Sustainability in Plant and Crop Protection

Series editor
Aurelio Ciancio, Sezione di Bari, Consiglio Nazionale delle Ricerche Istituto per la Protezione delle Piante, Bari, Italy

More information about this series at http://www.springer.com/series/13031

Sergei A. Subbotin • John J. Chitambar
Editors

Plant Parasitic Nematodes in Sustainable Agriculture of North America

Vol.2 - Northeastern, Midwestern and Southern USA

Editors
Sergei A. Subbotin
California Department of Food and Agriculture
Plant Pest Diagnostic Center
Sacramento, CA, USA

John J. Chitambar
California Department of Food and Agriculture
Plant Pest Diagnostic Center
Sacramento, CA, USA

ISSN 2567-9805 ISSN 2567-9821 (electronic)
Sustainability in Plant and Crop Protection
ISBN 978-3-319-99587-8 ISBN 978-3-319-99588-5 (eBook)
https://doi.org/10.1007/978-3-319-99588-5

Library of Congress Control Number: 2018965500

This Springer imprint is published by the registered company Springer Nature Switzerland AG
The registered company address is: Gewerbestrasse 11, 6330 Cham, Switzerland

Preface

Many changes globally affect the evolution of cropping systems today. There is hence a substantial need to periodically update our knowledge on the present and expected factors determining the performance of crops on continental scales. These two volumes on plant parasitic nematodes of North America are in line with this perspective. Both provide an impressive amount of updated information arising from one of the most technologically advanced agricultural system in the world. Topics include species composition, pathogenicity and losses, spatial distribution, and management approaches identified for most important nematode species. The volumes represent a rich source of information, also providing several historical reports and records, together with the description of main quarantine issues, related legislation, and adopted measures.

The chapters cover the whole continental range of geographic areas and crops, spanning from Mexico to Alaska. Although the same species are sometimes treated in different chapters, a *repetita juvant* approach has been considered necessary to provide a complete, detailed data source for the reader, including detailed geographic distribution patterns and incurred losses. The authors describe in fact the problems by regions, highlighting the different solutions that have been locally adopted and the main traits of the management approaches which have been identified and made available to farmers. These include, among others, use of rotation and resistant germplasm; nonhost or cover crops; agronomic management technologies; organic, integrated, or nematicide-based methods; as well as informations on the institutional initiatives aiming at the containment and exclusion of most threatening pests. All chapters have a stand-alone structure and represent a useful citation source.

In most intensive agricultural systems in the world, there is an increased need for new methods of nematode and other pest management, possibly with low environmental impacts, being sustainable in the long term. This view is today more necessary than ever due to the limitations in natural resources such as soil and water, the lack of traditional tools such as fumigants and nematicides, in part already banned or abandoned, or due to environmental issues. These factors have been considered by the authors, reporting prevalence data, updated quantitative estimations of losses, and data on the economic value of crops and products, in a broad regional context.

The two volumes result from the long-term work and experience of the authors, who represent a leading edge in the field of applied nematology, either for their experience or comprehensive research contributions. The volumes' compilation and production largely arose thanks to the careful and exhaustive coordination efforts that the editors, Sergei Subbotin and John J. Chitambar, deployed. Thanks to their excellent work, the readers will find a manual with complete source of information, literature data, and references, useful for any technical, teaching, and scientific need.

Bari, Italy

Aurelio Ciancio
SUPP Series Editor

Contents

1 **Plant Parasitic Nematodes of New England: Connecticut, Massachusetts and Rhode Island** 1
James A. LaMondia, Robert L. Wick, and Nathaniel A. Mitkowski

2 **Plant Parasitic Nematodes of New York, New Jersey and Pennsylvania** 27
George W. Bird, George S. Abawi, and James A. LaMondia

3 **Nematodes and Nematologists of Michigan** 57
George W. Bird and Fred Warner

4 **Nematodes of Agricultural Importance in Indiana, Illinois, Iowa, Missouri and Ohio** 87
Andreas Westphal, John J. Chitambar, and Sergei A. Subbotin

5 **Distribution and Importance of Plant Nematodes in Nebraska, Kansas and Colorado** 109
Timothy Todd and Thomas Powers

6 **Biology, Ecology and Management of Plant Parasitic Nematodes in Minnesota** 125
Senyu Chen and Zane J. Grabau

7 **Nematodes Important to Agriculture in Wisconsin** 157
Ann E. MacGuidwin

8 **Plant Parasitic Nematodes of North Dakota and South Dakota** 181
Guiping Yan and Richard Baidoo

9 **Management of Plant Parasitic Nematode Pests in Florida** 209
William Crow and Larry Duncan

10 Nematodes of Agricultural Importance in North and South Carolina 247
Weimin Ye

11 Plant Parasitic Nematodes of Virginia and West Virginia 277
Jonathan D. Eisenback

12 Plant Parasitic Nematodes of Tennessee and Kentucky 305
Ernest C. Bernard

13 Nematodes in Maryland and Delaware Crops 327
Ramesh R. Pokharel

14 Plant Parasitic Nematodes in Georgia and Alabama 357
Abolfazl Hajihassani, Kathy S. Lawrence, and Ganpati B. Jagdale

15 Important Plant Parasitic Nematodes of Row Crops in Arkansas, Lousiana and Mississippi 393
Travis R. Faske, Charles Overstreet, Gary Lawrence, and Terry L. Kirkpatrick

16 Plant Parasitic Nematodes of Economic Importance in Texas and Oklahoma 433
Terry A. Wheeler, Jason E. Woodward, and Nathan R. Walker

Index 453

Contributors

George S. Abawi New York State Agricultural Experiment Station, Cornell University, Geneva, NY, USA

Richard Baidoo Department of Plant Pathology, North Dakota State University, Fargo, ND, USA

Ernest C. Bernard Department of Entomology and Plant Pathology, University of Tennessee, Knoxville, TN, USA

George W. Bird Department of Entomology, Michigan State University, East Lansing, MI, USA

Senyu Chen Southern Research and Outreach Center, University of Minnesota, Waseca, MN, USA

John J. Chitambar California Department of Food and Agriculture, Plant Pest Diagnostic Center, Sacramento, CA, USA

William Crow IFAS, Department of Entomology and Nematology, CREC, University of Florida, Lake Alfred, FL, USA

Larry Duncan IFAS, University of Florida, IFAS, Department of Entomology and Nematology CREC, Lake Alfred, FL, USA

Jonathan D. Eisenback Department of Plant Pathology, Physiology and Weed Science, Virginia Tech, Blacksburg, VA, USA

Travis R. Faske Division of Agriculture, Lonoke Extension Center, Department of Plant Pathology, University of Arkansas, Lonoke, AR, USA

Zane J. Grabau Entomology and Nematology Department, University of Florida, Gainesville, FL, USA

Abolfazl Hajihassani Department of Plant Pathology, University of Georgia, Tifton, GA, USA

Ganpati B. Jagdale Department of Plant Pathology, University of Georgia, Athens, GA, USA

Terry L. Kirkpatrick Division of Agriculture, Southwest Research and Extension Center, Department of Plant Pathology, University of Arkansas, Hope, AR, USA

James A. LaMondia The Connecticut Agricultural Experiment Station, Windsor, CT, USA

Gary Lawrence Department of Biochemistry, Molecular Biology, Entomology and Plant Pathology, Mississippi State University, Starkville, MS, USA

Kathy S. Lawrence Department of Entomology and Plant Pathology, Auburn University, Auburn, AL, USA

Ann E. MacGuidwin Plant Pathology Department, University of Wisconsin, Madison, WI, USA

Nathaniel A. Mitkowski Department of Plant Sciences and Entomology, University of Rhode Island, Kingston, RI, USA

Charles Overstreet Department of Plant Pathology and Crop Physiology, Louisiana State University AgCenter, Baton Rouge, LA, USA

Ramesh R. Pokharel Maryland Department of Agriculture, Harry S Truman PKWY, Annapolis, MD, USA

Thomas Powers Department of Plant Pathology, University of Nebraska, Lincoln, NE, USA

Sergei A. Subbotin California Department of Food and Agriculture, Plant Pest Diagnostic Center, Sacramento, CA, USA

Timothy Todd Department of Plant Pathology, Kansas State University, Manhattan, KS, USA

Nathan R. Walker Entomology and Plant Pathology, Oklahoma State University, Stillwater, OK, USA

Fred Warner Department of Crop Soil and Microbial Sciences, Michigan State University, East Lansing, MI, USA

Andreas Westphal Department of Nematology, University of California Riverside, Kearney Research and Extension Center, Parlier, CA, USA

Terry A. Wheeler Texas A&M AgriLife Research, Lubbock, TX, USA

Robert L. Wick Stockbridge School of Agriculture, University of Massachusetts, Amherst, MA, USA

Jason E. Woodward Texas A&M AgriLife Extension Service, Lubbock, TX, USA

Guiping Yan Department of Plant Pathology, North Dakota State University, Fargo, ND, USA

Weimin Ye Nematode Assay Section, Agronomic Division, North Carolina Department of Agriculture and Consumer Services, Raleigh, NC, USA

Chapter 1
Plant Parasitic Nematodes of New England: Connecticut, Massachusetts and Rhode Island

James A. LaMondia, Robert L. Wick, and Nathaniel A. Mitkowski

1.1 Introduction

New England is a compact, northern region of the United States comprised of Connecticut, Rhode Island, Massachusetts, New Hampshire, Vermont and Maine. It has a small agricultural base compared to other regions of the country that benefit from longer growing seasons and more amenable soil types. New England states are not often associated with agriculture, but the economic value of agriculture is very important for these small states. A recent study (Lopez et al. 2017) estimated the 2015 economic impact of agriculture on the Connecticut economy to be $3.3–4 billion in direct sales, generating 21,000 jobs and approximately $800 million in wages. Sales of agricultural products in Massachusetts were over $490 million (Anon 2015) and were approximately $100 million in Rhode Island (Anon 2011). In addition, the green industry including landscaping and golf courses, adds significantly to both economic values and the quality of life in these states.

J. A. LaMondia (✉)
The Connecticut Agricultural Experiment Station, Windsor, CT, USA
e-mail: James.LaMondia@ct.gov

R. L. Wick
Stockbridge School of Agriculture, University of Massachusetts, Amherst, MA, USA
e-mail: rlwick@umass.edu

N. A. Mitkowski
Department of Plant Sciences and Entomology, University of Rhode Island, Kingston, RI, USA
e-mail: mitkowski@uri.edu

S. A. Subbotin, J. J. Chitambar (eds.), *Plant Parasitic Nematodes in Sustainable Agriculture of North America*, Sustainability in Plant and Crop Protection,
https://doi.org/10.1007/978-3-319-99588-5_1

1.2 Connecticut Agriculture

Agriculture has been important in Connecticut since colonial times and continues to be, with the total impact of agricultural industry in the state worth between $3.3 and 4 billion (Lopez et al. 2017). The green industry including ornamentals such as greenhouse, nursery, floriculture and sod production, accounts for 42% of the total agricultural products sold, nearly $500 million per year. Animal based production including dairy, poultry and cattle, accounts for about $340 million in sales and vegetable, fruit and cigar wrapper tobacco constitute 8, 7 and 6% of sales for an additional $200 million per year, respectively. Approximately 176,400 ha (12% of the land area) is classified as farmland. Much of the plant-based agriculture including ornamental production, vegetables and cigar tobacco, is located within the Connecticut River Valley running north-south through the center of the state (and north through Western Massachusetts). While only 23% of soils in Connecticut are classified as prime agricultural soils, 45% of the Connecticut River Valley and lowland soils are prime soils. These soils are the result of sedimentation from an ancient glacial lake and floodplain valley (Hartshorn and Colton 1967) and represent some of the most agriculturally productive soils in the state, New England, and the nation.

1.3 Massachusetts Agriculture

Massachusetts, the sixth smallest state has 20,305 km^3 of land. There are approximately 7800 farms with 210,400 ha under cultivation making agriculture worth about $492 million dollars annually. Ornamentals including sod, have the highest value, $144 million, followed by fruits and berries at $ 125 million; $69 million of which are from cranberries. Vegetables comprise the third highest market value at $81 million (USDA NASS 2012). In 1875 there were 14,549 farms in Massachusetts with 369,284 ha under cultivation (Census of Massachusetts 1876).

1.4 Rhode Island Agriculture

As the smallest state in the country, Rhode Island has a limited amount of agricultural production. Of 3140 km^3, approximately 20% of the state is comprised of the Providence area urban complex, in which 57% of the population resides. The total value of crop production in Rhode Island as of 2012 was approximately $49 million, ranking 49th in the nation (USDA NASS 2012). Greenhouse, nursery, floriculture and sod constitute the largest value at approximately $32.8 million. Within this group, turfgrass sod covers the largest area and averages 1214 ha annually, distributed among 15 farms. The mostly widely grown commodity group is forage grasses, with approximately 3318 ha in hay and other grains located on 285 farms (USDA

NASS 2012). Vegetable production constitutes the next largest commodity group at 890 ha distributed among 243 farms. The majority of crop producing farms in Rhode Island are less than 10 ha in size and are located throughout the rural and forested southern and western portions of the state. Surprisingly, Rhode Island was one of the few states to show an increase in agricultural production from 2007 to 2012, amounting to a 10% increase, and income from agritourism doubling to $1.4 million (USDA NASS 2012). Unfortunately, between 1981 and 2004, 25% of Rhode Island's prime farmland soils were converted to suburban or urban development and are no longer usable for agriculture (Turenne and Payne 2011).

1.5 Golf Course Industry

Turfgrasses are a significant agricultural commodity in all three states. The golf course industry in New England is worth approximately 10.6 billion dollars annually and there are more than 900 golf courses in the region. Massachusetts leads the New England states, where 377 golf courses generate about $5 billion and employ more than 57,000 people (Raub et al. 2015) (Table 1.1). While the sting nematode (*Belonolamius longicaudatus*) is typically considered the most damaging of turfgrass nematodes in the United States, it does not occur in northern climates. Consequently, nematode-related damage in northern golf course putting greens is frequently overlooked, even though multiple nematode genera are capable of causing severe turfgrass decline. The first significant survey of plant parasitic nematodes on golf course putting greens, from temperate regions in the United States, was undertaken in 1954. Researchers identified at least a dozen plant parasitic genera at variable levels from 41 putting greens throughout Rhode Island (Troll and Tarjan 1954). Although the study did not attempt to assign damage threshold numbers to populations of different genera, the researchers did notice observable turf declines in areas of extremely high *Tylenchorynchus claytoni*. As a final note, the authors stated, "*It had been assumed that plant nematodes were of only slight significance*

Table 1.1 Golf course statistics for New England States

State	Number of courses	Direct sales (dollars in millions)	Total value added (dollars in millions)	Total output (dollars in millions)
Connecticut	178	2473	1813	2853
Maine	140	1067	547	918
Massachusetts	377	4270	3157	4976
New Hampshire	113	1164	689	1098
Rhode Island	57	772	541	855
Vermont	69	928	356	609
All States	934	10,672	7102	11,308

After Raub et al. (2015)

in areas of the country subject to colder climates. It is hoped that surveys such as the one reported in this paper eventually will result in the abandonment of this fallacious view."

1.6 Historical Overview

In this chapter we will present some of the nematological problems that occur in Connecticut, Massachusetts and Rhode Island, with a historical perspective and overview of past and current management tactics. As agriculture including horticulture in New England is high-value and diverse, it is no surprise that nematode parasites of economically important plants are also diverse and can cause significant losses.

An early and informative study of the root knot nematode in Massachusetts can be found in Massachusetts Agricultural College Bulletin No. 55, *Nematode Worms* (Stone and Smith 1898). In this historical publication, the authors refer to the nematode as *Heterodera radicicola* since at the time, all root knot nematodes were considered to be the same species. The 68-page bulletin describes what nematodes are, symptoms produced in plants due to nematode infestations, histology of galls, life cycle of the nematode, and physical and chemical attempts to control the disease. Also included are some excellent drawings of different life stages of the nematode. They recognized the root knot nematode as the cause of decline in vegetable production. During the last decade of the 1800s, the value of vegetable crops propagated in glasshouses during the winter doubled and were worth $1,749,070 in 1895, according to a Massachusetts census of that year. However, root knot nematode often killed cucumber plants, and tomatoes were stunted and wilted, resulting in significant reduction and loss of crop growth. "*Realizing the impossibility of making definite recommendations to those seeking advice in the matter and feeling that the subject was one of great importance to the gardeners of Massachusetts, we finally undertook investigations, the results of which are contained in this bulletin*" (Stone and Smith 1898).

One of the first nematologists to work in New England, B. F. Lownsbery, conducted an extensive survey of plant parasitic nematodes on a wide variety of crops throughout the State of Connecticut from 1951 to 1953. The results were not published, but are summarized here. Plant parasitic nematodes were recovered from vegetables in 28 of 36 fields and included *Pratylenchus*, *Meloidogyne hapla* and *Tylenchorhynchus*. It was noted that *Pratylenchus* was associated with *Verticillium* wilt-affected plants. Tree and small fruit crops (16 farms) were affected by *Pratylenchus*, *Aphelenchoides*, *Meloidogyne hapla*, *Hoplolaimus*, *Mesocriconema* and *Xiphinema*. Ornamentals were positive for *Pratylenchus*, *Aphelenchoides*, *Meloidogyne hapla* and *Xiphinema* in 21 of 24 fields sampled. From over 300 tobacco fields sampled, the tobacco cyst nematode, *Globodera tabacum*, was found on one tobacco farm, while *Pratylenchus* and *Tylenchorhynchus* were present in several tobacco fields. Finally, in one of the first surveys of plant parasitic nematodes in turf in New England, *Tylenchorhynchus* and *Mesocriconema* were recovered from seven of nine golf course and turf farms. *Tylenchorhynchus* was noted as being associated with dead spots on golf greens. No comprehensive state-wide survey of nematode populations has been conducted in Connecticut since.

1.7 Root Lesion Nematode, *Pratylenchus penetrans*

Root lesion nematodes, primarily *Pratylenchus penetrans*, have been and continue to be the most commonly recognized nematode parasites of plants in the Northeast United States (Mai et al. 1960). Much of what we know about lesion nematodes in the Northeast was determined by coordinated research conducted in Connecticut, Illinois, Maryland, Massachusetts, New Jersey, New York, Ohio, Pennsylvania, Rhode Island and West Virginia as a part of the Regional Research Project (NE-64) and published in a collaborative Bulletin that outlined nematode biology and pathogenicity (Mai et al. 1977). The differences in lesion nematode impacts on crops that may exist between New England and other areas of the country may be due to the effects of lesion nematodes in interaction with certain fungal pathogens resulting in complex diseases. Lesion nematodes have been widely demonstrated to interact with vascular wilt pathogens (Rowe et al. 1985, 1987). Research in Connecticut has documented that *P. penetrans* can also interact with the fungal pathogen *Rhizoctonia fragariae* to increase the incidence and severity of cortical root rot in the strawberry black root rot disease complex (Fig. 1.1a) (LaMondia and Martin 1989; LaMondia 2003, 2004), which is a serious problem in strawberry replant situations or after several years in perennial plantings.

Lesion nematode populations fluctuate over time in response to strawberry growth and root biology (Szczygiel and Hasior 1972). LaMondia (2004) investigated the relation between strawberry root type, biomass, and nematode populations in roots and soil over time and determined that *Pratylenchus penetrans* primarily infected feeder and structural roots rather than perennial roots. In addition, *Rhizoctonia fragariae* was consistently isolated from both healthy and diseased perennial roots. Nematode feeding and movement directly resulted in cell damage and death. The indirect effects of lesion nematode infection were early periderm formation, initially seen as discoloration of the endodermis, followed by localized areas of secondary growth and cortical cell senescence and death. Weakened or dying cells resulting from the direct or indirect effects of *P. penetrans* were more susceptible to *R. fragariae*, and thereby, increased the extent of infection and cortical root rot.

Rhizoctonia fragariae and *P. penetrans* pathogens are widespread and common in strawberry plantings, making management of strawberry black root rot disease difficult, but necessary, in order to avoid serious losses. Martin (1988) was able to isolate *R. fragariae* from more than 70% of strawberry plants cultivated in commercial fields for more than 1 year. A survey of 41 commercial strawberry fields in Connecticut demonstrated that lesion nematodes occurred in greater than 75% of

Fig. 1.1 (**a**) Strawberry black root rot; (**b**) *Meloidogyne hapla* infected carrots; (**c**) *Meloidogyne hapla* infecting potato tuber; (**d**) *Meloidogyne hapla* infected *Hosta undulata*; (**e**) *Meloidogyne hapla* infected *Lobelia cardinalis*

sampled plants, especially in replanted fields, and that nearly half of growers were unaware of significant lesion nematode infestations in their fields. Stunted plants had nearly twice the *Pratylenchus* populations of adjacent healthier plants and populations ranged from undetectable to 2350/g root (LaMondia et al. 2005). Black root rot caused by co-infection of *R. fragariae* and *P. penetrans* can have severe economic consequences. An economic analysis of lesion nematode populations in *R. fragariae*-infested field soils was conducted based on a regression model

(DeMarree and Riekenberg 1998) using yield data with *P. penetrans* populations in small plots at the Connecticut Agricultural Experiment Station Valley Laboratory in Windsor, Connecticut. Based on 4 years of projected fruiting from a planting, strawberry profit, expressed as a percentage of gross sales, was predicted to be 33%, 30%, 18% or operation at a cumulative loss over four harvest years at initial densities of 0, 12, 50, or 125 *P. penetrans* per g root, respectively. Half of the samples recovered from surveyed growers' fields had populations in excess of 125 nematodes per g of root (LaMondia et al. 2005).

1.7.1 Management: Crop Rotation

Management of black root rot and lesion nematodes has historically relied on preplant soil fumigation. While fumigation had short-term effects that reduced nematode densities the next year and temporarily increased yields, sampling from fumigated fields still resulted in damaging lesion nematode populations (LaMondia et al. 2005). Rotation away from strawberry to unspecified crops reduced *R. fragariae* to about one third of that seen from continuous strawberry production (Martin 1988). A dense planting of small grains reduces broadleaf weeds, but the lesion nematode has a wide host range including most small grains (Mai et al. 1960) and rotation with grains has been associated with increased lesion nematode damage to potato (Florini and Loria 1990). Growers in Connecticut that rotated away from strawberry to small grains continued to observe poor strawberry growth and black root rot symptoms in the following crop. Rotation with cover crops that suppress nematodes such as 'Saia' oat, sorgho-sudangrass, *Rudbeckia hirta*, pearl millet '101' and 'Polynema' marigold can be effective (LaMondia 1999; LaMondia and Halbrendt 2003). Not all of these plants are suitable as they can be difficult to establish, may not compete well with weeds, or may be difficult to obtain. Strawberry growers in Connecticut who have had losses due to lesion nematodes and black root rot have reported that growing sorgho-sudangrass or millet before replanting strawberry greatly reduced lesion nematodes disease severity, especially after several cycles of rotation and strawberry. Additional rotational plant species need to be evaluated for non-host or antagonism efficacy against *P. penetrans*, the black root rot complex, seed availability, low cost and ease of establishment.

1.8 Northern Root Knot Nematode, *Meloidogyne hapla*

Root knot nematodes are some of the most important and damaging plant pathogens world-wide (Sasser and Carter 1985). The northern root knot nematode, *Meloidogyne hapla*, is common in Connecticut, Massachusetts and Rhode Island. It was relatively wide-spread in Connecticut in the 1951–1953 survey and continues to be a problem on a large number of crops due to its cold tolerance and wide host range.

Galls resulting from *M. hapla* infection are usually much smaller than the southern root knot species, but nonetheless may have a great economic impact. Most vegetable and fruit crops are hosts, and aboveground symptoms are often subtle. Plants can be stunted and grow and ripen unevenly with reduced yields and quality; for example, forking and galling on carrot (Fig. 1.1b) and lesions within the vascular ring of potato (Fig. 1.1c). Unlike some root knot nematode species, there are no resistant vegetable varieties available for *M. hapla* and nematode management has relied either on chemical controls (Gugino et al. 2006) or on rotation to a non-host plant such as a small grain crop.

High value nursery and greenhouse crops represent some of the most valuable components of agriculture in Connecticut, Massachusetts and Rhode Island. We have observed many ornamental plants, especially herbaceous perennials, to be infected with *M. hapla* (Fig. 1.1d, e) (LaMondia 1995c, 1996b). Not only may these plants be stunted, they may also have reduced winter survival and ornamental quality. Many of these infected plants are vegetatively propagated and their movement may result in distribution of the nematode to new previously uninfested areas. Once infected propagation stock is planted, the nematode will continue to spread in that field, garden or planting.

1.8.1 Management: Host Resistance

Management of the northern root knot nematode in ornamentals with the use of plant resistance has become very important, particularly in the absence of chemical control options. Resistance to *M. hapla* has been observed in many ornamental species (LaMondia 1995c, 1996b) and can aid in management in different ways. Inspection of incoming planting stock can be time consuming and expensive. Knowledge of *M. hapla* host status allows application of limited resources to the most likely host plant species to be infected. Some resistant plants such as *Rudbeckia fulgida* and *Aster novi-belgii,* can greatly reduce or eliminate *M. hapla* nematodes from potted nursery soil, garden beds or field soils in as little as 2–6 months, presumably due to both non-host and antagonistic effects against *M. hapla* (LaMondia 1997). This would be useful in controlling infestations in field-grown nurseries, landscapes and gardens after northern root nematodes have been introduced.

There may be instances when infected planting stock may be the only material available for a certain cultivar. *Meloidogyne hapla* juveniles typically infect roots at or near root tips (Christie 1936). This may explain why selective pruning of only the fibrous roots was successful in reducing *M. hapla* infection as well as the spread and establishment of *M. hapla* in propagation material from a known infested source (LaMondia 1997). This root-pruning sanitation is an alternative to heat treatment of propagation material. Heat treatment to kill *M. hapla* in roots is often difficult and may result in plant death.

1.8.2 Sustainable Management

Meloidogyne hapla infestation in soil may also be controlled by biofumigation, the incorporation of green manures such as Brassicas with high glucosinolate contents, which break down to nematicidal isothiocyanates (Halbrendt 1996). However, even the most effective biofumigant crops may be hosts of the northern root knot nematode, so there may be a danger of population increase if conditions are not suitable for biofumigation after green manure incorporation (LaMondia and Halbrendt 2003).

1.9 Tobacco Cyst Nematode, *Globodera tabacum*

While not generally recognized, tobacco has been grown as a high value agricultural crop in New England for a very long time and Connecticut and Massachusetts continue to have economically important tobacco producing areas in the Connecticut River Valley. Many high quality cigars are wrapped with tobacco leaves from Connecticut and Massachusetts. Native Americans in the Northeast grew *Nicotiana rustica* and tobacco was adopted along with corn as one of the first crops grown by European settlers, who first planted *N. rustica* but quickly switched to cultivation of the more palatable *Nicotiana tabacum*. Tobacco was important enough that Connecticut, settled in 1633, enacted legislation concerning tobacco by 1640. Tobacco was grown, not only for local consumption, but also for export, although exports were less than 10 tons per year until the end of the 1700s. Tobacco was primarily used in pipes until Colonel Israel Putnam, of revolutionary war fame, was credited with introducing cigars to Connecticut after a military expedition to Havana, Cuba in 1762. Cigars became popular and by 1810 numerous cigar factories had been established in and around the tobacco producing area of the Connecticut River Valley. Since that time, Connecticut tobacco has been grown and used almost exclusively for cigar production. Broadleaf cigar tobacco was introduced about 1833 as a new improved all-purpose strain that could be used for cigar filler, binder and wrapper and shade tobacco was developed in 1900 as a high quality cigar wrapper leaf (Jenkins 1925). Tobacco acreage was first officially recorded in Connecticut as approximately 2430–2840 ha during the US Civil War, and increased to over 12,140 ha in 1920 (Anderson 1953). Both broadleaf and shade tobacco continue to be grown in Connecticut as natural leaf cigar wrapper, with 1214–1618 ha of production.

In 1951, B. F. Lownsberry found a round cyst nematode to be the cause of a disease on shade tobacco in the Hazardville section of Enfield in Hartford County, Connecticut and subsequently described it as *Heterodèra tabacum* (Lownsbery and Lownsbery 1954) (Fig. 1.2a), which was later transferred to the genus *Globodera* as *G. tabacum* (Behrens 1975; Stone et al. 1983). A major concern at the time of its description was its morphological similarity to the potato cyst nematode *Globodera rostochiensis*, which had just recently been quarantined as a potato pest in New York

Fig. 1.2 (**a**) Tobacco cyst nematode *Globodera tabacum* females and males on roots; (**b**) Damage to shade tobacco due to *G. tabacum*; note treated areas surrounding the plot; (**c**) Fusarium wilt of broadleaf tobacco due to *Fusarium oxysporum* and *G. tabacum*

State in an effort to reduce or eliminate spread (Spears 1968). However, after further official evaluation, quarantine restrictions were not placed on Connecticut crops or acreage as the host ranges of *G. tabacum* and *G. rostochiensis* were demonstrated to be different (Lownsbery 1953; Harrison and Miller 1969) and morphological differences between the two nematodes demonstrated that the tobacco cyst nematode was a closely related, but new and distinct species (Lownsbery and Lownsbery 1954). The most important fact was that *G. tabacum* did not reproduce on potato and *G. rostochiensis* did not reproduce on *N. tabacum*. The host range of *G. tabacum* included all *N. tabacum* types tested as well as *N. rustica*. The ornamental tobacco species *N. alata*, *N. sanderae* and *N. longiflora* were not hosts. *Solanum nigrum*, eastern black nightshade, a common weed in the northeast and in tobacco fields, was shown to be the preferred host, with four to five times the number of cysts produced from the same amount of inoculum, in comparison to tobacco (Lownsberry 1953).

Once potential quarantine issues were resolved, further research demonstrated that *G. tabacum* significantly impacted tobacco growth and yields (Lownsbery and Peters 1955) (Fig. 1.2b). More recent research quantified yield losses by *G. tabacum* densities as low as 10–20% in soil at nematode densities over 50 J^2 per cm^3 soil and as high as 40–60% in shade and broadleaf tobacco at nematode densities of 500–1000 J^2 per cm^3 (LaMondia 1995a, 2002b). Tobacco cyst nematode increase in infested fields was greater in shade than broadleaf tobacco due to increased plant density and a longer growing season of over 100 days *versus* 75–80 days respectively. In addition, when *G. tabacum* occurred in combination with *Fusarium oxysporum* f.sp. *nicotianae*, broadleaf plants often died from Fusarium wilt before the nematodes could complete their life cycle (Fig. 1.2c). This not only seriously impacted the tobacco crop but also decreased cyst nematode populations (LaMondia 1992, 2015). Unlike broadleaf, shade tobacco cultivars are resistant to Fusarium wilt and the tobacco cyst nematode did not break that resistance. The introduction of wilt resistance genes to a new broadleaf tobacco release (C9) that has dominated subsequent commercial production (LaMondia and Taylor 1992) allowed broadleaf tobacco to be grown in wilt-infested fields, and also allowed increases in nematode populations to the point where damage thresholds were routinely exceeded.

The tobacco cyst nematode was initially found only on a single farm in Connecticut. The source of that infestation was unknown. Surveys, from 1951 to 1953, in 168 tobacco fields spread across all three tobacco producing counties (Hartford, Tolland and Middlesex) did not recover any additional tobacco cyst nematode infestations. Over a relatively short period of time, the tobacco cyst nematode infestation spread so that nearly 100% of the shade and broadleaf tobacco fields were infested by the 1980s. It is likely that the movement of soil from farm to farm on equipment and vehicles played a very important role in that spread.

The genus *Nicotiana* likely has its origin in South America and *Nicotiana tabacum* is a natural allopolyploid that has not been found in nature, being derived from the interspecific hybridization of the ancestors of *Nicotiana sylvestris* (maternal) and *Nicotiana tomentosiformis* (paternal) about 200,000 years ago (Leitch et al. 2008). The tobacco cyst nematode is also a likely native to the Andes of South America, similar to other round cyst nematodes, and now is world-wide in distribution (CAB International 2004; Bélair and Miller 2006). Genetic differences in nematode populations have been associated with tobacco farms operated by different companies in France, and it can therefore be assumed that both within-region and long-distance cyst spread has been unintentionally accomplished through human activities (Alenda et al. 2014).

1.9.1 Management: Chemical

For decades, tobacco cyst nematode management in Connecticut and Massachusetts relied almost exclusively on chemical controls: soil fumigation with 1, 3-Dichloropropene, ethylene dibromide, or methyl-isothiocyanate, oxamyl application as a non-fumigant nematicide in shade tobacco and either oxamyl or two or more years of rotation in broadleaf tobacco. Methyl bromide was not used as it had negative effects on tobacco quality characteristics, particularly, burn. Other tactics for managing nematode numbers involved crop root destruction immediately after harvest to kill nematodes which had not yet completed development in the roots

(LaMondia 2008) and trap cropping with plants that stimulated nematode hatch, again before reproduction could occur (LaMondia 1995b, 1996a).

1.9.2 Resistance

Breeding for resistance to tobacco cyst nematodes has been ongoing in Connecticut since the late 1980s. The incorporation of a single dominant-effect resistance gene, originally transferred to *N. tabacum* from *N. longiflora* (LaMondia 2002a) into an adapted and widely grown broadleaf variety, B2, has resulted in yield and leaf quality increases while reducing nematode populations (LaMondia 2012). Resistant plants stimulate cyst nematode hatch but the juveniles which infect roots do not establish viable feeding cells and do not reproduce. Cyst-nematode resistant shade tobacco lines are under development. An additional source of resistance to *G. tabacum* associated with black shank resistance has been documented to have a different inheritance and mode of action and can also be used in breeding programs (Johnson et al. 2009). Should a population of *G. tabacum* be able to reproduce on currently available single-gene resistance plants, the additional source of resistance will be available for continued management.

1.10 Nematodes on Turfgrasses in New England

Nematodes occur in all turfgrasses such as residential lawns, athletic fields, cemeteries, sod farms, school grounds and golf courses; however, in New England, damaging populations tend to occur primarily on golf course putting greens. Golf course greens in New England are particularly susceptible to nematode damage because of the intense utilization and management practices. This management results in shallow root systems due to low cutting heights, drought conditions and extreme soil compaction. While the authors have observed damage to golf course fairways and commercial sod farms from plant parasitic nematodes, these occurrences are the exception, not the rule.

New England has some of the oldest golf courses in the United States, several of them 100 or more years old. Most of the golf courses in the region have what are known as “pushup greens”, that is, greens formed by mounding-up field soil so that the greens surface is elevated from the approach and fairway. Top dressing with core aerification over the last 50 or more years has resulted in a cap of sandy soil (75–95% sand) 7–10 cm deep. Most turf-parasitic nematode populations are restricted to the sandy cap, although *Longidorus* can be found well below. New England greens are comprised mostly of *Poa annua* and *Agrostis stolonifera* with a few having mixtures that include *Agrostis canina.* Unlike field crops, golf greens are uniquely suited to propagate plant parasitic nematodes and very high populations often occur. Golf greens have a long season with a perennial host that forms a dense root system

Table 1.2 Occurrence and frequency of plant parasitic nematodes above damage threshold values from University of Rhode Island (URI) and University of Massachusetts (UMASS) sampling data (UMASS data derived from records of UMASS Extension Nematology Lab)

	URI 2003–2004		UMASS 2011–2017	
Nematode genus	Greens w/ nematode (%) (n = 114)[a]	Samples above threshold (%)[b, c]	Courses w/ nematode (%) (n = 692)	Samples above threshold (%)
Criconemoid species	97.4	7.0	57.6	1.8
Helicotylenchus	100.0	2.6	53.3	3.8
Tylenchorhynchus	100.0	35.1	95.4	24.4
Hoplolaimus	89.5	9.6	74.1	45.7
Heterodera juveniles	94.7	7.9	23.7	1.8
Meloidogyne J2's	50.0	1.8	29.8	6.8
Pratylenchus	–	–	27.1	5.9
Trichodorus	–	–	0.7	20
Other parasitic genera[d]	76.3	n/a	29.1	n/a

[a]URI sample was taken from 114 putting greens on 38 golf courses (Jordon and Mitkowski 2006
[b]Based on damage threshold data from Table 1.3
[c]Refers to the percent of total samples with nematode levels above damage threshold at any of the six sampling dates in 2003 and 2004
[d]Other genera include *Longidorus*, *Tylenchus*, *Paratylenchus* and *Xiphinema*

throughout the entire surface. This, coupled with the sandy texture and daily irrigation, make an ideal environment for nematode feeding and reproduction.

At least 12 genera of plant parasitic nematodes are found in golf course greens with *Tylenchorhynchus*, *Helicotylenchus*, *Hoplolaimus* and criconemoid nematodes most commonly encountered (Table 1.2). In 2003 and 2004, Jordon and Mitkowski (2006) undertook a sampling study of 114 putting greens from 38 different golf courses in Rhode Island, southern Connecticut and eastern Massachusetts. Golf courses were chosen based on previous history of nematode injury to turf. More currently, 2011–2017 data from the University of Massachusetts (UMASS) Extension Nematology Lab are also included in Table 1.2. While the percentages between the two data sets are different, the four most common turf-parasitic nematodes are the same. It should be noted that the University of Rhode Island (URI) sampling was focused on golf courses which were known to have high nematode populations, while the UMASS data is based on submissions from golf courses with suspected (but not necessarily confirmed) nematode issues at the time of submission. Additionally, sampling at both sites was conducted a decade apart from each other and the URI data is representative of a much narrower geographic area.

For the purposes of rapid diagnosis, turfgrass parasitic nematodes are typically only identified to genus, so the number and diversity of species encountered in the region are unclear. Within New England, *Tylenchorhynchus claytoni* and *T. dubius* have most commonly been reported (Troll and Tarjan 1954; Miller 1976; Blackburn

Table 1.3 Damage threshold levels for nematodes that parasitize turfgrasses (number of each genus/100 cm^3 soil)

Nematode genera	New England[a]	Other[b]
Criconemoids	1500	1500
Tylenchorhynchus	800	300
Hoplolaimus	400	150
Helicotylenchus	1500	600
Longidorus	100	–
Meloidogyne	500	100
Heterodera	500	–
Pratylenchus	100	150
Hemicycliophora	200	200
Trichodoroids	100	100
Xiphinema	–	200

[a]Developed by Robert Wick, PhD, University of Massachusetts
[b]From: Eric Nelson, PhD, Cornell University, Turfgrass Trends, Oct. 1995 (Nelson 1995)

et al. 1997). Recent ITS sequencing of *Tylenchorhynchus* nematodes from Massachusetts golf course greens confirms the presence of *T. claytoni* (N. Mitkowski, personal communication). However, *T. maximus* has been identified in New Jersey and is likely to be present in New England (Myers et al. 1992). As the name implies, *T. maximus* is noticeably longer than most species of the genus and unusually large *Tylenchorhynchus* individuals have been observed from different locations in the region over the past decade. *Tylenchorhynchus nudus* has been reported from turfgrasses in the Midwestern United States, but it is unclear if it is present along the east coast (Smolik and Malek 1972; Malek 1980). Malek (1980) reported *T. clarus* parasitizing creeping bentgrass in Ohio, but once again, it is unclear if it is present in New England.

To date, only a single species of *Hoplolaimus* has been reported in New England on golf course turfgrasses, *Hoplolaimus galeatus* (Troll and Tarjan 1954; Miller 1976) (Fig. 1.3a–c). Although the genus *Hoplolaimus* is not nearly as diverse as *Tylenchorhynchus*, with only 29 currently known species (Handoo and Golden 1992), ITS sequencing and morphological data have never identified any other species of *Hoplolaimus,* besides *H. galeatus,* parasitizing golf course turf in New England or elsewhere in the US (Lucas et al. 1978; Wick and Vittum 1988; Blackburn et al. 1997; Settle et al. 2006). While *Helicotylenchus* spp. are extremely common in New England turfgrass soils, the only attempt to identify populations of these nematodes to species was undertaken by Troll and Tarjan (1954) when they sampled 41 golf course putting greens in Rhode Island and identified *H. erythrinae* from half of the sampled greens. While prevalent, these nematodes rarely appear to reach high population levels, although Fushtey and McElroy (1977) did report significant numbers of *Helicotylenchus* spp. from different locations in Southern British Columbia. The most commonly reported species of *Helicotylenchus* found on turf in North America is *H. dihystera* (Sumner 1967; Lucas et al. 1978; Zeng et al. 2012). Davis et al. (1994b) reported *H. cornurus* from golf course putting greens in Chicago, IL

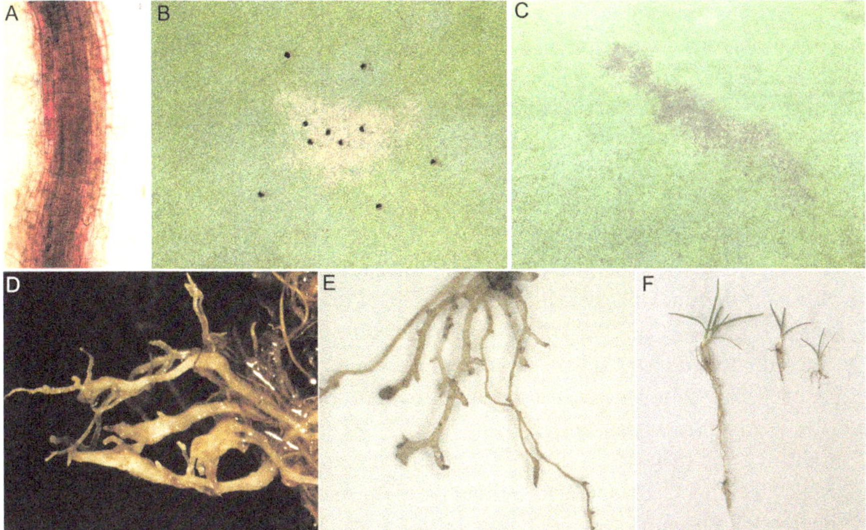

Fig. 1.3 (**a**) Four lance nematodes, *Hoplolaimus galeatus* in the bentgrass root; (**b**) This patch of dead turf on a golf green has more than 5000 lance nematodes per 100 cm^3 of soil. In surrounding patch the population is about 1500 lance nematodes per 100 cm^3 of soil; (**c**) This patch had 11,340 lance and 380 stunt nematodes/100 cm^3 soil; (**d**) Root-knot galls of grass roots caused by *Meloidogyne naasi*; (**e**) Galled and necrotic root tips of Kentucky bluegrass as the result of feeding injury by *Longidorus*, the needle nematode; (**f**) Healthy Kentucky bluegrass on the left, stunted galled roots as the result of about 50 *Longidorus* nematodes/100 cm^3 of soil

but the species has never been reported from additional locations. *Helicotylenchus pseudorobustus* has also been reported from New York State, a short distance from Connecticut (Feldmesser and Golden 1974). The most extensive list of *Helicotylenchus* found on turfgrasses comes from Kentucky bluegrass lawns, athletic fields and pastures in Wisconsin and includes *H. digonicus, H. dihystera* (=*H. nannus*), *H. erythrinae* (= *H. melancholicus*), *H. platyurus* and *H. microlobus* (Perry et al. 1959). Of the identified species, *H. digonicus* appeared to be the most pathogenic. As is the case with *Tylenchorhynchus,* distinctions between *Helicotylenchus* species can be subtle and there are currently 193 recognized species within the genus (Marais 2001).

Species of ring or criconemoid nematodes are very commonly found parasitizing turfgrasses. In the 1954 study by Troll and Tarjan, *Mesocriconema* (= *Criconemoides*) was reported from a single putting green in Rhode Island. While it is unclear which species of criconemoids are common in New England, in 1976, Miller reported *Criconemoides lobatum* from Connecticut, now recognized by some as *Mesocriconema rustica* (Ebsary 1991). No other reports of specific criconemoid taxa on turf from New England have been reported. Feldmesser and Golden reported *M. rustica* from West Point, NY in 1974, approximately 20 miles from the Connecticut border. *Criconema mutabile* has been reported from Rhode Island on unspecified hosts (Mai et al. 1960) and its ability to parasitize turf would make it a likely turfgrass pathogen in New England (Bernard 1980). *Mesoscriconema* spp. have been reported from turf in various locations in the USA (Bernard 1980; Zeng et al. 2012), and having been positively reported from New Jersey, it is likely present at some level in New England turfgrasses (Mai et al. 1960).

Meloidogyne naasi, the barley root knot nematode, commonly occurs in golf greens in New England and New York causing root galls (Fig. 1.3d). It was previously thought to be *M. graminis* (Rungrassamee et al. 2003). *Meloidogyne naasi* was first described in Great Britain (Franklin 1965) and later reported in Los Angeles and Orange Counties in California (Radewald et al. 1970). It has also been reported in Nevada, Oregon, Utah, Washington (McClure et al. 2012); Illinois, Kentucky, and Kansas (Michell et al. 1973). *Meloidogyne naasi* is not restricted to feeding on grasses; dicotyledonous hosts include alfalfa, clover, pea, soybean, chickweed and sugar beet. *Meloidogyne naasi,* either alone or in combination with *Tylenchorhynchus agri* or *Pratylenchus penetrans,* was very pathogenic to potted creeping bentgrass 'Toronto C-15' (Sikora et al. 1972). In another trial, inoculation of potted bentgrasses 'Toronto C-15 and 'Northmoor 9" with *M. naasi* resulted in a significant reduction of clipping weight 8 months after inoculation (Michell et al. 1973). Pathogenicity in golf greens has not been well established but circumstantial evidence suggests that several thousand juveniles/100 cm^3 of soil will result in compromised turf. In New England, this assessment must be made in the month of April when the juveniles are still in the soil. Regardless of the numbers in April, few juveniles will be seen throughout the growing season as there is only one hatch period per year. *Meloidogye naasi* is unique in that it requires a chilling period before the eggs will hatch. Diapause affects about 95% of the population so that in *M. naasi*-infested golf greens, a few juveniles can be recovered throughout the summer and fall. Depending on the year and location, high populations of *M. naasi* can be found between December and the end of April.

Pratylenchus penetrans has been reported from turfgrasses in New England (Troll and Tarjan 1954), but the species does not appear to be particularly aggressive on putting green turf. When *P. penetrans* was inoculated on the roots of greenhouse-grown *Poa annua*, very little damage was observed, even at the highest concentrations of 5000 nematodes/100 cm^3 soil (Bélair and Simard 2008). In fact, after 9 weeks, nematode concentrations had declined significantly from inoculation levels. Sikora et al. (1972) demonstrated significant pathogenicity of *P. penetrans* on *Agrostis palustris* but only in the presence of other pathogenic nematodes, particularly *Meloidogyne naasi.* While *P. penetrans* could reproduce on *A. palustris* in the absence of other plant parasitic nematodes, population levels had dropped significantly below inoculation concentrations after 6 months in the greenhouse and *A. paulstris* was described as a poor host for the nematode. Between 1992 and 1996, *Pratylenchus* spp. were only identified in an average of 7.2% of the approximately 2600 soil samples processed by the University of Rhode Island Turf Diagnostic Lab, with the vast majority of samples containing 40 or fewer nematodes/100 cm^3 soil. Roots were not extracted for *P. penetrans* in any of these cases. From 2011 to 2017, the University of Massachusetts (UMASS) Extension Nematology Lab reported 27.1% of submitted samples contained *P. penetrans* but only 5.9% of samples were above damage threshold levels (Table 1.2). The discrepancy between these two data sets is likely a result of the counting methodology used. The University of Rhode Island (URI) samples were all counted at a 1:20 dilution, meaning that if less than 20 nematodes of any type were present, they were statistically unlikely to be observed. The UMASS data set was derived by counting each nematode in an extracted sample. Interestingly, Qing et al. (1998) identified four species of *Pratylenchus* from Southern Ontario, an area relatively close to New England, but did not identify *P. penetrans*.

Longidorus spp. are relatively uncommon in turfgrass soils, but can cause significant damage even at low population levels. In 2006, the authors identified populations of *Longidorous* spp. attacking newly seeded Kentucky Bluegrass sod in Southern Maine, with damage resulting from only 50 nematodes/100 cm^3 soil. Hundreds of square feet of turf were killed and surviving seedlings had severely damaged root systems (Fig. 1.3e, f). Troll and Tarjan (1954) identified *L. sylphus* from 5 of 41 sampled Rhode Island golf courses. Although very little work has been undertaken on *Longidorus* on turf, *L. breviannulatus* was identified in Pennsylvania on creeping bentgrass greens (Forer 1977) and as few as 20 nematodes/100 cm^3 soil have been documented to cause severe damage to corn (Niblack 2003). *Longidorous elongatus* has also been observed to parasitize perennial ryegrass (*Lolium perenne*) under experimental conditions, but this grass is not used as putting surface, and while the grass was a host, no damage to affected plants was observed (Taylor 1967).

Plant parasitic nematodes vary widely in their virulence (Table 1.3), and their genus and number per given volume of soil needs to be considered when assessing their potential for damage. Damage threshold levels in Table 1.3 were not determined experimentally but were based on field observations and laboratory assays from golf courses in the New England region. As with all nematodes, the conditions in which turf-parasitic nematodes exist have a significant impact on their damage potential. Soil type (highly organic *vs.* sand *vs.* silt) and moisture content have the most significant impact on the success of turf-parasitic nematodes and can also affect the health of the turf grown in site-specific conditions. It has been shown that turf-parasitic nematodes are less prevalent in organically managed golf courses (Allan et al. 2015). Golf course age can also have a significant impact. In one study, older courses were shown to have higher nematode populations, with *Poa annua* and *Agrostis canina* being more susceptible to nematode increases, in general, than *A. palustris* (Jordan 2005). In addition, nematodes species can also play a role in virulence. While damage thresholds have been developed for the genus *Tylenchorhynchus,* no assumptions have been made as to the virulence of individual *Tylenchorhynchus* species. As a result, damage thresholds may be higher or lower than is appropriate as thresholds are applied to different *Tylenchorhynchus* species. Other extenuating circumstances such as geographic location, method of sample collection, time of the year, assay methods and prevailing environmental conditions, can affect the numbers and interpretation of threshold levels.

In addition to the difficulty of applying a single damage threshold to a variety of different potential environments, determining a threshold experimentally can be difficult and experimental conditions rarely correspond to real-world conditions on a golf course. As a consequence, turfgrasses in a greenhouse that can withstand very high populations of nematodes without showing symptoms may collapse when subjected to golf course traffic, drought and cutting heights. In addition, the determination of "damage" on golf turf is difficult to quantify because a clear and measurable yield is never achievable. Some researchers have chosen to examine clipping yield and rooting depth as a measure of nematode virulence but these parameters fluctuate throughout a growing season in cool-season turfgrasses and are often dependent on fertilization level and soil temperatures. The single most damaging nematode in the region is *Hoplolaimus*, which feeds both ecto- and endoparasitically on grass roots (Fig. 1.3a). Populations of 4000 nematodes/100 cm^3 of soil can result in dead patches of turf (Fig. 1.3b, c). Settle et al. (2006) utilized visual and multispectral radiometry to examine the effects of *H. galeatus* on *Agrostis palustris* "A-4". Researchers observed that damaging populations of *H. galeatus* ranged from 177 to 845 nematodes/100 cm^3 soil and that quality rating decreased 10% for each additional 400 nematodes counted (maximum counts of *H. galeatus* reached 1600 nematodes/100 cm^3 soil). Unfortunately, *H. galeatus* frequently burrows into roots and this study did not account for those nematodes, which cannot be removed via sugar flotation. When researchers examined the effect of turfgrass cultivar on nematode populations numbers and turfgrass injury, the error induced by varietal differences appeared to mask any possible correlation with nematode population levels.

Laughlin and Vargas (1972) observed significant reductions in foliar and root dry weight on both *Agrostis palustris* 'Toronto' and *Poa pratensis* 'Merion' in sand-based greenhouse trials using 500 and 1000 *Tylenchorhynchus dubius* nematodes/100 cm^3 soil. In similar greenhouse experiments, Davis et al. (1994a) were able to demonstrate pathogenicity on both *A. palustris* 'Penncross'and *P. annua* by *Tylenchorhynchus nudus*, indicated by reduced root mass of both species in the presence of approximately 1500 nematodes/100 cm^3 soil. This study also demonstrated that reduction in rooting was more severe on *A. palustris* than on *P. annua* and regression analysis suggested that as few as 120 *T. dubius* nematodes/100 cm^3 soil could potentially reduce root length by a centimeter, but the regression coefficient was very low ($R^2 = 0.31$). A study on the effect of *T. claytoni* on multiple cultivars of *A. palustris* indicated that cultivar had no effect on nematode reproduction. All six tested cultivars supported the nematode reproduction and visual decline in turfgrass quality was observed above 600 nematodes/100 cm^3 soil (Walker and Martin 2001).

As mentioned previously, *Longidorus* spp. were found to be very damaging in a turf farm with very sandy soil in Maine, in 2006. Populations as low as 30–50 nematodes/100 cm^3 of soil were capable of killing seedlings, where galled root tips and necrotic meristems were evident. In the 3rd edition of Couch's treatise on turfgrass diseases (Couch 1995), *Longidorous* spp. are listed at 20 nematodes/100 cm^3 soil, which approximates the experiences of the authors in 2006. Forer's (1977) work on *L. breviannulatus* also suggested a minimum number of 20 nematodes/100 cm^3 soil to cause observable damage, derived from an actual *A. palustris* putting green.

While these studies suggest that it is possible to obtain some level of accuracy in determining nematode thresholds, there is still a wide range of variability present. For example, while Laughlin and Vagas (1972) observed reductions in root and shoot weights at 500 *T. dubius* nematodes/100 cm^3 of soil, the majority of turfgrass putting greens that had not been treated with a nematicide frequently had at least 300–400 stunt nematodes/100 cm^3 of soil. Consequently, a damage threshold at this low level is meaningless and most samples will report as "over-threshold" even when turf may not appear symptomatic. For this reason, thresholds need to be used in conjunction with observed damage or a past history of observed damage. Because nematode virulence in turfgrasses is closely related to the overall health of the plant and prevailing environmental conditions, it is not advisable to use thresholds as a singular data point for determining whether control methods are warranted. Admittedly, when population numbers far exceed thresholds (such as the case of 39,000 stunt nematodes/100 cm^3 soil counted from a golf course putting green in Massachusetts in 2010), it becomes clear that nematode populations have exceeded whatever rational threshold is currently being employed and that treatment to reduce nematode numbers should be undertaken immediately.

1.10.1 Vertical and Horizontal Distribution of Nematodes in Golf Greens

Surveys to determine the vertical distribution of plant parasitic nematodes on golf greens were carried out during 1986 and 1987 on golf courses from all the New England states, except Rhode Island (Wick, Vittum and Swier, unpublished). As would be expected, considering the sandy cap of soil sitting on field soil, most of the nematodes were observed in the top 10 cm of soil, most particularly in the top 5 cm of soil (Fig. 1.4a). *Tylenchorhynchus, Helicotylenchus* and *Mesocriconema* were primarily in the top 5 cm of soil; *Hoplolaimus* was best represented throughout the top 10 cm but still mostly in the top 5 cm and *Longidorus* could be found even below 20 cm. The majority of the root system is present in the sandy cap of soil, which is about 10 cm deep. The roots remain in the top 5 cm for most of the growing season. Due to this stratification, soil samples taken to 5 cm may have twice the population than samples taken to 10 cm. The depth of soil sampling and root growth should be considered when interpreting nematode populations.

Nematodes in golf greens are distributed in a clumped pattern where the variation is always greater than the mean. The unequal distribution makes it difficult to show statistical differences in experimental plots. One practical consideration is the number of subsamples that can be taken per plot on a working golf green. For 20 plots, 200 subsamples are typical, but the sampled area must be in perfect condition once the sampling is finished. Golf usually proceeds within an hour after sampling, so time taken to collect samples is also a factor. In the following example, 16 cores were removed from a four-square meter plot and three genera of nematodes were

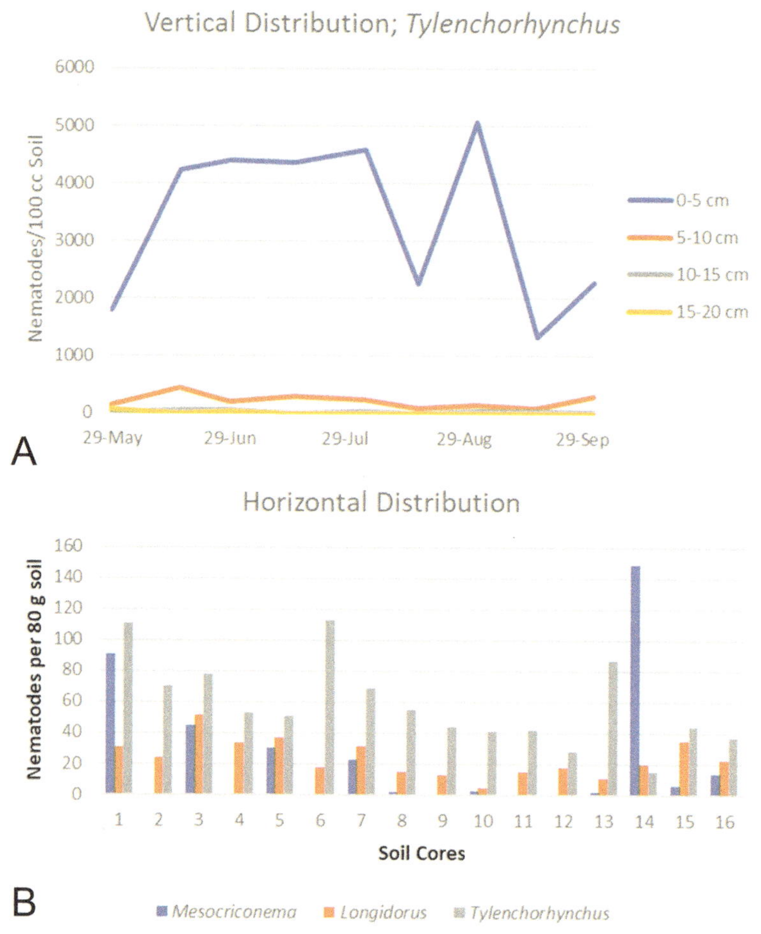

Fig. 1.4 (**a**) Vertical distribution of *Tylenchorhynchus* in a golf green; (**b**) Horizontal distribution of nematodes in a golf green

counted. To determine the least number of cores to estimate within 75% of the expected mean, the dispersion coefficient "k" was determined along with the mean, standard deviation and variance (Ferris 1990). For *Mesocriconema*, the minimum number of samples was about 200; for *Longidorus*, 18 and for *Tylenchorhynchus,* 16 (Fig. 1.4b). Note that in Fig. 1.4b, for *Mesocriconema,* 9 cores yielded less than 3 individuals, while 3 cores yielded 45, 91 and 149 individuals per core. The field distribution of *Mesocriconema* is expected to be highly clumped since the nematode is barely motile.

1.10.2 Sustainable Management of Nematodes in Golf Greens

Nematodes are particularly difficult to control in turfgrass because it is a perennial crop and no breeding for resistance to nematodes has been done. In addition, the thatch layer can make it difficult for materials to penetrate into the root zone by both its physical nature, and its ability to bind to chemicals. There has been interest in the application of chitin, ground up sesame plants and other materials, but it is difficult to effectively incorporate these materials into the root zone except for applications immediately following core aerification.

Currently, there are no effective intervention strategies to control nematodes in turfgrasses other than practices that enhance root growth and vigor. When damaging levels of nematodes are present, the height of cut should be raised and every effort to increase root growth should be undertaken. Root development occurs in the spring and fall and this is the favored time to encourage growth.

References

Alenda, C., Montarry, J., & Grenier, E. (2014). Human influence on the dispersal and genetic structure of French *Globodera tabacum* populations. *Infection Genetics and Evolution, 27*, 309–317.

Allan, E., Manter, D., & Jung, G. (2015). Comparison of nematode communities between organically and conventionally managed golf courses. *Phytopathology, 105*(3), S1.

Anderson, P. J. (1953). *Growing tobacco in Connecticut*. Connecticut Agricultural Experiment Station Bulletin 564. 110 pp.

Anon. (2011). *A vision for Rhode Island agriculture: Five-year strategic plan*. http://www.dem.ri.gov/programs/agriculture/documents/RI_agriculture_5yr_strategicplan.pdf

Anon. (2015). *Massachusetts agriculture*. http://www.mass.gov/eea/docs/agr/facts/snapshot-of-ma-agriculture.pdf

Behrens, E. (1975). *Globodera* Skarbilovic, 1959, eine selbstindige Gattung in der Unterfamilie Heteroderinae Skarbilovic, 1947 (Nematoda: Heteroderidae). Vortragstagung zuaktuellen Problemen der Phytonematologie am 29/5/1975 in Rostock. (pp. 12–26). Rostock: Manuskriptdruck der Vortrage

Bélair, G., & Miller, S. (2006). First report of *Globodera tabacum* infecting tobacco plants in Quebec, Canada. *Plant Disease, 90*, 527.

Bélair, G., & Simard, L. (2008). Effect of the root-lesion nematode (*Pratylenchus penetrans*) on annual bluegrass (*Poa annua*). Horticulture Research and Development Centre. *Phytoprotection, 89*, 37–39.

Bernard, E. C. (1980). *Identification, distribution and plant associations of plant-parasitic nematodes in Tennessee*. University of Tennessee, Agricultural Experiment Station Bulletin 594. 18 pp.

Blackburn, K., Alm, S. R., Yeh, T. S., & Dawson, C. G. (1997). High-pressure liquid injection of isazofos for management of *Hoplolaimus galeatus* and *Tylenchorhynchus dubius* infesting turfgrasses. *Journal of Nematology, 29*, 690–694.

CAB International. (2004). *Globodera tabacum*. In: *Distribution maps of plant diseases*. Map 929.

Census of Massachusetts. (1876). Prepared by Carroll D. Wright, Chief of the Bureau of Statistics of Labor. Albert wright, State Printer, Boston.

Christie, J. R. (1936). The development of root-knot nematode galls. *Phytopathology, 26*, 1–22.

Couch, H. (1995). *Diseases of Turfgrasses* (3rd ed.). Malabar: Krieger Publishing Company. isbn:0882750623.
Davis, R. F., Kane, R. T., Wilkinson, H. T., & Noel, G. R. (1994a). Population fluctuations of three nematode genera in putting greens in Northern Illinois. *Journal of Nematology, 26*, 522–530.
Davis, R. F., Wilkinson, H. T., & Noel, G. R. (1994b). Vertical distribution of three nematode genera in a bentgrass putting green in Central Illinois. *Journal of Nematology, 26*, 518–521.
DeMarree, A., & Riekenberg, R. (1998). Budgeting. In M. Pritts & D. Handley (Eds.), *Strawberry production guide for the Northeast, Midwest, and Eastern Canada* (pp. 118–131). Ithaca: Northeast Regional Agricultural Engineering Service.
Ebsary, B. A. (1991). *Catalog of the order* Tylenchida *(Nematoda)*. Ottawa, Canada, Agriculture Canada Publication 1869/B, 196 pp.
Feldmesser, J., & Golden, A. M. (1974). Bionomics and control of nematodes in a large turf area. *Journal of Nematology, 6*, 139.
Ferris, H. (1990). Sampling precision and reliability. In B. M. Zuckerman, W. F. Mai, & L. R. Krusberg (Eds.), *Plant nematology laboratory manual* (pp. 20–25). Amherst: The University of Massachusetts Agricultural Experiment Station.
Florini, D. A., & Loria, R. (1990). Reproduction of *Pratylenchus penetrans* on potato and crops grown in rotation with potato. *Journal of Nematology, 22*, 106–112.
Forer, L. B. (1977). *Longidorus breviannulatus* associated with a decline of bentgrass in Pennsylvania. *Plant Disease Reporter, 61*, 8.
Franklin, M. T. (1965). A root-knot nematode, *Meloidogyne naasi* n. sp., on field crops in England and Wales. *Nematologica, 11*, 79–86.
Fushtey, S. G., & McElroy, F. D. (1977). Plant parasitic nematodes in turfgrass in Southern British Columbia. *Canadian Plant Disease Survey, 57*, 54–56.
Gugino, B. K., Abawi, G. S., & Ludwig, J. W. (2006). Damage and management of *Meloidogyne hapla* using oxamyl on carrot in New York. *Journal of Nematology, 38*, 483–490.
Halbrendt, J. M. (1996). Allelopathy in the management of plant-parasitic nematodes. *Journal of Nematology, 28*, 8–14.
Handoo, Z. A., & Golden, A. M. (1992). A key and diagnostic compendium to the species of the genus *Hoplolaimus* Daday, 1905. *Journal of Nematology, 24*, 45–53.
Harrison, M. B., & Miller, L. I. (1969). Additional hosts of the tobacco cyst nematode. *Plant Disease Reporter, 53*, 949–951.
Hartshorn, J., & Colton, R. (1967). Geology of the southern part of glacial Lake Hitchcock and associated deposits. In P. Robinson (Ed.), *Guidebook to field trips*. (pp. 73–88). New England Intercollegiate Geological Conference, Amherst.
Jenkins, E. H. (1925). *Connecticut agriculture, history of connecticut*, (pp. 291–294). http://www.ct.gov/caes/lib/caes/documents/publications/a_history_of_connecticut_agriculture.pdf.
Johnson, C. S., Wernsman, E. A., & LaMondia, J. A. (2009). Effect of a chromosome segment marked by the Ph-p gene for resistance to *Phytophthora nicotianae* on reproduction of tobacco cyst nematodes. *Plant Disease, 93*, 309–315.
Jordan, K. S. (2005). *The ecology of plant -parasitic nematodes and their antagonists on golf course greens turf in southern New England*. Ph.D. Dissertation. Kingston: The University of Rhode Island.
Jordon, K. S., & Mitkowski, N. A. (2006). Population dynamics of plant-parasitic nematodes in golf course greens turf in southern New England. *Plant Disease, 90*, 501–505.
LaMondia, J. A. (1992). Predisposition of broadleaf tobacco to Fusarium wilt by early season infection with tobacco cyst nematodes. *Journal of Nematology, 24*, 425–431.
LaMondia, J. A. (1995a). Shade tobacco yield loss and *Globodera tabacum tabacum* population changes in relation to initial density. *Journal of Nematology, 27*, 114–119.
LaMondia, J. A. (1995b). Tobacco cyst nematode hatch and reproduction in response to tobacco, tomato, or black nightshade. *Journal of Nematology, 27*, 382–386.
LaMondia, J. A. (1995c). Response of perennial herbaceous ornamentals to *Meloidogyne hapla*. *Journal of Nematology, 27*, 645–648.

LaMondia, J. A. (1996a). Trap crops and population management of *Globodera tabacum tabacum*. *Journal of Nematology, 28*, 238–243.

LaMondia, J. A. (1996b). Response of additional herbaceous perennial ornamentals to *Meloidogyne hapla*. *Journal of Nematology, 28*(4S), 636–638.

LaMondia, J. A. (1997). Management of *Meloidogyne hapla* in herbaceous perennial ornamentals by sanitation and resistance. *Journal of Nematology, 29*, 717–720.

LaMondia, J. A. (1999). The effects of rotation crops on the strawberry pathogens *Pratylenchus penetrans*, *Meloidogyne hapla*, and *Rhizoctonia fragariae*. *Journal of Nematology, 31*, 650–655.

LaMondia, J. A. (2002a). Genetics of burley and flue-cured tobacco resistance to *Globodera tabacum tabacum*. *Journal of Nematology, 34*, 34–37.

LaMondia, J. A. (2002b). Broadleaf tobacco yield loss in relation to initial *Globodera tabacum tabacum* population density. *Journal of Nematology, 34*, 38–42.

LaMondia, J. A. (2003). Interaction of *Pratylenchus penetrans* and *Rhizoctonia fragariae* in strawberry black root rot. *Journal of Nematology, 35*, 17–22.

LaMondia, J. A. (2004). Strawberry black root rot. Feature article. *Advances in Strawberry Research, 23*, 1–10.

LaMondia, J. A. (2008). Early crop root destruction for management of tobacco cyst nematodes. *Journal of Nematology, 40*, 26–29.

LaMondia, J. A. (2012). Registration of B2 Connecticut broadleaf cigar-wrapper tobacco resistant to Fusarium wilt, Tobacco Mosaic Virus, cyst nematodes and blue mold. *Journal of Plant Registrations, 7*, 58–62.

LaMondia, J. A. (2015). Fusarium wilt of tobacco. *Crop Protection, 73*, 73–77.

LaMondia, J. A., & Halbrendt, J. M. (2003). Differential host status of rotation crops to dagger, lesion and root-knot nematodes. *Journal of Nematology, 35*, 349.

LaMondia, J. A., & Martin, S. B. (1989). The influence of *Pratylenchus penetrans* and temperature on black root rot of strawberry by binucleate *Rhizoctonia* spp. *Plant Disease, 73*, 107–110.

LaMondia, J. A., & Taylor, G. S. (1992). Registration of C8 and C9 Fusarium wilt resistant broadleaf tobacco germplasm lines. *Crop Science, 32*, 1066–1067.

LaMondia, J. A., Cowles, R. S., & Los, L. (2005). Prevalence and potential impact of soil-dwelling pests in strawberry fields. *HortScience, 40*(5), 1366–1370.

Laughlin, C. W., & Vargas, J. M. (1972). Pathogenic potential of *Tylenchorhynchus dubius* on selected turfgrass. *Journal of Nematology, 4*, 277–280.

Leitch, I. J., Hanson, L., Lim, K. Y., Kovarik, A., Chase, M. W., Clarkson, J. J., & Leitch, A. R. (2008). The ups and downs of genome size evolution in polyploid species of Nicotiana (Solanaceae). *Annals of Botany (London), 101*, 805–814.

Lopez, R. A., Boehm, R., Pineda, M., Gunther, P., Carstensen, F. (2017, September). *Economic impacts of Connecticut's agricultural industry: Update 2015*. Zwick Center for Food and Resource Policy Outreach Report No. 47. University of Connecticut.

Lownsbery, B. F. (1953). Host preferences of the tobacco cyst nematode (*Heterodera* sp.). *Phytopathology, 43*, 106–107.

Lownsbery, B. F., & Lownsbery, J. W. (1954). *Heterodera tabacum* new species, a parasite of solanaceous plants in Connecticut. *Proceedings of the Helminthological Society of Washington, 21*, 42–47.

Lownsbery, B. F., & Peters, B. G. (1955). The relation of the tobacco cyst nematode to tobacco growth. *Phytopathology, 45*, 163–167.

Lucas, L. T., Barker, K. R., & Blake, C. T. (1978). Seasonal changes in nematode densities on bentgrass golf greens in North Carolina. *Plant Disease Reporter, 62*, 373–376.

Mai, W. F., Crittenden, H. W., Jenkins, W. R. (1960). *Distribution of stylet-bearing nematodes in the Northeastern United States*. NJ. Agricultural Experiment Station Bulletin 795. 62 pp.

Mai, W. F., Bloom, J. R., & Chen, T. A. (1977). Biology and ecology of the plant-parasitic nematode *Pratylenchus penetrans*. Bulletin 815. In *Pennsylvania State University College of Agriculture*. Pennsylvania: University Park.

Malek, R. B. (1980). Population response to temperature in the subfamily Tylenchorhynchinae. *Journal of Nematology, 12*, 1–6.

Marais, M. (2001). *A monograph of the genus Helicotylenchus Steiner, 1945 (Nemata: Hoplolaimidae)*. Ph.D. Dissertation. University of Stellenbosch, Stellenbosch, South Africa.

Martin, S. B. (1988). Identification, isolation frequency, and pathogenicity of anastomosis groups of binucleate *Rhizoctonia* spp. from strawberry roots. *Phytopathology, 78*, 379–384.

McClure, M. A., Nischwitz, C., Skantar, A. M., Schmitt, M. E., & Subbotin, S. A. (2012). Root-knot nematodes in golf course greens of the western United States. *Plant Disease, 96*, 635–647.

Michell, R. E., Malek, R. B., Taylor, D. P., & Edwards, D. I. (1973). Races of the barley root-knot nematode, *Meloidogyne naasi* III. Reproduction and pathogenicity on creeping bentgrass. *Journal of Nematology, 5*, 41–44.

Miller, P. M. (1976). Effects of nematicides on nematode densities in turf in Connecticut. *Journal of Nematology, 10*, 122–127.

Myers, R. F., Wagner, R. E., & Halisky, P. M. (1992). Relationship between cultural factors and nematodes on merion Kentucky bluegrass. *Journal of Nematology, 24*, 205–211.

Nelson, E. (1995). Nematode disorders of turfgrass, how important are they? *Turfgrass Trends, 10*, 1–16.

Niblack, T. (2003). More details on corn nematodes. *The Bulletin*. University of Illinois Extension, 15 May 2003. bulletin.ipm.illinois.edu.

Perry, V. G., Darling, H. M., Thorne, G. (1959). *Anatomy, taxonomy and control of certain spiral nematodes attacking bluegrass in Wisconsin*. Bulletin of the Wisconsin Agricultural Experiment Station No. 207. 24 pp.

Qing, Y., Potter, J. W., & Gilby, G. (1998). Plant-parasitic nematodes associated with turfgrass in golf courses in southern Ontario. *Canadian Journal of Plant Pathology, 20*, 304–307.

Radewald, J. D., Pyeatt, L., Shibuya, F., & Humphrey, W. (1970). *Meloidogyne naasi*, a parasite of turfgrass in southern California. *Plant Disease Reporter, 54*, 940–942.

Raub, K., Campbell, B. L., Wallace, V., Henderson, J., Inguagiato, J., Rackliffe, D. (2015). *Economic impact of the turfgrass industry in New England*. University of Connecticut Publication. 43 pp. http://www.extension.uconn.edu/root/archives_32_4126732338.pdf

Rowe, R. C., Riedel, R. M., & Martin, M. J. (1985). Synergistic interactions between *Verticillium dahliae* and *Pratylenchus penetrans* in potato early dying disease. *Phytopathology, 75*, 412–418.

Rowe, R. C., Davis, J. R., Powelson, M. L., & Rouse, D. I. (1987). Potato early dying: Causal agents and management strategies. *Plant Disease, 71*, 482–489.

Rungrassamee, W., Wick, R. L., & Dicklow, B. (2003). Relationship of *Pasteuria* to root-knot and stunt nematodes. *Phytopathology, 93*, S135.

Sasser J. N., & Carter, C. C. (1985). Raleigh, NC: North Carolina State University Graphics: An advanced treatise on *Meloidogyne*, vol. I. Biology and Control. 422 pp.

Settle, D. M., Fry, J. D., Milliken, G. A., & Todd, T. C. (2006). Quantifying the effects of lance nematode parasitism in creeping bentgrass. *Plant Disease, 91*, 1170–1179.

Sikora, R. A., Taylor, D. P., Malek, R. B., & Edwards, D. I. (1972). Interaction of *Meloidogyne naasi, Pratylenchus penetrans,* and *Tylenchorhynchus agri* on creeping bentgrass. *Journal of Nematology, 4*, 162–165.

Smolik, J. D., & Malek, R. B. (1972). *Tylenchorhynchus nudus* and other nematodes associated with Kentucky bluegrass turf in South Dakota. *Plant Disease Reporter, 56*, 898–900.

Spears, J. F. (1968). *The golden nematode handbook, survey, laboratory, control and quarantine procedures*. US Department of Agriculture Handbook 353. 81 pp.

Stone, G. E., & Smith, R. E. (1898). *Nematode worms*. Hatch Experiment Station Massachusetts Agricultural College Bulletin No. 55. 69 pp.

Stone, A. R., Platt, H. M., & Khalil, L. F. (1983). Three approaches to the status of a species complex, with a revision of some species of *Globodera*. In A. R. Stone, H. M. Platt, & L. F. Khalil (Eds.), *Concepts in nematode systematics* (pp. 221–223). London: Academic.

Sumner, D. R. (1967). Nematodes in bluegrass. *Plant Disease Reporter, 51*, 457–460.

Szczygiel, A., & Hasior, H. (1972). Seasonal variations in population of plant parasitic nematodes in strawberry plantations. *Ekologia Polska, 20*, 1–16.

Taylor, C. E. (1967). The multiplication of *Longidorus elongatus* (de Man) on different host plants with reference to virus transmission. *Annals of Applied Biology, 59*, 275–281.

Troll, J., & Tarjan, A. C. (1954). Widespread occurrence of root parasitic nematodes in golf course greens in Rhode Island. *Plant Disease Reporter, 38*, 342–344.

Turenne, J., & Payne, M. (2011). *Rhode island prime farmland soil loss by development*. Publication of USDA NRCS. Warwick.

USDA NASS. (2012). *Census of agriculture, volume 1, chapter 1: State level data*. Rhode Island. Available at: https://www.agcensus.usda.gov/Publications/2012/Full_Report/Volume_1,_Chapter_1_State_Level/Rhode_Island/

Walker, N. W., & Martin, D. L. (2001). Effects of *Tylenchorhynchus claytoni* on creeping bent grass. *Phytopathology, 91*, S93.

Wick, R. L., & Vittum, P. J. (1988). Spatial and temporal distribution of plant parasitic nematodes in putting greens in the New England region. *Phytopathology, 78*, 1521.

Zeng, Y., Weimin, Y., Martin, S. B., Martin, M., & Tredway, L. (2012). Diversity and occurrence of plant-parasitic nematodes associated with golf course turfgrasses in North and South Carolina, USA. *Journal of Nematology, 44*, 337–347.

Chapter 2
Plant Parasitic Nematodes of New York, New Jersey and Pennsylvania

George W. Bird, George S. Abawi, and James A. LaMondia

2.1 Introduction

New York, New Jersey and Pennsylvania have diverse geologies and geographies including forests, rivers, mountains, lakes and associated rural, suburban and urban communities. Their plant agriculture also varies widely with numerous fruit, vegetable and agronomic crops grown on coarse-textured to fine-textured and organic soils. The population density of more than 40 million people allows for vibrant ornamental and recreational plant industries. Plant parasitic nematodes are known to be associated with most, if not all of these ecosystems. Some of these nematode species are key limiting factors of major economic significance and have a sound associated research base. Relatively little, however, is known about the majority of the species of plant parasitic nematodes that exist in New York, New Jersey and Pennsylvania ecosystems. The first report of a plant parasitic nematode in the region was *Meloidogyne* sp. identified in New Jersey by Halstead in 1891 (Mai 1995).

The Science of Nematology in New York, New Jersey and Pennsylvania is diverse. The Widely Prevalent Nematodes of the United States (WPNL, NY-2011, NJ-2014, PA-2015) lists 77, 52 and 49 species as being present in Pennsylvania, New Jersey and New York, respectively. While the discovery of *Globodera*

G. W. Bird (✉)
Department of Entomology, Michigan State University, East Lansing, MI, USA
e-mail: birdg@msu.edu

G. S. Abawi
New York State Agricultural Experiment Station, Cornell University, Geneva, NY, USA
e-mail: gsa@cornell.edu

J. A. LaMondia
Department of Agriculture, The Connecticut Agricultural Experiment Station, Windsor, CT, USA
e-mail: james.lamondia@ct.gov

S. A. Subbotin, J. J. Chitambar (eds.), *Plant Parasitic Nematodes in Sustainable Agriculture of North America*, Sustainability in Plant and Crop Protection,
https://doi.org/10.1007/978-3-319-99588-5_2

rostochiensis in New York in 1941 formed the basis of a strong nematology program at Cornell University (Mai and Lownsberry 1946), the January 23, 1956, organization of the Regional Research Project entitled, *Cultural and Chemical Control of Soil-Borne Pests* (NE-34) served as a major catalyst for nematology research and Extension in the region (Jenkins et al. 1963). The project was renewed and renamed on July 1, 1960, as the *Biology of Plant Parasitic and Soil Inhabiting Nematodes*. By 1963, there were six nematology courses taught among the three states; with nematodes being covered in seven other courses. In addition, 29 graduate students had been trained and *circa* 4000 h devoted to Extension programming. In 2017, this project, *Role of Plant parasitic Nematodes and Nematode Management in Biologically Based Agriculture* (NE-1640) is the oldest continuous Regional Research project in the U.S. Two of its many significant developments were the centrifugation-flotation technique for recovering nematodes from soil (Jenkins 1964) and pioneering nematode tissue culture research by Tiner (1961), both from Rutgers University. Pennsylvania State University provided early leadership for turfgrass nematology (Couch and Bloom 1959; Bloom 1961).

The objective of Plant Parasitic Nematodes of New York, New Jersey and Pennsylvania is to document the history, distribution, economic significance and management of the key species through summaries of published research. It is organized around sedentary endoparasites, migratory endoparasites and ectoparasites of root tissues and shoot system tissue parasites of major economic significance. Species presumed to be of less significance, but known to exist from various surveys, collections and the Widely Prevalent Nematodes of the United States database are included for these states. The topics of virus vector relationships, predisposition agents and nematode management are described in the appropriate parasitism sections.

2.2 Sedentary Endoparasites

2.2.1 Potato Cyst Nematode

The golden nematode, *Globodera rostochiensis* is a cyst-forming parasite of potato plants that originated in the center of diversity of potatoes in the Andes of South America (Mai 1977). The golden nematode has been widely distributed in many of the world's potato producing countries including much of Europe (Evans and Trudgill 1978), but this nematode and a closely related species, *Globodera pallida* are still of world-wide regulatory concern due to the severity of yield losses associated with infestations (Mai 1977). *Globodera rostochiensis* was first identified in the United States in New York State in July of 1941 when it was observed as the cause of severe stunting and yield loss in a Long Island Nassau County potato field (Cannon 1941). At that time, it had been present for years and was causing up to 70% yield loss in the affected area (Chitwood et al. 1942). Surveys of potato fields throughout Long Island demonstrated its spread, likely from that first infested field (Spears 1968). Quarantine measures were put into place by 1944 to restrict further spread. A new location, however, was discovered in Steuben County in upstate New York in 1967 (Mai 1977). Surveys of other potato producing states discovered small infestations in Delaware and New Jersey that were contained and permanently managed by taking the fields out of agricultural production.

A unique cooperative effort between the USDA/APHIS, USDA/ARS, New York State Department of Agriculture and Markets, Cornell University, New York State counties and towns and the New York State potato industry have successfully conducted a long-term program that combines regulatory and research efforts for survey and inspection, sanitation, and management. Management is designed to minimize the spread of cysts from infested fields and continually suppress nematode populations. For years, the golden nematode continued to be detected in additional locations in New York and it seemed as though regulators were chasing the pest. Sound sanitation and management systems developed by the research program, however, ultimately reduced nematode populations and limited spread. The long-term success of the program was demonstrated by cyst nematode population reduction to the point that potato fields were eventually deregulated. Beginning in 2010, some townships in New York were deregulated and released from the Golden Nematode Quarantine area (Kepich 2011). By 2015, it was reported that the Golden Nematode quarantine area had been reduced by 140,672 ha in nine counties to 2503 ha located in parts of eight New York counties (Kepich 2016).

While easily summarized, the Golden Nematode Regulatory Program in New York was a massive cooperative undertaking. Surveys are expensive and time-consuming. Quarantine and sanitation regulations were sometimes seen by growers as overly restrictive, and research progress required an integrated team approach and decades of effort (Mai and Spears 1954). The quarantine was initially put in place in 1944 and the first deregulation of potato fields still in production in quarantined areas did not happen until 2010. State and federal regulatory officials continue to conduct soil surveys and work toward deregulating additional areas which prove to be free of Golden Nematode infestations.

The Golden Nematode Program was deemed necessary as *G. rostochiensis* can be very damaging, and capable of causing total yield loss. The golden nematode can be easily spread to new locations as each cyst can contain hundreds of eggs which can survive desiccation and be easily moved in soil adhering to equipment, potato tubers, shoes or anything capable of moving soil. Each cyst can potentially transport a small population capable of starting a new infestation (Brodie 1993). Management of the nematode can be extremely difficult as a result of several unique features of its life cycle. Encysted juveniles within eggshells deposited in a new location may remain quiescent for years until stimulated to hatch by suitable environmental conditions and more importantly, recognition of host roots by the presence of an unknown chemical hatching stimulant, usually produced by potato plant roots (Fenwick and Widdowson 1959; Devine and Jones 2003). It may take years for nematode populations to reach damaging levels, even with continuous production of a suitable host. Few or no diagnostic symptoms are evident at low population densities. Even at low population densities, spread may occur as movement of a single cyst relocates many individuals and has the potential to start a new infestation in a new location (Brodie 1993).

Regulatory survey efforts concentrated on detecting new locations as early as possible (Spears 1968). Any field found to be infested and associated fields exposed to infestation were subject to quarantine regulations (Spears 1968). Population reduction procedures were initiated immediately. Control tactics and research efforts were not aimed at managing the pest to avoid yield losses as was done in many countries, but rather at eradication. Later, the long-term approach of continually reducing populations to maintain nematode population densities below detectable levels was used to reduce the probability of spread to fields in non-infested areas. One highly successful outcome of this intensive effort to reduce spread was that no potato grower in New York has experienced any level of yield loss due to the nematode since 1946 (Spears 1968; Brodie and Mai 1989).

2.2.1.1 Management Strategies

Management of the golden nematode initially focused on crop rotation away from host plants such as potato and tomato. Crop rotation alone did not allow an increase in nematode numbers, but was also relatively ineffective in reducing populations. Few juveniles hatched from cysts in the absence of a host and the hatch stimulation produced by host plant roots. Rotation required four or more years of cereals or other non-host crops to be effective (Mai and Lownsbery 1948; Mai and Harrison 1960). This was not practical for potato growers and the presence of solanaceous weeds during rotations could allow population increases and jeopardize control (Sullivan et al. 2007). Nematicides were included in the program in the 1940s. The soil fumigant dichloropropene-dichloropropane was quickly established as the most effective chemical product to reduce populations in fields where infestations were detected. Since fumigant nematicides are less effective near the soil surface where gases escape before nematode control can be achieved, fields were fumigated twice, 10 days apart, and the soil was turned between applications (Brodie and Mai 1989). Non-fumigant nematicides were investigated and incorporated as an additional tool in the management program in 1974 (Brodie 1980).

Breeding for plant resistance to the golden nematode in New York got its start when Ellenby (1954) identified a clone of *Solanum tuberosum* subsp. *andigena* that did not support reproduction and transferred that resistance into commercial *Solanum tuberosum* potatoes. This single dominant gene for resistance (H1 gene) was effective against the only pathotype present in New York at that time (Ro1). It was widely used in the Cornell potato breeding program to develop a range of varieties with resistance to *G. rostochiensis* Ro1. The first golden nematode resistant variety was released in 1966 as ‘Peconic’ (Peterson and Plaisted 1966). Since that time, over 20 varieties have been released in New York with the H1 gene for Golden Nematode resistance.

During the 1980s, concerns about environmental risks and contamination of groundwater with nematicides shifted the approach to management of the golden nematode away from chemical control to biological control, utilizing potato plant resistance and non-host crops. Plant resistance is very effective. Resistant plants stimulate nematode hatch, but do not allow feeding cell development, preventing reproduction of the nematode. A 4-year rotation cycle in soils with very low nematode populations (below 4 eggs/1 cm^3 of soil) consisting of two seasons of growing resistant potatoes followed by a non-host and then a susceptible potato crop, resulted in an overall population decline (Brodie 1996a; LaMondia and Brodie 1986). This strategy successfully reduced populations to the point where the nematode was eradicated from much of the quarantined acreage as stated above. The *G. rostochiensis* populations in the United States were believed to only consist of pathotype Ro1, which does not reproduce on potatoes with the H1 resistance gene, however, long-term exposure to that single source of resistance selected for an Ro2 pathotype, which was likely already present in certain locations (Belair and Simard 2009). The first indication of atypical reproduction on H1 potatoes in research plots was reported in 1996 (Brodie 1996b). Management of Ro2 pathotype was limited to the use of non-host crops until potato varieties with resistance to both Ro1 and Ro2 become available. A breeding line 'NY-140' has been demonstrated to carry resistance to both pathotypes (Kepich 2016). It is a promising source of resistance for future management of the golden nematode. In addition, it has been shown that populations of the golden nematode associated with Peconic, a potato variety resistant to the golden nematode, are very sensitive to soil disturbance through cultivation practices and desiccation.

2.2.2 *Sugar Beet Cyst Nematode*

The sugar beet cyst nematode, *Heterodera schachtii* (SBCN) has a broad host range. It consists of over 218 plant species, that are mostly members of the plant families Chenopodiaceae and Cruciferae (Steele 1965). This nematode is worldwide in distribution and is considered the major limiting factor in the production of sugar beets wherever they are grown. The SBCN was first reported in New York in 1961, causing damage to table beets grown for beet greens on a farm near Syracuse, New York (Mai 1961). The first observation of severe damage caused by the SBCN in a commercial field, however, was made in 1970 on table beets in a field near Lyon, New York (Mai et al. 1972). The occurrence and damage of the SBCN on several vegetable crops was well known and documented in numerous production areas in the United States and elsewhere (Lear et al. 1966; Olthof et al. 1974; Radewald et al. 1971). SBCN is also known to exist in Pennsylvania (WPNL 2015).

Results of an extensive survey conducted in 1970–1971 showed that the SBCN was distributed throughout the table beet- and cabbage-growing regions of Central and Western New York (Mai et al. 1972). The highest level of SBCN infestation level detected in a commercial field was 190 eggs and juveniles per 1.0 cm^3 soil. The practice of returning soil and debris (tare soil) to be spread on production fields from beet- and cabbage-processing plants undoubtedly contributed greatly to the wide spread distribution of this nematode. The tare soil returned to the farm usually included soil and debris from the grower(s) that made the previous delivery to the plant. Fortunately, this practice was voluntarily discontinued.

2.2.2.1 Crop Damage

Susceptible crops in addition to table beets and cabbage include cauliflower, broccoli, brussel sprouts, turnip, spinach, rutabagas and radish. In addition, the nematode feeds and reproduces on several weed hosts including shepherd's purse, dock, chickweed, hen nettle and wild radish. Above-ground symptoms of severe SBCN on table beets and cabbage are not diagnostic and appear initially as poor and uneven growth (Fig. 2.1a). Diagnostic symptoms of discoloration and reduced size can be observed on roots (Fig. 2.1b), in addition to white immature females on the root surface. The developing white females can be seen with the naked eye, especially on the red colored beet roots (Fig. 2.1c). Beet roots of severely infected plants are misshapen and small, whereas cabbage heads are loose, light in weight and smaller in diameter (Fig. 2.1d). The white females continue to develop on the root, become dirty white in appearance and turn brown, hard and lemon-shaped with up to 500 eggs within each cyst. The mature cysts fall into the soil when the roots die or decay, where the eggs within the cyst can survive for 7–8 years in the absence of a host. On its own, this nematode is able to move only a few centimeters per year, but

Fig. 2.1 (**a**) Poor and uneven growth of table beets in a commercial field with high infestation of *H. schachtii*; (**b**) Infected cabbage roots with *H. schachtii* (right) are discolored and smaller; (**c**) White immature females of *H. schachtii* on the surface of table beet roots; (**d**) Heavy infection of cabbage plants with *H. schachtii* result in the production of heads that are small, loose and light in weight (right)

distribution within and between production fields is accomplished through movement of contaminated soil and infected root tissues on farm equipment, animals, surface running water or in infected transplants.

Extensive damage to cabbage and table beet by the SBCN was demonstrated under greenhouse and field conditions in New York (Mai et al. 1972; Abawi and Mai 1977, 1980, 1983). The significant damage observed was negatively correlated with the initial soil population of the SBCN. The damage threshold density under field conditions to both total and marketable yields of table beet and cabbage was about 6–9 eggs/1 cm^3 soil. The latter affected both the quality and quantity of marketable yield of both crops. Significant damage to these and other vegetables have been reported from other regions in the United Stated and elsewhere (Lear et al. 1966; Radewald et al. 1971; Olthof et al. 1974).

2.2.2.2 Management Strategies

Cost-effective management of the SBCN dictates the employment of a multi-tactic strategy. Such a strategy might involve two or more of the following measures: sanitation practices, monitoring of soil population densities, crop rotation and use of chemical or biological control products. It is usually easier to prevent field contamination than to manage SBCN-infested field soils and those of other long-surviving pathogens. Thus, it is critical to use only certified transplants of host crops that are free of SBCN as well as other pathogens, if transplants are required for crop establishment. Also, farm equipment and tools used on infested or suspect soils should be washed thoroughly and dried before entering non-infested fields.

It is highly recommended that target planting sites are sampled and analyzed to establish a base line data and to follow-up with annual assessment to monitor the changes in the population density of SBCN. This information is needed as a decision tool for adjusting the length of the crop rotation with non-host crops or determining the appropriate control products to keep populations below the economic threshold density. In the absence of resistant commercial varieties of cabbage, table beets and other vegetables, crop rotation with non-hosts (corn, wheat, beans, cucurbits and many others) becomes of great importance in the management of SBCN. Results of an extensive crop rotation experiment on a commercial farm in New York demonstrated that populations of the SBCN increased with a non-host:host crop rotations of 1:2 and 1:1; whereas it decreased with those of 2:1–5:1 (Mai and Abawi 1980). The length of the rotation with non-host crops required to decrease the population of the SBCN below the damage threshold density is dependent on the level of soil infestation. In addition, the status of cover crops used should be determined and incorporated in the rotation scheme. Bio-fumigant crops incorporated as green manures might be helpful and their effects on the population of the SBCN, if any, should also be assessed. Furthermore, for any rotational scheme to be effective, weed hosts of the SBCN should be also controlled in order to prevent the survival of the nematode.

Numerous fumigant and non-fumigant nematicides were found to be effective in reducing the population and damage of the SBCN on cabbage and other hosts (Lear et al. 1966; Radewald et al. 1971; Keplinger et al. 1979; Abawi and Mai 1983). Pre-plant soil fumigants are more effective in light-textured soils as compared to their effectiveness in heavy-textured soils such as those prevailing in the cabbage and table beet producing areas of New York. Fenamiphos was found to be cost-effective and was registered for use on cabbage against this nematode in New York. Overall, it will be best to implement an integrated approach in the management of the SBCN and as a component of needed sustainable soil health management practices.

2.2.3 Other Cyst Nematodes

The soybean cyst nematode (*Heterodera glycines*) is a key limiting factor in U.S. soybean production. It was first detected in New Jersey more than two decades ago and more recently identified in Pennsylvania. The initial find of the soybean cyst nematode in New York was in Cayuga County in 2016 (Wang et al. 2017). The clover cyst nematode (*Heterodera trifolii*) is the only other species currently reported from the three states.

2.2.4 Root Knot Nematodes

Species of root knot nematodes are major pathogens of many crops in diverse plant families and are widely distributed throughout the United States and the world (Sasser and Carter 1985; Mitkowski and Abawi 2003a). Published reports on the root knot nematodes in New York and neighboring states date back to the late nineteenth century (Newhall 1942). However, only the northern root knot nematode (*Meloidogyne hapla*, NRKN) has been documented as being able to survive the characteristically low winter temperatures in New York (Mikowski et al. 2002; Mitkowski and Abawi 2003b) and likely in similar production areas, resulting in natural infections and damage to host crops the following growing season. Other warm temperature species of root knot nematodes (*M. incognita*, *M. javanica* and others) have been introduced on planting materials and observed causing damage during the growing season of introduction. Observations and standard recovery tests documented their failure to survive and cause damage to host crops in the following season. In addition to NRKN, *M. graminis* and *M. graminicola* are recorded in the WPNL (2015) for New York, *M. arenaria* and *M. incognita* (WPNL 2014) for New Jersey and *M. incognita* and *M. javanica* for Pennsylvania (WPNL 2011).

2.2.4.1 Damage and Losses

More than 550 crop and weed species are hosts to the NRKN. Significant variation has been shown to occur among and within field populations throughout New York, as determined by reproductive fitness and severity of galling on lettuce as well as by its nuclear and mitochondrial genomes (Mitkowski and Abawi 2003c). Over the years, this nematode has been observed infecting soybean, alfalfa, clovers, many vegetables and weed species (including dandelion, purslane, mallow and plantain) and others. Severe infections and damage by the NRKN have occurred frequently in recent years on various crops grown in New York including carrots, onions, lettuce, potatoes, and others grown on both organic and mineral soils (Abawi and Laird 1994; Viaene and Abawi 1996). Above-ground symptoms on host crops growing in heavily infested soils exhibit general stunting and uneven growth in a patchy field pattern (Widmer et al. 1999; Fig. 2.2a–c). Severely infected plants may also exhibit nutrient deficiency symptoms, delayed maturity, wilting and reduced marketable yield and quality. The latter symptoms are due largely to the reduced ability of severely infected root systems to absorb and transport water and nutrients. The diagnostic symptoms of root knot infection occur on roots in the form of galls that are large and distinct on lettuce (Fig. 2.2d, e) to smaller knots or root thickenings on onion roots (Fig. 2.2f).

Based on the characteristic galls produced on lettuce, an on-farm bioassay was developed for visually assessing soil infestation with this nematode and implemented as a decision management tool by producers and land managers in New York (Gugino et al. 2008). Other symptoms resulting from root knot nematode infection are extensive branching, stubby and forked roots (Fig. 2.2g). Most of the life cycle of this endoparasitic nematode is completed within the root tissues of its hosts. The life cycle of this nematode can be completed in as a few as 17 days, depending on the host and soil temperature. Spread of this nematode can be accomplished mainly through water, planting materials and infested soil adhering to farm equipment, humans and animals.

Under heavy soil infestation levels and favorable conditions, the northern root knot nematode is capable, and has caused, significant yield losses in quantity and quality of several crops grown in New York. For example, extensive losses impacting farm profitability have occurred on carrots, lettuce, onion and other crops, that necessitated the implementation of cost-effective management practices (Viaene and Abawi 1996; Gugino et al. 2006; Abawi et al. 2003). Carrots are among the most susceptible crops for extreme damage caused by NRKN. Symptoms of infection can be detected as early as 4 days after planting in infested soil (Slinger and Bird 1978). The damage threshold level of NRKN on commercial carrot varieties under field conditions was estimated to be <1 egg per 1.0 cm^3 soil (Gugino et al. 2006). Marketable yield losses as high as 45% were documented. Severely infected carrot roots are stubby, forked with numerous galls and not marketable. The damage threshold density of NRKN on lettuce was calculated to be one to two eggs per 1.0 cm^3 of soil (Viaene and Abawi 1996). Severely infected lettuce plants are small, loose and fail to produce marketable size heads. Similarly, heavily infected onion

Fig. 2.2 (**a**) Lettuce plants growing in organic soil heavily infested with *M. hapla*, exhibiting uneven and stunted growth; (**b**) Stunted growth of onion plants in a section of a commercial field with high infestation of *M. hapla*; (**c**) Extreme stunting and damage of *M. hapla* to carrot in a section of a commercial field left as a check for the standard application of Vydate against this nematode; (**d**) Distinct and rather large galls induced by *M. hapla* on lettuce roots; (**e**) Close-up of a rootlet with several galls and mature females with egg sacs stained red; (**f**) Root thickenings induced by *M. hapla* on onion roots; (**g**) Severely infected carrot roots with *M. hapla* exhibiting a stubby growth, forking and hairiness

plants produce smaller bulbs with thicker necks and are delayed in maturity. NRKN reduces the storability of onions and complicates harvest, increasing costs. Potatoes and other crops are affected similarly by reducing the marketable yield and quality. In addition, species of root knot nematodes are known to interact with other soil-borne pathogens such as *Fusarium* spp., resulting in increased disease severity and yield losses (Mai and Abawi 1987; Abawi and Chen 1998).

2.2.4.2 Management Strategies

A sustainable and cost-effective integrated pest management strategy (IPM) is the most effective approach for control of NRKN. Although only limited practical management options are available, there still exist numerous combinations of these practical tactics for the control of this nematode. In general, these are compatible with overall soil health management practices (Abawi and Widmer 2000). Unfortunately, none of the commercial varieties of major crops evaluated including carrot, onion, lettuce, potato and soybean, are resistant to the NRKN (Abawi and Ludwig 2003a, b; Gugino et al. 2006; Abawi et al. 2006). Fortunately, grain crops are not hosts of NRKN. Rotating out of host crops for two or more years with field or sweet corn, wheat or other grains will effectively manage this nematode, as long as, host weeds are also controlled.

Use of grain cover crops such as rye grain, annual ryegrass, barley, oats, tall fescue or sudangrass, in the rotation will contribute to the management of the NRKN. In addition, the incorporation of green manures of sudangrass cv. Trudan in moist and warm soils was found highly effective in reducing the population and damage of the NRKN (Viane and Abawi 1998). Suppression of the NRKN by sudangrass was attributed to its content of the cyanogenic glucoside dhurrin that upon biological decomposition in soil results in the production of hydrogen cyanide (HCN), a potent biocide, especially against nematode eggs (Widmer and Abawi 2000). Interestingly, cultivars and hybrids of sudangrass differed considerably in their suppressiveness against the NRKN. This was found to be closely correlated with the level of cyanogenic dhurrin in their cell walls (Widmer and Abawi 2002). Similarly, the incorporation of green manures of rapeseed, *Tagetes patula* (cvs. Jupiter, Polynema and Nema-Gone) and several accessions of white clover and flax were also found to be effective. Selected white clover and flax cultivars are also cyanogenic and have a similar suppressive mechanism through the production of HCN. However, the bio-fumigant effect of incorporated green manures of rapeseed and other cruciferous crops are due to the production of the nematicidal products, isothiocyanates (Halbrendt 1996).

In addition, the application of corn silage, brewery compost, chicken manures and other organic amendments have often been found to suppress the NRKN and plant parasitic nematodes in general. Thus, it is critical to carefully consider soil organic matter management, as it has major implications on the soil biology, crop productivity and sustainable soil health management (Abawi and Widmer 2000; Widmer et al. 2002).

Several fungi were found associated with eggs and juveniles of *M. hapla* including *Hirsutella rhossiliensis* and *Pochonia chlomydosporia*, which were found effective in reducing penetration and population of NRKN (Viaene and Abawi 1998a, b, 2000). Other nematophagous fungi and bacterial antagonists including *Pasteuria* and *Bacillus* spp., have been reported with activities against plant parasitic nematodes, but currently available commercial products are not widely used in New York.

Pre-plant soil fumigation with Telone-C or Vapam is available for nematode control in New York. Fumigation is effective when applied properly, but it is not widely used. Of non-fumigant nematicides, application of Vydate (Oxamyl) at planting as a spray incorporated or as an in-furrow treatment has been widely used, especially on carrots, onions and potatoes (Gugino et al. 2006, 2008). Seed treatment with Abamectin® was also found effective against NRKN in several trials on tomato, carrot and onions (Abawi et al. 2003; Abawi and Ludwig 2005). Overall, the adoption of multiple management tactics and the monitoring of the infestation levels of the NRKN are critical factors in the cost-effective management of this key pathogen and its damage to numerous crops in New York.

2.3 Migratory Endoparasites

2.3.1 Root Lesion Nematodes

Root lesion nematodes (*Pratylenchus* spp.) are widely distributed throughout the world and particularly in the northern temperate regions. A large number of species of lesion nematodes have been reported and are known to have a wide host range consisting of more than 400 plant species. The latter include fruits (apples, cherries, peaches, pears and other crops), grain crops (corn, wheat, rye, oat, barley and other crops), legumes (alfalfa, clover, vetch, soybean), vegetables (potato, bean, onion, cabbage, carrot, tomato and other crops) and many species of weed plants. The earliest reports on the occurrence of lesions nematodes in New York State were in 1956 (Mai and Parker 1956). It was reported that lesion nematodes caused losses on cherry, and their role in the replant problem on cherry and other fruits was also suggested (Mai and Parker 1956; Parker and Mai 1956a, b). The first report describing the relationships between *P. penetrans* on vegetables (Wando peas) and corn in New York was in 1960 and 1963 respectively (Dolliver et al. 1960, Miller et al. 1963). DiEdwardo (1961) documented the seasonal population variations of *P. penetrans* associated with strawberry, while Abu-Gharbieh et al. (1963) reported a relationship between root lesion nematodes and *Verticillium* and Heald (1963) documented root lesion parasitism of woody ornamentals. In addition, the collaborative Northeast Nematology Regional Research Project focused on the biology, ecology, culturing and management of lesion nematodes in the northeast region and beyond. The significant results of the latter were summarized in a bulletin authored by project participants (Mai et al. 1977). In addition, *P. neglectus* was detected in New York in 1997 (Timper and Brodie 1997).

2.3.1.1 Crop Damage and Losses

Today, lesion nematodes are widely distributed throughout the agricultural production regions in New York and neighboring states. In addition, population densities of lesion nematodes have increased significantly in recent years and the trend is continuing. Undoubtedly, the latter is due to the recent promotion and adoption of using grain and legume crops in long-term soil health management programs (Abawi and Widmer 2000). Numerous attempts over the years to characterize the populations of lesion nematodes on fruits, vegetables and other crops have shown that the primary species involved is *P. penetrans*, although with some morphological variability such as the shape and length of tail (Troccoli et al. 2003). Foliar symptoms on severely infected plants are of general poor and uneven growth, chlorosis and delay in maturity (Figs. 2.3a, b). Depending on the host, roots of severely infected plants exhibit poor development, discoloration and lack of adequate numbers of fibrous roots. Brown to black and narrow lesions may be visible on root surfaces at an early stage of infection of some hosts including soybeans and potatoes (Fig. 2.3c). Heavily infected young fruit trees often fail to produce good frame roots (Fig. 2.3d) and also generally lack functional feeder roots. In addition, shoot and root system growth of pear (Fig. 2.3e) and those of other fruits were shown to be drastically reduced under soil infestation with *P. penetrans* in greenhouse tests. The involvement of other soil-borne pathogens and saprotrophic organisms results in increased discoloration, rotting and death of roots. For positive diagnosis and damage assessment, it is critical to extract and confirm the identity of lesion nematodes in roots and associated soil. These nematodes survive as eggs and adults in roots of host crops or in soil. Distribution of lesion nematodes within and between production fields is mainly in infested soil on farm equipment, surface water or wind as well as in infected planting materials.

At high population densities, lesion nematodes are capable of causing significant losses in the quality and quantity of yield of both annual and perennials crops, especially in sandy soils. For example, maturity of onion was delayed and bulb weight was reduced by as low as 100 *P. penetrans*/100 cm^3 of soil. Most importantly, lesion nematodes, especially *P. penetrans*, are involved in classical disease complexes including the replant diseases of fruit trees (Fig. 2.3f), early dying of potato, and black root of strawberries (Mai and Abawi 1978; Abawi and Chen 1998; LaMondia 2004). Results from a survey of 27 strawberry farms in New York did not show a close correlation between populations of *P. penetrans* and poor root growth (Wing et al. 1995). This might be due to the different biology of black root rot in New York and the heavy and wet soil conditions of the farms sampled. However, lesion nematodes at even a very low number are capable of increasing infection and damage of other soil-borne pathogens and also non-pathogens on many crops.

Fig. 2.3 (**a**) Uneven growth of potato in a commercial field due to heavy infestation with *P. penetrans* (courtesy of W. F. Mai); (**b**) Delay in maturity of onions (green plants) due to high infestation with *P. penetrans*; (**c**) Close-up of the characteristically brown to black and narrow lesions of *P. penetrans* on soybean roots; (**d**) Infection of young trees with a high population of *P. penetrans* contribute to the production of poor frame roots (right); (**e**) Growth of pear seedlings in soil infested with different levels of *P. penetrans* (0; 5000; 10,000; and 20,000/13 cm clay pots under greenhouse conditions (Courtesy of W. F. Mai); (**f**) Uneven growth of 6-year old peach trees in a replant disease site in Western New York (Courtesy of W. F. Mai)

2.3.1.2 Management Strategies

Due to the wide and diverse hosts of lesion nematodes as well as the limited number of available resistant crop cultivars, it is a challenge to develop cost-effective management programs for these nematodes. The best approach requires implementation of multiple control measures on an as-needed basis. Thus, it is important to monitor the population densities of lesion nematodes on both annual and perennial crops, especially close to planting times.

There has been a considerable focus on the use of narrow or broad spectrum preplant fumigants for controlling lesion nematodes, especially on fruit crops (Arneson and Mai 1976; Mai and Abawi 1981; Mai et al. 1994). Reduction of lesion nematode populations was greater and significantly higher crop yields were achieved by broad-spectrum fumigants such as Vorlex (methyl isothiocyanate), as compared to the benefits resulting from using narrow spectrum nematicides such as Telone (1,3-dichloropropene). The benefits of soil fumigants were further enhanced by adjusting soil fertility, controlling weeds or establishing a ground cover with poor hosts for lesion nematodes such as perennial ryegrass and tall fescue. Also, the use of several non-fumigant nematicides including Fenamiphos, Carbofuran and Vydate, have been shown to be effective in reducing the populations of lesion nematodes on several crops including fruits and vegetables. In addition, the foliar application of Vydate was demonstrated to be effective in reducing the numbers of the lesion nematodes in roots and in soil around roots of fruit trees (Abawi and Mai 1975). The use of Vydate against the lesion nematode on onion, potatoes and other vegetables is also common in New York. Depending on the target crop and the production system, other management options that might be of interest are the addition of various organic amendments, incorporation of green manures of bio-fumigant crops, summer fallow and flooding, solarization, use of tolerant rootstocks of apples and other fruits, use of poor hosts (selected cultivars of ryegrass, oat, marigold, sesbania, alfalfa, and others) or resistant cultivars (marigold cv. Sparky and a wild oat cv. Saia) in a rotation, where available. Unfortunately, all commercial crop cultivars grown in New York, including those of onions, potatoes, beans, clovers and wheat tested under artificial inoculation in the greenhouse in New York, were found to be susceptible, although there were differences in their efficiencies as hosts to the population of *P. penetrans* used.

2.4 Ectoparasites

2.4.1 Dagger Nematodes

Dagger nematodes (*Xiphinema* spp.) are commonly found in temperate production areas of the world, around roots of perennial and annual crops as well as many weeds. In New York, Pennsylvania and other neighboring states, the species of dagger nematodes most reported belong to the *X. americanum*-group complex and *X. rivesi* (Forer and Stouffer 1982; Jaffee et al. 1987a, b; Molinari et al. 2004). In addition, populations of *Xiphinema* spp. from New York orchards expressed morphological morphological variations (Georgi 1988). In New York State, these nematodes were found primarily associated with roots of fruit trees (apple, pear, cherry and peach), small fruits (blueberry, raspberry and strawberry) and grapes. All life stages (eggs, juveniles and adults) can survive in soil and are spread by infested soil on farm equipment, water and planting materials. It has been suggested that only one generation per year is completed in northern temperate regions and that nematode population density is greatly impacted by soil temperature, moisture and texture. These large, ectoparasitic nematodes feed in the region just behind the root tips and can reach the vascular tissues with their long stylets. At extremely high populations, dagger nematodes are capable of causing numerous lesions, necrosis and destruction of feeder roots that at times result in the poor development of young plants. However, the main impact of dagger nematodes is their efficient ability to transmit several viruses. *Xiphinema* spp. are vectors of nepoviruses. These include tomato and tobacco ring spot viruses, *peach yellow bud mosaic virus* and other highly destructive virus pathogens of many crops. In New York, dagger nematodes were demonstrated to vector viruses that damage grapes, highbush blueberry and several fruit trees (Uyemoto et al. 1977; Fuchs et al. 2009).

Tomato ringspot virus (ToRSV) and *Tobacco ringspot virus* (TRSV) are efficiently transmitted from one host to another by *X. americanum*. Both viruses are located in different areas within the esophageal lumen (Brown et al. 1994, 1995; Wang et al. 2002). ToRSV has been identified in fruit trees in the region and infects a large number of small fruit, vegetable, ornamental and tree crops including blueberry and grape in New York as well as weeds. A large number of broadleaf orchard weed hosts of *X. americanum* serve as reservoirs of ringspot viruses. ToRSV can be seed transmitted in dandelion and acquired from infected plants by dagger nematode feeding (Mountain et al. 1983).

2.4.1.1 Management Strategies

Management of dagger nematodes and the nematode virus complex is best conducted with an integrated approach (Fuchs 2016). Virus-free certified planting stock is imperative. Tolerant or resistant cultivars or rootstocks should be used, if possible. Weed management in orchards is important as a means of reducing the presence and spread of virus reservoirs, and finally, reducing dagger nematode numbers is also critical, especially as disease incidence is not necessarily correlated with nematode numbers; individual nematodes may transmit virus. Limiting movement of soils between orchards or vineyards can also limit the distribution of viruliferous dagger nematodes. Management of dagger nematodes has historically relied on fumigant nematicides applied prior to orchard establishment and nonfumigant nematicides applied after planting (Halbrendt and Jing 1994; Bello et al. 2004; Halbrendt 2012).

More effective and practical nonchemical controls may include physical and biological tactics. While peach stem pitting has been known for a long time, its incidence increased when the cultural practices in orchards changed from mechanical cultivation of the entire orchard floor to eliminate weeds to establishing sod between rows and maintenance of weed-free strips within rows using herbicides to reduce compaction and soil erosion (Powell et al. 1982). Mechanical cultivation can reduce numbers of perennial weed hosts of nepo viruses such as dandelion, and also directly reduce population densities of dagger nematodes.

Biofumigation, as a form of allelopathy, has been demonstrated as an effective means of reducing the size of dagger nematode populations (Halbrendt and Jing 1994; Halbrendt 1996, 2012; Bello et al. 2004). Experiments in replanted orchard soils in Pennsylvania demonstrated that incorporation of a rapeseed rotation crop that released allelopathic nematicidal chemicals reduced dagger nematode numbers (Halbrendt 1996). Two rotations within a single year, an autumn planting of a winter rapeseed variety that was incorporated in the spring followed by a spring planting after 1 or 2 weeks that was incorporated in late summer, were as effective as a nematicide application (Halbrendt and Jing 1994; Jing 1994).

John Halbrendt's nematology research program at the Bigglerville Research Station of Pennsylvania State University was devoted to understanding the biology and control of dagger nematodes and virus diseases of peach trees and grape vines. Brown et al. (1994) demonstrated four *Xiphinema* spp. including three populations of the *X. americanum*-group and *X. rivesi*, as vectors of four North American nepoviruses. These included two strains of ToRSV, TRSV and *Cherry rasp leaf virus*. *Xiphinema rivesi*, however, was a more efficient vector than the other three species. Halbrendt's virus disease control research focused on prevention (removing virus reservoirs, testing for *Xiphinema*, evaluating site history and planting only certified virus-free vines or trees) and containment (removing symptomatic vines or trees and reducing the population density with chemical or biological procedures such as appropriate site preparation techniques). A significant amount of this research was pioneering work in respect to the potential roles of cover crops. In addition, it is important to note that fumigant- and non-fumigant-type nematicides are highly effective in reducing the numbers of dagger nematodes.

2.4.2 Stubby Root Nematodes

The nematode family Trichodoridae includes more than 95 species classified in five genera. The species currently recognized in the three states are all classified in the genera *Trichodorus, Paratrichodorus* and *Nanidorus*. The Widely Prevalent Nematode List (2014) includes *N. minor*, *P. nanus*, *P. pachydermus*, *P. porosus*, *T. aequalis* and *T. obscurus* for Pennsylvania, *P. nanus* and *P. pachydermus* for New Jersey and *N. minor* for New York. Species of the Trichodoridae vector short rod-shaped, single-stranded viruses (tobraviruses) that are severe pathogens of plants including potato which is widely grown in New York, Pennsylvania and New Jersey where the trichodoroid virus vectors are known to exist.

In the early 1960s, stubby root nematodes were a key limiting factor in onion production in organic soils in New York. Hoff and Mai (1962) documented the pathogenicity of *N. minor* on *Allium cepa*, and in 1967–1968, Bird and Mai published a series of articles on the embryogenesis, morphology, allometry and ecology of *N. minor* (Bird and Mai 1967a, b, 1968; Bird et al. 1968). They also published a numerical taxonomy of the family Trichodoridae and described a pointed tail stubby-root nematode, *Trichodorus acutus* (Bird 1967), from a population detected in the Cornell University Botanical Greenhouse. *Trichodorus acutus* has since been reported in nature in a number of localities worldwide.

2.4.3 Other Ectoparasites

The first report of ectoparasitic nematodes in New Jersey was made by Hutchinson et al. (1961). Bird and Jenkins (1964) identified 19 species of plant parasitic nematodes associated with *Vaccinium macrocaarpon* in cranberry bogs in New Jersey. They included *Atylenchus decalineatus, Mesocriconemas xenoplax, M. curvatum, Criconemoides* sp., *Ditylenchus* sp., *Helicotylenchus* sp., *Hemicycliophora similis, H. gracilis, Hemicycliophora* sp., *Hoplolaimus galeatus, Paratylenchus projectus, Rotylenchus uniformis, Scutellonema brachyurus, Tetylenchus joctus, Tylenchorhynchus dubius* and *T. maximus*. *Atylenchus decalineatus* was recovered from 80% of the 49 cranberry bogs included in the survey. *Atylenchus decalineaus* and *M. hapla* were the only two species that did not increase in population density in the associated parasitism studies and *M. curvatum, M. xenoplax,* and *H. similis* were pathogenic under greenhouse conditions. Jaffee et al. (1987a, b) reported on *M. curvatum* and *M. ornatum* associated with peach orchards in Pennsylvania.

Although the majority of the plant parasitic nematodes currently known to exist in New York, New Jersey and Pennsylvania are ectoparasites, relatively little is known about their specific biology and host parasite relationships in these states. Based on the Widely Prevalent Nematode List (2011, 2014, 2015) 77%, 65% and 51% of the species known to exit in New Jersey, Pennsylvania and New York, respectively, are ectoparasites. Species that are likely to be key pathogens in one or more of the three states, but not studied in detail include *Belonolaimus longicaudatus, Dolichodorus heterocephalus, D. marylandicus, Longidorus brevannulatus* and *L. elongatus*.

2.5 Shoot System Parasites

Seven species of shoot system tissue-feeding nematodes have been reported from New York, New Jersey or Pennsylvania. These include the stem and bulb nematode (*Ditylenchus dipsaci*), foliar nematodes (*Aphelenchoides ritzemabosi, A. fragariae, A. parietinus* and *A. besseyi*), seed gall nematodes (*Anguina* spp.) and potato rot nematode (*Ditylenchus destructor*). All of these can be highly destructive pathogens under the right conditions at specific locations. The stem and bulb nematode is a key pathogen and has been documented and researched extensively in New York. None of the other species have been documented or thoroughly researched in regards to their specific populations and relationships in the three states covered in this chapter.

2.5.1 Stem and Bulb Nematode

The stem and bulb (bloat) nematode, *Ditylenchus dipsaci,* is a destructive plant parasitic nematode of many crops, including onion, garlic and leek. It is widely distributed in temperate production regions of the world and has been known and studied extensively in Europe since 1877. In addition, this nematode occurs in many biological races with different host ranges and crop damage potential (Esquibet et al. 2003; Subbotin et al. 2005; Qiao et al. 2013). In addition to infecting garlics, onions, leeks, and chives, the garlic and onion race of *D. dipsaci* is reported to attack celery, certain varieties of peas and lettuce, hairy nightshade, Canada thistle, flower bulbs and several other plant species (Hooper and Southey 1978). The first report of the stem nematode in the United States was in 1929, when it was found damaging onions on a farm in Canastota, Madison County, New York, and again in 1939 on farms in Pine Island and Florida in Orange County, New York (Newhall et al. 1939; Newhall and Chitwood 1940). Severe infection and damage by the stem nematode to garlic was observed on a commercial farm in western New York in June 2010 (Abawi et al. 2011). The damage to garlic by the stem nematodes was reported from several other Northeastern states and Ontario, Canada and confirmed in 2011–2014 by the Nematode Diagnostic Service laboratory at Cornell University (Mountain 1957; Colett 2010; Johnson and Fuller 2012; Abawi pers. comm.).

Until the early 1960s, the stem nematode was widely distributed and caused economic losses to onions grown on organic soil throughout production regions in New York. The latter resulted in extensive research efforts to study the biology and management of this nematode (Lewis and Mai 1958, 1960; Mai et al. 1964; Smith and Mai 1964). Direct-seeding of onions was promoted and rapidly adopted by growers in the mid 1960s. This was done to avoid stem nematode damage as well as bacterial and fungal diseases that were associated with the use bulb sets at planting material. Since the use of true seeds of onions, damage by the stem nematode was rarely observed under commercial field conditions. A survey conducted shortly after the 2010 destructive outbreak of the stem nematode on garlic, clearly demonstrated that this nematode is widely distributed on garlic grown throughout New York (Abawi et al. 2011). It was recovered from garlic samples collected from 17 counties, with population densities as high as 3609/1 g of garlic tissue. The WPNL records *D. dipsaci* in Pennsylvania under the common name of alfalfa stem nematode (WPNL 2011), but not prevalent in New Jersey (WPNL 2015). A report by Pethybridge *et al.* (2016) confirmed the identity of the isolates recovered from infected garlic as *D. dipsaci* with extreme genetic uniformity. Only one isolate included in their study exhibited differences to those of *D. dipsaci* and only 97% similarity to *D. destructor*, thus it was labelled as *Ditylenchus* sp. The genetic uniformity of the characterized populations of the stem nematode suggested that a major introduction source was likely the cause of the latest infestation in garlic.

Vegetative propagation of garlic is continuing and seed exchanges and purchases among producers are of common occurrence. Unfortunately, early and light infestations of garlic bulbs by the stem nematode are symptomless. However, garlic plants severely infected by the stem nematode exhibit stunting, yellowing, collapse of leaves and premature dying. Infected bulbs initially show light discoloration, but later the entire bulb or individual cloves become dark brown, soft, sunken and light in weight. At later stages, infected bulbs may show cracks at the basal plate and various symptoms of decay resulting from the activities of other saprotrophic soil-borne organisms (Fig. 2.4a). Stem nematode infected onions and other hosts also show distinct swellings, twisting, and deformation of leaves, stems, bulbs and other foliar tissues. Severely infected seedlings and older plants may die before harvest (Fig. 2.4b).

2.5.1.1 Crop Damage and Losses

Infection and damage by the stem nematode significantly impacted onion production and profitability until the early 1960s. Currently, the stem nematode is a major constraint in garlic production. Yield losses as high as 100% have been observed in a few plantings. Symptomatic garlic bulbs are not marketable for fresh consumption

Fig. 2.4 (**a**) Mature garlic bulbs infected with *D. dipsaci* showing cracks and dry rotting of basal plates; (**b**) Pre-mature death of young garlic plants heavily infected with *D. dipsaci*; (**c**) Heavily infected garlic bulbs with *D. dipsaci* are not marketable and sorted out at harvest

(Fig. 2.4c) and those of infected lots, even at a low incidence, should not be sold for use as seeds. A good yield of garlic is about 8967 kg per hectare or higher and the price of garlic, although variable, is about $22 per kg or higher. Thus, even at a low percent of a yield loss, the impact of the stem nematode infection can be significant, especially to small-area garlic producers. About 30% of 400 garlic bulb samples obtained from garlic plantings throughout New York from June 2010 to early 2012, tested positive for the stem nematode. Interestingly, only about 10% of the garlic bulb samples analyzed in 2014 were found to be infected with the stem nematode. The latter might have been the results of the extensive outreach activities on the biology and management options of the stem nematode conducted in collaboration with personnel of Cornell Cooperative Extension, the Garlic Seed Foundation and garlic growers in New York and other states.

2.5.1.2 Management Strategies

Effective management of the stem nematode requires strict sanitation practices and the enforcement of quarantine regulations, in order to prevent the introduction of the nematode into production fields as well as the implementation of multiple control options. The latter includes the strict use of nematode-free seeds, hot water treatment of planting material, avoiding infested fields or treatment of soil with an appropriate product, practicing a proper crop rotation and the use of bio-fumigant cover crops (Lewis and Mai 1958; Dropkin 1989; Abawi and Moktan 2013). The wide host range of the stem nematode and its several biological races, however, makes its effective control difficult.

Infected planting materials are the major source of new infestation by the stem nematodes, thus only clean and stem nematode-free tested materials should be planted. Hot water treatment protocols of planting materials of garlic, onion and other crops, are available in the literature (Johnson and Lear 1965). However, hot water treatment should be considered only when clean planting materials are not available or when saving a valuable germplasm. Water temperatures reported to be effective against the stem nematode ranged from 38–49 °C, depending on the length of the soak period. Also, the efficacy of the hot water treatment was reported to improve with the addition of sodium hypochlorite, avermectin, formaldehyde, fungicides or other chemicals. Water temperature above 50 °C was reported to injure tissues of treated plant materials. The most common reported protocol for hot water treatment was a 20-min dip at 49 °C. In addition, clean planting materials should be planted only in stem nematode-free soil. It is critical to sample and analyze the soil of target planting sites for the presence of the stem nematode. Populations of the stem nematode as low as 10 per 500 cm^3 soil cause damage in many crops. Registered pre-plant soil fumigants (Telone®-C and Vapam® in NY) applied properly will control the stem nematode. Results of using non-fumigant nematicides (Vydate®, previously available for control of plant parasitic nematodes on onions, potatoes, carrots and other host crops in New York) have not been as affective as pre-plant fumigants against the stem nematode. Furthermore, practicing a long crop rotation

(3–4 years) out of susceptible hosts for the particular race(s) of the stem nematode is a highly effective management practice. For the onion and garlic race, rotating a site away from all *Allium* spp. (garlic, onion, leek, chives), celery, parsley, Shasta pea, salsify and other known hosts as well as controlling weed hosts (hairy nightshade and Canadian thistle) can be an important control tactic. Planting and incorporating green manures of known bio-fumigant crops (mustard, rapeseed, oilseed radish, sorghum-sudangrass hybrids and others) will also contribute to the management of the stem and other plant parasitic nematodes. However, the use of multiple and compatible management options is the best strategy to follow for the most effective and lasting control of the stem nematode.

2.5.2 Foliar Nematodes

Aphelenchoides ritzemabosi, *A. fragariae*, *A. parietinus* and *A. besseyi* have been reported from Pennsylvania; whereas, only *A. ritzemabosi*, *A. fragariae* and *A. parietinus* have been detected in New York and just *A. parietinus* in New Jersey. The early taxonomic history of these species includes a significant number of synonyms and confusion. The most common is *A. ritzemabosi*, the chrysanthemum foliar nematode. *A. fragariae* is the strawberry bud pathogen known as the spring crimp nematode. *Aphelenchoides besseyi* is the summer crimp nematode which also causes white tip of rice.

The chrysanthemum and spring crimp nematodes parasitize many herbaceous and woody plants. They feed on leaf mesophyll, resulting in necrotic tissue (blotches between veins) and non-functional apical meristems. The nematodes can move from plant to plant in thin films of moisture, splashing or rain/irrigation water or in infected plant material. *Aphelenchoides ritzemabosi* survives desiccation, but not extreme low temperatures. Use of nematode-free propagation materials and general sanitation procedures are the most appropriate management practices for these nematodes.

2.5.3 Potato Rot and Seed Gall Nematodes

Both the potato rot nematode (*Ditylenchus destructor*) and the wheat seed gall (*Anguina tritici*) are included in the 2014 WPN List for New York, but not for New Jersey or Pennsylvania. All three states have commercial potato industries, with New York's being the largest. While *D. destructor* is a regulatory species in some potato producing states, it has a relatively large host range including edible crops. *Ditylenchus destructor* causes severe necrosis of potato tuber tissue making infested tubers unmarketable. There are, however, no recorded detections of the potato rot nematode in New York potato production in recent years. Potato producing states with periodic detections of potato rot nematodes have highly developed and successful quarantine and certification programs that allow for continued export and certification of potatoes. Since potatoes are grown from tuber seed pieces, it is imperative for the seed to be pathogen-free. The overall U.S. potato seed certification programs allows farmers to obtain and plant high quality certified seed that is true to variety and pathogen free. Except for *Anguina tritici,* which has been eradicated from the USA, there are eleven recognized species of *Anguina*, each with its own biology and host range. It is highly likely that one or more of these species exist in New York, Pennsylvania or New Jersey.

2.6 Conclusions

New York, New Jersey and Pennsylvania have vibrant agricultural and human-living environment systems. These are inhabited by a diversity of plant parasitic nematodes. Throughout the years, strong research, academic instruction and extension-outreach programs in nematology evolved at Cornell University, Rutgers University and Pennsylvania State University. These institutions provided the information necessary to limit the detrimental impacts of these soil-borne organisms and contributed in significant ways to the evolution of the concept of both integrated pest management and sustainable agriculture. There were times, however, when nematology resources have been very limited at these three institutions. Ecosystems are dynamic and always changing. It is imperative that the Land Grant Institutions of New York, New Jersey and Pennsylvania provide highly significant future contributions towards the understanding of nematode biology, ecology and management in regards to the enhancement of overall human quality of life.

References

Abawi, G. S., & Chen, J. (1998). Concomitant pathogen and pest interaction. Chapter 7. In K. R. Barker, G. A. Pederson, & G. L. Windham (Eds.), *Plant-nematode interactions, monograph series book No. 36* (pp. 65–85). Madison: American Society of Agronomy 771 pp.

Abawi, G. S., & Laird, P. D. (1994). Increasing occurrence and damage of the northern root knot nematode on onions in New York State. *Phytopathology, 84*, 1370.

Abawi, G. S., & Ludwig, J. W. (2003a). Host efficiency of selected soybeans to *Meloidogyne hapla* and *Pratylenchus penetrans*. *Journal of Nematology, 35*, 3121.

Abawi, G. S., & Ludwig, J. W. (2003b). Evaluation of onion germplasm for resistance to *Meloidogyne hapla*. *Journal of Nematology, 35*, 321.

Abawi, G. S. & Ludwig, J. W.. (2005). *Nematodes on onions: New control products and soil bioassays*. Empire State Fruit and Vegetable Expo Proceedings, Cornell Coop. Ext., pp 107–110.

Abawi, G. S., & Mai, W. F. (1975). Effect of foliar applications of Oxamyl on movement of *Pratylenchus penetrans* in and outside host roots. *Plant Disease Report, 59*, 795–799.

Abawi, G. S., & Mai, W. F. (1977). Sugar beet cyst nematode: A threat to the red beet and cabbage industries in New York State. *New York's Food and Life Sciences, 10*, 10–13.

Abawi, G. S., & Mai, W. F. (1980). Effects of initial population densities of *Heterodera schachtii* on yield of cabbage and table beets in New York. *Phytopathology, 70*, 481–485.

Abawi, G. S., & Mai, W. F. (1983). Increase in cabbage yield by Fenamiphos treatment of uninfested and *Heterodera schachtii*-infested soils. *Plant Disease, 67*, 1343–1346.

Abawi, G. S., & Moktan, K. (2013). *Bloat nematode: A re-emerging and damaging pest of garlic and other crops* (pp. 168–170). Proceedings of the Mid-Atlantic Fruit and Vegetable Conference, Hershey, PA.

Abawi, G. S., & Widmer, T. L. (2000). Impact of soil health management practices on soilborne pathogens, nematodes and root diseases of vegetable crops. *Applied Soil Ecology, 15*, 37–47.

Abawi, G. S., Ludwig, J. W., Morton, H. V., & Hofer, D. (2003). Efficacy of abamectin as a seed treatment against *Meloidogyne hapla* and *Pratylenchus penetrans*. *Journal of Nematology, 35*, 321.

Abawi, G. S., Ludwig, J. W., & Gugino, B. K. (2006). Reproduction of *Meloidogyne hapla* on potato cultivars and its management with soil and foliar Vydate applications. *Phytopathology, 96*(S), 76 (Abstr.).

Abawi, G. S., Moktan, K., Stewart, C., Hoepting, C., & Hadad, R. (2011). Occurrence and damage of bloat nematode to garlic in New York. *Journal of Nematology, 43*, 223.

Abu-Gharbieh, W., Varney, E. H., & Jenkins, W. R. (1963). Relationships of meadow nematodes to *Verticillium* wilt of strawberries. *Phytopathology, 52*, 921.

Arneson, P. A., & Mai, W. F. (1976). Root diseases of fruit trees in New York State. VII. Costs and returns of preplant soil fumigation in a replanted apple soil. *Plant Disease Report, 60*, 1054–1057.

Bélair, G., & Simard, L. (2009). Potato cyst nematodes in Canada: Go forward with the science. *Fruit and Vegetable Magazine, 65*, 10–12.

Bello, A., Arias, M., López-Pérez, J. A., García-Álvarez, A., Fresno, J., Escuer, M., Arcos, S. C., Lacasa, A., Sanz, R., Gómez, P., Díez-Rojo, M. A., Piedra Buena, A., Goitia, C., de la Horra, J. L., & Martínez, C. (2004). Biofumigation, fallow, and nematode management in vineyard replant. *Nematropica, 34*, 53–64.

Bird, G. W. (1967). *Trichodorus acutus* n. sp. (Nematoda: Diptherophoroidea) and a discussion of allometry. *An Journal of Zoology*, 1201–1204.

Bird, G. W., & Jenkins, W. R. (1964). Nematodes associated with wild yam, Dioscorea sp., with special reference to the pathogenicity of *Meloidogyne incognita*. *Plant Disease Reporter, 46*, 858–860.

Bird, G. W., & Mai, W. F. (1967a). Factors influencing population densities of *Trichodorus christiei*. *Phytopathology, 57*, 1368–1371.

Bird, G. W., & Mai, W. F. (1967b). Morphometric and allometric variations of *Trichodorus christiei*. *Nematologica, 13*, 617–632.

Bird, G. W., & Mai, W. F. (1968). Numerical study of the growth and development of *Trichodorus christiei*. *Nematologica, 13*, 617–632.

Bird, G. W., Goodman, R. M., & Mai, W. F. (1968). Observations on the embryogenesis of *Trichodorus christiei* (Nematoda: Diptherophoroidea). *Canadian Journal of Zoology, 46*, 292–293.

Bloom, J. R. (1961). Nematodes and turf. *Golf Course Reptr, 29*, 12–17.

Brodie, B. B. (1980). Control of *Globodera rostochiensis* in relation to method and time of application of nematicides. *Journal of Nematology, 12*, 215–216.

Brodie, B. B. (1993). Probability of *Globodera rostochiensis* spread on equipment and potato tubers. *Journal of Nematology, 25*, 291–296.

Brodie, B. B. (1996a). Effect of initial nematode density on managing *Globodera rostochiensis* with resistant cultivars and nonhosts. *Journal of Nematology, 28*, 510–519.

Brodie, B. B. (1996b). *Golden nematode: A success story for biological control.* http://web.entomology.cornell.edu/shelton/cornell-biocontrol-conf/talks/brodie.html

Brodie, B. B., & Mai, W. F. (1989). Control of the golden nematode in the United States. *Annual Review of Phytopathology, 27*, 443–461.

Brown, D. J. F., Halbrendt, J. M., Jones, A. T., Vrain, T. C., & Robbins, R. T. (1994). Transmission of three North American nepoviruses by populations of four distinct species of the *Xiphinema americanum* group. *Phytopathology, 84*, 646–649.

Brown, D. J. F., Robertson, W. M., & Trudgill, D. L. (1995). Transmission of viruses by plant nematodes. *Annual Review of Phytopathology, 33*, 223–249.

Cannon, O. S. (1941). *Heterodera schachtii* found in a Long Island potato field. *Plant Disease Reporter, 25*, 408.

Chitwood, B. G., Clement, R. L., Morgan, R., & Tank, R. (1942). *Heterodera rostochiensis*, golden nematode of potatoes in New York State. *Plant Disease Reporter, 26*, 390–391.

Colett, M. 2010. Bulb and stem nematode menace again in garlic during 2010. Ministry of Agriculture Food and Rural Affairs, Ontario. http://www.omafra.gov.on.ca./english/crops/hort/news/hortmatt/.

Couch, H. B., & Bloom, J. R. (1959). Turfgrass disease control. *New York Turf-grass Bulletin, 64*, 248–250.

Devine, K. J., & Jones, P. W. (2003). Investigations into the chemoattraction of the potato cyst nematodes *Globodera rostochiensis* and *G. pallida* towards fractionated potato root leachate. *Nematology, 5*, 65–75.

DiEdwardo, A. A. (1961). Seasonal population variations of *Pratylenchus penetrans* and other nematodes associated with roots. *Phtytopathology, 50*, 633.

Dolliver, J. S., Clark, D. G., & Mai, W. F. (1960). Relationships of populations of *Pratylenchus penetrans* to vegetative growth of Wando peas. *Phytopathology, 50*, 239.

Dropkin, V. H. (1989). *Introduction to plant nematology* (p. 304). New York: Wiley.

Ellenby, C. (1954). Tuber forming species and varieties of the genus *Solanum* tested for resistance to the potato root eelworm *Heterodera rostochiensis* Wollenweber. *Euphytica, 3*, 195–202.

Esquibet, M., Grenier, E., Plantard, O., Andaloussi, F. A., & Gaubel, G. (2003). DNA polymprphism in the stem nematode *Ditylenchus dipsaci*: Development of diagnostic markers for normal and giant races. *Genome, 46*, 1077–1083.

Evans, K., D. L. Trudgill, 1978. Pest aspects of potato production. Part I. Nematode pests of potatoes. The potato crop, P. M. Harris, 1 6:440–469. London: Chapman and Hall.

Fenwick, D. W., & Widdowson, E. (1959). The emergence of larvae from free eggs of the potato root eelworm *Heterodera rostochiensis* (Woll.). *Annals of Applied Biology, 47*, 140–149.

Forer, L. B., & Stouffer, R. F. (1982). *Xiphinema* spp. associated with Tomato ringspot virus infection of Pennsylvania fruit crops. *Plant Disease, 66*, 735–736.

Fuchs, M. (2016). Virus transmission and grafting practices. *New York Fruit Quarterly, 24*, 25–27.

Fuchs, M., Abawi, G. S., Marsella-Herrick, P., Cox, R., Cox, K. D., Carroll, J. E., & Martin, R. R. (2009). Occurrence of Tomato ringspot virus and Tobacco ringspot virus in highbush blueberry in New York. *Journal of Plant Pathology, 92*, 451–459.

Georgi, L. L. (1988). Morphological variation in *Xiphinema* spp. from New York orchards. *Journal of Nematology* (1988), *20*, 47–57.

Gugino, B. K. J., Ludwig, W., Hen, P., & Abawi, G. S. (2006). *Greenhouse reactions of carrots cultivars to northern root knot nematode, 2003 and 2004, Biological and Cultural Control of Plant Diseases Report 21:V006*. St. Paul: APS. https://doi.org/10.1094/BC21.

Gugino, B. K., Ludwing, J. W., & Abawi, G. S. (2008). An on-farm bioassay for assessing *Meloidogyne hapla* infestations as a decision management tool. *Crop Protection, 27*, 785–791.

Halbrendt, J. M. (1996). Allelopathy in the management of plant parasitic nematodes. *Journal of Nematology, 28*, 8–14.

Halbrendt, J. M. (2012). *Managing nematodes and tomato ring spot virus in vineyards*. http://www.pawinegrape.com/uploads/PDF%20files/Meeting%20Presentations/IPM%202012/J%20 Halbrendt_GrapeNemaToRSVIPM. pdf

Halbrendt, J. M., & Jing, G. (1994). Nematode suppressive rotation crops for orchard renovation. *Acta Horticulturae, 363*, 49–56.

Heald, C. M., Jr. (1963). Pathogenicity of *Pratylenchus penetrans* to zinnia and gaden balsom. *Plant Disease Reporter, 47*, 260–271.

Hoff, J. K., & Mai, W. F. (1962). Pathogenicity of the stubby-root nematode to onion. *Plant Disease Reporter, 46*, 24–25.
Hooper, D. J., & Southey, J. F. (1978). *Ditylenchus*, *Anguina*, and related genera. In *Plant nematology* (Vol. I, pp. 78–97). London: Ministry of Agriculture, Fisheries and Food.
Hutchinson, M. T., Reed, J. P., Streu, H. T., DiEdwardo, A. A., & Schroeder, P. H. (1961). *Plant parasitic nematodes of New Jersey, Bulletin*. New Brunswick: New Jersey Agricultural Experimental Station, Rutgers: The State University.
Jaffee, B. B., Harison, M. B., Shaffer, R. L., & Strang, M. B. (1987a). Seasonal population fluctuation of *Xiphenema americanum* and *X. revesi* in New York and Pennsylvania orchards. *Journal of Nematology, 19*, 369–378.
Jaffee, B. A., Nyczepir, A. P., & Golden, A. M. (1987b). *Criconemella* spp. in Pennsylvania Peach Orchards with morphological observations of *C. curvata* and *C. ornate*. *Journal of Nematology, 19*, 420–423.
Jenkins, W. R. (1964). A rapid centrifugal-flotation technique for separating nematode from soil. *Plant Disease Reporter, 48*, 692.
Jenkins, W. R., Mai, W. F., Stessel, G. J. (1963). *A review of plant nematology in the northeastern United States, from 1956 to 1963 with an outlook for the future*. N.J. Agric. Expt. Sta. Bull. 805, 30 pp.
Jing, G. N. (1994). *Evaluation of the nematicidal properties of cruciferous plants for nematode management in replant orchards*. Ph.D. dissertation, The Pennsylvania State University, University Park.
Johnson, S. B., & D. Fuller. (2012). *Bloat nematode in Maine garlic*. University of Maine, Cooperative Extension Service, Orono. Bull. 1205:1–3.
Johnson, D. E., & Lear, B. (1965). Additional information regarding the hot water treatment of seed garlic cloves for the control of the stem and bulb nematode (*Ditylenchus dipsaci*). *Plant Disease Reporter, 49*, 898–899.
Kepich, D. (2011). *Golden nematode program update*. http://www.hort.cornell.edu/expo/proceedings/2011/Potatoes/Potato%20cyst%20nematode.pdf
Kepich, D. (2016). Golden nematode program update. http://www.hort.cornell.edu/expo/proceedings/2016/Potato.Golden%20nematode%20program%20updated.Kepich.pdf.
Keplinger, J. A., Mai, W. F., & Abawi, G. S. (1979). Efficacy of nonvolatile nematicides against the sugarbeet-cyst nematode on cabbage, 1978. *Fungicide and Nematicides Tests, 34*, 187–188.
LaMondia, J. A. (2004). Strawberry black root rot. *Advances in Strawberry Research, 23*, 1–10.
LaMondia, J. A., & Brodie, B. B. (1986). Effects of initial nematode density on population dynamics of *Globodera rostochiensis* on resistant and susceptible potatoes. *Journal of Nematology, 18*, 159–165.
Lear, B., Miyagawa, S. T., Johnson, D. E., & Atlee, C. B., Jr. (1966). The sugar beet nematode associated with reduced yields of cauliflower and other vegetable crops. *Plant Disease Report, 50*, 611–612.
Lewis, G. D., & Mai, W. F. (1958). Chemical control of *Ditylenchus dipsaci* Kuhn Filipjev in organic soils of southern New York. *Plant Disease Report, 42*, 1360–1363.
Lewis, G. D., & Mai, W. F. (1960). Overwintering and migration of *Ditylenchus dipsaci* in organic soils of southern New York State. *Phytopathology, 50*, 341–343.
Mai, W. F. (1961). Sugar beet nematode found in New York State. *Plant Disease Report, 45*, 151.
Mai, W. F. (1977). Worldwide distribution of potato cyst nematodes and their importance in crop production. *Journal of Nematology, 9*, 30–34.
Mai, W. F. (1995). *Plant nematology at Cornell* (p. 46). Ithaca: Department of Plant Pathology Publication, Cornell University.
Mai, W. F., & Abawi, G. S. (1978). Determining the cause and extend of apple, cherry and pear replant diseases under controlled conditions. *Phytopathology, 68*, 1540–1544.
Mai, W. F., & Abawi, G. S. (1980). Influence of crop rotation on spread and density of *Heterodera schachtii* on a commercial vegetable farm in New York. *Plant Disease, 64*, 302–305.

Mai, W. F., & Abawi, G. S. (1981). Controlling replant diseases of pome and stone fruits in northeastern United States by preplant fumigation. *Plant Disease, 65*, 859–864.

Mai, W. F., & Abawi, G. S. (1987). Interactions among root knot nematodes and *Fusarium* wilt fungi on host plants. *Annual Review of Phytopathology, 25*, 317–338.

Mai, W. F., & Harrison, M. B. (1960). *The golden nematode*. Cornell Ext. Bull. 870:1–32.

Mai, W. F., &Lownsberry, B.F. (1946). Progress report on experiments for the control of the golden nematode of potatoes. Department of Plant Pathology, Cornell University Mimeo.

Mai, W. F., & Lownsbery, B. F. (1948). Crop rotation in relation to the golden nematode population of the soil. *Phytopathology, 42*, 345–347.

Mai, W. F., & Parker, K. G. (1956). Evidence that the nematode *Pratylenchus penetrans* causes losses in New York State cherry orchards. *Phytopathology, 46*, 18.

Mai, W. F., & Spears, J. F. (1954). The golden nematode in the United States. *American Potato Journal, 31*, 387–396.

Mai, W. F., Lorbeer, J. W., Sherf, A. F. (1964). *Fumigating muck soils for nematode control*. Cornell Extension Bulletin 1133, 11.

Mai, W. F., Abawi, G. S., & Becker, R. F. (1972). Population levels of *Heterodera schachtii* in New York and damage to red table beet and cabbage under greenhouse conditions. *Plant Disease Report, 56*, 434–437.

Mai, W. F., Bloom, J. R., & Chen, T. A. (1977). *Biology and ecology of the plant parasitic nematode, Pratylenchus penetrans, Bulletin 815*. University Park: The Pennsylvania State University, College of Agriculture-AES 64 pp.

Mai, W. F., Merwin, I. A., & Abawi, G. S. (1994). Diagnosis, etiology and management of preplant disorders in New York cherry and apple orchards. *Acta Horticulturae, 363*, 33–41.

Mikowski, N. A., Van der Beek, J. G., & Abawi, G. S. (2002). Characterization of root knot nematode populations associated with vegetables in New York State. *Plant Disease, 86*, 840–847.

Miller, R. E., Boothroyed, C. W., & Mai, W. F. (1963). Relationship of *Pratylenchus penetrans* to roots of corn in New York. *Phytopathology, 53*, 313–315.

Mitkowski, N. A., & Abawi, G. S. (2003a). *Root knot nematodes. The plant health instructor*. Minneapolis: APS.

Mitkowski, N. A., & Abawi, G. S. (2003b). Genetic diversity of New York state *Meloidogyne hapla* populations determined by RAPDs and mitochondrial DNA. *Journal of Nemmatode Morphology and System, 5*, 191–202.

Mitkowski, N. A., & Abawi, G. S. (2003c). Reproductive fitness of populations of *Meloidogyne hapla* from New York State vegetable fields. *Nematology, 5*, 77–83.

Molinari, S., Lamberti, F., Duncan, L. W., Halbrendt, J., McKenry, M., Abawi, G. S., Magunacelaya, J. C., Crozzoli, R., Lemos, R., Nyczepir, A. P., Nagy, P., Robbins, R. T., Kotcon, J., Moens, M., & Brown, D. J. F. (2004). SOD polymorphism in *Xiphinema americanum*-group (Nematoda: Longidoridae). *Nematology, 6*, 867–876.

Mountain, W. B. (1957). Outbreak of the bulb and stem nematode in Ontario. *Canadian Plant Disease Survey, 37*, 62–63.

Mountain, W. L., Powell, C. A., Forer, L. B., & Stouffer, R. F. (1983). Transmission of *tomato ringspot virus* from dandelion via seed and dagger nematodes. *Plant Disease, 67*, 867–868.

Newhall, A. G. (1942). Chloropicrin and ethylene for root-knot nematode control. *Phytopathology, 32*, 626–630.

Newhall, A. G., & Chidwood, B. G. (1940). Onion eel-worm or bloat caused by the stem and bulb nematode *Ditylenchus dipsaci*. *Phytopathology, 30*, 390–400.

Newhall, A. G., Clement, R. L., Smith, I. D., Chitwood, B. G., & B. G. (1939). A survey of the occurrence of the bulb or stem nematode on onions in the state of New York. *Plant Disease Report, 23*, 291–292.

Olthof, T. A., Potterand, J. W., & Peterson, E. A. (1974). Relationships between populations of *Heterodera schachtii* and losses in vegetable crops in Ontario. *Phytopathology, 64*, 549–554.

Parker, K. G., & Mai, W. F. (1956a). Damage to tree fruits in New York by root lesion nematodes. *Plant Disease Report, 40*, 694–699.

Parker, K. G., & Mai, W. F. (1956b). The replant problem on cherry. *New York State Horticulture Society Proceeding, 101*, 298.

Peterson, L. C., & Plaisted, R. L. (1966). Peconic: A new potato variety resistant to the golden nematode. *American Potato Journal, 43*, 450–452.

Pethybridge, S. J., Gorny, A., Hoogland, T., Jones, L., Hay, F., Smart, C., & Abawi, G. S. (2016). Identification and characterization of *Ditylenchus* spp. populations from garlic in New York state, USA. *Trop Plant Pathology, 41*, 193–197.

Powell, C. A., Forer, L. B., & Stouffer, R. F. (1982). Reservoirs of *tomato ringspot virus* in fruit 14 orchards. *Plant Disease, 66*, 583–584.

Qiao, Y., Zaid, M., Badiss, A., Hughes, M., Celetti, J., & Yu, Q. (2013). Intra-racial genetic variation of *Ditylenchus dipsaci* from garlic in Ontario as revealed by random amplified polymorphic DNA analysis. *Canadian Journal of Plant Pathology, 35*, 346–353.

Radewald, J. D., Hall, B. J., Shibuya, F., & Nelson, J. (1971). Results of a preplant fumigation trial for the control of sugar-beet nematode on cabbage. *Plant Disease Report, 55*, 841–854.

Sasser, J. N., & Carter, C. C. (1985). *An advance treatise on Meloidogyne* (Vol. I, II). Raleigh: Dept. of Plant Pathology and Genetics, North Carolina State University Graphics and US-AID.

Slinger, L., & Bird, G. (1978). Ontogeny of *Daucus carota* infected with *Meloidogyne hapla*. *Journal of Nematology, 10*, 18–194.

Smith, J. J., & Mai, W. F. (1964). Host-parasite relationships of *Allium cepa* and *Meloidogyne hapla*. *Phytopathology, 55*, 693–697.

Spears, J. F. (1968). *The golden nematode handbook, survey, laboratory, control and quarantine procedures*. Agricultural Handbook No. 353. U. S. Department of Agricultural. 92 pp.

Steele, A. E. (1965). The host range of the sugar beet nematode, *Heterodera schachtii* Schmidt. *Journal American Society of Sugar Beet Technology, 13*, 573–603.

Subbotin, S. A., Madani, M., Krall, E., Sturhan, D., & Moens, M. (2005). Molecular diagnositcs, taxonomy, and phylogeny of the stem nematode *Ditylenchus dipsaci* species complex based on the sequences of the internal transcribed spacer-rDNA. *Phytopathology, 95*, 1308–1315.

Sullivan, M. J., Inserra, R. N., Franco, J., Moreno-Leheude, I., & Greco, N. (2007). Potato cyst nematodes: Plant host status and their regulatory impact. *Nematropica, 37*, 193–201.

Timper, P., & Brodie, B. B. (1997). First report of *Pratylenchus neglectus* in New York. *Plant Disease, 81*, 228.

Tiner, J. D. (1961). Cultures of the plant parasitic nematode genus *Pratylenchus* on sterile excised roots. II. A trap for collection of axenic nematodes and quantitative initiation of experiment. *Journal of Expirement Parasitology, 11*, 231–240.

Troccoli, A., Abawi, G. S., Ludwig, J. W., & Lamberti, F. (2003). Morpho-anatomical notes on populations of *Pratylenchus penetrans* and *P. crenatus* from New York State (USA). *Journal of Nematology, 35*(3), 367–368.

Uyemoto, J. K., Cummins, R. J., & Abawi, G. S. (1977). Virus and virus-like diseases affecting grapevines in New York vineyards. *American Journal of Enology and Viticulture, 28*, 131–136.

Viaene, N., & Abawi, G. S. (1996). Damage thresholds of *Meloidogyne hapla* to lettuce in organic soil. *Journal of Nematology, 28*, 537–545.

Viaene, N. M., & Abawi, G. S. (1998a). Management of *Meloidogyne hapla* on lettuce in organic soil with sudangrass as a cover crop. *Plant Disease, 82*, 945–952.

Viaene, N. M., & Abawi, G. S. (1998b). Fungi parasitic on juveniles and eggs of *Meloidogyne hapla* in organic soils in New York. *Supplement of the Journal of Nematology, 30*, 632–638.

Viaene, N. M., & Abawi, G. S. (2000). *Hirsutella rhosiliensis* and *Verticillium chlamydosporium* as biocontrol agents of the root knot nematode *Meloidogyne hapla* on lettuce. *Journal of Nematology, 32*, 85–100.

Wang, S., Gergerich, R. C., Wickizer, S. L., & Kim, K. S. (2002). Localization of transmissible and nontransmissible viruses in the vector nematode *Xiphinema americanum*. *Phytopathology, 92*, 646–653.

Wang, X., Bergstrom, G., Chen, S., Thurston, D., Handoo, Z., Hult, M., & Skantar, A. (2017). First report of the soybean cyst nematode, *Heterodera glycines*, in New York. *Plant Disease, 100*, 1957.

Widmer, T. L., & Abawi, G. S. (2000). Mechanism of suppression of *Meloidogyne hapla* and its damage by a green manure of Sudan grass. *Plant Disease, 84*, 562–568.

Widmer, T. L., & Abawi, G. S. (2002). Relationship between levels of cyanide in Sudangrass hybrids incorporated into soil and suppression of *Meloidogyne hapla*. *Journal of Nematology, 34*, 16–22.

Widmer, T. L., Ludwig, J. W., & Abawi, G. S. (1999). *The northern root knot nematode on carrot, lettuce, and onions in New York, New York's Food and Life Sciences Bulletin No. 156*. Geneva: NYSAES, Cornell University.

Widmer, T. L., Mitkowski, N. A., & Abawi, G. S. (2002). Soil organic matter and management of plant parasitic nematodes. 2002. *Journal of Nematology, 34*, 289–295.

Wing, K. B., Pritts, M. P., & Wilcox, W. F. (1995). Biotic, edaphic, and cultural factors associated with strawberry black root rot in New York. *Hortscience, 30*, 86–90.

WPNL. (2011, 2014, 2015). *Widely prevalent nematodes of the United States*. prevalentmnematodes.org

Chapter 3
Nematodes and Nematologists of Michigan

George W. Bird and Fred Warner

3.1 Introduction

The known glacial history of Michigan began about 2.4 million years ago (Gillespie et al. 1987). It involved six glaciations. The last was the Wisconsin Glacier. It retreated and the entire watershed was free of ice by 9000 years ago. Glaciation resulted in three major geological features. These include the Michigan Basin of the Lower Peninsula and eastern part of the Upper Peninsula, the southern margin of the Canadian Shield in the western part of the Upper Peninsula and the Great Lakes. Glaciation also resulted in a diversity of soils and local climates. This allowed for the pre-agricultural evolution of Eastern Deciduous, Spruce-Fir and Tall-Grass Prairie biomes. Today, Michigan farms produce more than 300 different commodities. These contribute $13 billion to the overall food and agriculture industry (MDARD 2017). They include agronomic crops, fruit, vegetables and ornamentals. In addition, Michigan has the largest state forest system in the USA. There are three national forests and more than eight hundred thousand hectares of private forest land. Plant parasitic nematodes are known to be associated with the vast majority, if not all of Michigan agricultural and forest systems. In addition, the roles of bacterial and fungal feeding nematodes have been documented. This would not have been possible without the diversity of nematologists that have studied the nematodes of Michigan. The objectives of this chapter are to: (1) summarize the history of Michigan nematology, (2) document the occurrence and distribution of plant

---

G. W. Bird (✉)
Department of Entomology, Michigan State University, East Lansing, MI, USA
e-mail: birdg@msu.edu

F. Warner
Department of Crop Soil and Microbial Sciences, Michigan State University, East Lansing, MI, USA
e-mail: fwnemalab@msu.edu

S. A. Subbotin, J. J. Chitambar (eds.), *Plant Parasitic Nematodes in Sustainable Agriculture of North America*, Sustainability in Plant and Crop Protection,
https://doi.org/10.1007/978-3-319-99588-5_3

parasitic nematode taxa known to be present in Michigan, (3) describe Michigan's contributions to understanding their biology and ecology and (4) outline the history and current state of integrated nematode management in the state.

3.2 Nematologists and Michigan Nematology

Michigan nematology began in 1910 with the arrival of Professor Ernst A. Bessey at Michigan Agricultural College. This was one year before he published his Nematological Classic entitled, *Root knot and Its Control*. The document includes a forward from the Honorable James Wilson, Secretary of the United States Department of Agricultural and William A. Taylor, Acting Chief of the Bureau of Plant Industry (Bessey 1911). It contains a list of the 480 species and subspecies known to be hosts of *Heterodera radicicola* (the 1910 taxonomic name for root knot nematodes). The nematode control section is divided into perennial and annual crops. The described practices included chemicals, fertilizers, flooding, drying, trap crops, steam, fallowing, non-susceptible crops and breeding for host-plant resistance. This was followed in 1915 with publication of Farmers Bulletin No. 648 entitled, *The Control of Root knot* (Bessey 1915). Prior to coming to Michigan, E. A. Bessey differentiated between summer crimp and spring crimp bud disease of strawberries. In 1942, J. R. Christie named *Aphelenchoides besseyi* (Christie 1942) in his honor. Bessey Hall is a constant reminder of E. A. Bessey's impact of the stature of biology at Michigan State University (Table 3.1).

In 1913, Margaret V. Cobb conducted a nematology survey of the Douglas Lake region of Michigan. Her findings included 12 known species and 11 new species (Cobb 1915). One of the new species was *Dolichodorus heterocephalus* (awl nematode), the first record of a plant parasitic nematode reported in Michigan. Seven additional species from the collection were described by N. A. Cobb (1914). In 1920, Professor Gerald Thorne made his first of several visits to Michigan to survey for *Heterodera schachtii*. This nematode, however, was not detected in Michigan until 1948 (Bockstahler 1950). In 1953, the Director of the Michigan Agricultural Experiment Station hired B. G. Chitwood to conduct a 6-month survey of Michigan nematodes (Chitwood 1953). This initiative included nematode surveys of vegetable, orchard, vineyard, berry, cover crop, nursery, florist and forest systems. In 1954, Dr. John Knierim was hired as Michigan State University's first full-time nematologist.

Throughout the years, a total of 22 professional nematologists have worked in Michigan (Table 3.1). This resulted in the training of a significant number of M.S. and Ph.D. students and their research forms a large portion of the knowledge base for this chapter. Most of these individuals have gone on to have successful careers in nematology. Michigan State University has offered both introductory and advanced courses in nematology, in addition to having nematology lectures included in plant pathology, horticulture, agronomic crop and soil science courses. Since the arrival of Dr. Charles Laughlin at Michigan State University in 1969, Extension

Table 3.1 Michigan Nematologists, positions and dates

Nematologist	Position	Dates
E. A. Bessey[a]	MSU, Professor, Chair, Dean	1910–1945
M. V. Cobb	University of Michigan Student	1913–1915
N. A. Cobb	USDA Nematologist	1914–1915
Gerald Thorne	MSU, Visiting Nematologist, Consultant	1920, 1962–1966
H.W. Bockstahler	USDA/ARS/Technician	1950
B. G. Chitwood	MSU, Visiting Nematologist	1953
John Knierim	MSU, Assistant Professor	1954–1980
Natalie Knobloch	MSU, Taxonomist and Diagnostician	1962–1978
Paul Wolley	MSU, Director, Nematology Program	1963–1968
Charles Laughlin	MSU, Associate Professor	1969–1973
John Davenport	MSU, Applied Research Technician	1972–2007
George Bird	MSU, Professor	1973-present
Lindy Rose	MSU, Nematode Diagnostician	1978–1981
Alma Elliott	MSU, Instructor	1979–1981
Loraine Graney	MSU, Nematode Taxonomist/Diagnostician	1982–1989
Linda Mansfield	MSU, Dis. Professor, Large Animal Clinic	1990-present
Fred Warner	MSU, Nematode Diagnostician	1990-present
Haddish Melakeberhan	MSU, Associate Professor	1994-present
Angie Tenney	MSU, Associate Diagnostician	1999-present
Todd Ciche	MSU, Assistant Professor	2006–2012
Jared Ali	MSU, Assistant Professor	2012–2015
Marisol Quintanilla	MSU, Applied Research and Extension Nematologist	2017-present
Kristin Poley	MSU, Applied Research Technician	2017-present
Jeff Shoemaker	MSU, Applied Research Technician	2017–2018

[a]Bessey Hall. There are three Bessey Halls at Big Ten Universities. The one at Michigan State University is named after E. A. Bessey (1877–1957), B.S., 1896, Univ. NE., M.S., 1898, Univ. NE. and Ph.D., Halle Univ., Germany (1904). The Bessey Halls at Iowa State University and the University of Nebraska are named after C. E. Bessey (1845–1915, E. A. Bessey's father), Michigan Agricultural College, Class of 1869, Horticulture; Professor, Iowa State University, Professor and Academic Dean, University of Nebraska.

nematology has been the primary focus of the program. Between 1962 and 1974, Michigan State University processed about 6000 extension samples for nematodes (Knobloch and Bird 1981). Ninety-four taxa of plant parasitic nematodes have been detected in Michigan (Table 3.2). In addition, at least ten formal nematode surveys were conducted in Michigan between 1913 and 2017. During the last 35 years, Michigan nematology has played an active leadership role in the evolution of the domains of integrated pest management, sustainable agriculture, sustainable-equitable development and soil health biology (Bird 2003; Bird and Smith 2013) (Table 3.3).

Table 3.2 Plant parasitic nematodes of Michigan: 1913–2018

Nematode species	Crop and plants	Reference
Aphelenchoides ritzemabosi	Chrysanthemums	Knierim (1963)
Atylenchus decalineatus	Blueberry	Tjepkema (1966)
Cactodera milleri	Lambs quarter	Graney and Bird (1990)
C. weissi	Smartweed	Chitwood (1953)
Criconema fimbriatum	Spruce	Knobloch and Bird (1981)
C. mutable	Unknown	Knobloch and Bird (1981)
C. permistum	Spruce	Knobloch and Bird (1981)
C. petasum	Unknown	Knobloch and Bird (1981)
C. princeps	Spruce	Knobloch and Bird (1981)
C. sphagni	White birch	Knobloch and Bird (1981)
Crossonema menzeli	Spruce	Knobloch and Bird (1981)
Ditylenchus dipsaci	Creeping phlox, onion	Schnabelrauch et al. (1981)
D. destructor	Potato	Chitwood (1953)
Dolichodorus heterocephalus	Beach grass	Cobb (1914)
Geocenamus longus	White birch	Knobloch and Bird (1981)
Gracilacus acicula	White birch, spruce	Knobloch and Bird (1981)
Helicotylenchus californicus	Willow	Knobloch and Bird (1981)
H. crenacauda	Willow, iris	Knobloch and Bird (1981)
H. digonicus	Clover, onion, potato	Chitwood 1953
H. platyurus	Phlox, onion	Knobloch and Bird (1981)
H. pseudorobustus	Phlox, willow	Knobloch and Bird (1981)
Hemicycliophora similis	Oak, clover	Chitwood (1953)
H. uniformis	Maple	Knobloch and Bird (1981)
H. vaccinium	Blueberry	Knobloch and Bird (1981)
H. vidua	Maple	Knobloch and Bird (1981)
Heterodera avenae	Wheat	Bernett (1986)
H. carotae	Carrots, Queen Anne's lace	Berney and Bird (1992)
H. glycines	Soybean	Warner and Golden (1987)
H. humuli	Hop	Warner et al. (2015)
H. orientalis	*Miscanthus* sp.	Warner and Handoo (pers. comm.)
H. pratensis	Turfgrass	Stouffer-Hopkins et al. (pers. comm.)
H. schachtii	Sugar beet, cabbage	Bockstaller (1950)
H. trifolii	Alfalfa	Brzeski and Laughlin (1971)
H. ustinovi	Bentgrass	Knobloch and Bird (1981)
Hirschmanniella gracilis	Beech, maple	Knobloch and Bird (1981)
Hoplolaimus galeatus	Maple, cherry	Chitwood (1953)
Lobocriconema thornei	Oak, maple	Knobloch and Bird (1978)
Longidorus breviannulatus	Corn	Corn extension samples
L. elongatus	Celery, onion	Knobloch and Bird (1981)
Meloidogyne arenaria	Celery, maple	Chitwood (1953)

(continued)

Table 3.2 (continued)

Nematode species	Crop and plants	Reference
M. hapla	Lettuce, celery, ornamentals	Chitwood (1953)
M. incognita	Greenhouse ornamentals	Knobloch and Bird (1981)
M. microtyla	Maple	Knobloch and Bird (1981)
M. naasi	Turfgrass	Knobloch and Bird (1981)
M. nataliei	Grape	Diamond and Bird (1994)
Merlinius brevidens	Sugar beet, onion	Knobloch and Bird (1981)
M. joctus	Ornamental nursery	Knobloch and Bird (1981)
M. macrodorus	Lily	Knobloch and Bird (1981)
M. tessellatus	Unknown	Knobloch and Bird (1981)
Mesocriconema axeste	Moss	Knobloch and Bird (1981)
M. curvatum	Grass, strawberry	Knobloch and Bird (1981)
M. ornatum	Unknown	Knobloch and Bird (1981)
M. reedi	Woods	Knobloch and Bird (1981)
M. serratum	Grass	Knobloch and Bird (1981)
M. simile	Peach, wormwood	Chitwood (1953)
M. xenoplax	Peach	Knobloch and Bird (1981)
Nacobbus batatiformis	Sugar beet	Knobloch and Bird (1981)
Nanidorus minor	Onion	Knobloch and Bird (1981)
Ogma cobbi	Willow, birch	Knobloch and Bird (1981)
O. octangularis	Maple	Chitwood (1953)
Paratrichodorus atlanticus	Unknown	Knobloch and Bird (1981)
P. pachydermus	*Dahlia* sp.	Knobloch and Bird (1981)
P. porosus	Potato	Knobloch and Bird (1981)
Paratylenchus hamatus	Celery, onion	Knobloch and Bird (1981)
P. projectus	Corn, alfalfa	Knobloch and Bird (1981)
Pratylenchoides laticauda	Mint	Knobloch and Bird (1981)
Pratylenchus crenatus	Corn, soybean, wheat	Knobloch and Bird (1981)
P. neglectus	Apple, Corn, soybeans, wheat	Chitwood (1953); Knobloch and Bird (1981)
P. penetrans	Fruits, vegetable crops	Knobloch and Bird (1981)
P. pratensis	Cherry	Chitwood (1953)
P. scribneri	Corn, soybean, wheat	Chitwood (1953)
P. vulnus	Cherry	Chitwood (1953)
Punctodera punctata	Turfgrass	Knobloch and Bird (1981)
Quinisulcius acti	Potato, corn	Knobloch and Bird (1981)
Q. acutus	Unknown	Knobloch and Bird (1981)
Q. capitatus	Unknown	Knobloch and Bird (1981)
Radopholus similis	*Miscanthus*	Warner (pers. comm.)
Rotylenchus buxophilus	Woody ornamentals	Knobloch and Bird (1981)
Rotylenchus robustus	Ornamental hedge	Knobloch and Bird (1981)
Trichodorus primitivus	Boxwood	Knobloch and Bird (1981)

(continued)

Table 3.2 (continued)

Nematode species	Crop and plants	Reference
T. proximus	Boxwood	Knobloch and Bird (1981)
T. similis	Turfgrass	Knobloch and Bird (1981)
Trophonema arenarium	*Spuria* sp.	Knobloch and Bird (1981)
Tylenchorhynchus agri	Unknown	Knobloch and Bird (1981)
T. clarus	Unknown	Knobloch and Bird (1981)
T. claytoni	Potato	Chitwood (1953)
T. dubius	Peach, pine, turfgrass	Chitwood (1953)
T. martini	Willow	Knobloch and Bird (1981)
T. maximus	Sugar beet	Knobloch and Bird (1981)
T. nudus	Turfgrass	Knobloch and Bird (1981)
T. parvus	Peach	Knobloch and Bird (1981)
Xenocriconemella macrodora	Wood lot	Knobloch and Bird (1981)
Xiphinema americanum	Elm, peach, apple, turfgrass	Chitwood (1953)
X. diversicaudatum	Greenhouse roses	G. Bird (pers. comm.)
X. rivesi	Grapes	Ramsdell et al. 1995

Table 3.3 Frequencies of detection and maximum counts per 100 cm^3 soil for plant-parasitic nematodes recovered from survey samples of turfgrasses collected in 2017 (n = 100) and 1993 (n = 106)

Nematode	Frequency of detection (%)		Maximum counts	
Year	2017	1993	2017	1993
Ring	97.0	69.7	6440	1400
Stunt	86.0	76.1	3280	880
Spiral	86.0	61.5	2160	2040
Root knot (j2)	22.0	16.5	300	55
Heterodera spp. (cyst)	21.0	10.1	141	41
Lance	19.0	22.0	330	399
Stubby root	8.0	0.9	100	1
Punctodera punctata (cyst)	7.0	0.0	17	0
Needle	3.0	3.7	50	1
Lesion	2.0	49.5	40	140
Sheath	2.0	6.4	460	60
Pin	0.0	17.4	0	99
Dagger	0.0	4.6	0	20

3.3 Plant Parasitic Nematodes

The ninety-four currently known taxa of plant parasitic nematodes in Michigan include sedentary endoparasites, migratory endoparasites, ectoparasites and virus vectors.

3.3.1 Cyst Nematodes

Michigan has eleven documented species of cyst nematodes. These include *Cactodera milleri, C. weissi, Heterodera avenae, H. carotae, H. glycines, H. humuli, H. pratensis, H. schachtii, H, trifolii, H. ustinovi (=H. iri)* and *Punctodera punctata*, as well as one tentatively identified species, *H. orientalis* (Handoo, USDA/ARS, pers. comm.). Many species of cyst nematodes are serious pathogens of agronomic crops. In Michigan, *Heterodera glycines* and *H. schachtii* are major limiting factors in the production of soybeans and sugar beets, respectively. *Heterodera carotae* can reduce carrot yields, but its impact has not been fully determined. Due to the number of cyst nematode species detected, Michigan is often referred to as the Cyst Nematode Capital of the U.S.

3.3.1.1 *Cactodera* spp.

The two species of *Cactodera* found in Michigan are of no agricultural importance. *Cactodera weissi* has existed in Michigan for at least 60 years and its type host is Pennsylvania smartweed, *Polygonum pennsylvanicum*. This weed is very abundant in the lower peninsula, hence we believe this nematode species is also widely distributed throughout this region. *Cactodera weissi* was first reported in Michigan in a 1971 (Brzeski, pers. comm.). It is found at an annual frequency of up to 1% in samples submitted to Michigan State University (MSU) Diagnostic Services. The type host for *C. milleri* is common lambsquarters, *Chenopodium album*. *Cactodera milleri* was described by Graney and Bird (1990). Other species of *Chenopodium* also serve as hosts for *C. milleri*.

3.3.1.2 *Heterodera avenae*

In a 1983, USDA/APHIS-sponsored a national cereal cyst nematode survey. *Heterodera avenae* was detected in a few locations in Tuscola County, Michigan. All of the sites had a similar production system history and ownership (Bernett 1986). A state-wide survey conducted soon after its initial discovery, revealed no additional detections although many economically significant hosts for this nematode grow in Michigan. Field trials in Michigan in 1986 indicated small grain yield losses can be associated with the presence of *H. avenae* (Bernett 1986). The farms with the original infestations were all managed with the same equipment. Although there are many hosts for this nematode in Michigan, no additional detections have been reported in the last three decades.

3.3.1.3 *Heterodera carotae*

The carrot cyst nematode, *Heterodera carotae,* was found in 1979 during a survey of organic soil (histosol) carrot/onion fields (Graney 1985). Results of surveys conducted in 1986 and 1988 to delineate the distribution of the carrot cyst nematode in Michigan indicated *H. carotae* was widely distributed in the major carrot production areas and had a frequency of detection of roughly 68% in the 43 fields surveyed (Berney and Bird 1992). *Heterodera carotae*, however, has never been detected in mineral soil carrot production systems in Michigan. In addition, *H. carotae* is often detected concomitantly with *Meloidogyne hapla,* so its impact on field-grown carrots is difficult to determine. For soil samples collected the fall prior to carrot, nematode control is recommended if *H. carotae* egg counts exceed 500/100 cm^3 soil. This threshold is essentially the same as that established by Oostenbrink (1972). Berney (1994) found that *H. carotae* had two root exudate mediated peaks of egg hatch. Hatch was common at 10 °C, complete at 15 °C and reduced at both 5 °C and 20 °C. No hatch occurred at 25 °C. Beginning in the early 1990s, however, much of the carrot production in Michigan began shifting from histosols to mineral soils.

3.3.1.4 Soybean Cyst Nematode, *Heterodera glycines*

The initial detection of the soybean cyst nematode (SCN), *H. glycines,* was in Gratiot County in April 1987. Random surveys of soybean fields were performed for SCN in 1992 and 1993 (Warner et al. 1994a). A statistically valid survey was conducted in 2010–2011 (Schumacker-Lott 2011). The Warner and Schumacher-Lott surveys indicate that slightly more than 50% of the 890 thousand Michigan soybean hectares are infested with *H. glycines*. In addition, Michigan has a SCN sampling program funded by the Michigan Soybean Promotion Committee. Over 22,000 samples have been submitted as of 2017, with 41 counties testing positive for SCN. The results covered 50 counties and indicated that on an annual basis, between 45% and 70% of the samples test positive for *H. glycines.*

Soybean cyst nematode is the most important plant pathogen of soybean in the U.S. If a grower opts to use an SCN-susceptible soybean variety on a site where SCN exists, 50% or greater yield loss can occur. Estimates in Michigan place yield loss at 5%, which costs growers about $40,000,000 annually. The Gratiott County location of the first Michigan detection was not harvested the previous year due to the low yield caused by SCN. At another site, bean yields were frequently below 70 kg/ha. In 1999–2000, Chen et al. (1995a) demonstrated both inter and intra-specific competition between *Glycines max* and *Chenopodium album* in the presence of *H. glycines*. Avendano (2003) conducted a comprehensive spatial distribution characterization in a Michigan soybean field. The nested design at 1-month intervals revealed a strong correlation between soil texture, pH, calcium and *H. glycines*. Bates (2006) reported that specific oilseed radish and Oriental mustard cultivars may have potential for use as trap crops for *H. glycines* and that some populations of this species appear to be aggressive in regard to PI 88788 as a source of resistance. In 2010 and 2011, Schumacher-Lott (2011) found significant greater yields in SCN-infested fields planted to PI 88788 and PI 437654-derived cultivars, compared to SCN-susceptible cultivars. Currently, HG and SCN Type testing are performed in the Diagnostic Lab at MSU (Warner et al. 2016). From 2014 to 2017, 97 SCN type tests were conducted. Approximately 95% of the SCN populations tested developed on the indicator line PI 88788 (SCN Type 2 populations), which is the source of resistance present in close to 98% of all SCN-commercially available SCN resistant varieties in maturity groups 0–3 soybeans. This a strong indication that a significant portion of Michigan *H. glycine* populations have become highly aggressive. A potential SCN trap crop blend of a trap crop legume, Wheeler rye and Maximus oilseed radish was tested in 2017–2018.

3.3.1.5 Hop Cyst Nematode, *Heterodera humuli*

The hop cyst nematode, *Heterodera humuli* was first detected in Michigan in 2012. It was found in a sample submitted to Diagnostic Services from an unthrifty hop planting. The site of the single detection of the hop cyst nematode yielded 241 *H. humuli* cysts/100 cm^3 soil. Michigan has a long history of hop production. It is highly probable that *H. humuli* exists in other hop yards.

3.3.1.6 Sugar Beet Cyst Nematode, *Heterodera schachtii*

The second plant parasitic nematode documented in Michigan was the sugar beet cyst nematode *Heterodera schachtii* (SBCN) (Bockstahler 1950). In a 1999 nematode survey of Michigan's sugar beet industry, Miller et al. (1999) found *Heterodera schachtii* widely distributed in six Michigan counties in the Thumb region (East Central Michigan). This nematode was also reported by Brzeski (pers. comm). in 1971. Three surveys for SBCN have occurred over the past 20 years: 1998, 2007 and 2012. The results have been similar in that SBCN occurs in 20–25% of samples collected from sugar beet fields. Michigan has a long history of sugar beet production, with roughly about 61,000 ha of beets grown annually. SBCN occurs in all of the major sugar beet producing areas and historically reduced beet yields 10,000–45,000 kg/ha. Muchena (1984) showed there were three generations of *H. schachtii* per year on *Brassica oleracea* cv. *Capitate*. Bates (2006), confirmed the potential of oilseed radish cvs Adagio and Colonel as trap crops for *H. schachtii*. Caswell's et al. (1986) model of *H. schachtii* remains one of, if not the most, comprehensive of all nematode simulation models.

3.3.1.7 Clover Cyst Nematode, *Heterodera trifolii*

The clover cyst nematode is detected at 5–20% in samples collected from forage legume fields in Michigan. It has never been considered an economic issue in Michigan in commercial production systems. This species was first reported to be present in an unpublished report by Brzeski (pers. comm.) in 1971. Relatively little is known about the biology and ecology of the clover cyst nematode in Michigan. In 2014, a greenhouse trial was conducted with alfalfa (Foregrazer), crimson clover, two varieties of red clover (Dynamite and Gallant), white clover (Domino), yellow sweet clover and rape (Dwarf Essex). Gallant red clover was the best host tested. *Heterodera trifolii* females and cysts were recovered from all of cultivars tested, but not from alfalfa or dwarf Essex rape.

3.3.1.8 *Heterodera ustinovi*

This species of cyst nematode occurs on golf greens where creeping bentgrass, *Agrostis stolonifera*, is grown. Detection of *Heterodera ustinovi* is 10–15% in samples submitted to Diagnostic Services from golf courses. In a 1993 survey of golf courses, *H. ustinovi* was recovered from about 10% of 106 samples. Creeping bentgrass is a good species for use on golf greens in temperate climates, but many greens are now dominated by annual bluegrass, *Poa annua*. *Heterodera ustonovi* prefers *Agrostis* sp., whereas, a second species of turfgrass nematodes, *Punctodera punctata*, prefers *Poa* plants. Most of the detections of *H. ustinovi* have occurred on golf courses and country clubs near metropolitan Detroit. Evidence suggests that nematode-infested sod was used for construction of the greens. Like many of the other plant parasitic nematodes associated with turfgrass, formal pathogenicity studies have not been conducted. Occurrence of this nematode, however, is usually associated with symptoms of foliar necrosis not attributed to other causes.

3.3.1.9 *Heterodera orientalis*

In the spring of 2000, a sample of *Miscanthus sinensis* was submitted to MSU Diagnostic Services. Numerous cysts were extracted from the soil. We attempted to identify these cysts to species using Mulvey and Golden's (1985) key to the cyst-forming genera and species of Heteroderidae. After two unsuccessful attempts, the cysts were sent to Dr. Z. Handoo (USDA-ARS, Beltsville, Maryland), who tentatively identified them as *Heterodera orientalis*. Unfortunately, we were not able to maintain a greenhouse culture of this nematode. The tentative identification, therefore, stands. We have not isolated this nematode from any other samples of grasses submitted for analyses since the initial detection.

3.3.1.10 *Punctodera punctata*

This species of cyst nematode was first identified in Michigan by Brzeski (pers. comm.) in 1971. White females were observed on the roots of Kentucky bluegrass, *Poa pratensis,* collected from a home lawn near Grand Rapids. The nematode was identified as *H. punctata*. Until recently, *P. punctata* had only been found associated with *P. pratensis* in Michigan. It was found in samples collected from home lawns and sod farms at a frequency of up to 5%. The first detection on a golf green was in 2012. *Punctodera punctata* is now recovered in about 5–10% of the samples collected from golf greens in MI. In a 1992 survey of golf courses, *P. punctata* was recovered from about 10% of 106 samples. Annual bluegrass now is the dominant grass species on many of golf greens, at least in the southern portion of the lower peninsula. While formal pathogenicity tests have not been conducted, anecdotal evidence suggests its feeding results in the development of necrotic symptoms on annual bluegrass golf greens.

3.3.2 Root Knot Nematodes

In his 1953 nematode survey, B. G. Chitwood identified twenty-eight taxa of plant parasitic nematodes including *Meloidogyne* spp. associated with vegetable and specialty crops (Table 3.2). *Meloidogyne hapla* is by far the most common of the four species of root knot nematodes currently recognized in Michigan. It is common (20–50%) in diagnostic samples from vegetables, brassicas, legumes, stone fruit, pome fruit, grapes and field-grown herbaceous perennials; infrequent (5–20%) on soybeans, dry beans, strawberry, raspberry, and field-grown woody ornamentals; rare (1–5%) on sugar beets and never (0%) on grains grasses/turf and blueberry.

Meloidogyne incognita and other *Meloidogyne* spp. are often associated with greenhouse crops and imported transplants. The fourth species, *M. naasi,* is not uncommon on turfgrass.

Under Michigan field and greenhouse environments, the northern root knot nematode, *M. hapla* and the southern root knot nematode, *M. incognita,* cause typical root galls, resulting in both necrotic and hypoplastic shoot system symptoms.

Slinger and Bird (1978) conducted a comprehensive study of the ontogeny of the carrot tap root in regards to pathogenesis by *M. hapla*. In addition to deformation of the tap root, plant maturity was delayed about 14 days. Kotcon (1979) found that both rotation crops and weeds impacted the population densities of predaceous nematodes in organic soil, but was not able to show a relationship between *M. hapla* and predaceous nematodes. Olsen (1984) demonstrated the benefits of having corn in crop rotations in *M. hapla* infested sites. MacGuidwin (1983) reported a single annual generation of *M. hapla* association with onions, and a negative linear relationship between mid-season *M. hapla* population density and onion bulb yield (MacGuidwin et al. 1987). The relationship generally resulted in yield losses less than those associated with other Michigan vegetable crops. *M. hapla* population development was similar in both *Glomus fasciculatum*-infected and non-mycorrhizal onion plants (MacGuidwin et al. 1985). The symbiont enhanced onion biomass in the absence, but not in the presence of *M. hapla*.

Meloidogyne nataliei (Michigan grape root knot nematode) is a highly unique taxon with a known global distribution limited to five townships in Southwest Michigan. This species has unique morphology, cytogenetics and biology (Golden et al. 1981). White females and egg masses are readily observable in November, under Michigan growing conditions. This species has a very limited host range (Diamond and Bird 1994) and a known distribution limited to a small geographical area in Michigan (Bird et al. 1994). "Studies of oogenesis and spermatogenesis revealed that *M. nataliei* is a diploid amphimictic species with four (n), relatively large chromosomes, and possibly with an XX (female)-XY (male) mechanism of sex determination. It differs considerably from all other amphimictic or meiotically parthenogenetic species of *Meloidogyne* which have 13–18 smaller chromosomes" (Triantaphyllou 1985). It is a species that needs to be studied in greater detail in regards to its overall relationship to the evolution of the Meloidogyninae.

3.3.3 Root Lesion Nematodes

Pratylenchus penetrans is considered the most common plant parasitic nematode in Michigan. While other *Pratylenchus* spp. exist, no recent survey at the species level has been undertaken. A highly aggressive population of *P. penetrans* exists in mineral soil in West Central Michigan. A 1988 survey of Michigan's potato industry found *Pratylenchus*, predominately *P. penetrans*, in more than 50% of the fields surveyed. There are about 80 described *Pratylenchus* spp. of which some are highly pathogenic, whereas, others have very little impact on host ontogeny. For several decades in Michigan, *P. penetrans* has been referred to as the "Penetrans Root Lesion Nematode". The 1980 study of *P. penetrans* associated with navy beans (*Phaseolus vulgaris*) demonstrated aggregate distribution of this nematode under field conditions. It also showed the variability among different plant cultivars in both nematode population development and plant symptoms associated with this host-parasite relationship. Olsen (1984) demonstrated the wide host range of *P. penetrans* associated with Michigan crops. As the predisposition agent for the Potato Early-Die Disease Complex, potato tuber yield losses are about 50% and range from 5500 to 22,500 kg/ha. Chen (1995) partitioned the below-ground potato biomass into eight components: seed piece, below-ground stem, stolons, basal roots, nodal roots, stolon roots, tuber roots and tubers in regards to *P. penetrans* population development. This species parasitized basal root, nodal root, stolon root, tuber root and stolon tissues. Basal root tissue was damaged as early as 21 days after planting and highly correlated with final tuber yield. Chen et al. (1995a, b) significantly reduced risk to the potato early-die disease complex and increased tuber yields with 2 years of rotation with a legume. Wernette (2011) studied the vertical distribution of *P. penetrans* and found it more common in the upper 30 cm of soil than at a

Pratylenchus penetrans Risk	*Verticillium dahliae* Risk					
	0	1	2	3	4	5
0	0	1	2	3	4	5
1	1	1	2	3	4	5
2	2	2	3	4	5	5
3	3	3	4	4	5	5
4	4	4	5	5	5	5
5	5	5	5	5	5	5

Risk	RLN / 1.0 g root +100 cc Soil	*V. dal*. Dil. Plate / g Soil	*V. dal*. Wet Siev. /10 g Soil
0	0	0	0
1	1-25	2	1-15
2	26-75	4	16-35
3	76-150	6	36-60
4	151-300	8-16	61-100
5	>300	>16	>100

Fig. 3.1 Michigan potato early-die disease complex risk matrix

Fig. 3.2 Carrot symptoms associated with an extremely aggressive population of *Pratylenchus penetrans*

30–60 cm soil depth. A potato early-die risk matrix is used in Michigan for making management recommendations in regards to this infectious disease (Fig. 3.1). The highly aggressive population of *P. penetrans* that exists in West Central Michigan is known to reduce marketable carrot yields by 50% (Fig. 3.2).

Elliott used a holistic approach to study the ecology of *P. penetrans* associated with navy beans. She detected it in aggregate distributions in 68% of Michigan bean fields, with a pathogenic relationship with cv. Sanilac. Cultivars Gratiot, Saginaw and Kentwood exhibited tolerance (Elliott and Bird 1985). Pathogenic severity and nematode population development was impacted by soil texture, moisture and temperature. Mycorrhizal associations appeared to be minimal in this system (Elliott et al. 1984a). Aldicarb provided effective control (Elliott et al. 1984b) and was used in this system for almost three decades.

Noling et al. (1984) used three population densities of *P. penetrans* and three population densities of the Colorado potato beetle (*Leptinotarsa decemlineata*) to study the joint action of these key limiting factors in potato production in Michigan. Root population densities of *P. penetrans* were significantly less in plants grown in the presence of *L. decemlineata*, compared to those maintained in the absence of this insect.

Studies on the joint impact of soil nutrition and *P. penetrans* in Michigan potato production began in 1980 and had a significant impact on the evolution of the concept of soil health and soil health biology in regards to overall Michigan agriculture (Vitosh et al. 1980; Bird and Smith 2013; Snapp et al. 2016). *Pratylenchus penetrans* was included among bacterial canker, nutrition, soil pH and winter injury, as factors associated with the decline of sweet cherry trees in Michigan (Melakeberhan et al. 1993). Melakeberhan et al. (1994) described the impact of *P. penetrans* on cherry rootstock growth and development and in a 1995 study, Melakeberhan et al. (1997) described the relationship between *P. penetrans* and the nutrition of *Prunus avium* rootstocks.

3.3.4 Pratylenchoides *spp.*

Historically, a species of the false root lesion nematode was commonly associated with mint production in organic soils in Michigan. It resulted in stunted plants and reductions in oil quantity. Because of this and other issues association with organic soil degradation, most of the Michigan mint industry has moved to mineral soils. *Pratylenchoides laticauda* has not been detected in Michigan in mineral soils.

3.3.5 Stubby Root Nematodes

In his 1953 nematode survey, B. G. Chitwood detected trichodorids associated with vegetable and specialty crops. While seven species classified in the Trichodoridae have been reported from Michigan (Table 3.2), their frequency of occurrence and population densities appear to have declined during the past two decades. Wernette (2011) studied the vertical distribution of *Paratrichodorus pachydermus*, a vector of Corky Ring Spot Disease of potato caused by *Tobacco rattle virus*. Its occurrence was more common at a soil depth of 30–60 cm, compared to a 0–30 cm soil depth.

3.3.6 Dagger Nematodes

The 1953 Michigan nematode survey by B. G. Chitwood detected *Xiphinema* sp. associated with Michigan orchards, vineyards, berries, cover crops, nurseries, florists and forests (Table 3.2). The 1966 nematode survey of Michigan cultivated blueberry plantings showed an association between *Xiphinema americanum* and *Necrotic Ring Spot Virus* symptoms (Tjepkema 1966). Other important virus diseases of Michigan crops associated with *X. americanum* as the vector include *Tomato Ring Spot Virus* disease of grapes, union necrosis of cherry and stem pitting of cherry (Ramsdell et al. 1995). *Xiphinema americanum* is commonly associated with tree fruit orchards, vineyards and other sites throughout Michigan. *Xiphinema rivesi* is also detected in these ecosystems on a less frequent basis. Both species serve as important vectors of *tomato ringspot virus* disease associated with apple, cherry and grape production. In addition, *Xiphinema diversicaudatum* was recently identified by the junior author of this chapter from a soil sample from roses grown in a Detroit greenhouse. It was previously detected from greenhouse rose samples in 1966.

3.3.7 Needle Nematodes

Longidorus elongatus is not uncommon in Michigan in both mineral and organic soils. It can be a serious problem in celery production. *Longidorus breviannulatus* is present, but limited to corn production in very coarse-textured sandy soils. Yield losses associated with this nematode can be extensive.

3.3.8 Other Ectoparasites

In his 1953 nematode survey, B. G. Chitwood detected criconematid species associated with Michigan orchards, vineyards, berries, cover crops, nurseries, florists and forests. The 1966 survey of 30 commercial blueberry farms by Tjepkema, detected *Atylenchus decalineatus*, *Nanidorus minor, Mesocriconema* spp., *Hemicliophora* spp., *Hoplolaimus galeatus*, *Tylenchorhynchus claytoni*, and *T. joctus*. Species of the Criconematinae, Paratylenchinae and stunt nematode taxa are currently common throughout Michigan agriculture and forest ecosystems. Both ring and stunt nematodes exist in a diversity of genera and species, whereas, known *Paratylenchus*

Table 3.4 Results of 32 years (1974–2015) of nematicide research for control of the potato early-die disease complex

Nematicide (rate and number of years of data)	Mean tuber yield
Non-treated control (32 years of data)	255 cwt/A
Metam (37.5 gal/A, 15 years of data)	387 cwt/A
Oxamyl (4.0 lbs. a.i/A, 11 years of data)	324 cwt/A
Ethoprop (6 years of data)	324 cwt/A
Telone (5 years of data	316 cwt/A
Aldicarb (17 years of data)	306 cwt/A

spp. are limited to *P. hamatus* and *P. projectus. Paratylenchus hamatus* is common in organic soils; whereas, *P. projectus* is common in mineral soils (Knobloch and Bird 1981b). *Lobocriconema thornei* was described by Knobloch and Bird (1978) from a forest location.

Michigan is home to approximately 900 golf courses. Two surveys of golf greens have been conducted; one in 1993 (Warner et al. 1994b) and another in 2017. Turfgrass species grown on golf greens are hosts to many genera of plant parasitic nematodes. At least 12 genera were identified without attempting to separate the stunt nematodes (*Merlinius, Quinsulcius* and *Tylenchorhynchus*) into their appropriate genera. Ring, spiral and stunt nematodes are, by a wide margin, the most frequently detected plant parasitic nematodes in golf green soil in Michigan (Table 3.4). The other genera/species of nematodes that are detected at frequencies >10% in turf samples from Michigan golf greens are *Hoplolaimus galeatus, Meloidogyne naasi,* and *Pratylenchus* spp.

3.3.9 Stem and Foliar Nematodes

In his 1953 nematode survey, B. G. Chitwood detected *Ditylenchus destructor* associated with vegetable and specialty crops. While known to be present in the past, the potato rot nematode, *Ditylenchus destructor* has not been detected in Michigan during the last 40 years. The stem and bulb nematode, *Ditylenchus dipsaci,* also known as the onion bloat nematode, is the most common shoot system tissue-feeding nematode in Michigan. It is frequently associated with herbaceous perennial ornaments. In the past, it was commonly associated with onions, and with garlic in more recent years. Schnabelrauch et al. (1980) reported evidence for four generations of *Ditylenchus dipsaci* associated with *Phlox subulata* during the first year of their study. This was followed by a significant population decline under field or storage conditions and only a single generation the following year. The chrysanthemum nematode, *Aphelenchoides ritzemabosi* is often associated with the greenhouse flower industry.

3.4 Management

At least 45 nematicides have been registered for use in Michigan (Table 3.5). In the late 1950s, halogenated hydrocarbon insecticides were used extensively on the Michigan State University (MSU) campus. This resulted in a serious robin-mortality problem. It also served as a catalyst for Rachel Carson's, land-mark book entitled, *Silent Spring* (Carson 1962). It resulted in a dynamic transdisciplinary team of MSU faculty working with scientists from Cornell University, University of California-Berkeley and Texas A&M University on the development of the philosophy and practices of Integrated Pest Management (IPM). While overall leadership for the MSU portion of this initiative came from the Department of Electrical Engineering and Systems Science, MSU nematology was responsible for the Extension-outreach component. In October of 1974, MSU hosted the Second U.S.A.-U.S.S.R. Symposium. It was entitled *Modeling for Pest Management: Concepts, Techniques and Applications* (Tummala et al. 1976). Volume 30 of BioScience (1980) was dedicated to Pest Management and included an article by Bird and Thomason (1980) entitled, *Pest Management, a Nematological Perspective.* This was expanded by Bird et al. (1985) in Volume II of Sasser's *Advanced Treatise on Meloidogyne*. Nematology continued to play an important role in IPM. In 1979–1980, Ivan Thomason of the Department of Nematology at the University of California-Riverside did a sabbatical leave at MSU. This resulted in development of the California State-Wide IPM Program. In addition, it became recognized that IPM had significant social and political attributes (Bird and Ikerd 1993). *The Integrated Pest Management Experience* (in) *Reform and Innovation of Science and Education Planning for the 1990 Farm Bill* was written from the view of a nematologist (Bird 1989). Much of the original U.S. IPM legislation is still in pace in 2017.

The IPM philosophy was incorporated into the Low Input Sustainable Agriculture (LISA) legislation of the 1985 Farm Bill which evolved into the Sustainable Agriculture Research and Education (SARE) legislation in the 1990 Farm Bill (Bird 1992). This was the stimulus for development of the highly popular Extension Bulletin entitled, *Michigan Field Crop Ecology*, which contains multiple chapters on nematodes (Cavigelli et al. 1998; Bird et al. 1998). The next steps included pioneering involvement in the soil health movement (Sanchez et al. 2003; Yao et al. 2005) and social aspects of conventional, alternative and organic agriculture systems (Francis et al. 2006; Kirschenmann and Bird 2006). The four fundamental strategies of Integrated Nematode Management (INM) include (1) exclusion/avoidance, (2) containment, (3) plant parasitic nematode population reduction and (4) do nothing. The current available tactics are essentially the same as described by Bessey in 1911. Recommendations of control of plant parasitic nematodes associated with fruit, vegetable and ornamental crops are published in MSU Extension Bulletins E-154 and E-312 (Bird and Warner 2015a, b). The objective of the Management Section of this chapter is to describe the current state of INM in Michigan.

Table 3.5 Nematicides marketed in Michigan 1973–2017

Common name	Active ingredient	Company
Aveo EZ	*Bacillus amyloliquefaciens* strain PTA-4838	Valent U.S.A. Corp.
Avicta Duo Corn	12.4% Abamectin; 28.1% Thiamethoxam	Syngenta
Avid 0.15EC	2.0% Abamectin	Syngenta
Avid 0.15EC	2.0% Abamectin	Sygenta
Basamid	99% Tetrahydro-3,5,-dimethyl-2H-1,3,5-thiadiazine-2-thione	BASF
BIOst Nematicide 100	94.46% *Burkholderia* spp. Strain A396 (heat killed)	Albaugh
Brom-O-Gas	96.75% Methyl bromide	Great Lakes Chemical Co.
ClandoSan 618	crustacean exoskeletons (10.4 lbs. N per 100 lbs. product)	IGENE Biotechnology, Inc.
Clariva	*Pasteuria nishiwazae*	Syngenta
Counter 20G	20% terbufos (OP)	AMVAC
Curfew EC	97.5% 1,3-dichloropropene	Dow AgroSciences, LLC
Nemagon 8.6 EC	1,2-Dibromo-3-chloropropane (8.6 lbs./gal)	Shell
DiTera DF	90% *Myrothecium verrucaria* strain AARC-0255 w/w	Valent Biosciences
Divanem 0.15 EC	2.0% Abamectin	Syngenta
Dursban 50W	50% chlorpyrifos (OP)	Dow AgroSciences, LLC
Dylox 80	80% Trichlorfon	Bayer
EarthMAX	4.2% humic acid	Harrell's
EDB	Ethylene-dibromide	Shell
Fumazone 70E	70% 1,2-Dibromo-3-chloropropane	Dow
Furadan	10% Carbofuran	FMC
ILevo	48.4% fluopyram	Bayer
Indemnify	34.5% fluopyram; 7.7% 1,2-propanediol	Bayer
Kontos	22.4% Spirotetramat	OHP, Inc.
K-PAM HL	54% potassium N-methyldithiocarbamate	AMVAC
Lorsban 15G	15% chlorpyrifos	Dow AgroSciences, LLC
Luna Tranquility		Bayer
Majestene	94.46% *Burkholderia* spp. Strain A396 (heat killed)	Marrone Bio Innovations
MeloCon WG	6% *Paecilomyces lilacinus* strain 251	Certis
Mocap 15G	15% Ethoprop (OP)	AMVAC
Mocap EC	69.6% Ethoprop (OP)	AMVAC
Movento	22.4% Spirotetramat	Bayer
Multiguard protect 90EC	90% furfural	Agriguard Co. LLC
NemaKILL	32% cinnamon oil; 8% clove oil; 15% thyme oil	Cisco

(continued)

Table 3.5 (continued)

Common name	Active ingredient	Company
NemaStrike	Tioxazafen, a disubstituted oxadiazole	Monsanto
Nematec	0.56% plant extract	Sci Protek, Inc.
Nematode control	Geraniol oil, egg powder and lecithin	Growers Trust
Nem guard gold	3.33% *Bacillus chitinosporus*	Agro Research International
Nimitz	40.0% fluensulfone	Adama
Nortica	5% *Bacillus firmis*	Bayer
Poncho/votivo	40.3% clothalandin; 8.1% *Bacillus firmis*	Bayer
Pylon	21.4% chlorfenapyr	OHP, Inc.
Sectagon	54% potassium N-methyldithiocarbamate	Tessenderlo Group
Telone C-17	81.2% 1,3-dichloropropene; 16.5% chloropicrin	Dow AgroSciences, LLC
Telone C-35	63.4% 1,3-dichloropropene; 34.7% chloropicrin	Dow AgroSciences, LLC
Telone II	97.5% 1,3-dichloropropene	Dow AgroSciences, LLC
Temik 15G	15% Aldicarb (carbamate)	Bayer
Thimet 20G	20% Phorate	AMVAC
Vapam HL	42% methyl dithiocarbamate	AMVAC
Velum	15.4% fluopyram; 22.2% imidacloprid	Bayer
Velum Prime	41.5% fluopyram	Bayer
Vorlex	1,3-D, 1,2-D and methyl isothiocyanate	Agrevo
Vydate L	24% Oxamyl (carbamate)	Dupont

3.4.1 Cysts Nematodes

In general, cyst nematodes have been more difficult to manage than other plant parasitic nematodes under Michigan conditions.

3.4.1.1 *Heterodera carotae*

Because of the extremely narrow host range of *Heterodera carotae*, risk of infestations can be avoided or reduced through rotations with non-host crops. Movement of the majority of the Michigan carrot acreage from high value organic soil used primarily for vegetable production, to mineral soils suitable for agronomic crop production, significantly increased the potential for rotations with non-host crops. Growers with a previously documented *H. carotae* problem or low marketable yields are encouraged to submit soil samples to MSU Diagnostic Services each fall before the next carrot crop. Carrot production systems with *H. carotae* in organic soil use both fumigant (Telone II and metam sodium) and non-fumigant nematicides (oxamyl) for control of this nematode.

3.4.1.2 *Heterodera glycines*

Crop rotation is recommended to reduce the risk of sites not infested with *H. glycines* from becoming infested and maintaining population densities below the damage threshold for this host-parasite relation. Soil analysis for *H. glycines* population dynamics determination is recommended on a 3-year basis. The length of any rotation should be based on the number of eggs/100 cm^3 soil. Soybeans should not be grown if SCN egg counts exceed 10,000/100 cm^3 soil. For predictive purposes, we estimate declines of 50% annually in SCN population densities in the presence of non-host crops. Producers should avoid growing SCN-susceptible soybean varieties if this nematode is detected in any soil sample. Use of soybean cultivars derived from the PI 88788 source of *H. glycines* resistance is common. Sites with yields less than 1350 kg/ha are expected with a susceptible cultivar, while, yields greater than 3000 kg/ha are expected with a resistant cultivar.

While growers are encouraged to rotate sources of *H. glycines* resistance, sources other than PI 88788 have not been readily available in recent years. This has resulted in an increase in highly aggressive populations, referred to as SCN or HG Types 1, or 1.2, reducing the yield potential enhancement of PI 88788-derived varieties (Warner et al. 2016). Soybean growers are experimenting with chemical, biological and plant health regulator seed treatments. These, however, are only designed for use with resistant varieties. Several new chemical nematicides for soil application are also in the development stage. To manage SCN, soybean producers grow SCN-resistant soybean varieties and rotate to non-host crops. The results of an SCN Type test can aid growers in selecting the best sources of SCN resistance found in commercial varieties. SCN Type testing provides growers information about the aggressiveness of their SCN populations. Most Type 2 populations are slightly or moderately aggressive at this time. A comprehensive analysis of SCN management is included in the 2018 book chapter entitled, "Role of Population Dynamics and Damage Thresholds in Cyst Nematode Management" (Bird et al. 2018). Michigan Farm Bureau, Michigan Agribusiness Association, Michigan soybean Promotion Committee and Michigan State University Extension have formed a Michigan SCN Resistance Management Coalition Partnership with more than twenty other states and eight industry partners. In addition to these partners, funding for this unique Coalition has been made available from the United Soybean Board and the North Central Soybean Research Program.

3.4.1.3 *Heterodera humuli*

While no research on hop cyst nematode management has been conducted in Michigan, Warner et al. (2015) were requested to write the nematode section of a Hop Production Bulletin. The following is a condensation of their recommendations. Avoidance/prevention is the key management strategy. This is primarily achieved through planting hop cyst nematode-free crowns. If a site does become infested, it is imperative that the nematodes are contained to that site, avoiding any activities that move soil and transport nematodes. Always be sure to clean equipment free from soil if working with soil infested with cyst nematodes. Mocap EC (ethoprop) is labeled for use on hop as a pre-plant and post-plant insecticide/nematicide. Cyst nematodes, however, can be difficult to control chemically and there is no information available to suggest that Mocap use results in population reductions of hop cyst nematodes as the product is recommended for insect control. Hop cultivars differ in their susceptibilities to hop cyst nematodes, but in general, most appear to support the nematodes quite well. Improving soil health can be beneficial because as the diversity and numbers of beneficial organisms in the soil increase, often the numbers and impacts of plant pathogens including nematodes, decrease. If hop cyst nematodes are detected in a hop yard, growers must then learn to optimize crop growth and yields in the presence of these nematodes because of the long-term persistence of cyst nematodes.

3.4.1.4 *Heterodera schachtii*

Historically, sugar beet cyst nematode management in Michigan was based on crop rotation. Since production is controlled by the sugar companies, beets were only allowed to be planted in non-infested sites once every 3 years. Fields with known *H. schachtii* infestations were limited to 5-year rotations. A few growers used soil fumigant (Telone, Vorlex) or non-fumigant (aldicarb) nematicides. Muchena and Bird (1987) evaluated the role of fenamiphos as a nemastat for control of *H. schachtii*. Subsequently, the crop rotations were shortened, resulting in increases in *H. schachtii* problem sites. Infested sites often yielded less than 25 tons per hectare, whereas, non-infested sites yielded double this amount. The development and availability of *H. schachtii* tolerant cultivars, in addition to other management changes, significantly enhanced beet yield potentials. Yields greater than 75 tons per hectares are not uncommon. It is recommended that growers sample their fields for SBCN either in the fall prior to a sugar beet crop or during an existing beet crop. Since the release of the first SBCN-tolerant (resistant) beet varieties in the mid to late 1990s, Michigan's sugar beet yields have increased significantly and awareness of SBCN is greatly elevated. Bird, Tylka and Zasada included an economic spreadsheet for SBCN decision-making in their 2018 book chapter entitled, "Role of Population Dynamics and Damage Thresholds in Cyst Nematode Management".

An additional important innovation is the use of a *H. schachtii* trap crop following wheat, pickles (Michigan grows pickles, not cucumbers) or peas, in the year prior to sugar beets. The trap crops are limited to *Raphanus sativus oleiferus* (oilseed radish) and are cultivar specific (e.g. Adagio, Colonel, Defender and Maximus). *Raphanus sativus longipineatus* (Daikon-type radish) is not a trap crop for *H. schachtii*. The *H. schachtii* trap crops attract second-stage juveniles. After root penetration, the nematode signals (cross-talk) for the plant to produce nurse cells. The plant fails to respond to the signal and the nematode dies without producing a next generation. The R-value (reproductive factor) for the *H. schachtii* trap crop ranges from about 0.01–0.10. Michigan sugar beet growers are also experimenting with biological seed treatments. The sugar beet industry has become one of the leaders in development of the concept of soil health.

3.4.1.5 *Heterodera ustinovi* and *Punctodera punctata*

Heterodera ustinovi has a strong preference for creeping bentgrass cv. Toronto. It does not develop well on bluegrasses and fescues. Rotation of grass species, however, is not an option for turf managers. Annual bluegrass is becoming more dominant on creeping bentgrass greens in Michigan, which does not favor *H. ustinovi.* This transition, however, has resulted in the presence of *Punctodera punctata* on these greens. A number of new nematicides including Divanem®, Indemnify® and Nimitz® have been registered for nematode control on turf. They have not, however, been evaluated for control of *H. ustinovi* and *P. punctata* under Michigan growing conditions. Experience indicates that cyst nematodes can be difficult to control with chemical nematicides. Obtaining a further understanding of the biology and parasitic habits of these two species may aid in the proper timing of nematicide applications.

3.4.2 *Root Knot Nematodes*

3.4.2.1 *Meloidogyne hapla*

Whenever possible, northern root knot nematode is managed through rotation with non-host crops such as corn, wheat or other small grain. Unfortunately, no resistant cultivars for *Meloidogyne hapla* susceptible crops are available. Soil fumigation with 1,3-D or metam is common for high cash value crops. 1,3-D, however, can be difficult to obtain. Most growers that use these fumigants are certified applicators and have their own application equipment. Methyl bromide is used on an emergency exemption basis in the production of field-grown herbaceous ornamental plants. This chemical is usually applied by a professional soil fumigation company. The most commonly used non-fumigant chemical nematicide is oxamyl. New supplies of this product, however, were not available for the 2016 and 2017 growing seasons. A number of new and old biological nematicides and plant health regulators are being evaluated by Michigan growers. Farms with highly susceptible *M. hapla* crops or problem sites are generally well aware of the situation and maintain formal or information crop yield, nematode management and soil sample records.

3.4.2.2 Michigan Grape Root Knot Mematode, *Meloidogyne nataliei*

In 1980, the Michigan grape root knot nematode became a state-mandated regulatory species for eradication. Known infestation sites were treated with shallow and deep high dosages of ethylene dibromide (EDB). Two decades later, EDB was detected in groundwater at the sites treated with this soil fumigant. In 2017, active populations of *Meloidogyne nataliei* were collected from at least two of the original infestation (eradication) sites.

3.4.2.3 *Meloidogyne incognita* and *M. naasi*

The southern root knot nematode is managed under greenhouse conditions through the use of soil sterilization, use of nematode-free propagation stocks and nematicides. Although *Meloidogyne incognita* has never been shown to survive Michigan winter conditions, Michigan potato enterprises have southern root knot nematode problems as far north as Indianapolis, Indiana. Additional climate change has the potential to allow for the over-wintering of *M. incognita* in the southern tier of Michigan agricultural counties. Nematicides are currently being evaluated for control of *M. naasi* in turf.

3.4.3 Root Lesion Nematodes

3.4.3.1 *Pratylenchus penetrans*

As the most common plant parasitic nematode in Michigan, control measures are required for a significant number of both annual and perennial crops. Before losing their registrations, use of EDB (Ethylene di-bromide) and DBCP (1,2-Dibromo-3-chloropropane) was common for management of *Pratylenchus penetrans*. Klonsky and Bird (1981) used a computer simulation to show that while the short-term economic loss of these fumigants in tree fruit production would be negligible, the long-term impact would depend on the availability of viable alternatives. Pre-plant applications of soil fumigants, primarily Telone II and metam sodium, are used where warranted for tree fruit, small fruit/grapes, vegetables and ornamentals. When aldicarb (Temik) was first registered for specific commodities, it immediately became the nematicide of choice. When its registrations were cancelled, there was a return to soil fumigation. The rate of 350 l/ha of metam sodium is used for potato early-die management. Some growers changed to other non-fumigant nematicides, predominately oxamyl and ethoprop. The results of 41 years of potato nematicide research clearly demonstrate the impact of these chemicals on tuber yield (Table 3.4).

3.4.4 Turfgrass Biological Control of Ecto and Migratory Endo-Parasitic Nematodes

In a turfgrass trial with *Heterorhabditis bacteriophora* applied through irrigation water, population densities of a mixture of ecto and endoparasitic nematodes were lower in the presence of the entomopathogenic nematode, compared to the absence of this biological control species (Smitley et al. 1992). Until recently, phenamiphos was the most common turf grass nematicide used in Michigan. A significant number of new chemical and biological products are currently being investigated.

3.4.5 Virus Vectors

3.4.5.1 *Xiphinema* spp.

In the presence of *Xiphinema americanum* and *X. rivesi*, new and replant orchard and vineyard sites undergo 1 or 2 years of soil preparation before planting. Ramsdell et al. (1983) evaluated the role of superimposed deep and shallow soil fumigation to control *X. americanum* and *peach rosette mosaic virus* re-infection in a "Concord" vineyard site. Historically, management has involved planting sudax as a cover crop. Recently, the recommendation has changed to Essex rape, because it is a poor to non-host for *X. americanum*. In addition, soil fumigants (Telone II and metam sodium) are used on a pre-plant basis. This is especially used in tart cherry orchards. *Xiphinema americanum* is the key target virus vector. Research has been funded by the Michigan tart cherry industry to find a replacement for soil fumigation. Cover crop blends that contain Essex rape appear promising.

3.4.5.2 *Paratrichodorus pachydermus*

Crop rotation and deep-shallow soil fumigation is used for control of *Paratrichodorus pachydermus* in locations known to be infested with this nematode. Growers have designed and built custom soil fumigant equipment designed to apply metam sodium at both 20 and 40-cm soil depths.

3.4.6 Shoot-System Nematodes

3.4.6.1 Stem and Bulb Nematode, *Ditylenchus dipsaci*

Both aldicarb and oxamyl reduced foliar symptoms of creeping phlox and population densities of *Ditylenchus dipsaci* during the first year after treatment, throughout winter storage and most of the second year (Schnabelrauch et al. 1981). In recent years, seeding, instead of planting bulb sets, is recommended for control of the bulb and stem nematode in onion systems. Crop rotation and use of nematode-free planting stock is essential for *D. dipsaci* management in garlic plantings. This nematode can be a key limiting factor in creeping phlox production. Crop rotation, use of nematode-free planting stock and soil fumigation are recommended and commonly used.

3.4.6.2 Foliar Nematode, *Aphelenchoides ritzemabosi*

Avoidance is by far the best strategy or foliar nematode management. Planting stock and planting media including field soil, should be free of *Aphelenchoides ritzemabosi*. This may require in-house quarantine to assure that crowns are nematode-free. Crop rotation, soil sterilization, foliar nematicides and hot water treatments can all be used as control tactics in specific situations. Hot water at 46 °C for 5–15 min is often adequate. Overhead watering should be avoided to prevent nematode dissemination to non-infested tissue.

3.5 Conclusions

Michigan's highly variable agriculture and nematodes (Table 3.2) evolved in conjunction with the state's great diversity of soils, local climates and talents of immigrant farmers. This fostered a leadership role in the evolution of the concepts of Integrated Pest Management, Sustainable Agriculture, Sustainable and Equitable Development and Soil Health. Today, large specialized farms are highly knowledgeable about nematodes and other associated production technologies. An emerging group of new small farms are having their initial experiences with both plant parasitic and other types of nematodes. In addition, there has been a significant increase in organic agriculture in Michigan (Bird 2017). These developments provide key challenges for nematologists, since the Nematodes of Michigan will always be evolving in response to the dynamics of their associated ecosystems.

References

Avendano, M. F. (2003). *Characterization of the spatial distribution of Heterodera glycines Ichinohe, 1955 (Nematoda), soybean cyst nematode in two Michigan fields*. Ph.D. Dissertation, Department of Entomology, Michigan State University, East Lansing. 217 pp.

Bates, C. L. (2006). *Bionomics and control of two Heterodera sp. in Michigan*. M.S. thesis, Department of Entomology, Michigan State University, East Lansing. 87 pp.

Bernett, B. (1986). *Effects of the oat cyst nematode on small grains in Michigan*. M.S. thesis, Department of Crop and Soil Science, Michigan State University, East Lansing. 65 pp.

Berney, M. F., (1994). *Study of the biology and ecology of Heterodera carotae (Jones, 1950) in Michigan*. Ph.D. Dissertation, Department of Entomology, Michigan State University, East Lansing, 155 pp.

Berney, M. F., & Bird, G. W. (1992). Distribution of *Heterodera carotae* and *Meloidogyne hapla* in Michigan carrot production. *Annals of Applied Nematology, 24*, 776–778.

Bessey, E. A. (1911). Root knot and its control. USDA/BPI, Bull. 217:1–89.

Bessey, E. A., 1915. *Farmers' bulletin: The control of root knot*. USDA/BPI, Bull. 648:1–19.

Bird, G.W. (1989). The integrated pest management experience. In *Reform and innovation of science and education planning for the 1990 farm bill* (pp. 31–41). 101st Congress S Prt 101–61.

Bird, G. W. (1992). *Sustainable agriculture and the 1990 farm bill. 1991–1992*. Proceedings of the Philadelphia Society for Promoting Agriculture. pp. 108–117.

Bird, G. W. (2003). Role of integrated pest management and sustainable development. In K. M. Maredia, D. Dakouo, & D. Mota-Sanchez (Eds.), *Integrated pest management in the global arena* (pp. 73–95). Cambridge, MA: CABI Pub. 512 pp.

Bird, G. W. (2017). The organic movement at MSU. In M. Kaufman & J. Christianson (Eds.), *The organic movement in Michigan* (pp. 70–80). Lansing: Michigan Organic Food and Farm Alliance. 209 pp.

Bird, G.W., Ikerd, J. (1993, September). Sustainable agriculture: A twenty-first-century system. *Annals of the American Academy of Political and Social Science, 529*, 92–102.

Bird, G., & Smith, J. (2013). Observations on the biology of organic orchard soils. *Acta Horticutural, 1001*, 287–293.

Bird, G. W., & Thomason, I. J. (1980). Pest management: A nematological perspective. *Bioscience, 30*, 670–674.

Bird, G. W., & Warner, F. (2015a). Nematode management. In *Michigan fruit management guide*. Michigan State University Extension Bulletin E-154.

Bird, G. W, & Warner, F. (2015b). Nematode control. In *Insect, disease and nematode control for commercial vegetables*. MI State Univ. Ext. Bull. E-312.

Bird, G. W., Tummala, R. L., & Gage, S. H. (1985). The role of systems science and data base management systems in nematology. In J. N. Sasser (Ed.), *Advanced meloidogyne treatise* (Vol. II, pp. 159–173). New York: Academic.

Bird, G. W., Diamond, C., Warner, F., & Davenport, J. (1994). Distribution and regulation of *Meloidogyne nataliei*. *Journal of Nematology, 26*, 727–730.

Bird, G. W., Berney, M. F., & Cavigelli, M. A. (1998). Soil ecology. In M. S. Cavigelli, S. R. Deming, L. K. Probyn, & R. R. Harwood (Eds.), *Michigan field crop ecology*, *Extension Bulletin E-2624* (pp. 12–16). East Lansing: Michigan State University, 92 pp.

Bird, G. W., Tylka, G. L., & Zasada, I. (2018). Role of population dynamics and damage thresholds in cyst nematode management. In R. Perry, M. Moens, & J. Jones (Eds.), *Cyst nematodes* (pp. 101–127). Wallingford: CAB International.

Bockstaller, H. W. (1950). *The sugar beet nematode in Michigan*. Proc. 6th Gen. Mtg. Sugar Beet Tech. pp. 479–480.

Brzeski, M. and C. W. Laughlin. (1971). Detection of *Heterodfera trifolii* in Michigan. Personal communications.

Carson, R. (1962). *Silent spring*. Cambridge, MA: Houghton Miffin.

Caswell, E. P., MacGuidwin, A. E., Miline, K., Nelson, C. E., Thomason, I. J., & Bird, G. W. (1986). A simulation model of *Heterodera schachtii* infecting *Beta vulgaris*. *Journal of Nematology, 18*, 512–519.

Cavigelli, M. S., Deming, S. R., Probyn, L. K., Harwood, R. R. (1998). Michigan field crop ecology. Michigan State University Extension Bulletin E-2624, East Lansing, 92 pp.

Chen, J. (1995). *Feature, function and nature of Pratlenchus penetrans and Verticillium dahliae interactions associated with Solanum tuberosum*. Ph.D. Disser. East Lansing: Department of Entomology, Michigan State University. 196 pp.

Chen, J., Bird, G. W., & Renner, K. A. (1995a). Influence of *Heterodera glycines* on interspecific and intraspecific competition associated with *Glycine max* and *Chenopodium album*. *Journal of Nematology, 27*, 63–69.

Chen, J., Bird, G. W., & Mather, R. (1995b). Impact of multi-year cropping regimes on Solanum tuberosum tuber yields in the presence of *Pratylenchus penetrans* and *Verticillium dahliae*. *Journal of Nematology, 27*, 654–660.

Chitwood, B. G. (1953). Plant parasitic nematode problems in Michigan. In N. A. Knobloch, G. W. Bird (Eds.), 1981. *Plant parasitic nematodes of Michigan: With special reference to the genera of the Tylenchorhynchinae (Nematoda)*, (pp. 6–8). Michigan Agricultural Research Station. Research Report 419:1–35.

Christie, J. R. (1942). A description of *Aphelenchoides besseyi*, n. sp. the summer dwarf nematode of strawberries with comments on the identity of *Aphelenchoides subtenuis* (Cobb, 1926) and *Aphelenchoides hodsoni* (Goodey, 1935). *Proceedings of the Helminthological Society of Washington, 9*, 82–84.

Cobb, N. A. (1914). Free-living fresh-water nematodes. *Transactions of the American Microscopical Society, 33*, 69–143.
Cobb, M. V. (1915). Some fresh-water nematodes of the Douglas Lake Region of Michigan, U.S.A. *Transactions of the American Microscopical Society, 34*, 21–47.
Diamond, C., & Bird, G. W. (1994). Observations on the host-range of *Meloidogyne nataliei*. *Plant Disease, 78*, 1050–1052.
Elliott, A. P., & Bird, G. W. (1985). Pathogenicity of *Pratylenchus penetrans* on navy beans (*Phaseolus vulgaris* L.). *Journal of Nematology, 17*, 81–85.
Elliott, A. P., Bird, G. W., & Safir, G. (1984a). Joint influence of *Pratylenchus penetrans* (Nematoda) and *Glomus fasciculatus* (Phycomyceta) on the ontogeny of *Phaseolus vulgaris*. *Nematropica, 13*, 111–119.
Elliott, A. P., Bird, G. W., Leavitt, R. A., & Rose, L. M. (1984b). Dynamics of aldicarb soil residues associated with *Pratylenchus penetrans* control in dry bean production. *Plant Disease, 68*, 873–874.
Francis, C., Poincelot, R., & Bird, G. (2006). *A new social contract: Developing and extending sustainable agriculture*. New York: Haworth Press, 367 pp.
Gillespie, R., Harrison, W. III, Grammer, M. (1987). Geology of Michigan and the Great Lakes. Cengage Learning. 39 pp.
Golden, A. M., Rose, L. M., & Bird, G. W. (1981). Description of *Meloidogyne nataliei* n. sp. (Nematoda: Meloidogynidae) from grape in Michigan with SEM observations. *Journal of Nematology, 13*, 393–400.
Graney, L. S. (1985). Observations on the morphology of *Heteroderae carotae* and *Heterodera avenae* in Michigan. *Journal of Nematology, 17*, 519.
Graney, L. S., & Bird, G. W. (1990). A review of the genus *Cactodera* with descriptions and comparative morphology of *Cactodera milleri n.* sp. and *Cactodera cacti* (Filipjev and Schuurmans, Stekhoven, 1941) Krall and Krall, 1978. *Journal of Nematology, 22*, 457–480.
Kirschenmann, F., & Bird, G. (2006). Future potential for organic farming: A question of ethics and productivity. In C. Francis, R. Poincelot, & G. Bird (Eds.), *A new social contract: Developing and extending sustainable agriculture* (pp. 307–324). New York: Haworth Press 367 pp.
Klonsky, K., Bird, G.W. (1981). An economic assessment of Michigan cherry production in relation to plant parasitic nematodes. MSU Ag. Econ. Rept. 401. 132 pp.
Knierim, J. (1963). Nematodes associated with crop plants in Michigan. *Agricultural Experiment Station Quarterly Bulletin, 46*, 254–262.
Knobloch, N., & Bird, G. W. (1978). Criconematidae habitats and *Lobocriconema thornei* n. sp. (Criconematidae: Nematoda). *Journal of Nematology, 10*, 61–70.
Knobloch, N., Bird, G.W. (1981). *Plant parasitic nematodes of Michigan; with special reference to the genera of the Tylenchorhynchidae*. Mich. State Univ. Agric. Expt. Res. Rept. 419. 35 pp.
Kotcon, J., (1979). Studies on the ecology of nematodes associated with vegetables grown in organic soils. M.S. thesis, East Lansing: Department of Entomology, Michigan State University. 124 pp.
MacGuidwin, A. E. (1983). Pathogenicity and ecology of *Meloidogyne hapla* associated with *Allium cepa*. Ph.D. Dissertation, East Lansing: Department of Entomology, Michigan State University. 237 pp.
MacGuidwin, A. E., Bird, G. W., & Safir, G. R. (1985). Influence of *Glomus fasciculatum* on *Meloidogyne hapla* infecting *Allium cepa*. *Journal of Nematology, 17*, 389–395.
MacGuidwin, A. E., Bird, G. W., Haynes, D. L., & Gage, S. H. (1987). Pathogenicity and population dynamics of *Meloidogyne hapla* associated with *Allium cepa*. *Plant Disease, 71*, 446–449.
MDARD. (2017). *Facts about Michigan agriculture*. https://www.michigan.gov/mdard/0,4610,7-125-1572-7775%2D%2D,00.html
Melakeberhan, H., Jones, A. L., Sobiczewski, P., & Bird, G. W. (1993). Factors associated with the decline of sweet cherry trees in Michigan: Nematodes, bacterial canker, nutrition, soil pH and winter injury. *Plant Disease, 77*, 266–270.

Melakeberhan, H., Bird, G. W., & Perry, R. (1994). Plant parasitic nematodes associated with cherry rootstocks in Michigan. *Journal of Nematology, 26*, 767–772.

Melakeberhan, H., Bird, G. W., & Gore, R. (1997). Impact of plant nutrition on *Pratylenchus penetrans* infection 381–388.on *Prunus avium* rootstocks. *Journal of Nematology, 29*, 381–388.

Miller, A. M., Warner, F. W., & Bird, G. W. (1999). Occurrence and distribution of *Heterodera* spp. in Michigan sugar beet production. *Journal of Nematology, 31*, 557.

Muchena, P. K. (1984). Host-parasite relationships and management of *Heterodera schachtii* associated with *Brassica oleracea* var. capitate L., M.S. thesis, Department of Entomology, Michigan State University, East Lansing. 191 pp.

Muchena, P. K., & Bird, G. W. (1987). Role of fenamiphos as a nemastat for control of *Heterodera schachtii* in cabbage production. *Plant Disease, 71*, 552–554.

Mulvey, R. H., & Golden, A. M. (1985). An illustrative key to the cyst forming genera and species of Heteroderidae in the western hemisphere with species morphometrics and distributions. *Journal of Nematology, 15*, 1–59.

Noling, J. W., Bird, G. W., & Grafius, E. J. (1984). Joint influence of *Pratylenchus penetrans* and *Leptinotarsa decemlineata* on *Solanum tuberosum* productivity and pest population dynamics. *Journal of Nematology, 16*, 230–234.

Olsen, H. C. (1984). Influence of rotation crops and management systems on *Pratylenchus penetrans* associated with *Solanum tuberosum* production. M.S. Thesis. East Lansing: Department of Entomology, Michigan Staste University. 58 pp.

Oostenbrink, M. (1972). Evaluation and integration of nematode control methods. In J. M. Webster (Ed.), *Economic nematology* (pp. 497–514). London: Academic.

Ramsdell, D. C., Bird, G. W., Gillett, J. M., & Rose, L. (1983). Superimposed deep and shallow soil fumigation to control *Xiphinema amercanum* and peach rosette mosaic virus reinfection in a "Concord" vineyard site. *Plant Disease, 67*, 625–627.

Ramsdell, D. C., Gillett, J. M., & Bird, G. W. (1995). Relative susceptibility of American grapevine scion cultivars and French hybrid rootstock and scion cultivars to infection and disease caused by peach rosette mosaic nepovirus. *Plant Disease, 79*, 154–157.

Sanchez, J. E., Edson, C. E., Bird, G. W., Whalon, M. E., Willson, T. C., Harwood, R. R., Kizilkaya, K., Nugent, J. E., Klein, W., Middleton, A., Loudon, T. L., Mutch, D. R., & Scrimger, J. (2003). Orchard nitrogen management influences soil and water quality and tart cherry yields. *Journal of the American Society for Horticultural Science, 128*, 277–284.

Schnabelrauch, L. S., Sink, K. C., Jr., Bird, G. W., & Laemmlen, F. F. (1980). Multiyear population dynamics of *Ditylenchus dipsaci* (Kuhn, 1857; Filipjev, 1937) associated with *Phlox subulata*. *Journal of Nematology, 12*, 203–207.

Schnabelrauch, L., Bird, G. W., Laemmlen, F. F., & Sink, K. C. (1981). Occurrence, symptomatology and control of *Ditylenchus dipsaci* associated with commercial production of *Phlox subulata*. *Plant Disease, 65*, 745–748.

Schumacker-Lott, L. (2011). Bionomics of *Heterodera glycines* and *Pratylenchus penetrans* associated with Michigan soybean production. M.S. thesis. 82 pp.

Slinger, L., & Bird, G. W. (1978). Ontogeny of *Daucus carota* infected with *Meloidogyne hapla*. *Journal of Nematology, 10*, 188–194.

Smitley, D. R., Warner, F. W., & Bird, G. W. (1992). Influence of irrigation and *Heterorhabditis bacteriophora* on plant parasitic nematodes in turf. *Journal of Nematology, 24*, 637–641.

Snapp, S., Tiemann, L., Rosenzweig, N., Brainard, D., Bird, G. (2016). Managing soil health for root and tuber crops. Michigan State University, East Lansing, Extension Bull 3343:1–10.

Tjepkema, J. P. (1966). The plant parasitic nematodes associated with cultivated blueberries in Michigan. M.S. thesis, Department of Entomology, Michigan State University, East Lansing. 45 pp.

Triantaphyllou, A. C. (1985). Gametogenesis and the chromosomes of *Meloidogyne nataliei*: Not typical of other root knot nematodes. *Journal of Nematology, 17*, 1–5.

Tumalla, R. L., Haynes, D. L., Croft, B. A. (1976). Modeling for pest management: Concepts, techniques and applications. Papers from the USA/USSR Symposium of Oct. 15–17, 1944, Michigan State University, East Lansing. 247 pp.

Vitosh, M. L., Noling, J. W., Bird, G. W., & Chase, R. W. (1980). The joint action of nitrogen and nematicides on *Pratylenchus penetrans* and potato yield. *American Potato Journal, 57*, 101–111.

Warner, F., Mather, R., Bird, G., & Davenport, J. (1994a). Nematodes in Michigan: distribution of *Heterodera glycines* and other plant parasitic nematodes in soybean. *Journal of Nematology, 26*, 720–726.

Warner, F. W., Davenport, J. F., & Bird, G. W. (1994b). *Nematodes in turfgrass research 1993*. 64th annual Michigan Turfgrass conference proceedings (pp 11–13).

Warner, F. W., Bird, G. W., Hay, F. S., Sirrine, J. R., Gent, D. H. (2015). Nematodes. *In Field guide for integrated pest management in Hops* (3rd edn., pp. 45–46). A Cooperative Publication Produced by Washington State University, Oregon State University, University of Idaho, and U.S. Department of Agriculture in cooperation with Michigan State University and Cornell University.

Warner, F. and M. Golden. 1987. USDA/ARS Confirmation of *Heterodera glycines* in Michigan. Personal communications.

Warner, F., Tenney, A., Bird, G. (2016). Current status of Michigan *Heterodera glycines* types. Proceedings of the 2016. Annual Meeting of the Society of Nematologists.

Wernette, L. (2011). Potato nematode research: With special reference to potato early-die, corky ringspot and soil enzymes. M.S. thesis, Department of Entomology, Michigan State University, East Lansing. 92 pp.

Yao, S., Merwin, I. A., Bird, G. W., Abawi, G. S., & Thies, J. E. (2005). Orchard floor management practices that maintain vegetative or biomass groundcover stimulate soil microbial activity and alter soil microbial community composition. *Plant and Soil, 271*, 377–389.

Chapter 4
Nematodes of Agricultural Importance in Indiana, Illinois, Iowa, Missouri and Ohio

Andreas Westphal, John J. Chitambar, and Sergei A. Subbotin

4.1 Agriculture in Indiana, Illinois, Iowa, Missouri and Ohio

These five states comprise portions of the southern part of the North Central Region of the U.S. They all are located at the southern portion of the Mississippi River Drainage Basin at surprisingly little elevation above sea level ranging from 200 to 300 m. Precipitation is distributed throughout the year allowing for efficient dryland farming on fertile soils that are largely glacially impacted in their origin. Except for Ohio, the states mostly receive summer rains, and total annual precipitation is declining from East to West. Agricultural production is focused on combine crops, foremost soybean, *Glycine max*, and corn, *Zea mays* (Table 4.1).

Indiana, Illinois, Iowa, Missouri and Ohio constitute the central part of the "corn belt" of the United States. Soybean and corn occupy large proportions of the entire acreage, and remaining lands are used for wheat, vegetable production, and minor areas for crops like fruit trees and vines. Based on this production pattern and climatic conditions of medium hard winters and mostly rainy summers, plant parasitic nematodes of the major crops are of greatest concern. The production emphasis on the two large-area combine crops results in narrow crop rotations, and market forces partially lead to monoculture cropping. Although plant parasitic nematodes are recognized in corn, foremost *Pratylenchus* sp. (Norton 1984), more attention is afforded for nematode problems in soybean. Production conditions are characterized by

A. Westphal (✉)
Department of Nematology, University of California Riverside, Kearney Research and Extension Center, Parlier, CA, USA
e-mail: andreas.westphal@ucr.edu

J. J. Chitambar · S. A. Subbotin
California Department of Food and Agriculture, Plant Pest Diagnostic Center, Sacramento, CA, USA
e-mail: john.chitambar@cdfa.ca.gov; sergei.subbotin@cdfa.ca.gov

S. A. Subbotin, J. J. Chitambar (eds.), *Plant Parasitic Nematodes in Sustainable Agriculture of North America*, Sustainability in Plant and Crop Protection,
https://doi.org/10.1007/978-3-319-99588-5_4

Table 4.1 Acreage of the large acre crops of Illinois, Indiana, Iowa, Missouri and Ohio in 2017 (in 1000,000 hectare)

State	Soybean	Corn
Illinois	4.3	4.5
Indiana	2.4	2.2
Iowa	4.0	5.4
Missouri	2.4	1.4
Ohio	2.1	1.4

https://quickstats.nass.usda.gov

maturity grouping of the crops. Flower induction is day length sensitive in soybean, limiting the cultivar pool to the maturity groups capable of producing high yields in this area. Similar restrictions apply for corn cultivars that are also limited by maturity groupings fitting in particular areas. To some extent, genetic resources are confined within these maturity groupings and are not easily exchanged with other production areas. This biological background increases the challenge of providing new cultivars fit for production in these states quickly and frequently.

4.2 Plant Parasitic Nematodes

4.2.1 Soybean Cyst Nematode, Heterodera glycines

In this region, the by-far most damaging and costly nematode parasite is the soybean cyst nematode (SCN), *Heterodera glycines* (Tylka and Marett 2017) (Fig. 4.1). While it is not fully understood if one or several introductions of the nematode to the U.S. occurred, one of the hypotheses includes the following: When soybean seed was first introduced to the U.S., plants lacked vigor and performance. To overcome this growth depression, soil was introduced from Asia, e.g. Japan where soybean had been cultivated and vigorously produced. The expected benefits and reasoning for this practice were strictly empirical at the time. Plants just performed more vigorously when amending fields with imported soil. From today's view, the material, probably unbeknownst, transferred rhizobium bacteria for the critical legume nodulation. This group of bacteria forms a symbiosis with the plants, in that the bacteria benefit from the plant host by obtaining photosynthates for nourishment while mineralizing atmospheric nitrogen that the plant can use for its growth. Using such type of soil amendments became standard practice to increase yields. Ignorant of other possible soil-borne culprits, the story goes that the soybean cyst nematode was also introduced with such inoculum soil. But the first official report of soybean cyst nematode in the U.S. was made in 1954 in North Carolina in a field where soybean was grown after several years of growing flower bulbs of planting material imported from Japan (Winstead et al. 1955). The current consensus is that the soybean cyst nematode has continuously spread throughout soybean production areas while early after its discovery, some discussions persisted that the nematode would be endemic to the U.S. (Noel 1992).

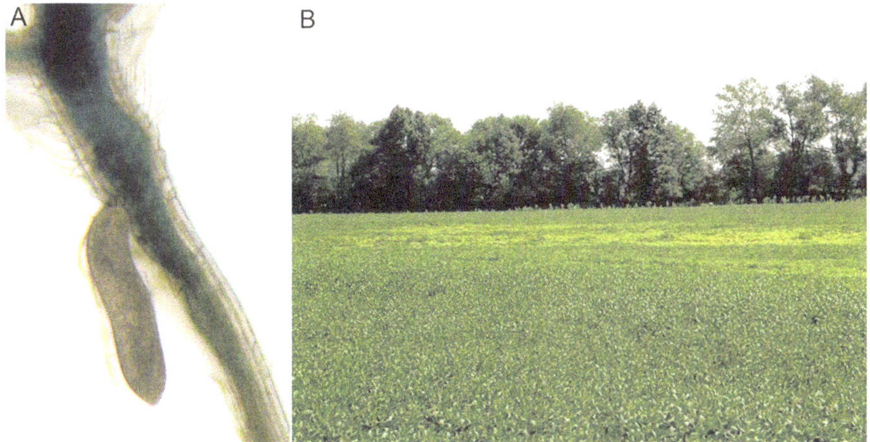

Fig. 4.1 (**a**) Plant root infected with soybean cyst nematode; note the vascular swelling of the developing nematode feeding site (syncytium) (Credit: Xiaoli Guo, Division of Plant Sciences and Bond Life Sciences Center, University of Missouri). (**b**) Field view of a soybean field with patches of yellow and stunted soybean caused by soybean cyst nematode (Credit: Purdue University)

Surveys in the different states, confirmed the wide distribution of SCN in Missouri, (Niblack et al. 1994) and Ohio (Riedel and Golden 1988), however, its spread throughout the region is now fully recognized (Niblack 2005). For years, collaborative efforts of nematologists, extension specialists and others in the respective states have documented its spread and damage potential. Over the years, the soybean cyst nematode has excelled in being one of the most important plant pests of soybean on an annual basis (Allen et al. 2017; Koenning and Wrather 2010; Wrather and Koenning 2009).

A soil-borne problem like soybean cyst nematode requires multiple management approaches (Niblack 2005). Utility of some practices may be reduced because of environmental/climatic conditions and economic forces. In the here-covered states, economic returns of the soybean and corn crops clearly favor their production over alternative crops. The most effective vegetation period from April to September is used for growing these cash crops, thereby minimizing opportunities for alternative or cover crops.

Based on these production constraints and interactions with other soil-borne maladies, a strong research focus is on the development of resistant soybean germplasm, and the search continues (Arelli et al. 2015). Crop rotation and cover cropping are investigated along with clean field strategies, and so involve removal of alternate hosts during the soybean crop and outside the vegetation period of the cash crop. The use of naturally occurring nematode population density regulation has received noteworthy attention, but further comprehensive studies are indicated to implement their use. Chemical seed and in-furrow treatments find interest but are limited because of the typically small margins of the return to investment for such strategies. In detail the following aspects are discussed:

4.2.1.1 Interactions of *Heterodera glycines* with Soil-Borne Fungi on Soybean

In the 1990s, the new symptomology of the so-called "sudden death syndrome" (SDS) of soybean was observed in soybean fields. Depending on the epidemiology, symptoms are most often observed after onset of reproductive stages. The etiology was traced to *Fusarium virguliforme* (Aoki et al. 2003), formerly *F. solani* f. sp. *glycines* (Roy et al. 1997). SDS typically occurs at the beginning of reproductive stages of the soybean crop and is evidenced in the field as varying areas of premature defoliation. Single leaves initially have interveinal chlorosis and later necrosis before these toxin-induced symptoms result in leaf abortion (Westphal et al. 2008). Early in the discovery of the disease the interrelationship of the fungal pathogen with *H. glycines* was discovered (McLean and Lawrence 1993). This interrelationship was later described as a truly synergistic disease complex (Xing and Westphal 2013). In contrast to nematode damage that can be conspicuous and allow for apparently normal growth but severely reduced yields, SDS symptoms are obvious and trigger "catastrophic" fears when plants in infected fields rapidly and prematurely defoliate (Westphal et al. 2008). Though yield losses can be extreme, they may not be as large as the symptomology suggests if the disease is occurring late in the season. A similar interactive disease complex was also described for *H. glycines* and *Phialophora gregata* in the development of brown stem rot (Tabor et al. 2003). If the diseases are likely to occur in the same region, proper diagnostic is essential to take the proper remedial actions (Tabor et al. 2018).

4.2.1.2 Host Plant Resistance Including Considerations for Virulence Differences of *Heterodera glycines*

Use of host plant resistance against *H. glycines* has been recognized as one of the key options in managing these soil-dwelling worms. Large efforts have been made to find sources of resistance to this parasite. This resulted in the discovery of several resistant sources. Challenges remained because host plant resistance was detected in soybean lines with otherwise undesirable agronomic characteristics. For example, the seed color was dark in some of the lines, or the plant lacked overall vigor. This required additional back-crossing efforts to high yielding soybean lines. This lengthy process was only partially overcome, and initially resistant cultivars experienced the so-called "yield drag". A phenomenon described as yield inferiority of a resistant cultivar compared to the susceptible high-yield cultivar under non-infested conditions. Experimentally, this can be demonstrated when *H. glycines*-resistant and susceptible cultivars are exposed to different nematode infestation levels (Koenning 2000). This characteristic of early resistant cultivars slowed adoption of the resistant cultivars.

Shortly after the introduction of resistant lines, occasional failures under *H. glycines*-infested conditions were observed. Plant damage and nematode reproduction were detected in single fields while the same cultivars fared well in other infested fields. This led to the hypothesis that *H. glycines* populations vary in virulence. It became obvious that this variability was encumbering the utility of the new resistant cultivars. Thus, a system for capturing this variability was developed. Initially, a race system was developed that used arbitrary classification schemes of nematode reproduction on resistance sources of soybean (Riggs and Schmitt 1988). Nematode reproduction was measured on the resistant lines and compared to a susceptible standard. An arbitrary 10% reproduction compared to reproduction on the susceptible was used as cut-off for calling the nematode-host plant interaction resistant. Based on such classification, a race was assigned to specific nematode populations (Riggs and Schmitt 1988). Because it was difficult to quickly translate the race information into useful information for cultivar choice, and because of other shortcomings of the system due to the high variability of nematode reproduction, an improved classification system was needed. In the HG-type system, population notifications were greatly improved (Niblack 1992b). For example, the indexing gives unambiguous notification on what resistance source a population of *H. glycines* can overcome, and the seed description clearly indicates the utility of the cultivar. In this latter system, similar classification cut-offs are used but the index gives a direct lead to what sources of resistance are ineffective against certain field populations (Niblack et al. 2002). Both classification systems were first introduced to describe field populations for pure biological observations, but especially the modernized system does allow for guidance in the cultivar choice when the soybean cyst nematode is present in specific fields. These very practical considerations also feed back into the development of new genetic resources to generate broader breeding strategies. For example, the generation of broader germplasm has been reported to allow for efficient future selection efforts (Cianzio et al. 2018)

Knowledge of the specific virulence pattern of *H. glycines* is important because this nematode is not only spread in different virulence groups but can also change its virulence pattern when observed over periods of decades. For example, resistance in the source PI88788 has been excessively used for generating soybean cultivars with resistance to *H. glycines* while few other sources were used for decades (Tylka 2017). In a comparison of virulence patterns of *H. glycines* of historical data and those of the early 2000s, populations had changed in their capacity to infect soybean lines with resistance derived from PI88788 (Niblack et al. 2008). This observation, confirmed in Missouri (Howland et al. 2016) and Iowa (McCarville et al. 2017), illustrated the need to use different sources of resistance, and that the simple recommendation to at least rotate the resistant cultivars. Even if alternate cultivars were based on the same PI88788 resistance, thus probably insufficient to avoid the selection pressure for higher virulence on this resistance source, some benefit of the supporting genetics was surmised. Hypothetically, this overuse of one resistance source on large areas of nematode populations favored nematode populations of virulence patterns that can overcome the resistance of PI88788 (Niblack et al. 2008). Other resistance sources have more side-effects possibly of less agronomically desirable traits. The soybean line 'Hartwig' derived from PI437654 has high levels of resistance to multiple *H. glycines* populations (Anand 1992). Careful breeding experiments of Hartwig with the highly susceptible

'Williams 82' coupled with molecular work, identified molecular markers (Faghihi et al. 1995), and methods for molecular tracing during breeding have been patented (Vierling et al. 2000). The promise of these lines is the lack of undesirable traits not being transferred during the breeding efforts. Lines developed with this technology were marketed as CystX cultivars. Success of this technology is still evaluated.

Great hope is set on understanding the infection process more comprehensively, and to find novel ways to interfere with the infection process. A recent summary has been given on respective improvements of understanding the infection process of *H. glycines,* and other sedentary plant parasitic nematodes could contribute in the long-term to improve methods for nematode management (Hewezi and Baum 2017).

4.2.1.3 Cultural Methods for Managing *Heterodera glycines*

Host plant resistance is one of the cornerstones of management of *Heterodera glycines*. Use of crop rotation and cover cropping has received some attention. In many areas of these states, options for crop rotations are minimal because soybean and corn comprise the majority of the acreage. A rotation to the non-host corn is aimed at taking advantage of the natural decline of the population densities when reproduction cannot occur. Presumably, spontaneous hatch and the reduction by natural enemies reduces the population densities of *H. glycines*. Unless the soil has extraordinarily high activity of nematode antagonist (see discussion below on nematode suppressive soil) these decline rates are often insufficient to reduce nematode population densities below economic threshold levels. One or 2 years of corn are insufficient to reduce nematode population densities below threshold levels (Tylka 2016). Accordingly, a rotation scheme of rotating resistant soybean with corn, then a susceptible soybean, again corn, and then back to resistant soybean has been proposed. Anecdotally, the resistant soybean reduces the population densities together with the 1 year of the non-host corn to protect the high-yielding susceptible cultivar. It is surmised that the interspersing of the susceptible line reduces the selection pressure on the nematode population thereby preserving the desirable resistance characteristic. This is a wide-spread recommendation though with little data foundation. Some benefit for production has been documented where winter wheat can be produced and harvested early enough to allow for a soybean crop right after wheat harvest. While this "double-cropping" is a strategy mostly feasible in the southern counties of the respective states, it appears that the incidence of SDS is reduced and the per-acre productivity increase (Von Qualen et al. 1989).

Because changes to the overwhelming crop rotation of alternating soybean with corn annually are difficult to implement, other agronomic practices have found research interest. For example, the use of cover crops that may have benefits for the cropping system. A co-cropping was employed to overcome the challenges of limited growing periods outside the production cycle of the cash crops (Chen et al. 2006). Winter covers may have potential but options are limited for specific cover crop species in the northern counties of the states after harvest of the soybean or corn crop after the middle of September (Villamil et al. 2006). In concert with using plantings of potentially nematode-antagonistic plants, the avoidance in permitting volunteer vegetation as alternate hosts has been intensively studied in multiple states. Reports of very high reproductive rates of *H. glycines* under controlled greenhouse conditions (Venkatesh et al. 2000), and the positive confirmation of reproduction of *H. glycines* on fall volunteer purple henbit (Creech et al. 2007) illustrated the interest in these plants, and subsequently, a variety of greenhouse, field and microplot experiments were conducted to investigate the role of this volunteer vegetation (Wong and Tylka 1994). Even after several years of differential weed control, no differences of nematode population densities were found (Mock et al. 2012). In summary, these projects demonstrated that not field hygiene, though important in preventing transfer of infested soil from infested fields into non-infested fields, but weed management in the fallow period for nematode management strategies, often plays a minor role.

In large-acre crops, tillage is another obvious agronomic practice that could be altered if it was beneficial to the production strategy. Minimum, and especially no-tillage practices have shown potential to suppress population increases of *H. glycines* (Westphal et al. 2009). But multiple factors go into decisions of tillage operations. Even within the five states, agroecological conditions may vary sufficiently to modify how tillage intensity affects nematode population densities and other soil-borne pests and diseases. For example, no-tillage was foremost beneficial to reduce *H. glycines* population densities in Indiana but was detrimental in Southern Illinois where no-tillage increased the severity of SDS. One is to speculate that the winter freezes in the north impact the soil edaphon differently than the constantly non-frozen conditions in the southern counties of Illinois. So, careful implementation at the area of interest is indicated.

4.2.1.4 Biological, Chemical Control and Suppressive Soils for Managing *Heterodera glycines*

Cyst nematodes persist in the soil for many years before the contents of the dead but protective female body are depleted. During this time, the cysts plus content are exposed to a multitude of soil organisms that can feed on them. Early work by Carris and Glawe has accumulated a wonderful pictorial account of fungi found in these nematode propagules (Carris and Glawe 1989). Similarly, work in Arkansas had found a sterile hyphomycete fungus that later was found to be related to *Dactylella oviparasitica*, a very effective female parasite of *H. schachtii* which is a close relative of *H. glycines* (Kim et al. 1998; Yang et al. 2012). Because of the large-acre set-up of these crops, the interest of studying suppressive soils has constantly increased. There are various forms of soil suppressiveness of which the most studied is "… where a pathogen establishes at first, causes damage, and then diminishes with continued culture of the crop" (Cook and Baker 1983). Hypothetically, this type of suppressiveness is microbially incited, and thus it may be possible to manipulate the soil environment for evolution of this natural population density regulation. In studies in Indiana, such soil suppressiveness developed in random fields that had come out of soybean-corn rotations when they were mono-cultured to susceptible soybean (Westphal and Xing 2011). Comparisons were made between non-treated and methyl-bromide fumigated plots both infested with the SDS pathogen. Over the time course of 5 years, *H. glycines* populations increased several-fold in pre-plant fumigated plots and remained low in the non-treated plots illustrating some biological reduction of nematode reproduction and concomitant reduction of SDS severity. In these field experiments, SDS also was suppressed. The suppressive phenomenon only developed when at least *H. glycines* was naturally present at initiation of the experiment. In such growing contexts, nematode population densities probably were kept low by microbial antagonists. A program in the neighboring State of Minnesota has examined microbial effects and tested the effects of tillage and crop sequence on the dynamics of soil suppressiveness (Chen 2007). Unfortunately, these early studies were not followed up thoroughly, and the use of novel sequencing approaches to describe the soil microbiome was only initiated a decade later and at rather random patterns (Srour et al. 2017).

In studies at the University of Illinois, the obligate nematode parasite *Pasteuria nishizawae* was discovered under soybean monoculture. Transfer studies with soil provided the proof of concept that this bacterial nematode parasite was effectively reducing cyst nematode population densities (Atibalentja et al. 1998). This obligate bacterial nematode parasite was successfully grown in culture in the early 2000s, and subsequently commercialized. This allowed developing *P. nishizawae* into a biological seed treatment. Based on the expense and the biology of the nematode and the bacterium, seed treatments have been the preferred choice (Anon 2018). Other control options rely on seed treatments with Avicta (abamectin), Votivo (a mixture of a chemical and *Bacillus firmus*), and Clariva (*P. nishizawae*) (Tylka 2016). Older chemistries of in-furrow treatments have mostly left the market place partially because of regulation.

4.2.2 Root Knot Nematodes, Meloidogyne *spp.*

4.2.2.1 Root Knot Nematodes *Meloidogyne* Species on Soybean and Corn

Root knot nematodes (RKN) have a much broader host range than soybean cyst nematodes, suggesting that much greater problems could be expected to occur in the States of Indiana, Illinois, Iowa, Missouri and Ohio (Fig. 4.2). At the same time, root knot nematodes cause more prevalent damage in sandy soils. There is good evidence that *Meloidogyne* spp. prefer pore spaces provided by such soil texture. Many soils of this region are fine-textured and may, for that reason, be less prone for root knot nematode infestations. Although corn and soybean are hosts for different *Meloidogyne* spp., reports on RKN problems in soybean and corn in the five states are rare. The species of *Meloidogyne* in these states is probably *M. incognita*. Because of the overwhelming importance of soybean cyst nematode, only few deep-going surveys have been conducted to see if other species also occur. Just a few surveys have been conducted in Illinois and Indiana (Allen et al. 2005; Kruger et al.

Fig. 4.2 Soybean infected with root knot nematodes. Note the infestation hot spot surrounded by more vigorously growing soybean

2007) in contrast to challenges with RKN in southern states, e.g., Georgia, where large screening efforts of soybean lines were established for years and new sources for resistance were identified (Harris et al. 2003). More recent work demonstrated the damage potential on corn in southern states (Bowen et al. 2008). Against previous general perception of monocot plants being poor hosts to RKN species, corn can be a supportive host of RKN. Maybe because of its hybrid vigor and confusion with other soil-borne issues, RKN are rarely diagnosed. As both soybean and corn are hosts for RKN but rarely examined for their responses, problems can occur when these field crops are cropped in rotation with vegetable crops. Such problems were frequently observed in Southern Indiana where rotations of corn, soybean, and watermelon are frequent, and more comprehensive evaluations of crop sequences are necessary when the valuable cash crop is to be grown in a Midwest crop sequence where soybean and corn may be susceptible to root knot nematodes (Westphal 2011). In the southwest of Indiana, a defined area of sandy soil is used for specialized vegetable production. Foremost, watermelon is the crop of choice in this area. In this region *Meloidogyne* spp. were easily detected in a survey (Kruger et al. 2007). Choice of resistant cultivars of soybean (Kruger et al. 2008) was proposed as a mitigation strategy in rotations with watermelons where all current cultivars are susceptible and sensitive to RKN infection.

4.2.2.2 *Meloidogyne* spp. on Vegetable Crops

In vegetable crops, because of the higher per acre crop value, chemical remedies are used against the soil-dwelling culprits. For example, soil fumigation is often used in preparation of watermelon plantings. The unstable weather typical for this area of the Midwest can make such applications challenging, and has led to variable results when cool and wet soils are treated with 1,3-dichloropropene (1,3-D). After the loss or use restrictions of carbamates and organophosphate resulting from the Food Quality Protection Act (FQPA), novel chemistries are just slowly gaining registration in vegetable crops. These new chemistries have very different modes of action for suppressing nematodes, but are much less toxic to the user and the environment.

Cover cropping against root knot nematodes was of some success in Europe. There, similar strategies for the management of the sugar beet cyst nematode are implemented. In these systems, the nematodes are attracted to the host plant roots. They penetrate these roots, and then fail to reproduce if the host plant is resistant to the particular nematode species. The cover crop serves as a trap crop because this process leads to active reduction of the nematode population density. Cultivar choice is critical to ensure the reduction of nematode numbers versus an unwanted increase of population densities if the nematodes were to reproduce on the cover crop. Cover cropping of nematode-reducing plants in Midwest vegetable production has not been widely adopted, and instead the moderately supportive host rye is often grown in non-crop periods. When planted in fall, this cereal crop is used for erosion control during winter and left as windbreaks when the majority of the field is prepared for watermelon seedbeds. Investigations have demonstrated the risk for nematode increases under this cover crop, resulting in a treatment need that potentially could be avoided by cover cropping with nematode-resistant plant species (Westphal et al. 2006).

4.2.3 Root Lesion Nematodes, **Pratylenchus** *spp.*

Lesion nematodes are more often a problem on perennial plants, both woody and herbaceous. Several *Pratylenchus* species occur in the five states: *Pratylenchus penetrans, P. alleni, P. crenatus, P. hexincisus, P. neglectus* and *P. scribneri* (Brown et al. 1980; Anon 1999). More than 50% samples collected in Ohio contained *Pratylenchus* spp. (Wilson and Walker 1961). Over 80% of the corn and soybean fields sampled in Ohio are infested with lesion nematodes. *Pratylenchus penetrans* is most often found in nurseries, orchards, and strawberry fields in Illinois. In the Eastern United States, *P. penetrans* has been responsible for severe decline and replant failure in many cherry, apple, and peach orchards, and is one of the most common species of plant parasitic nematodes found on corn in the United States. *Pratylenchus penetrans* can affect the host directly or through interactions with other organisms in disease complexes such as those involving fungi, including *Fusarium* spp., *Rhizoctonia solani, Verticillium albo-atrum* and *Verticillium dahliae*. The two other *Pratylenchus, P. hexincisus* and *P. scribneri* are also commonly associated with field crops in Illinois. *Pratylenchus scribneri* also caused yield losses of potato in Ohio (Wheeler and Riedel 1994). These two species often occur in mixed populations and have been associated with damage and yield loss in corn and soybeans. Host suitability studies indicated that corn and potato were good hosts for both *Pratylenchus* species, whereas alfalfa and red clover were non-hosts. Wheat and rye were better hosts for *P. hexincisus* than *P. scribneri*, whereas sorghum, soybean, tomato and white clover were better hosts for *P. scribneri* (Anon 1999).

For the management of root lesion nematode, several procedures have been recommended. (i) Maintain optimum growing conditions. The greatest damage by lesion nematodes occurs to plants under stress. (ii) Crop rotation. Despite the wide range of hosts for lesion nematodes, crop rotation can provide control in some instances. The use of crop rotation depends on the species involved and the economic feasibility of such rotations. (iii) Treat propagation material with heat. A hot water treatment is an effective method of eradicating lesion nematodes from the roots of transplants. Temperatures of 45–55 °C sustained for 10–30 min are commonly used. Before making a large-scale treatment, treat several plants of each variety to make sure that heat damage will be, at most, minimal. (iv) Treat the soil with dry or moist heat. Nematodes are killed by exposure to temperatures of 40–52 °C, depending on the species. Aerated steam is the most efficient method but baking small quantities of soil in an oven at 82 °C for 30 min or 71 °C for 60 min is also effective, especially for the homeowner who needs a small quantity of soil. (v) Apply nematicides. The use of chemical fumigants to control lesion nematodes can be effective and economical, especially where high-value crops are involved. Preplant fumigation with nematicides may be necessary in order to control replant and other lesion nematode diseases in orchards, nurseries, strawberry beds and other areas (Anon 1999). Seed treatments containing the nematicide abamectin in combination with fungicides can also reduce root infection by root lesion nematodes (da Silva et al. 2016).

4.2.4 Foliar Nematode, Aphelenchoides fragariae

Foliar nematode causes serious damage to alfalfa, strawberries, *Lamium*, and many ornamentals in nursery and landscape enviroments and during the last decades, has emerged as an important pest of hosta and ornamentals in North America (Jagdale and Grewal 2002). In the United States, the cultivation and production of hosta is a multimillion-dollar industry and nurseries grow and sell over 1000 selections representing 10 different hosta species and their hybrids (Grewal and Jagdale 2001). As hosta foliage offers a variety of leaf shapes, textures and colors, there is a growing concern among growers and nursery managers about leaf damage caused by foliar nematodes. This nematode was found in many nurseries in Ohio and Illinois (Noel and White 1994; Grewal and Jagdale 2001). Many studies on the biology and management of this nematode have been made in Ohio by Heinlein (1982), Grewal and Jagdale (2001), Jagdale and Grewal (2002, 2004, 2006), An et al. (2017). These nematodes infect young leaves through stomata and feed on the mesophyll cells, causing large sections of the leaf to become chlorotic and subsequently turn necrotic.

The necrotic lesions are usually bounded by large veins. In the Midwest, typical symptoms of foliar nematodes on hosta can first be observed in July (Grewal and Jagdale 2001). Jagdale and Grewal (2006) showed that foliar nematode overwinters as juveniles and adults, but not as eggs in soil, dry infected leaves, and dormant crowns, but not in the roots. These authors also revealed that survival of *A. fragariae* in soil and dormant buds was influenced by the location of plants. It has been found that higher numbers of nematodes survived in soil collected from a polyhouse than those collected from plants held under polythene cover or plants in the bare ground in a home garden.

Control of foliar nematodes has been problematic as most of the standard chemical nematicides have been banned by the United States Environmental Protection Agency (US EPA) under the implementation of FQPA. Some effective nematicides and fumigants including methyl bromide have been phased out due to their broad spectrum toxicity and the threat to the degradation of the ozone layer (An et al. 2017). Use of hot water as a preventive treatment to manage foliar nematodes may provide an environmentally safe alternative to nematicides. Jagdale and Grewal (2004) evaluated the effects of hot and boiling water on growth parameters of hosta and ferns and concluded that the application of a 90 °C water drench as a preventive treatment in autumn or spring could prove effective in reducing foliar nematode infection of hosta without affecting plant vigor.

Many efforts have been made to search for alternative products for management of foliar nematodes affecting ornamental and horticultural plants. Jagdale and Grewal (2002) tested a biological agent, *Pseudomonas cepacia*, two plant products (clove extract and Nimbecidine) and twelve chemical pesticides for the control of *A. fragariae*. They found that only diazinon EC, trichlorfon SP, oxamyl GR and ZeroTol™ were effective in reducing nematode population in soil and leaves. However, diazinon EC, trichlorfon SP and oxamyl GR were banned by the US Environmental Protection Agency. ZeroTol™ was therefore suggested as a useful product for managing foliar nematodes in soil, however, it did not provide acceptable levels of control in the leaves because of limited contact with nematodes and its short persistence (Jagdale and Grewal 2002).

An et al. (2017) used a three-stage approach to evaluate 24 products for their potential to control foliar nematodes *Aphelenchoides fragariae* in hosta. Out of the 24 products screened, Pylon (24% chlorfenapyr) and Nemakill (32% cinnamon oil, 8% clove oil, 15% thyme oil mixture) showed the highest nematicidal activity against foliar nematodes in the aqueous suspension, soil drench, and leaf-disc spray assays even at the low concentrations (20-fold dilution) tested. These two products were significantly more effective than ZeroTol™ in the leaf disc assays. There was no evidence of phytotoxicity from Pylon and Nemakill regardless of being applied through spray or soil drench. The authors concluded that these two products have potential for foliar nematode management and can be recommended for further field evaluations.

4.2.5 *Stem and Bulb Nematode,* **Ditylenchus dipsaci**

An infestation of the stem nematode was discovered in onion fields in Cook County, Illinois, in 1954 (Edwards and Taylor 1963). It has been demonstrated that snap bean, soybean, pea and tomato were excellent and good host for this nematode. Weed hosts may also be of importance in survival of *D. dipsaci* in the field. Of the weeds demonstrated to be capable of functioning as hosts in this study, *Hibiscus trionum, Solanum nigrum*, and *Polygonum persicaria* were observed to be infected with *D. dipsaci* in infested onion fields. The stem nematode is also a key pest of garlic. Although garlic is not a major crop in Ohio, this crop is grown in diversified vegetable production systems. In July 2013, diseased garlic bulbs were received from a grower in Lorain County, OH, from a field with wide symptom distribution. Bulbs were discolored, exhibited splitting, and had basal plate damage including reduced roots. Nematodes were extracted for examination by placing bulb slices in water (Testen et al. 2014).

4.2.6 *Needle Nematode,* **Longidorus breviannulatus**

Longidorus breviannulatus is one of the most damaging nematodes in corn in highly sandy soils (Malek et al. 1980). It causes root pruning of seedlings resulting in severe stunting and greatly reduced yields. The nematode has been found in most sandy fields examined in the central and southwestern counties of Iowa where corn has been grown continuously for a few years (Norton 1989). This species was also detected in association with the perennial grasses *Miscanthus* spp. plot and corn in Havana, Illinois (Ye and Robbins 2004; Mekete et al. 2009). The occurrence of this species was associated with severe damage to the fibrous root system, including stunting and necrosis of grasses.

4.2.7 *Spiral Nematodes,* **Helicotylenchus** *spp.*

Spiral nematodes were the most frequently found plant parasitic nematode (Table 4.2), present in 77% of the soil samples collected from corn in Iowa. The spiral nematode also was present at the highest maximum population density (2340 nematodes per 100 cm^3 soil) and with the greatest mean population density (87 per 100 cm^3 soil) of all nematode genera identified in the samples. A damage threshold of 500–1000 nematodes per 100 cm^3 soil were proposed (Tylka et al. 2011). *Helicotylenchus* spp. were found in Ohio in 14.6% samples (Wilson and Walker

Table 4.2 Some plant parasitic nematodes reported from agricultural fields, pastures and golf courses in Indiana, Illinois, Ohio, Iowa and Missouri

Species	States	Crop or plants	References
Aphelenchoides fragariae	Illinois, Indiana, Iowa	Strawberry	Logsdon et al. (1968)
A. ritzemabosi	Illinois, Indiana, Iowa	Ornamentals	Logsdon et al. (1968)
Discocriconemella inarata	Iowa	Prairies	Powers et al. (2010)
Ditylenchus destructor	Indiana	Potato	Logsdon et al. (1968)
Helicotylenchus cornurus	Illinois	Bentgrass	Davis et al. (1994a)
H. digonicus	Iowa	Prairies	Norton and Schmitt (1978)
H. dihystera	Missouri, Iowa	Soybean, prairies	Norton and Schmitt (1978) and Niblack (1992a)
H. platyurus	Illinois	Peach	Walters et al. (2008)
H. peudorobustus	Missouri, Illinois, Indiana, Iowa	Soybean, peach, corn	Ferris et al. (1971), Norton (1977), Lawn and Noel (1986), Niblack (1992a), and Walters et al. (2008)
Heteroanguina graminophila	Iowa, Ohio	*Calamagrostis canadensis*	Norton et al. (1987)
Heterodera glycines	Missouri, Iowa, Indiana, Illinois, Ohio	Soybean	Anon (1984), Niblack et al. (1993), and Willson et al. (1996)
H. trifolii	Missouri	Soybean	Niblack (1992a)
H. ustinovi	Ohio	Grasses	Joseph et al. (2018)
Hoplolaimus galeatus	Iowa, Illinois, Indiana, Missouri	Prairies, cotton, corn	Logsdon et al. (1968), Norton and Hinz (1976), Norton and Schmitt (1978), and Wrather et al. (1992)
Longidorus breviannulatus	Iowa, Illinois	Corn	Malek et al. (1980) and Norton et al. (1982)
Meloidogyne hapla	Missouri, Iowa, Illinois, Indiana	Soybean	Logsdon et al. (1968) and Niblack (1992a)
M. incognita	Missouri, Illinois	Soybean, cotton	Anon (1984), Niblack (1992a), and Wrather et al. (1992)
M. naasi	Illinois	Grasses	Michell et al. (1973)
Mesocriconema curvatum	Illinois, Iowa	Bentgrass	Logsdon et al. (1968) and Davis et al. (1994a)
M. xenoplax	Illinois, Indiana, Iowa	Peach	Logsdon et al. (1968), Reis et al. (1979), and Walters et al. (2008)
Nanidorus minor	Iowa	Corn	Norton et al. (1982)
Paratrichodorus allius	Ohio	Corn	Lopez-Nicora et al. (2014)

(continued)

Table 4.2 (continued)

Species	States	Crop or plants	References
Paratylenchus neoamblycephalus	Ohio	Soybean	Ankrom et al. (2017)
P. projectus	Missouri, Illinois, Indiana	Soybean, peach	Ferris et al. (1971), Lawn and Noel (1986), Niblack (1992a), and Walters et al. (2008)
Pratylenchus agilis	Missouri	Soybean	Niblack (1992a)
P. crenatus	Ohio	Potato, corn	Brown et al. (1980)
P. hexincisus	Missouri, Iowa	Soybean, corn	Norton and Hinz (1976) and Niblack (1992a)
P. neglectus	Ohio	Potato	Brown et al. (1980)
P. penetrans	Ohio, Missouri, Illinois, Indiana	Potato, soybean, peach, peppermint	Mai et al. (1977), Brown et al. (1980), Niblack (1992a, b), Wheeler and Riedel (1994), and Walters et al. (2008)
P. scribneri	Missouri, Ohio, Illinois	Soybean, potato, cotton	Lawn and Noel (1986), Niblack (1992a), Wrather et al. (1992), and Wheeler and Riedel (1994)
P. thornei	Ohio	Potato	Brown et al. (1980)
P. vulnus	Missouri, Illinous	Cotton, peach	Wrather et al. (1992) and Walters et al. (2008)
Rotylenchulus reniformis	Missouri	Cotton	Wrather et al. (1992)
Quinisulcius acutus	Iowa, Missouri, Illinois, Indiana	Corn, soybean	Ferris et al. (1971), Norton (1989), and Niblack (1992a)
Tylenchorhynchus annulatus	Illinois	Peach	Walters et al. (2008)
T. agri	Iowa, Illinois	Corn, soybean	Norton (1989)
T. claytoni	Iowa, Missouri, Illinois	Corn, soybean, peach	Norton (1989), Niblack (1992a), and Walters et al. (2008)
T. maximus	Iowa, Missouri	Lowland prairies, turf, soybean	Norton and Schmitt (1978), Norton (1989), and Niblack (1992a)
T. martini	Illinois	Soybean	Lawn and Noel (1986)
T. nudus	Illinois, Iowa	Bentgrass, corn, prairies	Norton and Schmitt (1978), Norton (1989), and Davis et al. (1994a)
Xiphinema americanum sensu lato	Missouri, Iowa, Illinois, Indiana, Ohio	Soybean, prairies, peach, alfalfa, red clover	Logsdon et al. (1968), Ferris et al. (1971), Norton and Schmitt (1978), Norton et al. (1982), Niblack (1992a), and Walters et al. (2008)
X. chambersi	Iowa	Prairies	Norton and Schmitt (1978)

1961) and also in almost all corn fields surveyed in Ohio. This genus was found in 94% of the fields sampled, at a mean population density of 90 nematodes per 100 cm^3 soil. It has been concluded that the spiral nematode was not a major threat to corn production in most Ohio corn fields. However, further research will be needed to determine how damaging this plant parasitic nematode genus really is under conditions in Ohio (Niblack 1992a)

4.2.8 Stunt Nematodes, **Tylenchorhynchus** *spp.*

Taylor et al. (1963) sampled 26 putting greens from six golf courses in Illinois and determined that *Tylenchorhynchus* spp. were the most abundant plant parasitic nematodes present. An unidentified *Tylenchorhynchus* sp. was also recovered from all putting greens, with an average 284 nematodes recovered per 125 g of soil. Nematodes of the genus *Tylenchorhynchus* were also recovered from 22% of putting greens sampled in an Ohio study (Safford and Riedel 1976). It has been shown in experimental study that *T. nudus* suppresses root growth of both bentgrass and annual bluegrass (Davis et al. 1994b, c). *Quinisulcius acutus* and *T. martini* were found in soybean fields in Illinois and Indiana (Ferris et al. 1971).

4.3 Conclusion

In summary, nematode problems exist in the herein discussed states. Reports include isolated challenges of specialty crops. Most vulnerable are the large-acre crops soybean and corn, in which relatively narrow genetic bases lead to large production areas with crop cultivars of similar susceptibilities. Examining these highly intensive production systems illustrates how ensuring sustainability of production systems relies on vigilant and constant search for alternative production strategies and genetic improvements. Management strategies already in use against nematode threats to crop production in the five states represent on-going, integrated, sustainable approaches to nematode management and due to restrictions on nematicides, the use of biological and cultural alternatives will continue to challenge future development of sustainable nematode management practices in the region.

References

Allen, J. B., Bond, J. P., & Schmidt, M. E. (2005). Incidence of *Meloidogyne incognita* and development of resistant soybean germplasm in Illinois. *Plant Health Progress*. https://doi.org/10.1094/PHP-2005-0606-01-RS.

Allen, T. W., Bradley, C. A., Sisson, A. J., Byamukama, E., Chilvres, M. I., Coker, C. M., Collins, A. A., Damincone, J. P., Dorrance, A. E., Dufault, N. S., Esker, P. D., Faske, T. R., Giesler, L. J., Grybauskas, A. P., Hershman, D. E., Hollier, C. A., Isakeit, T., Jardine, D. J., Kelly, H. M., Kemerait, R. C., Kleczewski, N. M., Koenning, S. R., Kurle, J. E., Malvick, D. K., Markell, S. G., Mehl, H. L., Mueller, D. S., Mueller, J. D., Mulrooney, R. P., Nelson, B. D., Newman, M. A., Osborne, L., Overstreet, C., Padgett, G. B., Phipps, P. M., Price, P. P., Sikora, E. J., Smith, D. L., Spurlock, T. N., Tande, C. A., Tenuta, A. U., Wise, K. A., & Wrather, J. A. (2017). *Soybean yield loss estimates due to diseases in the United States and Ontario, Canada, from 2010 to 2014*. doi: https://doi.org/10.1094/PHP-RS-16-0066.

An, R., Karthik, N. K., & Grewal, P. (2017). Evaluation of botanical and chemical products for the control of foliar nematodes *Aphelenchoides fragariae*. *Crop Protection, 92*, 107–113.

Anand, S. C. (1992). Registration of 'Hartwig' soybean. *Crop Science, 32*, 1069–1070.

Ankrom, K. E., Lopez-Nicora, H., Niblack, T. L., & Lindsey, L. E. (2017). First report of a pin nematode (*Paratylenchus neoamblycephalus*) from soybean in Ohio. *Plant Disease, 101*, 1330–1331.

Anon. (1984). *Distribution on plant parasitic nematode species in North America*. D.C. Norton, chairman, Nematode Geographical Distribution Committee. Hyattsville, MD, Society of Nematologists.

Anon. (1999). Lesion nematodes. *Report on Plant Disease*. Department of Crop Sciences, University of Illinois at Urbana-Champaign, RPD 1103.

Anon. (2018). *Clariva pn seed treatment*. http://www.syngenta-us.com/seed-treatment/clariva-pn. Accessed 27 Mar 2018.

Aoki, T., O'Donnell, K., Homma, Y., & Latanzi, A. R. (2003). Sudden-death syndrome of soybean is caused by two morphologically distinct species within the *Fusarium solani* species complex – *F. virguliforme* in North America and *F. tucumaniae* in South America. *Mycologia, 95*, 660–684.

Arelli, P., Mengistu, A., Nelson, R., Cianzio, S., & Vuong, T. (2015). New soybean accessions evaluated for reaction to *Heterodera glycines* populations. *Crop Science, 55*, 1236–1242.

Atibalentja, N., Noel, G. R., Liao, T. F., & Gertner, G. Z. (1998). Population changes in *Heterodera glycines* and tis bacterial parasite *Pasteuria* sp. in naturally infested soil. *Journal of Nematology, 30*, 81–92.

Bowen, K. L., Hagan, A. K., Campbell, H. L., & Nightendale, S. (2008). Effect of southern root-knot nematode (*Meloidogyne incognita* race 3) on corn yields in Alabama. *Plant Health Progress*. https://doi.org/10.1094/PHP-2008-0910-01-RS.

Brown, M. J., Riedel, R. M., & Rowe, R. C. (1980). Species of *Pratylenchus* associated with *Solanum tuberosum* cv Superior in Ohio. *Journal of Nematology, 12*, 189–192.

Carris, M., & Glawe, D. A. (1989) *Fungi colonizing cysts of Heterodera glycines*. USDA Bulletin 786, University of Illinois, Urbana.

Chen, S. (2007). Suppression of *Heterodera glycines* in soils from fields with long-term soybean monoculture. *Bicontrol Science and Technology, 17*, 125–134.

Chen, S., Wyse, D. L., Johnson, G. A., Porter, P. M., Stetina, S. R., Miller, D. R., Betts, K. J., Klossner, L. D., & Haar, M. J. (2006). Effect of cover crops alfalfa, red clover, and perennial ryegrass on soybean cyst nematode population and soybean and corn yields in Minnesota. *Crop Science, 46*, 1890–1897.

Cianzio, S. R., Arelli, P. R., Swaminathan, S., Lundeen, P., Gebhart, G., Rivera-Velez, N., Guilherme, S. R., Soares, I. O., Diers, B. W., Knapp, H., Westgate, M., & Hudson, M. E. (2018). Genetically diverse soybean cyst nematode–resistant full-sib soybean germplasm lines AR4SCN, AR5SCN, AR6SCN, AR7SCN, and AR8SCN. *Journal of Plant Registrations, 12*, 124–131.

Cook, R. J., & Baker, K. F. (1983). *The nature and practice of biological control of plant pathogens* (539 pp). St. Paul: The American Phytopathological Society.

Creech, J. E., Webb, J. S., Young, B. G., Bond, J. P., Harrison, S. K., Ferris, V. R., Faghihi, J., Westphal, A., & Johnson, W. G. (2007). Development of soybean cyst nematode on henbit (*Lamium amplexicaule*) and purple deadnettle (*Lamium purpureum*). *Weed Technology, 21*, 1064–1070.

da Silva, M. P., Tylka, G. L., & Munkvold, G. P. (2016). Seed treatment effects on maize seedlings coinfected with *Fusarium* spp. and *Pratylenchus penetrans*. *Plant Disease, 100*, 431–437.

Davis, R. F., Wilkinson, H. T., & Noel, G. R. (1994a). Vertical distribution of three nematode genera in a bentgrass putting green in Central Illinois. *Journal of Nematology, 26*, 518–521.

Davis, R. F., Noel, G. R., & Wilkinson, H. T. (1994b). Pathogenecity of *Tylenchorhynchus nudus* to creeping bentgrass and annual bluegrass. *Plant Disease, 78*, 169–173.

Davis, R. F., Wilkinson, H. T., & Noel, G. R. (1994c). Root growth of bentgrass and annual bluegrass as influenced by coinfection with *Tylenchorhynchus nudus* and *Magnaporthe poae*. *Journal of Nematology, 26*, 86–90.

Edwards, D. I., & Taylor, D. P. (1963). Host range of an Illinois population of the stem nematode (*Ditylenchus dipsaci*) isolated from onion. *Nematologica, 9*, 305–312.

Faghihi, J., Vierling, R. A., Halbrendt, J. M., Ferris, V. R., & Ferris, J. M. (1995). Resistance genes in a 'Williams 82' × 'Hartwig' soybean cross to an inbred line of *Heterodera glycines*. *Journal of Nematology, 27*, 418–421.

Ferris, V. R., Ferris, J. M., Bernard, R. L., & Probst, A. H. (1971). Community structure of plant parasitic nematodes related to soil types in Illinois and Indiana soybean fields. *Journal of Nematology, 3*, 399–408.

Grewal, P. S., & Jagdale, G. B. (2001). Biology and management of foliar nematodes. *The Hosta Journal, 32*, 64–66.

Harris, D. K., Boerma, H. R., Hussey, R. S., & Finnerty, S. L. (2003). Additional sources of soybean germplasm resistant to two species of root knot nematode. *Crop Science, 43*, 1848–1851.

Heinlein, M. A. (1982). *Symptomatology and host range of Aphelenchoides fragariae (Ritzema-Bos, 1890) Christie 1932 on the Gesneriaceae*. M.Sc. thesis. The Ohio State University, Columbus, Ohio, 53 pp.

Hewezi, T., & Baum, T. J. (2017). Chapter twelve – communication of sedentary plant parasitic nematodes with their host plants. In *Advances in botanical research* (Vol. 82, pp. 305–324).

Howland, A., Nathan, M., & Mitchum, M. G. (2016). The distribution and management practices of *Heterodera glycines* in Missouri in 2015. *Journal of Nematology, 348*, 377–378.

Jagdale, G. B., & Grewal, P. S. (2002). Identification of alternatives for the management of foliar nematodes in floriculture. *Pest Management Science, 58*, 451–458.

Jagdale, G. B., & Grewal, P. S. (2004). Effectiveness of a hot water drench for the control of foliar nematodes, *Aphelenchoides fragariae*, in floriculture. *Journal of Nematology, 36*, 49–53.

Jagdale, G. B., & Grewal, P. S. (2006). Infection behavior and overwintering survival of foliar nematodes, *Aphelenchoides fragariae*, on hosta. *Journal of Nematology, 38*, 130–136.

Joseph, S., Akyazi, F., Habteweld, A. W., Meteke, T., Creswell, T., Ruhl, G. E., & Faghihi, J. (2018). First report of cyst nematode (*Heterodera iri*) in Ohio. *Plant Disease, 102*, 1042.

Kim, D. G., Riggs, R. D., & Correll, J. C. (1998). Isolation, characterization, and distribution of a biocontrol fungus from cysts of *Heterodera glycines*. *Phytopathology, 88*, 465–471.

Koenning, S. R. (2000). Density-dependent yield of *Heterodera glycines*-resistant and susceptible cultivars. *Journal of Nematology, 32*, 502–507.

Koenning, S. R., & Wrather, J. A. (2010). Suppression of soybean yield potential in the continental United States by plant diseases from 2006 to 2009. *Plant Health Progress*. https://doi.org/10.1094/PHP-2010-1122-01-RS.

Kruger, G. R., Xing, L. J., Santini, J. B., & Westphal, A. (2007). Distribution and damage caused by root-knot nematodes on soybean in southwest Indiana. *Plant Health Progress*. https://doi.org/10.1094/PHP-2007-1031-01-RS.

Kruger, G. R., Xing, L. J., LeRoy, A., & Westphal, A. (2008). *Meloidogyne incognita* resistance in soybean under Midwest conditions. *Crop Science, 48*, 716–726. https://doi.org/10.2135/cropsci2007.04.0196.

Lawn, D. A., & Noel, G. R. (1986). Field interrelationships among *Heterodera glycines, Pratylenchus scribneri*, and three other nematode species associated with soybean. *Journal of Nematology, 18*, 98–106.

Logsdon, C. E., Taylor, D. P., Bergeson, G. B., Ferris, J. M., Norton, D. C., Dickerson, O. J., Knierim, J. A., MacDonald, D. H., Jenkins, L., Schuster, M. L., Pepper, E. H., Wilson, J. D., Malek, R. B., & Darling, H. M. (1968). *Nematology in the North Central Region, 1956–1966*. Special Report 58, 20 pp.

Lopez-Nicora, H. D., Mekete, T., Sekora, N., & Niblack, T. L. (2014). First report of the stubby-root nematode (*Paratrichodorus allius*) from a corn field in Ohio. *Plant Disease, 98*, 1164.

Mai, W. F., Bloom, J. R., & Chen, T. A. (1977). *Biology and ecology of the plant parasitic nematode Pratylenchus penetrans*. Bulletin, Pennsylvania State University, Agricultural Experiment Station 815, 66 pp.

Malek, R. B., Norton, D. C., Jacobsen, B. J., & Acosta, N. (1980). A new corn disease caused by *Longidorus breviannulatus* in the Midwest. *Plant Disease, 64*, 1110–1113.

McCarville, M. T., Marett, C. C., Mullaney, M. P., Gebhart, G. D., & Tylka, G. L. (2017). Increase in soybean cyst nematode virulence and reproduction on resistant soybean varieties in Iowa from 2001 to 2015 and the effect on soybean yields. *Plant Health Progress*. https://doi.org/10.1094/PHP-RS-16-0062.

McLean, K. S., & Lawrence, G. W. (1993). Interrelationship of *Heterodera glycines* and *Fusarium solani* in sudden death syndrome of soybean. *Journal of Nematology, 25*, 434–439.

Mekete, T., Gray, M. E., & Niblack, T. L. (2009). Distribution, morphological description, and molecular characterization of *Xiphinema* and *Longidorous* spp. associated with plants (*Miscanthus* spp. and *Panicum virgatum*) used for biofuels. *GCB Bioenergy, 1*, 257–266.

Michell, R. E., Malek, R. B., Taylor, D. P., & Edwards, D. I. (1973). Races of the barley root knot nematode, *Meloidogyne naasi*. I. Characterization by host preference. *Journal of Nematology, 5*, 41–44.

Mock, V. A., Creech, J. E., Ferris, V. R., Faghihi, J., Westphal, A., Santini, J. B., & Johnson, W. G. (2012). Influence of winter annual weed management and crop rotation on soybean cyst nematode (*Heterodera glycines*) and winter annual weeds: Years four and five. *Weed Science, 60*, 634–640.

Niblack, T. L. (1992a). *Pratylenchus, Paratylenchus, Helicotylenchus*, and other nematodes on soybean in Missouri. *Supplement to Journal of Nematology, 24*(4S), 738–744.

Niblack, T. L. (1992b). Chapter 7: The race concept. In R. D. Riggs & J. A. Wrather (Eds.), *Biology and management of the soybean cyst nematode* (pp. 73–86). St. Paul: APS Press.

Niblack, T. L. (2005). Soybean cyst nematode management reconsideration. *Plant Disease, 89*, 1020–1026.

Niblack, T. L., Heinz, R. D., Smith, G. S., & Donald, P. A. (1993). Distribution, density, and diversity of *Heterodera glycines* in Missouri. *Supplement to Journal of Nematology, 25*(4S), 880–886.

Niblack, T. L., Heinz, R. D., Smith, G. S., & Donald, P. A. (1994). Distribution and diversity of *Heterodera glycines* in Missouri. *Journal of Nematology, 25*, 880–886.

Niblack, T. L., Arelli, P. R., Noel, G. R., Opperman, C. H., Orf, J. H., Schmitt, D. P., Shannon, J. G., & Tylka, G. L. (2002). A revised classification scheme for genetically diverse populations of *Heterodera glycines*. *Journal of Nematology, 34*, 279–288.

Niblack, T. L., Colgrove, A. L., Colgrove, K., & Bond, J. P. (2008). Shift in virulence of soybean cyst nematode is associated with use of resistance from PI 88788. *Plant Health Progress*. https://doi.org/10.1094/PHP-2008-0118-01-RS.

Noel, G. R. (1992). Chapter 1: History, distribution and economics. In R. D. Riggs & J. A. Wrather (Eds.), *Biology and management of the soybean cyst nematode* (pp. 1–14). St. Paul: APS Press.

Noel, G. R., & White, D. (1994). *Hosta:* a new host record for *Aphelenchoides fragariae*. *Plant Disease, 78*, 924.

Norton, D. C. (1977). *Helicotylenchus pseudorobustus* as a pathogen on corn, and its densities on corn and soybean. *Iowa State Journal of Research, 51*, 279–285.

Norton, D. C. (1984). Maize nematode problems. *Plant Disease, 67*, 253–256.

Norton, D. C. (1989). Plant parasitic nematodes in Iowa. *Journal of Iowa Academy of Sciences, 96*, 24–32.

Norton, D. C., & Hinz, P. (1976). Relationship of *Hoplolaimus galeatus* and *Pratylenchus hexincisus* to reduction of corn yields in sandy soils in Iowa. *Plant Disease Report, 60*, 197–200.

Norton, D. C., & Schmitt, D. P. (1978). Community analyses of plant parasitic nematodes in the Kalsow prairie, Iowa. *Journal of Nematology, 10*, 171–176.

Norton, D. C., Dunlap, D., & Williams, D. D. (1982). Plant parasitic nematodes in Iowa: Longidoridae and Trichodoridae. *Proceedings of the Iowa Academy of Sciences, 89*, 15–19.

Norton, D. C., Cody, A. M., & Gabel, A. W. (1987). *Subanguina calamagrostis* and its biology in *Calamagrostis* spp. in Iowa, Ohio, and Wisconsin. *Journal of Nematology, 19*, 260–262.

Powers, T. O., Harris, T., Higgins, R., Sutton, L., & Powers, K. S. (2010). Morphological and molecular characterization of *Discocriconemella inarata*, an endemic nematode from North American Native tallgrass prairies. *Journal of Nematology, 42*, 35–45.

Riedel, R. M., & Golden, A. M. (1988). First report of *Heterodera glycines* on soybean in Ohio. *Plant Disease, 72*, 363.

Ries, S. M., Noel, G. R., & Doll, C. C. (1979). Illinois peach orchard nematode survey. *Transactions of the Illinois State Horticultural Society, 112*, 22–24.

Riggs, R. D., & Schmitt, D. P. (1988). Complete characterization of the race scheme for *Heterodera glycines*. *Journal of Nematology, 20*, 392–395.

Roy, K. W., Hershman, D. E., Rupe, J. C., & Abney, T. S. (1997). Sudden death syndrome of soybean. *Plant Disease, 81*, 1100–1111.

Safford, J., & Riedel, R. M. (1976). *Criconemoides* species associated with golf course turf in Ohio. *Plant Disease, 60*, 405–408.

Srour, A. Y., Gibson, D. J., Leandro, L. F. S., Malvick, D. K., Bond, J. P., & Fahoury, A. M. (2017). Unraveling microbial and edaphic factors affecting the development of sudden death syndrome in soybean. *Phytobiomes, 1*, 91–101.

Tabor, G. M., Tylka, G. L., Behm, J. E., & Bronson, C. R. (2003). *Heterodera glycines* infection increases incidence and severity of brown stem rot in both resistant and susceptible soybean. *Plant Disease, 87*, 655–661.

Tabor, G. M., Leandro, L., & Robertson, A. (2018). *Brown stem rot and sudden death syndrome: Can you tell them apart?* Iowa State University – Integrated Crop Management, online: https://crops.extension.iastate.edu/brown-stem-rot-and-sudden-death-syndrome-can-you-tell-them-apart. Accessed 13 Apr 2018.

Taylor, D. P., Britton, M. P., & Hechler, H. C. (1963). Occurrence of plant parasitic nematodes in Illinois golf greens. *Plant Disease Report, 47*, 134–135.

Testen, A. L., Walsh, E. K., Taylor, C. G., Miller, S. A., & Lopez-Nicora, H. D. (2014). First report of bloat nematode (*Ditylenchus dipsaci*) infecting garlic in Ohio. *Plant Disease, 98*, 859.

Tylka, G. (2016). Integrated management of *Heterodera glycines* in the midwestern United States. *Journal of Nematology, 48*, 377–378.

Tylka, G. (2017, November 3). *Over 1,000 SCN-resistant soybean varieties – all but 29 have PI 88788 resistance*. Iowa State University Extension and Outreach Integrated Crop Management News.

Tylka, G. L., & Marett, C. C. (2017). *Known distribution of the soybean cyst nematode, Heterodera glycines, in the United States and Canada, 1954 to 2017*. doi: https://doi.org/10.1094/PHP-05-17-0031-BR.

Tylka, G. L., Sisson, A. J., Jesse, L. C., Kennicker, J., & Marett, C. C. (2011). Testing for plant parasitic nematodes that feed on corn in Iowa 2000–2010. Online. *Plant Health Progress, 5*. doi: https://doi.org/10.1094/PHP-2011-1205-01-RS.

USDA-NASS. (2018). *Area planted for all purposes, yield and production for soybean and corn* quickstats. Washington, DC: USDA-National Agricultural Statistics Service (NASS).

Venkatesh, R., Harrison, S. K., & Riedel, R. M. (2000). Weed hosts of soybean cyst nematode (*Heterodera glycines*) in Ohio. *Weed Technology, 14*, 156–160.

Vierling, R. A., Faghihi, J., Ferris, V. R., & Ferris, M. (2000). *Methods for conferring broad-based soybean cyst nematode resistance to a soybean line*. United States Patent number 6,096,944.

Villamil, M. B., Bollero, G. A., Darmoday, R. G., Simmons, F. W., & Bullock, D. G. (2006). No-till corn/soybean systems including winter cover crops: Effects on soil properties. *Soil Science Society of America Journal, 70*, 1936–1944.

Von Qualen, R. H., Abney, T. S., Huber, D. H., & Schreiber, M. M. (1989). Effects of rotation, tillage, and fumigation on premature dying of soybean. *Plant Disease, 73*, 740–744.

Walters, S. A., Bond, J. P., Russell, J. B., Taylor, B. H., & Handoo, Z. A. (2008). Incidence and influence of plant parasitic nematodes in southern Illinois peach orchards. *Nematropica, 38*, 63–74.

Westphal, A. (2011). Sustainable approaches to the management of plant parasitic nematodes and disease complexes. *Journal of Nematology, 43*, 122–125.

Westphal, A., & Xing, L. J. (2011). Soil suppressiveness against the disease complex of the soybean cyst nematode and sudden death syndrome of soybean. *Phytopathology, 101*, 878–886.

Westphal, A., Xing, L. J., & Egel, D. S. (2006). *Use of cover crops for management of root knot nematodes in cucurbits*. Annual International Research Conference on Methyl Bromide Alternatives and Emissions Reductions, presentation 22–1 to 22–3, Orlando, FL, November 5–9, 2006.

Westphal, A., Abney, T. S., Xing, L. J., & Shaner, G. (2008). Sudden death syndrome of soybean. *The Plant Health Instructor*. https://doi.org/10.1094/PHI-I-2008-0102-01.

Westphal, A., Xing, L. J., Pillsbury, R., & Vyn, T. J. (2009). Effects of tillage on population densities of *Heterodera glycines*. *Field Crops Research, 113*, 218–226.

Wheeler, T. A., & Riedel, R. M. (1994). Interaction among *Pratylenchus penetrans, P. scribneri*, and *Verticillium dahlia* in the potato early dying disease complex. *Journal of Nematology, 26*, 228–234.

Willson, H. R., Riedel, R. M., Eisley, J. B., Young, C. E., Jasinski, J. R., Wheeler, T. A., Kauffman, P. H., Pierson, P. E., & Stuart, M. C. (1996). Distribution of *Heterodera glycines* in Ohio. *Supplement to Journal of Nematology, 28*(4S), 599–603.

Wilson, J.D., & Walker, J.T. (1961). *An inventory of stylet-bearing nematodes in Ohio*. Ohio Agricultural Experimental Station, Wooster, Ohio. Special Circular 97, 10 pp.

Winstead, N. N., Skotland, C. B., & Sasser, J. N. (1955). Soybean-cyst nematode in North Carolina. *Plant Disease Report, 39*, 9–11.

Wong, A. T. S., & Tylka, G. L. (1994). Eight nonhost weed species of *Heterodera glycines* in Iowa. *Plant Disease, 78*, 365–367.

Wrather, A., & Koenning, S. (2009). Effects of diseases on soybean yield sin the United States 1996 to 2007. *Plant Health Progress*. https://doi.org/10.1094/PHP-2009-0401-01-RS.

Wrather, J. A., Niblack, T. L., & Milam, M. R. (1992). Survey of plant parasitic nematodes in Missouri cotton fields. *Supplement to Journal of Nematology, 224*(4S), 779–782.

Xing, L. J., & Westphal, A. (2013). Synergism in the interaction of *Fusarium virguliforme* and *Heterodera glycines* in sudden death syndrome of soybean. *Journal of Plant Diseases and Protection, 120*, 209–217.

Yang, J. I., Benecke, S., Jeske, D. R., Rocha, F. S., Smith Becker, J., Timper, P., Becker, J. O., & Borneman, J. (2012). Population dynamics of *Dactylella oviparasitca* and *Heterodera schachtii*: Toward a decision model for sugar beet planting. *Journal of Nematology, 44*, 237–244.

Ye, W., & Robbins, R. T. (2004). Cluster analysis of *Longidorus* species (Nematoda: Longidoridae), a new approach in species identification. *Journal of Nematology, 36*, 207–219.

Chapter 5
Distribution and Importance of Plant Nematodes in Nebraska, Kansas and Colorado

Timothy Todd and Thomas Powers

5.1 Introduction

Nematological research in the Central Great Plains historically focused on management of nematodes in sugar beets and irrigated corn fields, primarily through the use of granular nematicides and soil fumigants. With the introduction of the soybean cyst nematode *Heterodera glycines* into eastern soybean production areas in the 1980s, emphasis largely shifted to using resistant varieties and crop rotation to address the new threat. Recent developments in the Central Great Plains, including diminishing effectiveness of host resistance in soybean, new developments in seed treatments for nematode control in corn and soybean, and grower adoption of cover crops, has led to a renewed focus on integrated pest management (IPM) practices.

5.2 Economically Important Crop Production

Colorado, Kansas and Nebraska compose the Central Great Plains' breadbasket of North America. Each state ranks in the top ten winter wheat and sorghum producing states in the U.S. (Table 5.1), collectively accounting for nearly 30% of U.S. winter wheat production and approximately 45% of U.S. grain sorghum production. Corn and soybean are the next most important crops in the region, representing approximately 16% of U.S. corn production and 10% of U.S. soybean production.

T. Todd
Department of Plant Pathology, Kansas State University, Manhattan, KS, USA
e-mail: nema@ksu.edu

T. Powers (✉)
Department of Plant Pathology, University of Nebraska, Lincoln, NE, USA
e-mail: tpowers1@unl.edu

S. A. Subbotin, J. J. Chitambar (eds.), *Plant Parasitic Nematodes in Sustainable Agriculture of North America*, Sustainability in Plant and Crop Protection,
https://doi.org/10.1007/978-3-319-99588-5_5

Table 5.1 Crop production in the Central Great Plains

	Corn				Winter wheat			
	Hectarage[a]		Production[b]		Hectarage[a]		Production[b]	
Colorado	0.40	0.01	3.47	0.01	0.83	0.06	1.80	0.04
Kansas	1.59	0.05	12.24	0.04	3.55	0.26	8.62	0.21
Nebraska	3.72	0.11	38.17	0.12	0.53	0.04	1.49	0.04
Total	5.71	0.16	53.88	0.16	4.91	0.37	11.90	0.29
	Soybean				Sorghum			
	Hectarage[a]		Production[b]		Hectarage[a]		Production[b]	
Colorado	–	–	–	–	0.09	0.04	0.15	0.02
Kansas	1.53	0.05	3.28	0.03	1.03	0.43	3.81	0.42
Nebraska	2.04	0.06	6.82	0.07	0.05	0.02	0.22	0.02
Total	3.57	0.11	10.09	0.11	1.17	0.48	4.18	0.46

[a]Millions of hectares and proportion of U.S. total
[b]Millions of metric tons and proportion of U.S. total

Additional major agricultural commodities in the region include hay production and rangeland, which each represent approximately 10% of U.S. totals.

5.3 Geological Characteristics

In order to understand the crops that are grown and the practices employed to manage pests and pathogens in the Great Plains, it is useful to have some insight into the past. The deep, rich soils of the Central Great Plains are a legacy of ancient geological events. The uplift of the Rocky Mountains, which began more than 50 million years ago, created a rain shadow effect that initiated the transformation of the ancient plant and animal communities. As a result of this uplift, together with a climatic cooling during the Cenozoic era, tropical forests were converted to grasslands as the winters became dry and cold, and summers hot and wet. Grasses, well adapted for the seasonal extremes in temperature and rainfall, diversified in the mid-Miocene, as did a multitude of mammals that specialized in browsing on the graminaceous forage. Rivers flowing eastward from the Rocky Mountains carried sand, silt and clay in their winter meltwater, building the deep layers of the soil, and the glacial winds of the Pleistocene Ice Ages deposited additional layers of sand and loess. These legacy events shaped the landscape that European settlers encountered in their expansion to western territories. Rainfall decreases by 50% from the eastern to the western borders of Kansas and Nebraska. In the eastern counties, the annual rainfall of 63–75 cm supports dryland corn and soybeans. In the west, irrigation is essential for corn production and winter wheat has been the preferred crop. Most of the productive agricultural land of both states overlies the Ogallala Aquifer, the largest source of groundwater on the North American continent. The availability of this water for irrigation, although highly regulated in Nebraska, has allowed for the intensive cultivation of vast areas of land that would otherwise be marginally productive.

5.4 Nematological Problems

5.4.1 Root Lesion Nematodes, Pratylenchus *spp.*

The widespread distribution of root lesion nematodes (RLN), *Pratylenchus* spp., along with their well-documented damage potential, warrant the designation of this genus as the most economically important plant parasitic nematodes in the Central Great Plains (Table 5.2). *Pratylenchus* spp. are among the most frequently encountered parasitic taxa in corn and wheat fields in the region, with an observed prevalence of approximately 80% for both crops (Todd et al. 2014; Jackson-Ziems et al., unpubl.). Most populations of *Pratylenchus* associated with crops in the Central Great Plains belong to a small number of parthenogenetic morphospecies including *P. neglectus* (Fig. 5.1) and *P. scribneri* on corn, and *P. neglectus* and *P. thornei* on wheat (Powers and Todd unpubl.).

5.4.1.1 Importance

Yield loss estimates for corn, on a per nematode basis, range from 0.014% based on numbers of *P. penetrans* per 100 cm^3 soil + root fragments (MacGuidwin and Bender 2016) to 0.001% based on numbers of *P. neglectus* (Todd and Oakley 1996b) and *P. scribneri* (Smolik and Evenson 1987) per g root. Yield loss estimates for winter wheat, on a per nematode basis, similarly range from 0.01% to 0.001% for numbers of *P. neglectus* per g root (Smiley et al. 2004; Todd et al. 2014), but approach 1% for numbers of *P. thornei* per 100 cm^3 soil (Armstrong et al. 1993).

Table 5.2 Economically important nematodes in Central Great Plains agriculture

Economically important taxa[a]	Primary agronomic hosts	References
Pratylenchus neglectus	Corn, wheat	Todd and Oakley (1996b) and Todd et al. (2014)
P. thornei	Corn, wheat	Powers and Todd (unpublished) and Thorne (1961)
P. cf. alleni	Corn, potato	Powers and Todd (unpublished)
P. scribneri		
Heterodera glycines	Soybean	Todd et al. (1995) and Tylka and Marett (2014)
Belonolaimus sp.	Corn	Cherry et al. (1997), Kerr and Wysong (1979), and Todd (1989)
Longidorus breviannulatus	Corn	Powers (unpublished)
Heterodera schachtii	Sugar beet	Thorne (1961)
Nacobbus aberrans	Sugar beet	Schuster and Thorne (1956)
Ditylenchus dipsaci	Onion, alfalfa	Pokharel et al. (2009)

[a]Listed in order of importance based on distribution

Fig. 5.1 *Pratylenchus neglectus* is probably the most widespread agronomically important nematode in the Great Plains region

Based on regional surveys, the authors currently estimate Central Great Plains losses to root lesion nematodes at 3–4% for corn and 1.5–2% for wheat.

5.4.1.2 Management

Commercial cultivars with high levels of resistance to root lesion nematodes are currently unavailable for the major crops in the Central Great Plains region, although quantitative variation in both corn and wheat germplasm is well documented. In the absence of deployable resistance, management of root lesion nematodes has historically relied on nematicide applications and crop rotation. More recent cropping system approaches have also incorporated cover crops for nematode management, in addition to their other agronomic benefits. Each of these management strategies will be discussed in detail below. Regardless of which strategies are deployed, however, the economics of nematode management in crops such as corn and wheat are often marginal, necessitating a holistic approach that considers added ecological and agronomic benefits.

Variability among corn inbred lines and commercial hybrids in the reproductive potential of root lesion nematodes was identified decades ago, with the best performing cultivars supporting fewer than 10% of the population densities supported by the worst performing cultivars (Norton 1983; Smolik and Wicks 1987). One study specifically identified lines with high levels of the antibiotic 2,4-dihydroxy-7-methoxy-1,4-benzoxazin-3-one (DIMBOA) present in corn and other grasses, as exhibiting reduced host suitability (Melton and Simcox 1987). Information on differential corn cultivar suitability is available primarily for *P. hexincisus* and *P. scribneri*, and the responses of other *Pratylenchus* species and of modern corn cultivars remain to be determined.

Similar ranges in host suitability to *P. neglectus* and *P. thornei* have been reported among winter wheat cultivars in the Pacific Northwest (PNW), and numerous quantitative trail loci for resistance have been identified from commercial cultivars, landrace lines and wheat relatives (Smiley and Nicol 2009). Additionally, wheat cultivars with tolerance to these two RLN species have been deployed effectively in

Australia (Thompson et al. 2008; Vanstone et al. 2008). Less is known about the response of winter wheat cultivars to RLN populations in the Great Plains region, but a recent evaluation of 200 synthetic wheat lines produced by CIMMYT (International Maize and Wheat Improvement Center), along with 10 commercial winter wheat cultivars from Kansas and the PNW, identified 2 synthetic wheat lines with resistance to *P. thornei* that was better than the least susceptible commercial cultivar (Gaynor 2015). No synthetic wheat lines with better resistance to *P. neglectus* than any of the commercial cultivars were identified, and no correlation was observed between cultivar responses to *P. thornei* and *P. neglectus*.

Chemical control of RLN in winter wheat is impractical due to a lack of registered compounds and marginal economic viability. The situation is somewhat different for corn however, and numerous studies have demonstrated that effective RLN control can be achieved with soil-applied carbamate and organophosphate nematicides. Nonetheless, yield responses are erratic, ranging from negligible to nearly 30% improvement across Kansas environments (Lengkeek et al. 1981; Todd and Oakley 1995, 1996a). Declining product availability, along with new developments in protectant seed treatments, have led to reduced dependence on soil-applied nematicides in recent years. Several commercial seed protectant products are currently available for nematode control in corn, but ongoing research in Kansas suggests that yield responses will be similarly erratic.

Crop rotation represents a promising approach to RLN management in the Central Great Plains, but its full potential has yet to be realized, largely due to insufficient knowledge of species-specific host preferences. The situation is further complicated by the under-characterized intraspecific variability among RLN populations and the difficulty in obtaining a reliable species designation. Current research in Kansas and Nebraska is focusing on resolving these inadequacies. The relative host status of the major agricultural crops in the region including corn, sorghum, soybean and wheat have been characterized for reference RLN populations representing four of the most commonly encountered morphospecies (Fig. 5.2a). These data clearly suggest that RLN populations are not as polyphagous as generally assumed, and that with sufficiently detailed information, crop rotation can be a reliable management strategy. Intraspecific variability in host preferences among RLN populations within the region remains to be characterized, and this is a goal of several ongoing research initiatives sponsored by USDA National Institute of Food and Agriculture Multistate Research projects.

Research has identified a limited number of suitable rotation crops for RLN species associated with corn and wheat production in the Central Great Plains. Barley, durum wheat and field pea have been identified as relatively poor or resistant hosts for *P. thornei* and *P. neglectus* populations in Australia and the PNW (Hollaway et al. 2000; Taylor et al. 2000; Smiley et al. 2013). Sorghum and soybean can be added to the list of poor hosts for Kansas populations of *P. neglectus* (Fig. 5.2a). Alfalfa and sorghum have previously been identified as poor hosts for *P. scribneri* populations in Kansas (Todd 1991), and current research confirms sorghum as a poor host and adds soybean and wheat to the list of poor hosts (Fig. 5.2a).

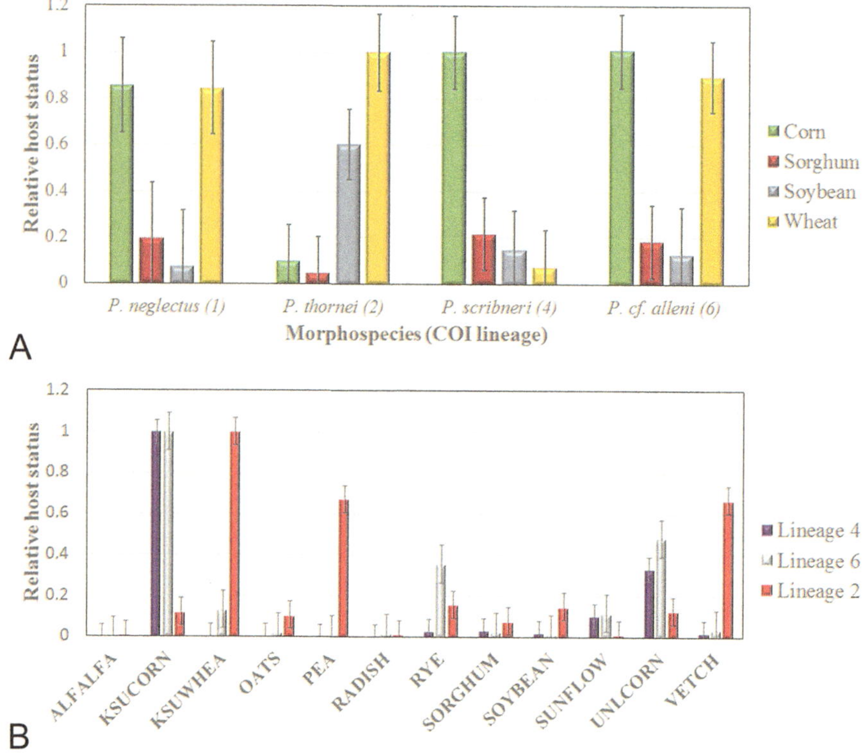

Fig. 5.2 (**a**) Relative host status of corn, sorghum, soybean and wheat for Kansas *Pratylenchus* populations grouped by morphospecies and mitochondrial COI lineage. Each crop was represented by two genotypes. Error bars represent standard errors of mean relative host status; (**b**) Relative host status of selected cover crops for Kansas *Pratylenchus* populations identified by mitochondrial COI lineage. Error bars represent standard errors of mean relative host status

Limited information on RLN population responses to cover crops is currently available for the Central Great Plains and the results of studies outside of the region do not support many generalizations. Grasses, including graminaceous cover crops, for example, are often good to moderate hosts for RLN species, yet pearl millet appears to be a poor host for *P. brachyurus* and *P. penetrans* (Bélair et al. 2002; Inomoto and Asmus 2010). Many other cover crops are also moderate to good hosts for *P. penetrans*, and can lead to increased nematode pressure and damage on ensuing host crop species (Abawi and Widmer 2000; Kimpinski et al. 2000). In a recent Midwest cover crop study, a radish cover crop increased *P. penetrans* population densities during both the cover crop phase and the subsequent carrot crop production phase (Grabau et al. 2017). Even nematode-suppressive cover crops such as *Brassica* spp., can fail to suppress RLN populations because the desired effect is offset by a favorable host status (Potter et al. 1999; Taylor et al. 2000).

Preliminary results from Kansas *Pratylenchus* populations suggest that the host status of cover crops varies unpredictably across nematode species, but pea, vetch and rye (in addition to corn and wheat) represent significant potential for population increase in some cases (Fig. 5.2b). Reliable information on the host status of cover crops in the Central Great Plains continues to be a critical need. For the present however, cover crop recommendations in the region must continue to be based on agronomic benefits other than nematode management.

5.4.2 *False Root Knot Nematode,* Nacobbus abberans *and Sugar Beet Cyst Nematode,* Heterodera schachtii *on Sugar Beet*

The westernmost counties of Nebraska retain what used to be a large area of sugar beet production just east of the Rocky Mountains foothills. In the early 1950s soil surveys of sugar beet fields in Western Nebraska discovered the false root knot nematode, *Nacobbus abberans*, in combination with *Heterodera schachtii* and *Meloidogyne hapla* (Thorne and Schuster 1956). It was suggested that *Nacobbus* alone was an economically important nematode based on stand reduction and loss of plant vigor (Schuster and Thorne 1956). Thorne (1952) speculated that nematodes may be related to the closing of several beet-sugar factories. The sugar beet isolate of *Nacobbus* was considered sufficiently distinct enough to deserve separate species status by Thorne and Schuster (1956) and was named *Nacobbus batatiformis*. It has since been synonymized with *N. abberans* (Sher 1970). Whether or not damage by this nematode indirectly led to the decline of the industry in the region, there is only occasional evidence of involvement of *Nacobbus* in present day sugar beet production in Nebraska. Halting the mandated practice of returning tare soil from the processing facility back to the production fields may have reduced the spread of the nematode (Gray et al. 1992; Wilson 2013). Since 1980, the primary nematode concern has been the sugar beet cyst nematode (SBCN) *Heterodera schachtii*. In the 1980s and 1990s, it was routine to sample production fields, and based on an economic threshold of 2.8 eggs/cm^3, treat fields that exceeded that level with Aldicarb/Temik™. When ground water contamination became a concern within the production area, Telone® II became the fumigant of choice. At the same time, management approaches developed 100 years earlier were reconsidered. Trap crops and biofumigation with oil radish and yellow mustards were tested to reduce SBCN populations (Smith et al. 2004). Interestingly oil radish and mustard trap crops had no effect on reducing levels of *Nacobbus*. In 2017, over 18,500 ha of sugar beets were grown in Nebraska. It is difficult to determine the percent of sugar beet production acreage that is specifically managed for nematode control, but records from the 1999 sugar beet crop in Nebraska estimates 8% of the production was chemically treated for nematode management (Wilson 2013).

5.4.3 *Corn Needle Nematode,* Longidorus breviannulatus *and Sting Nematode,* Belonolaimus *sp.*

Approximately one quarter of the total area of Nebraska is covered by the largest expanse of grass-stabilized sand dunes in the Western Hemisphere. The Nebraska Sandhills, deposited by periglacial winds 8–10,000 years ago, cover nearly 50,000 km^2. Some of this region has been converted to agriculture. It is not surprising that the corn needle nematode (*Longidorus breviannulatus*) and sting nematodes (*Belonolaimus* species) have been discovered in cornfields within the Sandhills region (Kerr and Wysong 1979; Powers unpubl.). Both nematodes require soils with a high percentage of sand for their existence (Robbins and Barker 1974; Malek et al. 1980). Both nematodes also exhibit extensive vertical migration, resulting in erratic distribution in the soil profile (MacGuidwin 1989; Todd 1989). These nematodes are most readily observed at early stages of corn development when stunted seedlings occur in patches that are half the height of uninfected sites. These seedlings, particularly in the case of *Longidorus*, display a "rat tail" root structure as all lateral roots cease development. Stunted plants may achieve a height equivalent to uninfected plants as the season progresses, although it is believed that grain fill is affected. There are no studies that quantitatively measure the yield impact of these nematodes in the Great Plains. However, both nematodes are routinely mentioned in extension circulars that address nematode management in corn.

Sand dunes represent a much less extensive area of Kansas, but they are frequently associated with the Arkansas River, which runs through the southwestern and south central regions of the state. Populations of one or more unique and undescribed *Belonolaimus* species can be found wherever intensive corn production coincides with these soils (Cherry et al. 1997). Distributions are extremely localized and sporadic both within fields and across regions.

5.4.3.1 Importance

The economic threshold for *Belonolaimus* species is typically set at the detection limit, reflecting the extremely high damage potential of these nematodes. Total corn yield loss is predicted for population densities as low as 30 nematodes per 100 cm^3 soil (Todd 1989). Nonetheless, the restricted distribution of these nematodes (<1% of Kansas and Nebraska corn production hectarage) limits their economic importance overall.

Longidorus breviannulatus exhibits similar damage potential, with population densities greater than 100 nematodes per 100 cm^3 soil capable of killing corn seedlings in the U.S. Midwest (Malek et al. 1980).

5.4.3.2 Management

It is likely that needle and sting nematodes are responsible for a large percentage of nematicide applications to corn production fields in the Great Plains. Damage from these nematodes is conspicuous, as is the response to treatment (Todd 1989). Nematicide efficacy is rarely sufficient, however, to achieve normal yield potential in severely affected areas.

Non-host options for crop rotation are limited, particularly for *Belonolaimus* species. Sorghum, soybean and wheat all support reproduction by Kansas populations of sting nematodes, but alfalfa produces population declines that are comparable to those under fallow conditions (Todd 1991). The host range of *Longidorus breviannulatus* appears to be restricted to graminaceous crops, with leguminous crops such as soybean and alfalfa serving as non-hosts (Malek et al. 1980).

5.4.4 *Stem and Bulb Nematode,* Ditylenchus dipsaci

Colorado has the fifth highest storage onion production in the U.S. (Pokharel et al. 2009). Nematodes are considered among the major limiting constraints of onion production, which is compounded by the constraints applied to exported onions with regards to stem and bulb nematode, *Ditylenchus dipsaci* -free produce. The actual economic impact of the nematode on onion production in Colorado has not been determined.

5.4.5 *Soybean Cyst Nematode,* Heterodera glycines

The soybean cyst nematode (SCN) is a relatively recent introduction in Kansas and Nebraska, having first been identified in 1985 and 1986, respectively (Sim and Todd 1986; Powers and Wysong 1987). Nonetheless, as of 2014, SCN had been reported from 56 counties in Kansas and 58 counties in Nebraska (Tylka and Marett 2014; Brodrick, pers. com.). Nearly all of the known infested counties are located in Eastern and Central Kansas and Nebraska, where soybean production is concentrated. Based on a 2010–2012 survey of soybean production fields in Kansas, less than 15% of the soybean acreage in the state is currently infested, although the prevalence in several counties exceeds 50% (Todd unpubl.).

5.4.5.1 Importance

Yield loss for susceptible soybean cultivars in Kansas has been estimated to be approximately 9% for each 1000 eggs per 100 cm^3 soil present at planting (Todd et al. 1995). An average egg density of 1200 eggs per 100 cm^3 soil was observed for infested soybean fields in Kansas in the 2010–2012 survey, but as SCN prevalence was estimated at ~15%, this represents a statewide yield loss of only ~1.5% (Todd unpubl.). Since most commercial soybean cultivars exhibit some resistance to SCN, actual losses in the region are likely to be substantially lower, although there is evidence of increasing virulence among SCN populations on the most widely deployed source of resistance (Faghihi et al. 2010).

The economic impact of SCN in the Central Great Plains is exacerbated by its interaction with two important fungal diseases of soybean: charcoal rot, caused by *Macrophomina phaseolina*, and sudden death syndrome, caused by *Fusarium virguliforme*. Increased severities of both diseases have been observed in the presence of *H. glycines* (Todd et al. 1987; Winkler et al. 1994; Brzostowski et al. 2014), and soybean yield loss mitigation in the region frequently requires strategies for the simultaneous management of nematode and fungal pathogens.

5.4.5.2 Management

As for U.S. soybean hectarage generally, host resistance has been widely deployed to manage SCN populations in Kansas and Nebraska. Resistant soybean cultivars limit SCN population increase and typically yield better than susceptible cultivars at nematode-infested sites (Todd et al. 1995; Wang et al. 2000; Long and Todd 2001; Donald et al. 2006). Resistance nevertheless presents two major challenges to sustainable management: (1) a narrow basis for resistance in commercial soybean germplasm and (2) virulence diversity and plasticity among SCN populations (i.e. HG Types). An estimated 97% of soybean cultivars in the North Central U.S. derive their resistance from a single source – PI 88788 (Faghihi et al. 2010). This dependence on a single source of resistance has selected for widespread virulence among SCN populations, with the frequency of virulent populations positively correlated with the length of resistance deployment (Faghihi et al. 2010). Surveys in Kansas have determined that 65% of SCN populations in the state display levels of reproduction (>30% of the reproduction on a susceptible cultivar) on PI 88788-derived cultivars that would limit their effectiveness (Todd unpubl.). Resistance sources that display universally high levels of resistance to Kansas SCN populations, such as PI 437654, are available, but these have not been widely deployed.

Crop rotation continues to play a vital role in SCN management in the region. Non-host crops such as corn, sorghum, and wheat provide extended periods without *H. glycines* reproduction, but significant reductions in nematode population densities generally require two or more years without a susceptible soybean crop (Long and Todd 2001; Todd et al. 2003). There is a limit to the effectiveness of non-host crops, however, as nematode survival rates are density dependent and approach 100% at low population densities. Rapid population increases on subsequent susceptible soybean cultivars further limit the effectiveness of crop rotation for SCN management. Rotation with non-host crops is, therefore, most effective when combined with host resistance.

Nematicides have historically played a minor role in SCN management in the Central Great Plains, largely because of the availability of host resistance. Nonetheless, economic levels of control have been demonstrated with traditional soil-applied nematicides (Todd et al. 1987), and commercial seed protectants have recently become an important area of research and development.

5.4.6 Other Nematodes

Rangeland and hay production represent significant land uses in the Central Great Plains, and the nematode assemblages associated with these systems, particularly those in the tallgrass prairie, have been extensively studied (Darby et al. 2013; Orr and Dickerson 1966; Olson et al. 2017; Todd 1996; Todd et al. 2006). Plant parasitic taxa are dominated by species of *Helicotylenchus* including *H. digonicus*, *H. platyurus* and *H. pseudorobustus*. *Mesocriconema* spp. and *Xiphinema americanum* are also abundant in tallgrass prairie. While the importance of these species in terms of their abundance and prevalence is well-characterized, however, their impact on agriculture remains to be established.

The lance nematode *Hoplolaimus galeatus*, along with one or more species of *Helicotylenchus*, *Mesocriconema*, and *Tylenchorhynchus* are common inhabitants of creeping bentgrass putting greens in the region, but damage has been documented only for *H. galeatus* (Settle et al. 2007; Todd and Tisserat 1990). Recommendations for management of this nematode are complicated by the observation that damage thresholds vary significantly with environment and cultural practices. Research suggests that putting greens can generally tolerate relatively high levels of feeding pressure as long as cultural practices are optimized for overall turf quality.

Other taxa prevalent on agronomic crops in the region include *Merlinius brevidens*, *Quinisulcius acutus*, and *Paratylenchus projectus* on wheat (Todd et al. 2014), and a number of poorly characterized *Helicotylenchus* and *Tylenchorhynchus* spp. on corn (Todd unpubl.). Insufficient evidence of damage exists to warrant management of any of these species.

5.4.7 Sustainable Nematode Management in the Great Plains

The nature and scale of production agriculture in the Great Plains has necessitated sustainable approaches on economic considerations alone. When commodity prices are low, profit margins are negligible and the standard management options of crop rotation and resistant cultivars prevail. Cover crops are increasingly viewed as an integral component of sustainability throughout the region, with recognized benefits for soil erosion and soil improvement, but water and pest relationships still need to be evaluated. Uncertainties due to climate change, water availability and exotic pest introductions pose significant challenges to the future sustainability of agriculture in the Great Plains, requiring, in the words of Lewis Carroll's Red Queen, "all the running you can do to keep in the same place" (Carroll 1871). Agricultural practices will need to adapt to these challenges, but sustainable systems in the future, as in the past, will continue to rely on established integrated pest management approaches augmented by new research results.

References

Abawi, G. S., & Widmer, T. L. (2000). Impact of soil health management practices on soilborne pathogens, nematodes and root diseases of vegetable crops. *Applied Soil Ecology, 15*, 37–47.

Armstrong, J. S., Peairs, F. B., Pilcher, S. D., & Russell, C. C. (1993). The effect of planting time, insecticides, and liquid fertilizer on the Russian wheat aphid (Homoptera: Aphididae) and the lesion nematode (*Pratylenchus thornei*) on winter wheat. *Journal of the Kansas Entomological Society, 66*, 69–73.

Bélair, G., Fournier, Y., Dauphinais, N., & Dangi, O. P. (2002). Reproduction of *Pratylenchus penetrans* on various rotation crops in Quebec. *Phytoprotection, 83*, 111–114.

Brzostowski, L. F., Schapaugh, W. T., Todd, T. C., Little, C. R., & Rzodkiewicz, P. A. (2014). Effect of host resistance to *Fusarium virguliforme* and *Heterodera glycines* on sudden death syndrome disease severity and soybean seed yield. *Plant Health Progress, 15*, 1–8. https://doi.org/10.1094/PHP-RS-13-0100.

Carroll, L. (1871). *Through the looking-glass and what Alice found there*. Chapter 2.

Cherry, T., Szalanski, A. L., Todd, T. C., & Powers, T. O. (1997). The internal transcribed spacer region of *Belonolaimus* (Nemata: Belonolaimidae). *Journal of Nematology, 29*, 23–29.

Darby, B. J., Todd, T. C., & Herman, M. A. (2013). High-throughput amplicon sequencing of rRNA genes requires a copy number correction to accurately reflect the effects of management practices on soil nematode community structure. *Molecular Ecology*. Online. https://doi.org/10.1111/mec.12480.

Donald, P. A., Pierson, P. E., St. Martin, S. K., Sellers, P. R., Noel, G. R., MacGuidwin, A. E., Faghihi, J., Ferris, V. R., Grau, C. R., Jardine, D. J., Melakeberhan, H., Niblack, T. L., Stienstra, W. C., Tylka, G. L., Wheeler, T. A., & Wysong, D. S. (2006). Assessing *Heterodera glycines*-resistant and susceptible cultivar yield response. *Journal of Nematology, 38*, 76–82.

Faghihi, J., Donald, P. A., Noel, G., Welacky, T. W., & Ferris, V. R. (2010). Soybean resistance to field populations of *Heterodera glycines* in selected geographic areas. *Plant Health Progress*. Online. https://doi.org/10.1094/PHP-2010-0426-01-RS.

Gaynor, R. C. (2015). *Genomic selection for Kansas wheat*. PhD Dissertation, Kansas State University, Manhattan, KS. 45 pp.

Grabau, Z. J., Maung, Z. T. Z., Noyes, D. C., Baas, D. G., Werling, B. P., Brainard, D. C., & Malekeberhan, H. (2017). Effects of cover crops on *Pratylenchus penetrans* and the nematode community in carrot production. *Journal of Nematology, 49*, 114–123.

Gray, F. A., Francl, G. D., & Kerr, F. D. (1992). *Sugar beet nematode*. University of Wyoming Cooperative Extension Service Circular B-975.

Hollaway, G. J., Taylor, S. P., Eastwood, R. F., & Hunt, C. H. (2000). Effect of field crops on population densities of *Pratylenchus neglectus* and *P. thornei* in Southeastern Australia; Part 2: *P. thornei*. *Supplement Journal Nematology, 32*, 600–608.

Inomoto, M. M., & Asmus, G. L. (2010). Host status of graminaceous cover crops for *Pratylenchus brachyurus*. *Plant Disease, 94*, 1022–1025.

Kerr, E. D., & Wysong, D. S. (1979). Sting nematode, *Belonolaimus* sp., in Nebraska. *Plant Disease Report, 63*, 506–507.

Kimpinski, J., Arsenault, W. J., Gallant, C. E., & Sanderson, J. B. (2000). The effect of marigolds (*Tagetes* spp.) and other cover crops on *Pratylenchus penetrans* and on following potato crops. *Supplement Journal Nematology, 32*, 531–536.

Lengkeek, V. H., Sanden, G. E., & Lash, L. D. (1981). Efficacy of nematicides in corn in western Kansas. *Transactions of the Kansas Academy of Science, 84*, 209–215.

Long, J. H., & Todd, T. C. (2001). Effects of crop rotation and cultivar resistance on seed yield and the soybean cyst nematode in full-season and double-cropped soybeans. *Crop Science, 41*, 1137–1143.

MacGuidwin, A. E. (1989). Abundance and vertical distribution of *Longidorus breviannulatus* associated with corn and potato. *Journal of Nematology, 21*, 404–408.

MacGuidwin, A. E., & Bender, B. E. (2016). Development of a damage function model for *Pratylenchus penetrans* on corn. *Plant Disease, 100*, 764–769.

Malek, R. B., Norton, D. C., Jacobsen, B. J., & Acosta, N. (1980). A new corn disease caused by *Longidorus breviannulatus* in the Midwest. *Plant Disease, 64*, 1110–1113.

Melton, T. A., & Simcox, K. D. (1987). Effects of 2, 4-dihydroxy-7-methoxy-2H 1,4-benzoxazin-3(4H)-one (DIMBOA) in corn lines on reproduction on *Pratylenchus hexincisus*. *Journal of Nematology, 19*, 543.

Norton, D. C. (1983). Maize nematode problems. *Plant Disease, 67*, 253–256.

Olson, M., Harris, T., Higgins, R., Mullin, P., Powers, K., Olson, S., & Powers, T. O. (2017). Species delimitation and description of *Mesocriconema nebraskense* n. sp. (Nematoda: Criconematidae), a morphologically cryptic, parthenogenetic species from North American grasslands. *Journal of Nematology, 49*(1), 42–68.

Orr, C. C., & Dickerson, O. J. (1966). Nematodes in true prairie soils of Kansas. *Transactions of the Kansas Academy of Science, 69*, 317–334.

Pokharel, R. R., Larsen, H., Hammon, B., Gourd, T., & Bartolo, M. (2009). *Plant parasitic nematodes, soil and root health of Colorado onion fields* (pp. 31–38). Colorado State University agricultural experiment Station technical report TR09-12.

Potter, M. J., Vanstone, V. A., Davies, K. A., Kirkegaard, J. A., & Rathjen, A. J. (1999). Reduced susceptibility of *Brassica napus* to *Pratylenchus neglectus* in plants with elevated root levels of 2-phenylethyl glucosinolate. *Journal of Nematology, 31*, 291–298.

Powers, T. O., & Wysong, D. S. (1987). First report of the soybean cyst nematode (*Heterodera glycines)* in Nebraska. *Plant Disease, 71*, 1146.

Robbins, R. T., & Barker, K. R. (1974). The effects of soil type, particle size, temperature, and moisture on reproduction of *Belonolaimus longicaudatus*. *Journal of Nematology, 6*, 1–6.

Schuster, M. L., & Thorne, G. (1956). Distribution, relation to weeds, and histology of sugar beet root galls caused by *Nacobbus batatiformis* Thorne and Schuster. *Journal of the American Society of Sugar Beet Technologists, 9*, 193–197.

Settle, D. M., Fry, J. D., Milliken, G. A., Tisserat, N. A., & Todd, T. C. (2007). Quantifying the effects of lance nematode parasitism in creeping bentgrass. *Plant Disease, 91*, 1170–1179.

Sher, S. A. (1970). Revision of the genus *Nacobbus* Thorne and Allen, 1944 (Nematoda: Tylenchoidea). *Journal of Nematology, 2*, 228–235.

Sim, T., & Todd, T. C. (1986). First field observation of the soybean cyst nematode in Kansas. *Plant Disease, 70*, 603.

Smiley, R. W., & Nicol, J. M. (2009). Nematodes which challenge global wheat production. In B. F. Carver (Ed.), *Wheat science and trade* (pp. 171–187). Ames: Wiley-Blackwell.

Smiley, R. W., Merrifield, K., Patterson, L. M., Whittaker, R. G., Gourlie, J. A., & Easley, S. A. (2004). Nematodes in dryland field crops in the semiarid Pacific Northwest United States. *Journal of Nematology, 36*, 54–68.

Smiley, R. W., Machado, S., Gourlie, J. A., Pritchett, L. C., Yan, G. P., & Jacobsen, E. E. (2013). Effects of crop rotations and tillage on *Pratylenchus* spp. in the semiarid Pacific Northwest United States. *Plant Disease, 97*, 537–546.

Smith, H. J., Gray, F. A., & Koch, D. W. (2004). Reproduction of *Heterodera schachtii* Schmidt on resistant mustard, radish, and sugarbeet cultivars. *Journal of Nematology, 36*, 123–130.

Smolik, J. D., & Evenson, P. D. (1987). Relationship of yields and *Pratylenchus* spp. population densities in dryland and irrigated corn. *Annals of Applied Nematology, 1*, 71–73.

Smolik, J. D., & Wicks, Z. W. (1987). Reproduction of *Pratylenchus hexincisus* and *P. scribneri* in corn inbreds. *Annals of Applied Nematology, 1*, 29–31.

Taylor, S. P., Hollaway, G. J., & Hunt, C. H. (2000). Effect of field crops on population densities of *Pratylenchus neglectus* and *P. thornei* in Southeastern Australia; part 1: *P. neglectus*. *Supplement Journal Nematology, 32*, 591–599.

Thompson, J. P., Owen, K. J., Stirling, G. R., & Bell, M. J. (2008). Root lesion nematodes (*Pratylenchus thornei* and *P. neglectus*): A review of recent progress in managing a significant pest of grain crops in northern Australia. *Australasian Plant Pathology, 37*, 235–242.

Thorne, G. (1952). *Control of the sugar beet nematode*. U.S. Dept. Agr. Farmers' Bull. 2054, 18 pp.

Thorne G. (1961). *Principles of nematology* (553 pp.). New York: McGraw-Hill.

Thorne, G., & Schuster, M. L. (1956). *Nacobbus batitiformis*, n. sp. (Nematoda: Tylenchidae), producing galls on the roots of sugar beets and other plants. *Proceedings of the Helminthological Society of Washington, 23*, 128–134.

Todd, T. C. (1989). Population dynamics and damage potential of *Belonolaimus* sp. on corn. *Supplement Journal Nematology, 21*, 697–702.

Todd, T. C. (1991). Effect of cropping regime on populations of *Belonolaimus* sp. and *Pratylenchus scribneri* in sandy soil. *Supplement Journal Nematology, 23*, 646–651.

Todd, T. C. (1996). Effects of management practices on nematode community structure in tallgrass prairie. *Applied Soil Ecology, 3*, 235–246.

Todd, T. C., & Oakley, T. R. (1995). Evaluation of nematicides for lesion nematode control in corn, 1994. *Fungicide and Nematicide Tests, 50*, 188.

Todd, T. C., & Oakley, T. R. (1996a). Evaluation of liquid Furadan for lesion nematode control in corn, 1995. *Fungicide and Nematicide Tests, 51*, 173.

Todd, T. C., & Oakley, T. R. (1996b). Seasonal dynamics and yield relationships of *Pratylenchus* spp. in corn roots. *Supplement Journal Nematology, 28*, 676–681.

Todd, T. C., & Tisserat, N. A. (1990). Occurrence, spatial distribution, and pathogenicity of some phytoparasitic nematodes on creeping bentgrass putting greens in Kansas. *Plant Disease, 74*, 660–663.

Todd, T. C., Pearson, C. A. S., & Schwenk, F. W. (1987). Effect of *Heterodera glycines* on charcoal rot severity in soybean cultivars resistant and susceptible to soybean cyst nematode. *Annals of Applied Nematology, 1*, 35–40.

Todd, T. C., Schapaugh, W. T., Long, J. H., & Holmes, B. (1995). Field response of soybean in maturity groups III–V to *Heterodera glycines* in Kansas. *Supplement Journal Nematology, 27*, 628–633.

Todd, T. C., Long, J. H., & Oakley, T. R. (2003). Density-dependent multiplication and survival rates in *Heterodera glycines*. *Journal of Nematology, 35*, 98–103.

Todd, T. C., Powers, T. O., & Mullin, P. G. (2006). Sentinel nematodes of land-use change and restoration in tallgrass prairie. *Journal of Nematology, 38*, 20–27.

Todd, T. C., Appel, J. A., Vogel, J., & Tisserat, N. A. (2014). Survey of plant parasitic nematodes in Kansas and eastern Colorado wheat fields. *Plant Health Progress, 15*, 112–117.

Tylka, G. L., & Marett, C. C. (2014). *Distribution of the soybean cyst nematode, Heterodera glycines, in the United States and Canada: 1954 to 2014*. doi:https://doi.org/10.1094/PHP-BR-14-0006.

Vanstone, V. A., Hollaway, G. J., & Stirling, G. R. (2008). Managing nematode pests in the southern and western regions of the Australian cereal industry: Continuing progress in a challenging environment. *Australasian Plant Pathology, 37*, 220–234.

Wang, J., Donald, P. A., Niblack, T. L., Bird, G. W., Faghihi, J., Ferris, J. M., Grau, C., Jardine, D. J., Lipps, P. E., MacGuidwin, A. E., Melakeberhan, H., Noel, G. R., Pierson, P., Riedel, R. M., Sellers, P. R., Stienstra, W. C., Todd, T. C., Tylka, G. L., Wheeler, T. A., & Wysong, D. S. (2000). Soybean cyst nematode reproduction in the north central United States. *Plant Disease, 84*, 77–82.

Wilson, R. G. (2013). *Sugar beet production guide*. University of Nebraska Extension EC 156. Pub. University of Nebraska-Lincoln. 244pp.

Winkler, H. E., Hetrick, B. A. D., & Todd, T. C. (1994). Interactions of *Heterodera glycines*, *Macrophomina phaseolina*, and mycorrhizal fungi on soybean in Kansas. *Supplement to Journal of Nematology, 26*, 675–682.

Chapter 6
Biology, Ecology and Management of Plant Parasitic Nematodes in Minnesota

Senyu Chen and Zane J. Grabau

6.1 Introduction

Agriculture is a major part of the economy in Minnesota which is located in the Northcentral United States with a total of more than 10.5 million hectares of farmland. Corn (*Zea mays*) and soybean (*Glycine max*) are two major crops grown in Minnesota with 3.5 and 2.8 million hectares grown in 2012, respectively (Vilsack and Clark 2014). The state is the country's largest producer of sugar beet (*Beta vulgaris*), sweet corn and green pea (*Pisum sativum*) for processing. Other crops include wheat (*Triticum aestivum*), oats (*Avena sativa*), barley (*Hordeum vulgare*), sorghum (*Sorghum bicolor*), rye (*Secale cereale*), dry edible bean (*Phaseolus angularis*), canola (*Brassica napus*), sunflower (*Helianthus annuus*), flax (*Linum usitatissimum*), potato (*Solanum tuberosum*), sweet potato (*Ipomoea batatas*), strawberry (*Fragaria x ananassa*) and vegetable crops (Vilsack and Clark 2014). In Southern Minnesota, corn-soybean annual rotation is the most common cropping system. Wheat, oats, barley, canola, sunflower, dry bean and potato are mainly grown in the northern areas. A number of plant parasitic nematodes have been found in Minnesota fields and orchards associated with various agricultural crops and other plants (Table 6.1). However, only a few of them have been investigated for their economic importance in Minnesota. The soybean cyst nematode (SCN, *Heterodera glycines*) is the most important nematode, and has been the focus of research in the state. Lesion nematodes (*Pratylenchus* spp.) are probably the second-most important group of plant-parasitic nematodes and there is some research of its biology and

S. Chen (✉)
Southern Research and Outreach Center, University of Minnesota, Waseca, MN, USA
e-mail: chenx099@umn.edu

Z. J. Grabau
Entomology and Nematology Department, University of Florida, Gainesville, FL, USA
e-mail: zgrabau@ufl.edu

S. A. Subbotin, J. J. Chitambar (eds.), *Plant Parasitic Nematodes in Sustainable Agriculture of North America*, Sustainability in Plant and Crop Protection,
https://doi.org/10.1007/978-3-319-99588-5_6

Table 6.1 Plant parasitic nematodes reported to cause economic damage in Minnesota crop fields and orchards

Nematode species	Associated crop[a]	Reference
Aphelenchoides spp.	Ornamentals	Taylor et al. (1958) and Taylor and Schleder (1959)
Criconemoides sp.	Alfalfa, barley, corn, soybean, wheat	Taylor et al. (1958) and Taylor and Schleder (1959)
Ditylenchus dipsaci	Garlic	Mollov et al. (2012)
Gracilacus sp. (*G. marylandicus* like)	Wheat	MacDonald (1979)
Helicotylenchus pseudorobustus	Corn, wheat	MacDonald (1979)
Helicotylenchus spp.	Alfalfa, apple, barley, corn, flax, pea, soybean, wheat	Taylor et al. (1958), Taylor and Schleder (1959), and Wallace and MacDonald (1979)
Heterodera glycines	Soybean, dry bean	MacDonald et al. (1980) and Yan et al. (2017)
H. trifolii	Pea	Taylor et al. (1958)
Hoplolaimus spp.	Alfalfa, corn, flax, pea, soybean	Taylor et al. (1958), Taylor and Schleder (1959), and Wallace and MacDonald (1979)
Hoplolaimus galeatus	Apple, wheat	MacDonald (1979) and Wallace and MacDonald (1979)
Meloidogyne hapla	Strawberry	Crow and MacDonald (1978)
Meloidogyne sp.	Alfalfa, barley, flax, corn	Taylor et al. (1958) and Taylor and Schleder (1959)
Paratrichodorus allius	Sugar beet	Yan et al. (2016)
Paratylenchus projectus	Corn, flax, oats, soybean, strawberry, wheat	Crow and MacDonald (1978) and MacDonald (1979)
Paratylenchus spp.	Apple, alfalfa, barley, corn, flax, oats, pea, soybean, rye	Wallace and MacDonald (1979)
Pratylenchus hexincisus	Alfalfa, corn, barley, flax, oats, pea, rye, soybean, wheat	MacDonald (1979), Taylor et al. (1958), and Taylor and Schleder (1959);
P. scribneri	Alfalfa, corn, flax, soybean	Palmer and MacDonald (1974) and Taylor et al. (1958)
P. penetrans	Alfalfa, flax, soybean, corn	Palmer and MacDonald (1974), Taylor et al. (1958), and Wallace and MacDonald (1979)
P. pratensis	Alfalfa, flax, soybean	Taylor et al. (1958)
P. tenuis	Rye, strawberry	Crow and MacDonald (1978)
Pratylenchus spp.	Corn, soybean	Grabau and Chen (2016a, b)
Trichodorus spp.	Alfalfa, corn, flax, oats, soybean, wheat	Taylor et al. (1958) and Taylor and Schleder (1959)
Trophurus minnesotensis	Sugar beet	Caveness (1958)

(continued)

Table 6.1 (continued)

Nematode species	Associated crop[a]	Reference
Tylenchorhynchus spp.	Alfalfa, apple, barley, corn, flax, oats, pea, soybean, wheat	Taylor et al. (1958), Taylor and Schleder (1959), MacDonald (1979), and Wallace and MacDonald (1979)
Xiphinema americanum	Barley, corn, oats, rye, soybean, raspberry, wheat	Crow and MacDonald (1978), MacDonald (1979), Taylor et al. (1958), Taylor and Schleder (1959), and Grabau and Chen (2016a, b)

[a]This table lists hosts that are reported to be associated with the given nematode. For many nematodes, host range and pathogenicity on any particular crop is not well defined

management in Minnesota. The emphasis of this review will be on the soybean cyst nematode and lesion nematodes. A brief summary of other potential important plant parasitic nematodes is also included.

6.2 Soybean Cyst Nematode, *Heterodera glycines*

6.2.1 Infestation History and Distribution

In Minnesota, the soybean cyst nematode was first detected in 1978 in Faribault County on the southern border with Iowa (MacDonald et al. 1980). Since then, the nematode has been found in several counties from south to north, and by 2016, its presence had been confirmed in most (68) soybean-growing counties in the state (Fig. 6.1). In a survey conducted in 2013, the highest percentage of samples infested with SCN was observed in Southwestern (89.7%) and Southcentral (88.2%) Minnesota. The frequency of occurrence decreased from south to north, and in Northwestern Minnesota only 19% of samples were positive for SCN infestation (Table 6.2).

6.2.2 Host Crops and Economic Importance

Approximately 400 plant species have been reported as hosts of soybean cyst nematode, of which 260 species are leguminous plants in the family Fabaceae. Other families that contain good or moderately good hosts include Brassicaceae, Plantaginaceae, Caryophyllaceae, Solanaceae, Lamiaceae, and Asteraceae. Fortunately, few crops are hosts of SCN. Besides soybean, which is the most important host crop, SCN host crops grown in Minnesota also include common bean (*Phaseolus vulgaris*), peas, field pennycress (*Thlaspi arvense*), hairy vetch (*Vicia villosa* var. *villosa*), lupine (*Lupinus* spp.) and mungbean (*Vigna radiata*).

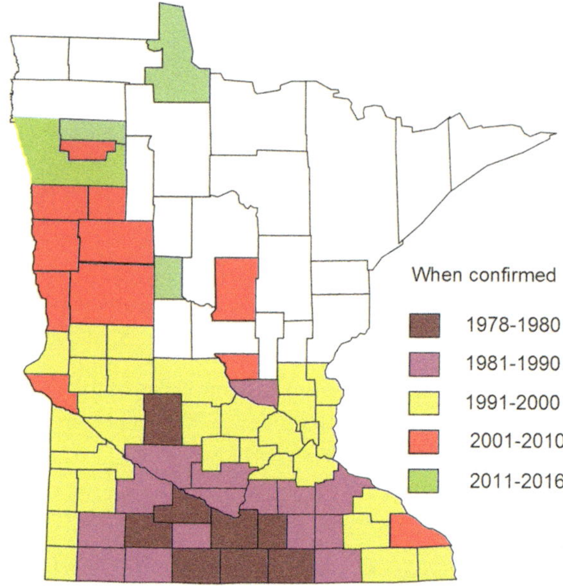

Fig. 6.1 Minnesota counties with confirmed infestation of the soybean cyst nematode by 2016

Table 6.2 Frequency infestation and abundance of the soybean cyst nematode in soil samples from soybean fields by region in Minnesota in 2013

Region	Number of samples collected	Samples infested	% samples infested	Average eggs/100 cm³ soil
Southcentral	34	30	88.2	1853
Southwest	39	35	89.7	2323
Central	26	19	73.1	3228
Southeast	17	12	70.6	928
Westcentral	42	27	64.3	1545
Northwest	42	8	19.0	2948
Total	200	131	65.5	2210

Soybean cyst nematode is the most important pathogen of soybean in most soybean-growing countries and regions and Minnesota is not an exception. Soybean hecterage is the second largest for any crop, following corn, in Minnesota, which is the third largest soybean-producing state in the United States. Although accurate estimation of soybean yield loss to the SCN is difficult to obtain, SCN damage to soybean has well been documented. Damage of soybean by SCN is greater in sandy soil than in clay soil. In some heavily infested sandy fields, SCN can cause complete soybean yield loss. In Minnesota, most soybean fields have clay or loamy soil with high organic contents. In such fertile soil, sometimes there are no obvious symptoms even when infested with a high SCN population density. Yield reduction of soybean is related to initial SCN egg population density. In a field study across

Southern Minnesota, susceptible soybean cultivars produced more than 28% less seed yield compared to SCN-resistant cultivars, when initial population densities were more than 5000 eggs/100 cm^3 soil (Chen et al. 2001a). Actual soybean yield loss may be higher because SCN can cause some yield loss in resistant cultivars.

Common bean (dry bean) is a good host of SCN. Some kidney bean cultivars are similar or even better SCN hosts than soybean (Poromarto and Nelson 2009). Minnesota is one of the top states for common bean production. Severe damage of common bean plants has been observed in commercial fields in Minnesota (Yan et al. 2017). Most common beans are grown in the northern parts of the state, especially in the Red River Valley in Northwestern Minnesota where SCN was detected recently. The extent of SCN infection of common bean in fields has not been investigated, but the increasing importance of SCN in common bean production in Minnesota is expected with the spread of the nematode in Northern Minnesota.

Pea is grown mainly as a canning crop in Minnesota, which is the top pea-producing state in the United States. Although good reproduction of SCN on pea has been reported, most pea cultivars are highly resistant or immune to SCN (Riggs and Hamblen 1962; Zhang 1995).

Field pennycress is an oilseed cover crop in Minnesota. It has been reported as a moderately good host (Poromarto et al. 2015; Venkatesh et al. 2000). The impact of SCN on pennycress has not been studied. Since pennycress is proposed to be planted in corn-soybean production systems, its impact on the SCN population and soybean production needs to be evaluated.

6.2.3 Life Cycle, Population Dynamics and Important Soil Factors

The length of the SCN life cycle is typically about 4 weeks, depending on geographic location, soil temperature and nutritional conditions. In Southern Minnesota, SCN can complete three to four generations during a soybean-growing season. In Central to Northern Minnesota, the nematode is likely to complete only three generations due to a shorter growing season and colder conditions.

Soybean cyst nematode population density is affected by a number of environmental factors as well as host status. The most important environmental factor is the temperature. Optimal soil temperatures are 24 °C for egg hatch, 28 °C for root penetration, and 28–31 °C for juvenile and adult development. Little or no development takes place either below 15 °C or above 35 °C (Schmitt and Noel 1984). Seasonal changes in SCN population densities vary in different geographic locations. SCN can survive at −20 °C (Hu et al. 2016), so the freezing temperatures of Minnesota winters are probably not detrimental to the nematode. During April, after the soil has thawed and temperatures have increased, second-stage juveniles (J2) start to hatch from eggs. After soybean is planted, J2 hatch increases due to chemical stimulants released from soybean roots. Egg population density in soil declines gradually due to the hatch of J2 until late June to early July, when the females of the first generation become mature and produce eggs, and egg population densities start to increase in SCN-susceptible soybean. From late July or early August to the end of the season, SCN egg population densities increase rapidly (Fig. 6.2a). At harvest, egg population densities in susceptible soybean can range from a few thousand to tens of thousands per 100 cm^3 of soil. Average annual reduction of egg population density in non-host corn plots is about 50% (Fig. 6.2b). It takes about 5 years to lower the egg population density from 10,000 to approximately 300 eggs/100 cm^3 of soil (Chen et al. 2001b), a level at which there is limited or no damage to soybean.

Soil properties affect SCN population densities and damage in Minnesota. Soybean cyst nematode population densities are generally positively correlated with soil pH and damage to soybean is greater at high pH because SCN infection exacerbates iron deficiency chlorosis (Chen and Miller 2002; Grabau et al. 2017; Rogovska et al. 2009). Soil texture varies in Minnesota and SCN population development is generally favored by sandy soil (Perez-Hernandez and Giesler 2014).

Agriculture soil ecosystems are complex and organisms in soil affect each other. Natural antagonistic organisms of soybean cyst nematode are one of the most inves-

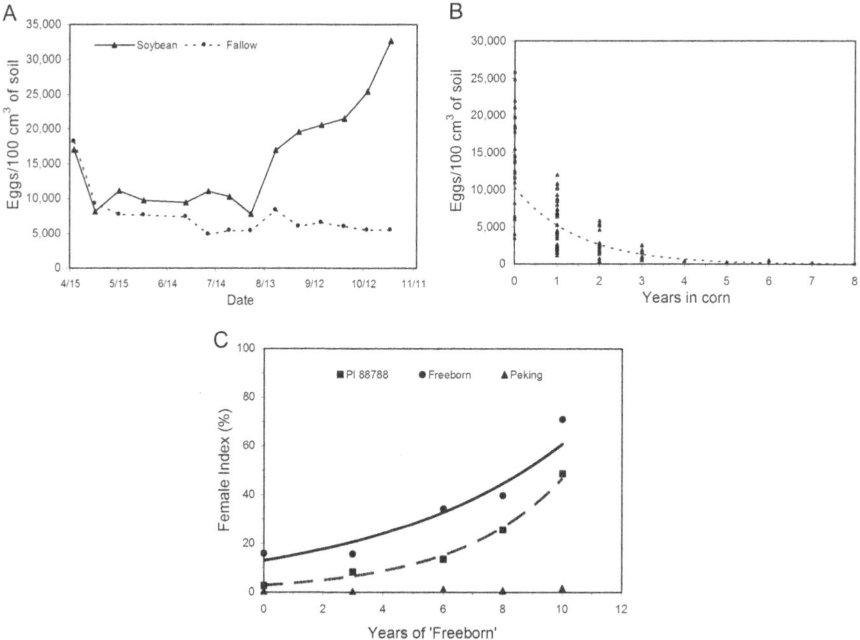

Fig. 6.2 (**a**) Seasonal change of egg population density of the soybean cyst nematode in plots planted with susceptible soybean and in fallow in a Minnesota field in 2008; (**b**) Relationship of soybean cyst nematode egg population density at harvest with number of years of corn following SCN-susceptible soybean during 1996–2004 in a field in southern Minnesota; (**c**) Relationship between reproduction potential (Female Index) of the soybean cyst nematode on 'PI 88788', 'Freeborn', and 'Peking' after the use of the SCN-resistant soybean Freeborn for various years during 1996–2007 in a Minnesota field infested with an original population of HG Type 0 (race 3) (Zheng and Chen 2011)

tigated groups among nematode species. Numerous fungi and bacteria have been isolated from SCN females, cysts, eggs and juveniles, and some of them have been demonstrated to have significant roles in regulating SCN population densities in soils (Chen and Dickson 2012; Chen 2004). Soils suppressive to SCN have been found in a number of locations in the world. Suppression of SCN in some fields in China (Ma and Liu 2000), Japan (Nishizawa 1986), Korea (Kim et al. 1998) and the United States (Noel et al. 2007) have been attributed to parasitism by the endospore-forming bacterium *Pasteuria nishizawae.* In contrast, *Hirsutella rhossiliensis* and *H. minnesotensis* may play an important role in suppressing SCN population densities in Minnesota soybean systems (Chen 2007a; Chen and Reese 1999; Liu and Chen 2000). Another Minnesota study of microbial communities in suppressive soil demonstrated that a number of fungal and bacterial taxa may contribute to suppression of SCN population density (Hu et al. 2017).

6.2.4 Changes of HG Types

The variability of SCN virulence is described by HG Type schemes, in which the virulence phenotypes of SCN populations are determined by the number of females that develop on seven indicator lines as compared with a susceptible soybean cultivar (Niblack et al. 2002). For example, a population with HG Type 1.2 can reproduce on resistant line 1 (Peking) and line 2 (PI 88788), while HG Type 0 cannot reproduce on any of the seven resistant lines. PI 88788 is the source of resistance in most (approximately 95%) commercial SCN-resistant cultivars available in the region, while Peking is available in some cultivars. SCN-resistant soybean cultivars have been used for more than two decades in Minnesota. At the present time, most of soybean fields in Southern and Central Minnesota are planted to SCN-resistant cultivars. The impact of the use of SCN resistant cultivars on the virulence phenotypes of SCN populations was investigated in field plot experiments and statewide surveys. In a field plot experiment, SCN reproduction potential on the resistant soybean cultivar Freeborn and its resistance source PI 88788 increased with increasing years of growing the cultivar (Fig. 6.2c). After 5 years, the population changed from the original HG Type 0 (race 3) to a population that was able to overcome the resistance of PI 88788 (FI > 10; HG Type 2.5.7). After 10 years, Freeborn, that had been moderately resistant (FI $\approx$ 15) to the original population became susceptible (FI > 60) to the resulting SCN population (Zheng and Chen 2011). Four statewide surveys of SCN virulence phenotypes were conducted in 1997–1998, 2002, 2007–2008, and 2013. Before 2002, most SCN populations in Minnesota were HG Type 0- (race 3) (Zheng et al. 2006), after 2008, most of SCN populations in Minnesota were HG Type 2.5.7 (race 1) (Table 6.3), which can overcome PI 88788.

Table 6.3 Percentage of soybean cyst nematode populations from Minnesota with Female Index more than 10 on the indicator soybean lines

Soybean lines	1997–1998	2002	2007–2008	2013
Peking	3.4	1.1	12.1	12.4
PI 88788	13.6	17.0	72.6	64.0
PI 90763	3.4	0	7.3	3.4
PI 437654	2.1	0	0	0
PI 209332	3.7	14.9	71.8	65.2
PI 89722		0	6.5	1.1
PI 548316		33.3	91.1	98.9

6.2.5 *Damage and Disease Interactions*

Soybean cyst nematode infection causes damage to plants not only physically by penetrating and moving through the roots, but also physiologically by altering the metabolism of root cells surrounding the nematode. These modified root cells called syncytia, produce the nutrients needed for the nematode's growth and development. Infection by SCN can also induce secondary infection by one or more microbial pathogens resulting in a disease complex (Bond and Wrather 2004). As a result, function of soybean roots is reduced, and the soybean plant may show nutrient deficiency symptoms.

"Yellow dwarf" is an appropriate description for symptoms that are commonly caused by SCN. When soybean plants are severely infected, the plants become stunted, canopy development is impaired, and leaves may become chlorotic depending on soil and weather conditions. In Minnesota, iron-deficiency chlorosis (IDC) is a common problem that may be induced or made more severe by SCN infection in high pH soil (Fig. 6.3a). Similarly, SCN may induce potassium deficiency symptoms in soils with low potassium levels. SCN populations are not evenly distributed throughout fields. Areas of severely affected and symptomatic soybean plants are often round or elliptical in shape. Those heavily infested areas are often elongated in the direction of tillage due to localized mechanical spread of cysts by tillage equipment. These uneven distributions are often observed in a field where the nematode was recently introduced, and in fields with various soil types.

Soybean cyst nematode infection may limit nodulation by nitrogen-fixing bacteria (Lehman et al. 1971). Because SCN damages roots and limits nutrient uptake by the soybean plants, iron, potassium, and nitrogen deficiencies may increase in severity. Severely infected plants may die before flowering, especially during dry years in soils with poor water-holding capacity.

Belowground symptoms include dark-coloured roots, poorly developed root systems, and reduced nodule formation. SCN infection may increase susceptibility of plants to microbial pathogens by altering plant metabolism or by creating wounds

Fig. 6.3 (**a**) The interaction of SCN with high soil pH resulting in iron-deficiency chlorosis. Pot on left of each image was infested with 10,000 eggs/100 cm^3 of soil, and pot on right had no SCN; (**b**) SCN-susceptible soybean cultivar Pioneer 92B05 (left) and resistant soybean cultivar Prairie Brand PB-2183NRR (right) in plots infested with high SCN egg population densities (17,400 eggs/100 cm^3 of soil left, 10,500 right)

for other pathogens to enter the plant. Several important diseases including brown stem rot, sudden death syndrome, and other fungal root rots of soybean, are associated with or increased in severity by the presence of SCN (Bond and Wrather 2004; Tabor et al. 2003; Westphal et al. 2014).

6.2.6 Management

6.2.6.1 Host Resistance

Resistant cultivars can effectively reduce SCN population density and increase plant growth and yield (Fig. 6.3b). Therefore, use of host resistance is the major strategy for managing SCN. Many SCN-resistant cultivars in Maturity Groups II and I and a few in Maturity Group 0 have been developed and are available for use in Minnesota. Performance of a resistant cultivar in an SCN-infested field depends on the type and level of resistance in hosts, the type and level of virulence of SCN, initial soybean cyst nematode population densities, and environments.

Numerous soybean germplasm lines have been reported to be resistant to SCN, but only a few lines have been used for breeding SCN-resistant cultivars (Shannon et al. 2004). Most (about 95%) commercial SCN-resistant cultivars available in Minnesota are developed from the single source of resistance PI 88788 and a few from Peking. A number of quantitative trait loci (QTL) have been found associated with SCN resistance in different germplasm lines (Concibido et al. 2004; Guo et al. 2006; Vuong et al. 2011). The most important SCN-resistance QTLs are the *rhg 1* on chromosome 18 and *Rhg 4* on chromosome 8 (Mitchum 2016). Both molecular and SCN population studies showed that Peking and PI 88788 carried distinct types of resistance. SCN resistance was found to be mainly due to the copy number variation of genes in the *rhg 1* locus (Cook et al. 2012, 2014). In SCN-susceptible soybean Williams 82, there is only one copy of *rhg 1* genes, but in the resistant soybean there is more than one copy. In the Peking-type of resistance, there are three copies, and in the PI 88788-type of resistance there are more than six copies. Not only does the copy number differ, but also the sequences of different copies of repeats may be different (Lee et al. 2015). The copy and DNA sequence variations may confer different types and levels of SCN resistance, but the pattern of SCN resistance by soybean cultivars to different SCN populations cannot be explained by only these copy number and types in the soybean (Lee et al. 2015). Peking-type resistance also involves the *Rhg 4* allele, but there is no *Rhg 4* in PI 88788-type resistance. In PI 88788-type resistance, the three genes GmSNAP18, GmAAT, and GmWI12 at the *rhg 1-b* allele simultaneously contribute to the SCN resistance, while in Peking-type resistance the *rhg1-a* GmSNAP18 is sufficient for resistance to SCN (Liu et al. 2017). An important QTL found on the chromosome 10 explained a large portion of SCN resistance in PI 567516C that is different from the resistance in PI 88788 and Peking (Lian et al. 2014; Vuong et al. 2010). In addition, many other genes in the genome of soybean including genes in SCN-susceptible cultivars, may affect SCN reproduction (Bao et al. 2014).

Repeated use of the same resistant cultivar or continuous use of cultivars with the same resistance source may eventually lead to SCN populations that can overcome resistance from the common source. Consequently, soybean cultivars with resistance genes from different sources should be alternated to slow down changes in HG Type composition and maintain effectiveness of resistant cultivars. In a long-term field plot experiment in a field initially infested by an SCN population HG Type 0-, monoculture of a PI 88788-derived cultivar selected a population that can overcome resistance in PI 88788, but did not increase Female Index on Peking; similarly, monoculture

Table 6.4 Effect of monoculture of SCN-resistant cultivars on the virulence of soybean cyst nematode (SCN) populations

Monoculture		PI 88788		Peking		PI 437654	
6 years (2003–2009)							
	Susceptible	10.9	bc	0.6	c	0.9	
	R1 (PI 88788 SR)	29.6	a	0.5	c	1.7	
	R2 (Peking SR)	9.0	c	20.3	a	0.1	
	R3 (PI 437654 SR)	20.3	ab	10.4	b	0.1	
11 years (2003–2014)							
	Susceptible	12.0	d	0.8	c	0	b
	R1 (PI 88788 SR)	60.5	a	1.0	c	0	b
	R2 (Peking SR)	12.2	d	42.1	a	2.6	a
	R3 (PI 437654 SR)	29.2	bc	21.7	b	0.1	b

R1 is SCN-resistant 'Latham EX547 RR N' with PI 88788 source of resistance (SR); R2 is SCN-resistant '91M90' with Peking SR; R3 is SCN-resistant 'AR5084' with PI347654 SR.

of a Peking-derived cultivar selected a population that can overcome resistance in Peking, but did not increase Female Index on PI 88788 (Table 6.4). This study demonstrated that the two types of SCN resistance have distinct impact on SCN population development. Rotation of the soybean cultivars with these two sources of resistance is a good choice for SCN management. However, in a number of fields in Minnesota, SCN populations (HG Type 1.2-) can overcome both PI 88788 and Peking sources of resistance. Diversification of source of resistance is needed for long-term effectiveness of using SCN resistance in management.

In the past, resistant cultivars produced less yields than susceptible cultivars when both were grown in the absence of soybean cyst nematode (Fig. 6.4). This is probably due to linkage of SCN-resistance genes with yield-depression genes (Kopisch-Obuch et al. 2005; Mudge et al. 1996). Although the yield potential of resistant cultivars has been improved, and some elite resistant cultivars have fairly high yield, continuous efforts are needed to breed SCN-resistant cultivars that do not carry yield-depression genes. Also, it is important to use SCN resistance in an integrated management program that includes other strategies such as crop rotation and biological control.

6.2.6.2 Crop Rotation and Other Crop Managements

Crop rotation is used not only to reduce SCN population density, but also to benefit general crop management. Many non-host crops such as barley, corn, oat, potato, sorghum, sunflower, canola and wheat, and poor hosts such as alfalfa, pea, and sugar beet are grown in Minnesota. They can be included in a crop rotation to reduce SCN population densities.

The number of years of non-host crops needed to effectively lower SCN population density depends on many factors including crop species, initial SCN population

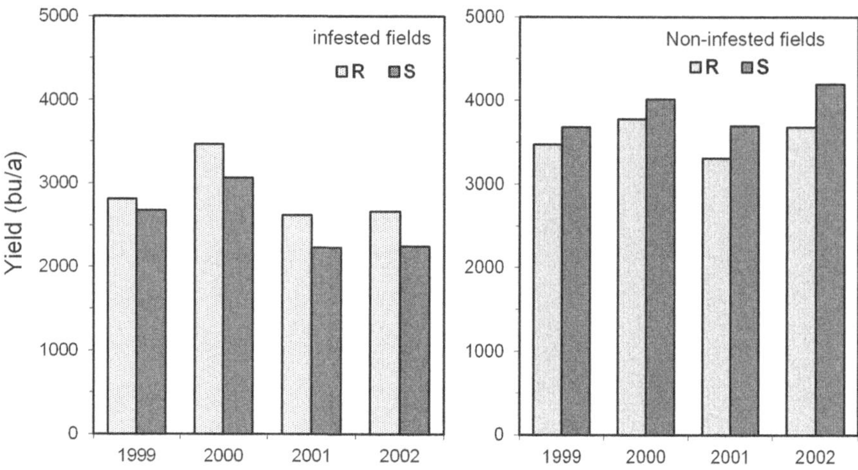

Fig. 6.4 Average yield of top ten SCN-resistant cultivars (R) and top ten susceptible cultivars (S) in SCN-infested and non-infested fields in Minnesota

density, climate, and soil biotic and abiotic factors that affect nematode mortality. In Minnesota, SCN survives well during winter, and with high populations (>10,000 eggs/100 cm^3 soil) after a susceptible soybean, it may take as long as 5 years, depending on soil environments, of non-host or poor-host crops to reduce the SCN population to a low level (e.g., 200–500 eggs/100 cm^3 soil) that will have no or little damage to a susceptible soybean cultivar (Chen 2011; Chen et al. 2001b). Some leguminous crops such as pea and sunn hemp, are poor hosts that produce SCN-hatch stimulants, and are more effective in lowering SCN population density than monocots including corn and wheat (Warnke et al. 2006). Soybean is a major crop in Minnesota, and in most cases soybean is annually rotated with corn. The short period of rotation with corn or other non-host crops is not long enough to lower the SCN egg population density below a damage level. Consequently, inclusion of SCN-resistant soybean cultivars in the crop rotation is essential for successful management of SCN in the Minnesota crop production systems. Resistant cultivars should not be used when initial SCN egg population density is too high, (e.g., >10,000 eggs/100 cm^3 of soil) because a high density can cause a significant yield loss even to a resistant cultivar. Fields with SCN population densities at or above 10,000 eggs/100 cm^3 of soil should be planted to a non-host crop for one or more years until the population densities drop below that level. If SCN egg population density is reduced sufficiently by the rotation of non-hosts and resistant cultivars, which generally takes 5 years in Minnesota, a susceptible soybean can be used. Susceptible cultivars may help avoid or slow down the development of SCN populations that may overcome resistance (Noel and Edwards 1996). In some fields, because the soil is suppressive to SCN (Bao et al. 2011; Chen 2007a), 3 years of non-host and SCN-resistant soybean may be sufficient to reduce the SCN population to a low level, and susceptible soybean can be considered.

Cover crops are used to reduce soil erosion, improve soil fertility, increase biological diversity, and reduce disease and pest pressures. Cover crops can be planted in rotation, or inter-seeded with primary crops, or during off-crop season (e.g., winter) for managing a number of plant-parasitic nematodes (Duncan and Noling 1998). Population densities of plant-parasitic nematodes are reduced because the cover crops are non-hosts or poor hosts, produce allelochemicals that are toxic or inhibitory, provide niches for antagonistic flora and fauna, and/or trap nematodes (Grabau and Chen 2014; Kloepper et al. 1992; Wang et al. 2002). An ideal cover crop should have multiple modes of action in suppressing nematodes. In a Minnesota field experiment, the cover crops alfalfa, red clover and perennial rye grass had little impact on SCN population density (Chen et al. 2006).

It is important to maintain proper soil fertility. Appropriate cultural practices such as maintenance of good soil fertility, may enhance plant growth, increase tolerance of plants and minimize yield loss to SCN. A 2-year study demonstrated that application of swine manure increased yield of susceptible soybean in an SCN-conducive soil, but no yield increase was observed in SCN-suppressive soil (Bao et al. 2013). In contrast to the Southern USA where tillage generally reduce SCN population density (Donald et al. 2009; Koenning et al. 1995), in Minnesota, tillage has no or limited effect on SCN egg population density (Chen 2007b). In fact, conventional tillage may improve early season root development, and reduce damage to soybean caused by SCN.

6.2.6.3 Biological Control

Commercial application of biological control agents has potential in the integrated management of the soybean cyst nematode. A number of fungal and bacterial organisms have been tested in the greenhouse and fields for control of the SCN and some of them have demonstrated promising potential (Chen and Dickson 2012; Chen 2004; Chen and Liu 2005; Liu and Chen 2005). There is no widely accepted commercial biological control agent for SCN management, although some commercial seed coating products that contain biocontrol agents are commercially available. Much more research from public institutions and commitment from industries are needed for development of commercial biological agents for control of the SCN and other nematodes. Commercial biological control agents must be evaluated in local fields to confirm their efficacy.

In a field environment, soybean cyst nematode is subjected to attack by a wide range of natural enemies including fungi, bacteria, predacious nematodes, insects, mites and other microscopic soil animals. The species and activities of natural antagonists vary in different fields. In some soybean fields in Minnesota, high percentages (more than 60%) of the SCN second-stage juveniles are parasitized by the fungi *Hirsutella minnesotensis* and/or *H. rhossiliensis* (Fig. 6.5) (Chen and Reese 1999; Liu and Chen 2000). Some other well-known nematode-parasitic fungi such as *Pochonia*, *Exophiala*, *Clonostachys*, *Trichoderma* and *Purpureocillium* were detected in a suppressive soil and apparently associated with SCN suppression in

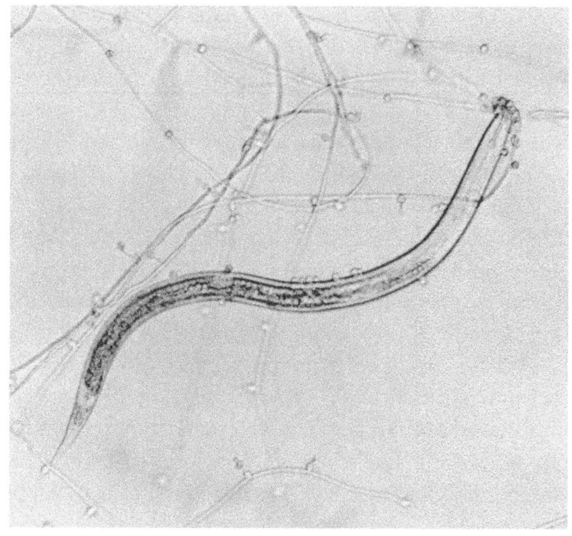

Fig. 6.5 A soybean cyst nematode second-stage juvenile parasitized by the fungus *Hirsutella minnesotensis* (Chen and Reese 1999)

the field (Hu et al. 2017). Introduction of one or more organisms that play important roles in regulating SCN population into a field where there are no such organisms, or enhancement of these organisms where there is low population density, may be another strategy of using biological agents for SCN management.

6.2.6.4 Chemical Control

Some nematicides are registered for use in soybean. A few nematicides are effective in lowering SCN population density, but their performance depends on many soil and environmental factors including soil type, rainfall, soil moisture, temperature, and soil microbial activities. Use of nematicides adds significantly to production costs and does not guarantee increased yields. Economics, as well as environmental and personal health concerns, should be considered before using nematicides. For these reasons, nematicides are not commonly recommended for SCN management. Recently, some seed coating products that contain chemical nematicides have become commercially available in the USA. The efficacy of these products needs to be evaluated in local fields.

6.3 Root Lesion Nematodes, *Pratylenchus* spp.

6.3.1 Species and Their Associated Crops

The genus *Pratylenchus* contains more than 70 species (Castillo and Vovlas 2007; Subbotin et al. 2008), most of which have wide host ranges. At least eight species of *Pratylenchus* have been reported from Minnesota (Table 6.1) in fields associated with different crops including corn, soybean, flax, alfalfa, wheat, barley, oats, rye and strawberry (Palmer and MacDonald 1974; Taylor et al. 1958; Taylor and Schleder 1959; Wallace and MacDonald 1979).

6.3.2 Frequency of Occurrence and Population Dynamics

Lesion nematodes are frequently found in crop fields in Minnesota. In a state-wide survey in organically farmed fields during midseason from August to September 2006, lesion nematodes were found in 65% of samples with an average population density across all samples of 68 nematodes/100 cm^3 soil (Chen et al. 2012). In another survey conducted in late July to early August in 2013, 81.5% of soil samples near-randomly collected from soybean fields across the soybean-growing counties in the state were positive for lesion nematode infestation with an average population density of 22 nematodes/100 cm^3 soil. In a long-term corn-soybean rotation study in Southern Minnesota, *Pratylenchus* sp. soil population density increased with number of years in corn from 1 to 5 years following a 5-year soybean monoculture, and decreased with number of years in soybean from 1 to 5 years following a 5-year corn monoculture (Fig. 6.6). In corn-soybean annual rotation, soil population density of lesion nematodes was similar before planting and at midseason (2 months after planting) in both corn and soybean phases (Fig. 6.7). In the corn phase, in annual rotation with soybean, lesion nematode soil population density increased from midseason to harvest ending with an average population density of

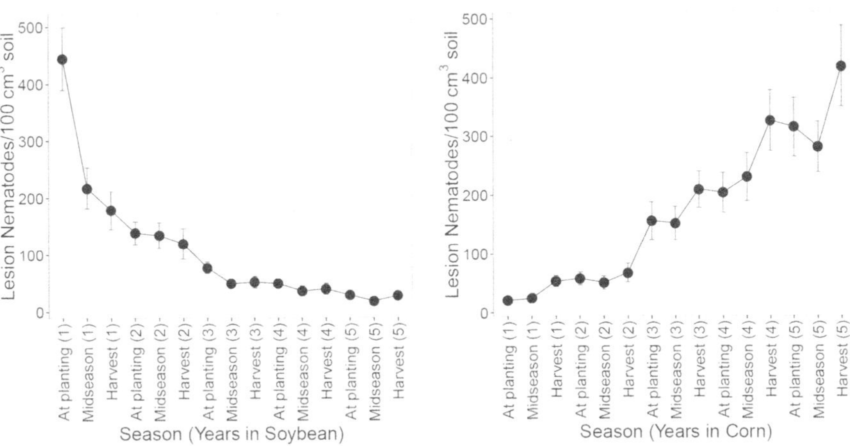

Fig. 6.6 Lesion nematode soil abundance through seasons in different crop phases in a long-term corn-soybean crop rotation experiment where 5 years of soybean (or corn) are followed by 5 years of corn (or soybean). Treatments are staggered so that each phase is present each year. Data points represent the average for the given crop phase over 5 years (2010–2014) and 4 replicates (n = 20). Midseason is approximately 2 months after planting

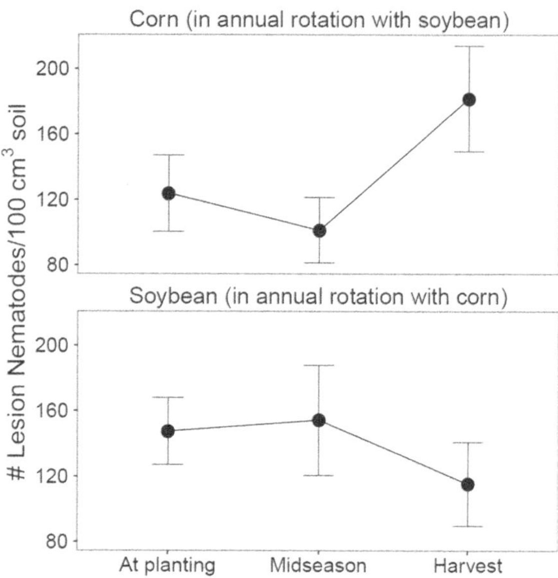

Fig. 6.7 Lesion nematode soil abundance through the growing season corn and soybean phases of corn-soybean annual rotation at a long-term crop rotation experiment site. Data points represent the average for the given crop phase over 5 years (2010–2014) and 4 replicates (n = 20). Midseason is approximately 2 months after planting

180 lesion nematodes/100 cm^3 soil (Fig. 6.7). In the soybean phase of corn-soybean annual rotation, population density decreased from midseason to harvest (Fig. 6.7), ending with an average population density of 120 lesion nematodes/100 cm^3 soil (Grabau and Chen 2016a, b).

6.3.3 Damage and Yield Loss

Lesion nematodes are considered the third most important groups of plant parasitic nematodes following the root knot and cyst nematodes (Jones et al. 2013). Lesion nematodes can cause damage to hosts on its own and through interaction with microbial pathogens. The migratory endoparasitic habits can cause severe damage to host plant roots. The nematodes can enter and exit the roots multiple times creating numerous root wounds for secondary microbial infection, and a single nematode can damage several roots. The damage to crops by lesion nematodes is further manifested by secondary infection of microbial pathogens.

Yield loss caused by lesion nematodes is widely recognized for a number of crops, but accurate estimation of yield loss to the nematodes is difficult (MacGuidwin and Bender 2016). There is limited research of the effect of lesion nematodes on crop yields in Minnesota. Grabau and Chen (2016a) demonstrated that corn yield was increased 3–11% by nematicide application, depending on crop sequence, in a long-term crop rotation study with lesion and spiral (*Helicotylenchus* sp.) nematode infestations. They also demonstrated that corn yield was negatively related with lesion nematode population density, and lesion nematode explained a greater proportion of variation in corn yield than spiral nematode population density. Lesion nematode is known to enhance fungal invasion into corn roots, increase root rot disease, and decrease corn productivity in some corn-producing regions (Egunjobi et al. 1986; Jordaan et al. 1987; da Silva et al. 2017). However, in a study from Minnesota, *Fusarium moniliforme* alone decreased corn root and shoot weights more than when the fungus was combined with either *P. scribneri* or *P. penetrans*, probably due to confounding factors (Palmer and MacDonald 1974).

Grabau and Chen (2016b) did not detect soybean yield loss by the lesion nematode in a long-term crop rotation study in a field that was concomitantly infested by the soybean cyst nematode. Lesion nematodes may be important in alfalfa (Thies et al. 1992) and strawberry in Minnesota.

Research from other states and regions may be used for reference in determining yield loss to lesion nematode in Minnesota crop production systems. In 1970–1980s, Norton and his team in Iowa demonstrated that corn yield was negatively related with plant-parasitic nematodes, mainly the lesion nematodes (Norton 1983; Norton et al. 1978). Nematicide trials in South Dakota also showed yield reduction of corn by *P. hexincisus* and *P. scribneri* (Smolik and Evenson 1987). In a recent study in Wisconsin, yield loss of corn to *P. penetrans* was estimated at 3.79% in a total of six field-years during 2008–2010 (MacGuidwin and Bender 2016).

Relatively little is known about the extent of damage and yield loss from lesion nematodes in other crops in Minnesota. However, based on research in other states and countries, lesion nematodes can cause extensive damage to wheat, potato, and soybean (Ross et al. 1967; Schmitt and Barker 1981; Bowers et al. 1996; Morgan et al. 2002; Smiley and Machado 2009), plants with which lesion nematodes are associated in Minnesota. Worldwide, lesion nematode infection is known to interact with *Verticillium dahliae* to cause early dying of potato (Rowe and Riedel 1984; Riedel et al. 1985; Rotenberg et al. 2004). Early dying of potato occurs in Minnesota (Johnson 1988), but the role of lesion nematodes in this disease in Minnesota is not well-documented. Similarly, interaction between lesion nematodes and fungal pathogens including *Rhizoctonia* spp. and *Fusarium* spp. among others, is known to increase wheat damage in some locations (Benedict and Mountain 1956; Taheri et al. 1994) but the extent to which this disease complex occurs in Minnesota wheat is unclear. In summary, several important lesion nematode species are present in Minnesota, and further studies are needed to determine their importance in Minnesota crop productions.

6.3.4 Management

There are limited studies of management of lesion nematodes in Minnesota cropping systems. Castillo and Vovlas (2007) did extensive review of management strategies for lesion nematodes. Presented here are brief discussions of some potential options for management of lesion nematodes in the crops in Minnesota.

6.3.4.1 Resistance

Use of host resistance is always a major strategy in management of plant parasitic nematodes if resistance exists in a host. Studies of host resistance to lesion nematodes throughout the world have increased rapidly in recent years. Resistance has been found in several crops that are grown in Minnesota including wheat (Sheedy and Thompson 2009; Smiley et al. 2014), barley (Galal et al. 2014), potato (Davis et al. 1992, 2004), corn (Smolik and Wicks 1987) and soybean (Acosta et al. 1979). In Minnesota, host resistance is not a common technique for managing lesion nematodes because commercial cultivars adapted to the Minnesota climate are not resistant to lesion nematodes or their level of resistance or tolerance is unknown. Resistance to *P. penetrans* has been found in the alfalfa breeding lines in the University of Minnesota (Thies et al. 1994), and the University of Idaho (Hafez et al. 2007). Two germplasms lines MNGRN2 (PI536526) and MNGRN4 (PI536527) were released in 1989 from Minnesota (Barnes et al. 1990).

6.3.4.2 Crop Rotation

Crop rotation is difficult for lesion nematode management because of their broad host ranges and concomitant infestation of multiple lesion nematode species in many fields. Nevertheless, successful use of crop rotation for management of lesion nematodes has been reported in other states using crops such as millet and marigold for *P. penetrans* (Ball-Coelho et al. 2003; Bélair et al. 2006; MacGuidwin et al. 2012) or alfalfa for *P. scribneri* (Todd 1991). Accurate identification of nematode species in a field is critical for effective use of crop rotation. Host status and susceptibility of crops to various important plant parasitic nematode species in fields need to be determined and the effectiveness of the crops in reducing the nematode population densities evaluated in the local environments. Corn and soybean are the only economically viable crops throughout much of Minnesota, which also restricts crop rotation efficacy. In a long-term crop rotation study, soybean crop reduced abundances of lesion nematodes (Fig. 6.7) and increased corn yield in the following year (Grabau and Chen 2016a), suggesting that corn-soybean crop rotation may reduce some populations of lesion nematodes.

6.3.4.3 Chemical Control

Chemical control with nematicides were widely used for management of plant parasitic nematodes, but has been reduced during the past two decades due to environmental concerns (Marban-Mendoza and Manzanilla-López 2012; Rich et al. 2004). In addition, chemical control is generally costly. Consequently, chemical control is mainly used in high value crops when host resistance is not available and other methods including cultural and biological methods, are ineffective. In Minnesota, most crops do not fit this description and broad-acreage chemical control for nematodes is used infrequently in the state. Nevertheless, chemical control is still a viable option for management of lesion nematodes in some cropping systems (Castillo and Vovlas 2007). In potato, fumigants such as metam sodium and 1,3-dichloropropene, are effectively used to manage lesion nematodes and early dying disease and increase marketable tuber yield (McKeown et al. 2001; Olthof 1989; Pasche et al. 2014). In recent years, a number of nematicidal seed treatments have become available for nematodes including lesion nematodes. By reducing the amount of nematicidal material applied, these seed treatments are economically viable for low-value crops and are common in corn, soybean, and other crops. The efficacy of the seed treatments needs to be confirmed in local environments before it can be relied on for consistent nematode management.

6.3.4.4 Soil Amendments

Soil amendments may affect soil fauna and flora and improve soil health. Some soil organic amendments and green manures have been successfully used to lower lesion nematode populations and increase crop yields in other states. Examples include mushroom compost or straw mulch to reduce *V. dahliae* and *P. penetrans* population densities and potato damage (LaMondia et al. 1999), acidified swine manure to control *Pratylenchus* spp. (Mahran et al. 2008), and biofumigation of soil with Brassicas for control of *P. penetrans* in potato production (Al-Rehiayani and Hafez 1998). In Minnesota field studies, liquid swine manure application did not significantly affect lesion nematode populations but increased corn yield (Bao et al. 2013; Grabau et al. 2017). More research is needed to determine and optimize efficacy of other organic amendments for managing lesion nematodes in Minnesota crop production systems.

6.3.4.5 Biological Control

Soil nematodes including lesion nematodes are subjected to attack by all kinds of organisms including fungi, bacteria, microscopic predators and plants through parasitism or antibiotic effect. There are a number of reported studies on the biological control of lesion nematodes (Castillo and Vovlas 2007). For examples, Samac and Kinkel (2001) found that some strains of *Streptomyces* isolated from scab-suppressive soil in Minnesota suppressed *P. penetrans* population density in alfalfa, probably by inducing host resistance. In a field trial on strawberry in Massachusetts, a strain of *Streptomyces costaricanus* effectively suppressed *P. penetrans* in strawberry (Dicklow et al. 1993). *Pasteuria*, an endospore-forming bacterium, has been commonly found on *Pratylenchus* (Chen and Dickson 1998). It has been recently found on *Pratylenchus* spp. in Minnesota potato (Oliveira et al. 2015) and corn fields, but its role in regulating *Pratylenchus* populations has not been fully studied. Endophytic bacteria such as *Microbacterium esteraromaticum* and *Kocuria varians*, isolated from the nematode antagonistic plant species African (*Tagetes erecta*) and

French marigold (*T. patula*), were effective in suppressing population density of *P. penetrans* in soil planted to potato (Sturz and Kimpinski 2004). The nematode-endoparastic fungi *Hirsutella rhossiliensis* reduced the number of *P. penetrans* entering roots by 25% (Timper and Brodie 1994). Some endophytic fungi such as *Trichoderma* can suppress lesion nematode population density (Miller and Anagnostakis 1977). These studies demonstrated that there are a wide range of microbial antagonists that may play roles in suppressing lesion nematode populations, and some of them may have potential as commercial biocontrol agents. There are a number of commercial biological control agents available for management of plant parasitic nematodes including lesion nematodes (Chen and Dickson 2012), but none of them has been widely accepted. Recently, some seed treatments containing biological agent(s) are available for control of nematodes in corn, mainly *Pratylenchus*. The efficacy of the seed treatments with biological agents needs to be fully evaluated before they can be widely accepted.

6.4 Other Nematodes

6.4.1 *Stem and Bulb Nematode,* Ditylenchus dipsaci

In Minnesota, *Ditylenchus dipsaci* has been detected on garlic (*Allium sativum*) in a few counties (Mollov et al. 2012). *Ditylenchus dipsaci* is an obligate plant parasite that can feed on both upper plant parts including leaves, stem, flower and seeds, and lower parts including roots, bulbs and tubers. The nematode prefers cool temperatures, but the optimum temperature for development varies depending on different populations and hosts. For example, the generation time on yellow pea was 24, 18 and 22 days at 17, 22 and 27 °C, respectively (Hajihassani et al. 2017). *Ditylenchus dipsaci* is highly pathogenic and can cause devastating damage to many crops such as sugar beet, alfalfa, red clover, oat, potato, corn, strawberry, garlic, onion (*Allium cepa*) and horticultural plants such as narcissus (*Narcissus*), hyacinth (*Hyacinthus*), and tulip (*Tulipa*) (Sturhan and Brzeski 1991). Its prevalence and economic importance in Minnesota warrants further studies.

6.4.2 *Dagger Nematodes,* Xiphinema *spp.*

Dagger nematodes are commonly found in crop fields in Minnesota. Two species, *Xiphinema americanum* and *X. chambersi* have been reported, but their identities may need to be confirmed. The dagger nematodes are large ectoparasites with a long life-cycle from several months to 2 years. They feed on root tips and induce root galls on some hosts. The nematodes are considered moderately pathogenic to host crops. However, there is limited study of their importance in Minnesota agriculture. In a long-term corn and soybean crop rotation study, *Xiphinema americanum* population densities were small in monoculture of either crop or annual rotation of the two crops (Fig. 6.8a), and did not cause detectable yield loss to corn and soybean (Grabau and Chen 2016a, b). Norton et al. (1978) reported a negative correlation between corn yield and dagger nematode population density. Some studies have suggested that *Xiphinema americanum* reproduction is lower on corn than soybean in the field with populations reduced when corn is continuously

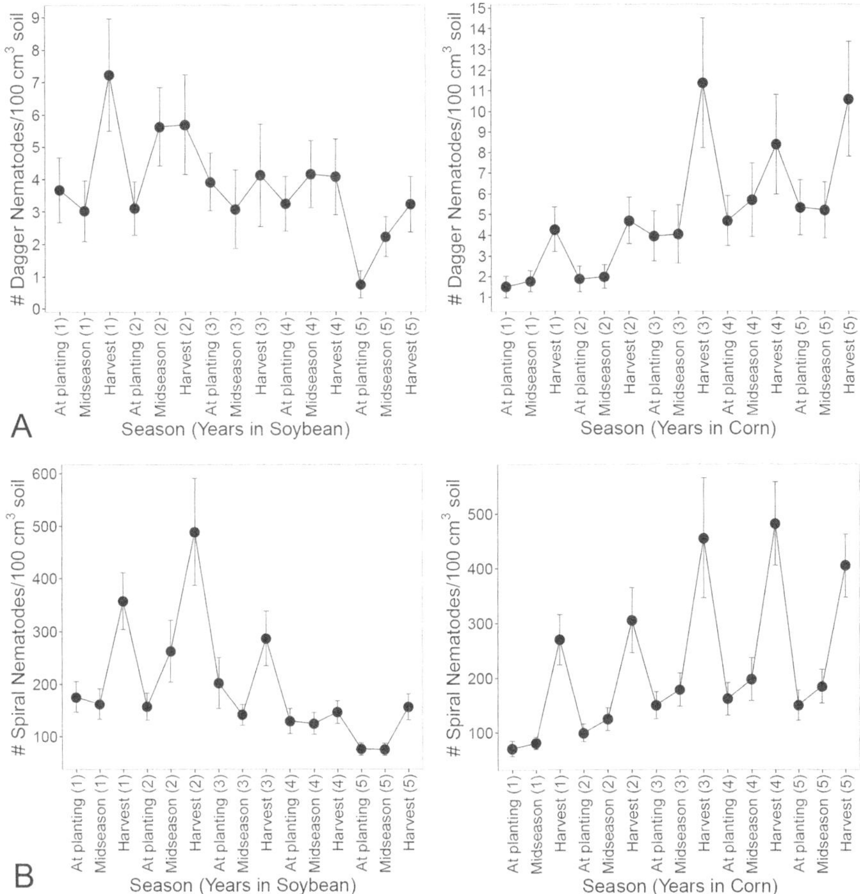

Fig. 6.8 (**a**) Dagger nematode soil abundance through seasons in different crop phases in a long-term corn-soybean crop rotation experiment where 5 years of soybean (or corn) are followed by 5 years of corn (or soybean). Treatments are staggered so that each phase is present each year. Data points represent the average for the given crop phase over 5 years (2011–2015) and 4 replicates (n = 20). Midseason is approximately 2 months after planting; (**b**) Spiral nematode soil abundance through seasons in different crop phases in a long-term corn-soybean crop rotation experiment where 5 years of soybean (or corn) are followed by 5 years of corn (or soybean). Treatments are staggered so that each phase is present each year. Data points represent the average for the given crop phase over 5 years (2010–2014) and 4 replicates (n = 20). Midseason is approximately 2 months after planting

cropped (Evans et al. 2007; Ferris and Bernard 1971) although one study in Southern Minnesota found that *X. americanum* populations reached very high levels when corn was planted for 10 years (MacDonald 1979). Outside of Minnesota, dagger nematodes can cause yield loss to soybean when population density is high, especially in sandy soil (Schmitt and Noel 1984), but damage by *Xiphinema* to most crops in Minnesota is unknown. *Xiphinema* is an effective virus vector, and its role of virus vector is probably more important than its feeding. For example, soybean severe stunt (SSS), caused by the *Soybean severe stunt virus* (SSSV), is a soil-borne virus disease and vectored by *X. americanum*. If both pathogens exist in a soybean field, it may cause severe damage to soybean (Weldekidan et al. 1992; Evans et al. 2007), but this disease has not yet been reported in Minnesota. *Xiphinema* may be more important on some perennial crops and fruit trees due to its long life-cycle that can be completed on the perennial host plants.

6.4.3 Stubby Root Nematodes, **Trichodorus** *spp. and* **Paratrichodorus** *spp.*

Stubby-root nematodes from the genera *Trichodorus* and *Paratrichodorus* were reported in Minnesota crop fields (Table 6.1). Yan et al. (2016) reported *Paratrichodorus allius* in a Minnesota sugar beet field. Stubby root nematodes can feed on entire root systems, but generally feed on root tips and root hairs. The growth of root tips is stopped and occasionally slightly swollen by the nematode feeding at the zone of elongation and apical meristem. Lateral roots can be produced at feeding sites, but are immediately attacked by the nematodes which results in stunted, branched, stubby roots (Decraemer 1991). Stubby root nematodes can be problematic in some crops in Minnesota, especially in sandy soils (Taylor et al. 1958; Taylor and Schleder 1959). Their role as virus vectors is important as virus-induced damage is associated with stubby root nematodes in a number of crops. The nematodes can transmit *tobacco rattle virus* (TRV), *pea early browning virus* (PEBV) and *pepper ringspot virus* (PEPRSV). *Tobacco rattle virus* has been found on potato in Minnesota (Gudmestad et al. 2008). This virus is effectively vectored by *P. allius* or other stubby root nematodes, and causes potato corky ringspot disease. Stubby root nematodes can be an important pest in potato production in the state.

6.4.4 Foliar Nematodes, **Aphelenchoides** *spp.*

Foliar nematodes have been reported in Minnesota. Foliar nematodes infect the emerging shoots of germinating plants and feed ectoparasitically on leaf primordia. Several species of *Aphelenchoides* have been recognized as important agriculture pests including *A. ritzemabosi*, *A. fragariae* and *A. besseyi*. The former two species prefer cool temperature and are widely spread in the temperate regions. Although the species of *Aphelenchoides* in Minnesota have not been identified, damage of horticultural plants such as hosta (*Hosta* sp.), zinnia (*Zinnia* sp.) and peony (*Paeonia lactiflora*) has been observed. The importance of *Aphelenchoides* on field crops such as strawberry and alfalfa warrants further investigation.

6.4.5 Root Knot Nematodes, Meloidogyne *spp.*

Root knot nematodes, as a group, are the most important plant parasitic nematodes worldwide (Jones et al. 2013), but they are not frequently found in Minnesota. Crow and MacDonald (1978) investigated plant parasitic nematodes in strawberry plantations and found high population densities of *Meloidogyne hapla* associated with strawberry crop, and poorly growing plants were observed in an area infested with a high density of 1154 *M. hapla* J2/100 cm^3 soil, thereby suggesting the importance of the nematode in strawberry production in Minnesota. Damage of carrots by *M. hapla* was also observed in some fields in Minnesota.

6.4.6 Spiral Nematodes, Helicotylenchus *and* Rotylenchus *spp.*

Spiral nematodes of the genera *Helicotylenchus* and *Rotylenchus* have been reported in Minnesota (Table 6.1). In a recent survey, *Helicotylenchus* was the most abundant and prevalent vermiform nematode genus present in organically farmed crop fields (Chen et al. 2012), and *H. pseudorobustus* is probably the most predominant species in corn fields in Minnesota (MacDonald et al. 1978). Most spiral nematodes have low pathogenicity to host plants, but high nematode abundance can result in significant yield loss. Infection by *Helicotylenchus dihystera* significantly suppressed plant growth of millet (Villenave and Duponnois 1998). In some fields in Minnesota, the nematode population densities can be high (>1000 nematodes/100 cm^3 soil), which probably cause some yield loss. In a long-term crop rotation study (Grabau and Chen 2016a, b), *H. pseudorobustus* population densities increased during the growing season, particularly from midseason to harvest, through all 5 years in corn monoculture. In that study, spiral nematode abundances peaked at about 500 spiral nematodes/100 cm^3 soil (Fig. 6.8b). In the same study, spiral nematode abundances increased during the growing season in initial years in soybean monoculture, but declined by years 4–5 in monoculture to 150 nematodes/100 cm^3 soil (Fig. 6.8b). Nematicide application also increased corn yield 3–11% depending on crop sequence. Spiral nematode abundance was negatively correlated with corn yield, but lesion nematode abundance explained a greater proportion of variation in corn yield.

6.4.7 *Stunt Nematodes,* Tylenchorhynchus *spp.,* Merlinius *spp. and* Trophurus *spp.*

Stunt nematodes are among the most common nematodes in Minnesota (Table 6.1). In a survey of nematodes in organically farmed fields, stunt nematodes were the second most frequently encountered group of nematodes across the state, and the most frequently found nematodes in Northwestern Minnesota (Chen et al. 2012). Most stunt nematode species are not considered important in agriculture, but high population densities of some species have been reported to cause damage to crops. For example, *Tylenchorhynchus claytoni,* which has been found in Minnesota, caused significant yield loss of soybean in a microplot experiment (Ross et al. 1967). *Trophurus minnesotensis* was reported to associate with sugar beet in Minnesota, but its economic importance has not been determined (Caveness 1958).

6.5 Nematode Management as Part of Sustainable Agriculture in Minnesota

Increasing emphasis has been placed on sustainable agriculture, which is defined as production that sustains crop productivity, economic viability of farms, and natural resources over the long term. Based on this definition, nematode management practices in Minnesota are already very sustainable. SCN is the most economically important nematode in Minnesota and is perhaps the only target of active management practices. Resistant soybean cultivars and corn-soybean crop rotation are the primary practices used to manage SCN and both expend few natural resources while improving crop productivity in an economical manner. However, there are also ways in which these practices are not sustainable. The SCN resistance in most commercial cultivars is derived from a common genetic source and resistance-breaking SCN is developing in Minnesota making resistant cultivars ineffective against some SCN populations. For resistant cultivars to continue to be effective for managing SCN in the future, new sources of resistance will need to be incorporated into high-performing commercial soybean cultivars.

Similarly, crop rotation to manage SCN is based primarily on a single non-host crop, corn. A short rotation such as corn-soybean is relatively resource-intensive and is only economically sustainable while corn and soybean prices are high. However, nematode management is not a limiting factor in Minnesota crop rotations as many non-hosts of SCN are known and could be used effectively. The sustainability of agriculture in Minnesota could be improved further by incorporating alternative, environmentally friendly nematode management practices into crop production. For example, manure application can help reduce SCN damage and increase soil organic matter while cycling animal waste. Growing cover crops that help reduce nematode populations could also reduce soil erosion. In summary, Minnesota nematode management practices are relatively sustainable, but practices could be adopted to improve their sustainability.

References

Acosta, N., Malek, R. B., & Edwards, D. I. (1979). Susceptibility of soybean cultivars to *Pratylenchus scribneri*. *Journal of Agriculture of the University of Puerto Rico, 63*, 103–110.

Al-Rehiayani, S., & Hafez, S. (1998). Host status and green manure effect of selected crops on *Meloidogyne chitwoodi* race 2 and *Pratylenchus neglectus*. *Nematropica, 28*, 213–230.

Ball-Coelho, B., Bruin, A. J., Roy, R. C., & Riga, E. (2003). Forage pearl millet and marigold as rotation crops for biological control of root-lesion nematodes in potato. *Agronomy Journal, 95*, 282–292.

Bao, Y., Neher, D. A., & Chen, S. Y. (2011). Effect of biocides and soil disturbance on nematode community and extracellular enzyme activity in soybean cyst nematode suppressive soil. *Nematology, 13*, 687–699.

Bao, Y., Chen, S. Y., Vetsch, J., & Randall, G. (2013). Soybean yield and *Heterodera glycines* responses to liquid swine manure in nematode suppressive soil and conducive soil. *Journal of Nematology, 45*, 21–29.

Bao, Y., Vuong, T., Meinhardt, C., Tiffin, P., Denny, R., Chen, S., Nguyen, H. T., Orf, J. H., & Young, N. D. (2014). Potential of association mapping and genomic selection to explore PI88788 derived soybean cyst nematode resistance. *The Plant Genome, 7*. https://doi.org/10.3835/lantgenome2013.11.0039.

Barnes, D. K., Thies, J. A., Rabas, D. L., Nelson, D. L., & Smith, D. M. (1990). Registration of two alfalfa germplasms with field resistance to the root-lesion nematode. *Crop Science, 30*, 751–752.

Bélair, G., Dauphinais, N., Fournier, Y., Dangi, O. P., & Ciotola, M. (2006). Effect of 3-year rotation sequences and pearl millet on population densities of *Pratylenchus penetrans* and subsequent potato yield. *Canadian Journal of Pant Pathology, 28*, 230–235.

Benedict, W. G., & Mountain, W. B. (1956). Studies on the etiology of a root rot of winter wheat in southwestern Ontario. *Canada Journal Botany, 34*, 159–174.

Bond, J., & Wrather, J. A. (2004). Interactions with other plant pathogens and pests. In D. P. Schmitt, J. A. Wrather, & R. D. Riggs (Eds.), *Biology and management of the soybean cyst nematode* (pp. 111–129). Marceline: Schmitt and Associates of Marceline.

Bowers, J. H., Nameth, S. T., Riedel, R. M., & Rowe, R. C. (1996). Infection and colonization of potato roots by *Verticillium dahliae* as affected by *Pratylenchus penetrans* and *P. crenatus*. *Phytopathology, 86*, 614–621.

Castillo, P., & Vovlas, N. (2007). *Pratylenchus (Nematoda: Pratylenchidae): Diagnosis, biology, pathogenecity and management*. Leidin: Brill.

Caveness, F. E. (1958). *Clavaurotylenchus minnesotensis*, n. gen., n. sp. (Tylenchinae: Nematoda) from Minnesota. *Proceedings of the Helminthological Society of Washington, 25*, 122–124.

Chen, S. Y. (2004). Management with biological control. In D. P. Schmitt, J. A. Wrather, & R. D. Riggs (Eds.), *Biology and management of the soybean cyst nematode* (pp. 207–242). Marceline: Schmitt and Associates of Marceline.

Chen, S. Y. (2007a). Suppression of *Heterodera glycines* in soils from fields with long-term soybean monoculture. *Biocontrol Science and Technology, 17*, 125–134.

Chen, S. Y. (2007b). Tillage and crop sequence effects on *Heterodera glycines* and soybean yields. *Agronomy Journal, 99*, 797–807.

Chen, S. (2011). *Soybean cyst nematode management guide*. University of Minnesota Extension. Web/URL: http://www1.extension.umn.edu/agriculture/soybean/soybean-cyst-nematode/. Access date 25 Nov 2017.

Chen, Z. X., & Dickson, D. W. (1998). Review of *Pasteuria penetrans*: Biology, ecology, and biological control potential. *Journal of Nematology, 30*, 313–340.

Chen, S., & Dickson, D. W. (2012). Biological control of plant-parasitic nematodes. In R. H. Manzanilla-López & N. Marbán-Mendoza (Eds.), *Practical plant nematology* (pp. 761–811). Guadalajara: Colegio de Postgraduados and Mundi-Prensa, Biblioteca Básica de Agricultura.

Chen, S. Y., & Liu, X. Z. (2005). Control of the soybean cyst nematode by the fungi *Hirsutella rhossiliensis* and *Hirsutella minnesotensis* in greenhouse studies. *Biological Control, 32*, 208–219.

Chen, S., & Miller, D. (2002). Effects of soil pH on *Heterodera glycines* reproduction and soybean chlorosis. *Nematology, 4*, 251.

Chen, S. Y., & Reese, C. D. (1999). Parasitism of the nematode *Heterodera glycines* by the fungus *Hirsutella rhossiliensis* as influenced by crop sequence. *Journal of Nematology, 31*, 437–444.

Chen, S. Y., Porter, P. M., Orf, J. H., Reese, C. D., Stienstra, W. C., Young, N. D., Walgenbach, D. D., Schaus, P. J., Arlt, T. J., & Breitenbach, F. R. (2001a). Soybean cyst nematode population development and associated soybean yields of resistant and susceptible cultivars in Minnesota. *Plant Disease, 85*, 760–766.

Chen, S. Y., Porter, P. M., Reese, C. D., & Stienstra, W. C. (2001b). Crop sequence effects on soybean cyst nematode and soybean and corn yields. *Crop Science, 41*, 1843–1849.

Chen, S. Y., Wyse, D. L., Johnson, G. A., Porter, P. M., Stetina, S. R., Miller, D. R., Betts, K. J., Klossner, L. D., & Haar, M. J. (2006). Effect of cover crops alfalfa, red clover, and perennial ryegrass on soybean cyst nematode population and soybean and corn yields in Minnesota. *Crop Science, 46*, 1890–1897.

Chen, S. Y., Sheaffer, C. C., Wyse, D. L., Nickel, P., & Kandel, H. (2012). Plant-parasitic nematode communities and their associations with soil factors in organically farmed fields in Minnesota. *Journal of Nematology, 44*, 361–379.

Concibido, V. C., Diers, B. W., & Arelli, P. R. (2004). A decade of QTL mapping for cyst nematode resistance in soybean. *Crop Science, 44*, 1121–1131.

Cook, D. E., Lee, T. G., Guo, X. L., Melito, S., Wang, K., Bayless, A. M., Wang, J. P., Hughes, T. J., Willis, D. K., Clemente, T. E., Diers, B. W., Jiang, J. M., Hudson, M. E., & Bent, A. F. (2012). Copy number variation of multiple genes at *rhg1* mediates nematode resistance in soybean. *Science, 338*, 1206–1209.

Cook, D. E., Bayless, A. M., Wang, K., Guo, X. L., Song, Q. J., Jiang, J. M., & Bent, A. F. (2014). Distinct copy number, coding sequence, and locus methylation patterns underlie *rhg1*-mediated soybean resistance to soybean cyst nematode. *Plant Physiology, 165*, 630–647.

Crow, R. V., & MacDonald, D. H. (1978). Phytoparasitic nematodes adjacent to established strawberry plantations. *Journal of Nematology, 10*, 204–207.

da Silva, M. P., Tylka, G. L., & Munkvold, G. P. (2017). Seed treatment effects on maize seedlings coinfected with *Rhizoctonia solani* and *Pratylenchus penetrans*. *Plant Disease, 101*, 957–963.

Davis, J. R., Hafez, S. L., & Sorensen, L. H. (1992). Lesion nematode suppression with the Butte potato and relationships to *Verticillium wilt*. *American Potato Journal, 69*, 371–383.

Davis, J. R., Pavek, J. J., & Corsini, D. L. (2004). Potato genotypes, a tool for managing soilborne pathogens – A summary. *Acta Horticulture, 635*, 93–100.

Decraemer, W. (1991). Stubby root and virus vector nematodes: *Trichodorus*, *Paratrichodorus*, *Allotrichodorus*, and *Monotrichodorus*. In W. R. Nickle (Ed.), *Manual of agricultural nematology* (pp. 587–626). New York: Marcel Dekker.

Dicklow, M. B., Acosta, N., & Zuckerman, B. M. (1993). A novel *Streptomyces* species for controlling plant-parasitic nematodes. *Journal of Chemical Ecology, 19*, 159–173.

Donald, P. A., Tyler, D. D., & Boykin, D. L. (2009). Short- and long-term tillage effects on *Heterodera glycines* reproduction in soybean monoculture in west Tennessee. *Soil tillage research, 104*, 126–133.

Duncan, L. W., & Noling, J. W. (1998). Agricultural sustainability and nematode integrated pest management. In K. R. Barker, G. A. Pederson, & G. L. Windham (Eds.), *Plant and nematode interactions* (pp. 251–287). Madison: American Society of Agronomy.

Egunjobi, O. A., Norton, D. C., & Martinson, C. (1986). Interaction of *Pratylenchus scribneri* and *Helminthosporium pedicellatum* in the etiology of corn root rot. *Phytoparasitica, 14*, 287–295.

Evans, T. A., Miller, L. C., Vasilas, B. L., Taylor, R. W., & Mulrooney, R. P. (2007). Management of *Xiphinema americanum* and soybean severe stunt in soybean using crop rotation. *Plant Disease, 91*, 216–219.

Ferris, V. R., & Bernard, R. L. (1971). Crop rotation effects on population densities of ectoparasitic nematodes. *Journal of Nematology, 3*, 119–122.

Galal, A., Sharma, S., Abou-Elwafa, S. F., Sharma, S., Kopisch-Obuch, F., Laubach, E., Perovic, D., Ordon, F., & Jung, C. (2014). Comparative QTL analysis of root lesion nematode resistance in barley. *Theoretical and Applied Genetics, 127*, 1399–1407.

Grabau, Z. J., & Chen, S. Y. (2014). Efficacy of organic soil amendments for management of *Heterodera glycines* in greenhouse experiments. *Journal of Nematology, 46*, 267–274.

Grabau, Z. J., & Chen, S. Y. (2016a). Determining the role of plant-parasitic nematodes in the corn-soybean crop rotation yield effect using nematicide application: I. corn. *Agronomy Journal, 108*, 782–793.

Grabau, Z. J., & Chen, S. Y. (2016b). Determining the role of plant-parasitic nematodes in the corn-soybean crop rotation yield effect using nematicide application: II. soybean. *Agronomy Journal, 108*, 1168–1179.

Grabau, Z., Vetsch, J., & Chen, S. (2017). Effects of fertilizer, nematicide, and tillage on plant-parasitic nematodes and yield in corn and soybean. *Agronomy Journal, 109*, 1651–1662. https://doi.org/10.2134/agronj2016.09.0548.

Gudmestad, N. C., Mallik, I., Pasche, J. S., & Crosslin, J. M. (2008). First report of *tobacco rattle virus* causing corky ringspot in potatoes grown in Minnesota and Wisconsin. *Plant Disease, 92*, 1254.

Guo, B., Sleper, D. A., Lu, P., Shannon, J. G., Nguyen, H. T., & Arelli, P. R. (2006). QTLs associated with resistance to soybean cyst nematode in soybean: Meta-analysis of QTL locations. *Crop Science, 46*, 595–602.

Hafez, S. L., Palanisamy, S., & Miller, D. R. (2007). Reaction of twenty-five alfalfa breeding lines to the lesion nematode *Pratylenchus penetrans*. *Journal of Nematology, 39*, 75.

Hajihassani, A., Tenuta, M., & Gulden, R. H. (2017). Influence of temperature on development and reproduction of *Ditylenchus weischeri* and *D. dipsaci* on yellow pea. *Plant Disease, 101*, 297–305.

Hu, W. M., Chen, S. Y., & Liu, X. Z. (2016). Effect of temperature treatment on survival of *Heterodera glycines* and its associated fungi and bacteria. *Nematology, 18*, 845–855.

Hu, W., Samac, D. A., Liu, X., & Chen, S. (2017). Microbial communities in the cysts of soybean cyst nematode affected by tillage and biocide in a suppressive soil. *Applied Soil Ecology, 119*, 396–406 https://doi.org/10.1016/j.apsoil.2017.07.018.

Johnson, K. B. (1988). Modeling the influences of plant infection-rate and temperature on potato foliage and yield losses caused by *Verticillium dahliae*. *Phytopathology, 78*, 1198–1205.

Jones, J. T., Haegeman, A., Danchin, E. G. J., Gaur, H. S., Helder, J., Jones, M. G. K., Kikuchi, T., Manzanilla-Lopez, R., Palomares-Rius, J. E., Wesemael, W. M. L., & Perry, R. N. (2013). Top 10 plant-parasitic nematodes in molecular plant pathology. *Molecular Plant Pathology, 14*, 946–961.

Jordaan, E. M., Loots, G. C., Jooste, W. J., & De-Waele, D. (1987). Effects of root-lesion nematodes (*Pratylenchus brachyurus* Godfrey and *P. zeae* Graham) and *Fusarium moniliforme* Sheldon alone or in combination, on maize. *Nematologica, 33*, 213–219.

Kim, D. G., Lee, Y. K., & Lee, J. K. (1998). Spatial distribution of *Pasteuria nishizawae* attacking *Heterodera glycines*. *Korean Journal Plant Pathology, 14*, 710–713.

Kloepper, J. W., Rodriguez-Kabana, R., McInroy, J. A., & Young, R. W. (1992). Rhizosphere bacteria antagonistic to soybean cyst (*Heterodera glycines*) and root-knot (*Meloidogyne incognita*) nematodes: Identification by fatty acid analysis and frequency of biological control activity. *Plant and Soil, 139*, 75–84.

Koenning, S. R., Schmitt, D. P., Barker, K. R., & Gumpertz, M. L. (1995). Impact of crop rotation and tillage system on *Heterodera glycines* population density and soybean yield. *Plant Disease, 79*, 282–286.

Kopisch-Obuch, F. J., McBroom, R. L., & Diers, B. W. (2005). Association between soybean cyst nematode resistance loci and yield in soybean. *Crop Science, 45*, 956–965.

LaMondia, J. A., Gent, M. P. N., Ferrandino, F. J., Elmer, W. H., & Stoner, K. A. (1999). Effect of compost amendment or straw mulch on potato early dying disease. *Plant Disease, 83*, 361–366.

Lee, T. G., Kumar, I., Diers, B. W., & Hudson, M. E. (2015). Evolution and selection of *rhg1*, a copy-number variant nematode-resistance locus. *Molecular Ecology, 24*, 1774–1791.
Lehman, P. S., Huisingh, D., & Barker, K. R. (1971). The influence of races of *Heterodera glycines* on nodulation and nitrogen-fixing capacity of soybean. *Phytopathology, 61*, 1239–1244.
Lian, L., Wang, F., Denny, R., Schaus, P., Young, N., Orf, J., & Chen, S. (2014). Soybean cyst nematode resistance in PI 567516C soybean: A promising new source of resistance. *Journal of Nematology, 46*, 193(Abstr).
Liu, X. Z., & Chen, S. Y. (2000). Parasitism of *Heterodera glycines* by *Hirsutella* spp. in Minnesota soybean fields. *Biological Control, 19*, 161–166.
Liu, S. F., & Chen, S. Y. (2005). Efficacy of the fungi *Hirsutella minnesotensis* and *H. rhossiliensis* from liquid culture for control of the soybean cyst nematode *Heterodera glycines*. *Nematology, 7*, 149–157.
Liu, S. M., Kandoth, P. K., Lakhssassi, N., Kang, J. W., Colantonio, V., Heinz, R., Yeckel, G., Zhou, Z., Bekal, S., Dapprich, J., Rotter, B., Cianzio, S., Mitchum, M. G., & Meksem, K. (2017). The soybean GmSNAP18 gene underlies two types of resistance to soybean cyst nematode. *Nature Communications, 8*. https://doi.org/10.1038/ncomms14822.
Ma, R., & Liu, X. Z. (2000, August, 26–28) Detection of *Hirsutella* spp. and *Pasteuria* sp. from second-stage juveniles of *Heterodera glycines* in China. In G. H. Zhou & H. F. Li (Eds.), Proceedings: The first Asian conference on plant pathology (p. 205). Beijing: China Agricultural Science Press.
MacDonald, D. H. (1979). Plant-parasitic nematodes associated with field crops grown in monoculture in Minnesota. *Journal of Nematology, 11*, 306.
MacDonald, D. H., Pierce, A. R., & Mansager, P. A. (1978). Plant parasitic nematodes in Minnesota corn fields. *Minnesota Agricultural Extension Service Special Report, 12*, 81–87.
MacDonald, D. H., Noel, G. R., & Lueschen, W. E. (1980). Soybean cyst nematode, *Heterodera glycines*, in Minnesota. *Plant Disease, 64*, 319–321.
MacGuidwin, A. E., & Bender, B. E. (2016). Development of a damage function model for *Pratylenchus penetrans* on Corn. *Plant Disease, 100*, 764–769.
MacGuidwin, A. E., Knuteson, D. L., Connell, T., Bland, W. L., & Bartelt, K. D. (2012). Manipulating inoculum densities of *Verticillium dahliae* and *Pratylenchus penetrans* with green manure amendments and solarization influence potato yield. *Phytopathology, 102*, 519–527.
Mahran, A., Conn, K. L., Tenuta, M., Lazarovits, G., & Daayf, F. (2008). Effectiveness of liquid hog manure and acidification to kill *Pratylenchus* spp. in soil. *Journal of Nematology, 40*, 266–275.
Marban-Mendoza, N., & Manzanilla-López, R. H. (2012). Chemical and non-chemical tactics to control plant-parasitic nematodes. In R. H. Manzanilla-López & N. Marbán-Mendoza (Eds.), *Practical plant nematology* (pp. 729–759). Guadalajara: Colegio de Postgraduados and Mundi-Prensa, Biblioteca Básica de Agricultura.
McKeown, A. W., Potter, J. W., & Straiton, E. (2001). Yield of 'Superior' potatoes (*Solanum tuberosum*) and dynamics of root-lesion nematode (*Pratylenchus penetrans*) populations following "nematode suppressive" cover crops and fumigation. *Phytoprotection, 82*, 13–23.
Miller, P. M., & Anagnostakis, S. (1977). Suppression of *Pratylenchus penetrans* and *Tylenchorhynchus dubius* by *Trichoderma viride*. *Journal of Nematology, 9*, 182–183.
Mitchum, M. G. (2016). Soybean resistance to the soybean cyst nematode *Heterodera glycines*: An update. *Phytopathology, 106*, 1444–1450.
Mollov, D. S., Subbotin, S. A., & Rosen, C. (2012). First report of *Ditylenchus dipsaci* on garlic in Minnesota. *Plant Disease, 96*, 1707.
Morgan, G. D., MacGuidwin, A. E., Zhu, J., & Binning, L. K. (2002). Population dynamics and distribution of root lesion nematode (*Pratylenchus penetrans*) over a three-year potato crop rotation. *Agronomy Journal, 94*, 1146–1155.
Mudge, J., Concibido, V., Denny, R., Young, N., & Orf, J. (1996). Genetic mapping of a yield depression locus near a major gene for soybean cyst nematode resistance. *Soybean Genetics Newsletter, 23*, 175–178.

Niblack, T. L., Arelli, P. R., Noel, G. R., Opperman, C. H., Orf, J. H., Schmitt, D. P., Shannon, J. G., & Tylka, G. L. (2002). A revised classification scheme for genetically diverse populations of *Heterodera glycines*. *Journal of Nematology, 34*, 279–288.

Nishizawa, T. (1986). On a strain of *Pasteuria penetrans* parasitic to cyst nematodes. *Revue de Nematologie, 9*, 303–304.

Noel, G. R., & Edwards, D. I. (1996). Population development of *Heterodera glycines* and soybean yield in soybean-maize rotations following introduction into a noninfested field. *Journal of Nematology, 28*, 335–342.

Noel, G. R., Bauer, S., & Atibalentja, N. (2007). Soybean cyst nematode populations suppressed by *Pasteuria nishizawae*. *Journal of Nematology, 39*, 82.

Norton, D. C. (1983). Maize nematode problems. *Plant Disease, 67*, 253–256.

Norton, D. C., Tollefson, J., Hinz, P., & Thomas, S. H. (1978). Corn yield increases relative to nonfumigant chemical control of nematodes. *Journal of Nematology, 10*, 160–166.

Oliveira, C. J., Grabau, Z., Samac, D. A., & Chen, S. Y. (2015). *Pasteuria* sp. endospore attachment to *Pratylenchus* species. *Journal of Nematology, 47*, 261.

Olthof, T. H. A. (1989). Effects of fumigant and nonfumigant nematicides on *Pratylenchus penetrans* and yield of potato. *Journal of Nematology, 21*, 645–649.

Palmer, L. T., & MacDonald, D. H. (1974). Interaction of *Fusarium* spp. and certain plant parasitic nematodes on maize. *Phytopathology, 64*, 14–17.

Pasche, J. S., Taylor, R. J., David, N. L., & Gudmestad, N. C. (2014). Effect of soil temperature, injection depth, and metam modium mate on the management of *Verticillium wilt* of potato. *American Journal of Potato Research, 91*, 277–290.

Perez-Hernandez, O., & Giesler, L. J. (2014). Quantitative relationship of soil texture with the observed population density reduction of *Heterodera glycines* after annual corn rotation in Nebraska. *Journal of Nematology, 46*, 90–100.

Poromarto, S. H., & Nelson, B. D. (2009). Reproduction of soybean cyst nematode on dry bean cultivars Adapted to North Dakota and Northern Minnesota. *Plant Disease, 93*, 507–511.

Poromarto, S. H., Gramig, G. G., Nelson, B. D., & Jian, S. (2015). Evaluation of weed species from the Northern Great Plains as hosts of soybean cyst neamtode. *Plant Health Progress, 16*, 23–28. https://doi.org/10.1094/PHP-RS-14-0024.

Rich, J. R., Dunn, R. A., & Noling, J. W. (2004). Nematicides: Past and present uses. In Z. X. Chen, S. Y. Chen, & D. W. Dickson (Eds.), *Nematology, advances and perspectives* (pp. 1179–1200). Beijing: Tsinghua University Press and CAB International.

Riedel, R. M., Rowe, R. C., & Martin, M. J. (1985). Differential interactions of *Pratylenchus crenatus*, *P. penetrans*, and *P. scribneri* with *Verticillium dahliae* in potato early dying disease. *Phytopathology, 75*, 419–422.

Riggs, R. D., & Hamblen, M. L. (1962). Soybean-cyst nematode host studies in the family Leguminosae. *Arkansas Agricultural Experiment Station Report Series, 110*, 1–17.

Rogovska, N. P., Blackmer, A. M., & Tylka, G. L. (2009). Soybean yield and soybean cyst nematode densities related to soil pH, soil carbonate concentrations, and alkalinity stress Index. *Agronomy Journal, 101*, 1019–1026.

Ross, J. P., Nusbaum, C. J., & Hirschmann, H. (1967). Soybean yield reduction by lesion, stunt, and spiral nematodes. *Phytopathology, 57*, 463–464 Abstr.

Rotenberg, D., MacGuidwin, A. E., Saeed, I. A. M., & Rouse, D. I. (2004). Interaction of spatially separated *Pratylenchus penetrans* and *Verticillium dahliae* on potato measured by impaired photosynthesis. *Plant Pathology, 53*, 294–302.

Rowe, R. C., & Riedel, R. M. (1984). Synergistic interactions between *Verticillium dahliae* and *Pratylenchus* species in potato early dying disease. *Phytopathology* 1984, *74*, 845.

Samac, D. A., & Kinkel, L. L. (2001). Suppression of the root-lesion nematode (*Pratylenchus penetrans*) in alfalfa (*Medicago sativa*) by *Streptomyces* spp. *Plant and Soil, 235*, 35–44.

Schmitt, D. P., & Barker, K. R. (1981). Damage and reproductive potentials of *Pratylenchus brachyurus* and *Pratylenchus penetrans* on soybean. *Journal of Nematology, 13*, 327–332.

Schmitt, D. P., & Noel, G. R. (1984). Nematode parasites of soybean. In W. R. Nickle (Ed.), *Plant and insect nematodes* (pp. 13–59). New York: Marcel Dekker.

Shannon, J. G., Arelli, P. R., & Young, L. D. (2004). Breeding for resistance and tolerance. In D. P. Schmitt, J. A. Wrather, & R. D. Riggs (Eds.), *Biology and management of the soybean cyst nematode* (pp. 155–180). Marceline: Schmitt and Associates of Marceline.

Sheedy, J. G., & Thompson, J. P. (2009). Resistance to the root-lesion nematode *Pratylenchus thornei* of Iranian landrace wheat. *Australasian Plant Pathology, 38*, 478–489.

Smiley, R. W., & Machado, S. (2009). *Pratylenchus neglectus* reduces yield of winter wheat in dryland cropping systems. *Plant Disease, 93*, 263–271.

Smiley, R. W., Gourlie, J. A., Yan, G. P., & Rhinhart, K. E. L. (2014). Resistance and tolerance of landrace wheat in fields infested with *Pratylenchus neglectus* and *P. thornei*. *Plant Disease, 98*, 797–805.

Smolik, J. D., & Evenson, P. D. (1987). Relationship of yields and *Pratylenchus* spp. population densities in dryland and irrigated corn. *Annals Applied Nematology, 1*, 71–73.

Smolik, J. D., & Wicks, Z. W. I. I. (1987). Reproduction of *Pratylenchus hexincisus* and *P. scribneri* in corn inbreds. *Annals Applied Nematology, 1*, 29–31.

Sturhan, D., & Brzeski, M. W. (1991). Stem and bulb nematodes, *Ditylenchus* spp. In W. R. Nickle (Ed.), *Manual of agricultural nematology* (pp. 423–464). New York: Marcel Dekker.

Sturz, A. V., & Kimpinski, J. (2004). Endoroot bacteria derived from marigolds (*Tagetes* spp.) can decrease soil population densities of root-lesion nematodes in the potato root zone. *Plant and Soil, 262*, 241–249.

Subbotin, S. A., Ragsdale, E. J., Mullens, T., Roberts, P. A., Mundo-Ocampo, M., & Baldwin, J. G. (2008). A phylogenetic framework for root lesion nematodes of the genus *Pratylenchus* (Nematoda): Evidence from 18S and D2-D3 expansion segments of 28S ribosomal RNA genes and morphological characters. *Molecular Phylogenetics and Evolution, 48*, 491–505.

Tabor, G. M., Tylka, G. L., Behm, J. E., & Bronson, C. R. (2003). *Heterodera glycines* infection increases incidence and severity of brown stem rot in both resistant and susceptible soybean. *Plant Disease, 87*, 655–661.

Taheri, A., Hollamby, G. J., Vanstone, V. A., & Neate, S. M. (1994). Interaction between root lesion nematode, *Pratylenchus neglectus* (Rensch 1924) Chitwood and Oteifa 1952, and root rotting fungi of wheat. *New Zealand Journal Crop Horticulture, 22*, 181–185.

Taylor, D. P., & Schleder, E. G. (1959). Nematodes associated with Minnesota crops. II. Nematodes associated with corn, barley, oats, rye, and wheat. *Plant Disease Report, 43*, 329–333.

Taylor, D. P., Anderson, R. V., & Haglund, W. A. (1958). Nematodes associated with Minnesota crops. I. Preliminary survey of nematodes associated with alfalfa, flax, peas, and soybeans. *Plant Disease Report, 42*, 195–198.

Thies, J. A., Barnes, D. K., Rabas, D. L., Sheaffer, C. C., & Wicoxson, R. D. (1992). Seeding date, carbofuran, and resistance to root-lesion nematode affect alfalfa stand establishment. *Crop Science, 32*, 786–792.

Thies, J. A., Basigalup, D., & Barnes, D. K. (1994). Inheritance of resistance to *Pratylenchus penetrans* in alfalfa. *Journal of Nematology, 26*, 452–459.

Timper, P., & Brodie, B. B. (1994). Effect of *Hirsutella rhossiliensis* on infection of potato by *Pratylenchus penetrans*. *Journal of Nematology, 26*, 304–307.

Todd, T. C. (1991). Effect of cropping regime on populations of *Belonolaimus* sp. and *Pratylenchus scribneri* in sandy soil. *Journal of Nematology, 23*, 646–651.

Venkatesh, R., Harrison, S. K., & Riedel, R. M. (2000). Weed hosts of soybean cyst nematode (*Heterodera glycines*) in Ohio. *Weed Technology, 14*, 156–160.

Villenave, C., & Duponnois, R. (1998). Influence of soil microorganisms on the reproduction of *Helicotylenchus dihystera* and its pathogenicity to millet. *Nematologica, 44*, 195–206.

Vilsack, T., & Clark, Z. F. (2014). *2012 census of agriculture – Minnesota state and county data*. USDA National Agricultural Statistics Service. https://www.agcensus.usda.gov/Publications/2012/Full_Report/Volume_1,_Chapter_1_State_Level/Minnesota/mnv1.pdf. Access date: 7/20/2017.

Vuong, T. D., Sleper, D. A., Shannon, J. G., & Nguyen, H. T. (2010). Novel quantitative trait loci for broad-based resistance to soybean cyst nematode (*Heterodera glycines* Ichinohe) in soybean PI 567516C. *Theoretical and Applied Genetics, 121*, 1253–1266.

Vuong, T. D., Sleper, D. A., Shannon, J. G., Wu, X., & Nguyen, H. T. (2011). Confirmation of quantitative trait loci for resistance to multiple-HG types of soybean cyst nematode (*Heterodera glycines* Ichinohe). *Euphytica, 181*, 101–113.

Wallace, M. K., & MacDonald, D. H. (1979). Plant-parasitic nematodes in Minnesota apple orchards. *Plant Disease Report, 63*, 1063–1067.

Wang, K. H., Sipes, B. S., & Schmitt, D. P. (2002). *Crotalaria* as a cover crop for nematode management: A review. *Nematropica, 32*, 35–57.

Warnke, S. A., Chen, S. Y., Wyse, D. L., Johnson, G. A., & Porter, P. M. (2006). Effect of rotation crops on *Heterodera glycines* population density in a greenhouse screening study. *Journal of Nematology, 38*, 391–398.

Weldekidan, T., Evans, T. A., Carroll, R. B., & Mulrooney, R. P. (1992). Etiology of soybean severe stunt and some properties of the causal virus. *Plant Disease, 76*, 747–750.

Westphal, A., Li, C. G., Xing, L. J., Mckay, A., & Malvick, D. (2014). Contributions of *Fusarium virguliforme* and *Heterodera glycines* to the disease complex of sudden death syndrome of soybean. *Plos One, 9*. https://doi.org/10.1371/journal.pone.0099529.

Yan, G. P., Khan, M., Huang, D., Lai, X., & Handoo, Z. A. (2016). First report of the stubby root nematode *Paratrichodorus allius* on sugar beet in Minnesota. *Plant Disease, 100*, 1022.

Yan, G. P., Plaisance, A., Chowdhury, I., Baidoo, R., Upadhaya, A., Pasche, J., Markell, S., Nelson, B., & Chen, S. (2017). First report of the soybean cyst nematode *Heterodera glycines* infecting dry bean (*Phaseolus vulgaris* L.) in a commercial field in Minnesota. *Plant Disease, 101*, 391.

Zhang, D. S. (1995). Paulownia and pea as two additional hosts of the soybean cyst nematode (*Heterodera glynices*). *Acta Phytopathologica Sinica, 25*, 275–278.

Zheng, J. W., & Chen, S. Y. (2011). Estimation of virulence type and level of soybean cyst nematode field populations in response to resistant cultivars. *Journal Entomology and Nematology, 3*, 37–43.

Zheng, J. W., Li, Y. H., & Chen, S. Y. (2006). Characterization of the virulence phenotypes of *Heterodera glycines* in Minnesota. *Journal of Nematology, 38*, 383–390.

Chapter 7
Nematodes Important to Agriculture in Wisconsin

Ann E. MacGuidwin

7.1 Agriculture in Wisconsin

Agriculture in Wisconsin is very diverse and important to the U.S. economy (Table 7.1). The state is a leader in the dairy industry and the production of corn for silage. Wisconsin produces more snap beans for processing, cranberries and ginseng than any other state and ranks among the top five states for forage, potatoes and processing vegetables (USDA NASS 2017). The state contributes to the forest products industry with significant acreage devoted to silviculture. Organic agriculture is very prominent and Wisconsin is outranked only by California for the number of organic farms.

About 40% of Wisconsin's land mass is devoted to agriculture. The top five commodities in terms of acreage are corn, soybean, hay, wheat and oat (WAS 2017; USDA NASS 2017). In terms of sales value, the ranking changes to corn, soybean, hay, potato and cranberry. The majority of corn and soybean are grown in the south and north central regions of the state on alfisols with a good potential for productivity. Potatoes and processing vegetables are concentrated in the central region of the state on irrigated sandy soils. Cranberry production is focused in Westcentral Wisconsin.

7.2 Nematology in Wisconsin

Root knot nematode on ginseng was the first nematology project in Wisconsin (McClintock 1914), but it was the discovery of *Ditylenchus destructor* in seed potatoes in 1953 that established the need for nematode expertise in Wisconsin. In 1953, Dr. H. Darling, Director of the Wisconsin Seed Certification Program at the University

A. E. MacGuidwin (✉)
Plant Pathology Department, University of Wisconsin, Madison, WI, USA
e-mail: aem@plantpath.wisc.edu

S. A. Subbotin, J. J. Chitambar (eds.), *Plant Parasitic Nematodes in Sustainable Agriculture of North America*, Sustainability in Plant and Crop Protection,
https://doi.org/10.1007/978-3-319-99588-5_7

Table 7.1 Agricultural crops of Wisconsin

Crop	Harvested hectares[a]	U.S. rank
Corn grain	1,186,000	8
Soybean	866,000	12
Hay	506,000	13
Corn for silage	356,000	1
Wheat	69,000	18
Oat	34,000	4
Potato	27,000	3
Snap bean	25,000	1
Sweet corn	22,000	3
Pea	9,100	3
Cranberry[b]	8,500	1
Cabbage	2,400	2
Cucumber	2,200	6
Carrot	1,800	3
Peppermint[b]	1,250	5

[a]Estimates for 2017 (USDA NASS 2017)
[b]Estimates for 2016 (WAS 2017)

of Wisconsin (UW), went to the Netherlands to study nematology with Drs. M. Oostenbrink and J. W. Seinhorst. Dr. Darling was joined by G. Thorne, who was appointed to the UW faculty in 1956. *Ditylenchus destructor* became the focus of nematology research at UW Wisconsin during this period (Faulkner and Darling 1961; Smart and Darling 1963; Anderson and Darling 1964; Darling 1959). Also at that time, the Wisconsin Department of Agriculture Trade and Consumer Protection (WI DATCP) began a zero-tolerance regulatory program that quarantined fields for seed potato production if *D. destructor* was detected.

With nematology expertise in place, other nematode-plant associations in Wisconsin were studied: *Helicotylenchus* spp. on bluegrass (Perry 1958), *Xiphinema* spp. on spruce (Griffin and Darling 1964; Griffin and Barker 1966), pine (Krebill et al. 1968) and strawberry (Perry 1958), *Pratylenchus penetrans* on potato and corn (Dickerson et al. 1964) and *Meloidogyne* (=*Hypsoperine*) *ottersoni* on reed Canary grass (Webber and Barker 1967; Thorne 1969). Drs. K. Barker and V. Dropkin were in the Department briefly after Thorne's retirement in 1961, but by the time of their departure Dr. Darling had resumed research on seed potatoes and a focused nematology program did not resume until 1983 with the hire of Dr. A. MacGuidwin. Nematode problems that arose in the interim included the soybean cyst nematode *Heterodera glycines* and increased pressure from *P. penetrans*. The Wisconsin DATCP began surveying the state for cyst nematodes of regulatory importance in 1982; an activity that continues to this day.

In 1984, a nematode diagnostic service was established at the UW Department of Plant Pathology. Samples submitted for nematode testing and research projects revealed a wide array of plant parasitic nematodes in the state. Currently, more than

90% of soil samples are positive for *Pratylenchus*, making it the most common genus of plant parasitic nematodes in Wisconsin. Other widely prevalent genera in order of incidence are *Helicotylenchus*, *Paratylenchus* and *Xiphinema*. Common genera associated with particular soil types or hosts include *Heterodera*, *Tylenchorhynchus*, *Hoplolaimus*, *Paratrichodorus* and criconematid genera. Genera that are relatively rare in the agricultural soils of Wisconsin include *Meloidogyne*, *Longidorus* and *Ditylenchus*. The five genera with the greatest economic impact to Wisconsin's most valuable crops are *Pratylenchus*, *Heterodera*, *Meloidogyne*, *Longidorus* and *Ditylenchus*.

7.3 Root Lesion Nematode, *Pratylenchus* spp.

7.3.1 Impact to Wisconsin

The widespread occurrence and extensive host range of *Pratylenchus* spp. make the root lesion nematode the most important nematode pest in Wisconsin. Population densities of *Pratylenchus* spp. are too low in most fields to induce significant yield loss, but nearly every field has the potential for problems to develop. Yield loss has been documented in Wisconsin for *P. penetrans* on potato (MacGuidwin et al. 2012) and corn (MacGuidwin and Bender 2016) and there is anecdotal evidence of damage for many crops. *Pratylenchus* spp. have been recovered from UW diagnostic and research samples for grain (corn, soybean), cereals (wheat, oat), vegetables (potato, pea, sweet corn, carrot, onion, pepper, cabbage, green bean), forage (alfalfa, corn), fruit (cranberry, cherry, apple), ornamental (peony, daisy, marigold, boxwood), cover (rye, sorghum, radish, kale, mustard, rapeseed) and specialty (ginseng, hops, mint) crops. The host range of *P. penetrans* extends well beyond the diagnostic samples received in Wisconsin and includes many weeds (Bélair et al. 2007) present in the state. The majority of infestations have not been identified, but *P. penetrans*, *P. scribneri* and *P. neglectus* are some of the most common species in the state. *Pratylenchus penetrans* has the greatest incidence and impact on specialty crops, while all species are associated with grain and forage crops.

7.3.2 Life History in Wisconsin

There are multiple overlapping generations of *Pratylenchus* spp. in Wisconsin. All life stages overwinter and the roots of annual crops become infected before shoots emerge. Dead roots from previous crops shelter nematodes, so nematode assays that include an incubation step provide the best estimate of initial nematode population densities (MacGuidwin and Bender 2012). Sampling to predict the damage potential of root lesion nematodes in Wisconsin is typically done in the fall. A study on the vertical distribution of *P. scribneri* in the soil profile supports this practice.

There was no evidence of downward migration in fields planted with corn or potato in September or October, indicating an "escape in time" strategy to survive Wisconsin winters (MacGuidwin and Stanger 1991). Estimates of overwinter survival rates for *P. scribneri* over a 3-year period averaged 41% for corn and 25% for potato, with most of the mortality occurring in the soil before soil froze (MacGuidwin and Forge 1991).

7.3.3 Interactions with Other Pathogens

The potato early dying disease caused by the interaction of *P. penetrans* and *Verticillium dahliae*, is a major constraint to potato production in Central Wisconsin (Fig. 7.1). Pioneering research that revealed this disease interaction in Ohio (Martin et al. 1982) was confirmed for potato varieties and conditions common to Wisconsin (Kotcon et al. 1985; MacGuidwin and Rouse 1990b; Morgan et al. 2002b). Very low population densities of these pathogens cause economic loss when the infestations are concomitant, rendering the majority of potato fields vulnerable to reduced yield and quality. The combined effects for yield loss are additive when one or both of the pathogens are present at high population densities. Details of the interaction have been elusive at the molecular level, but it has been demonstrated that the nematode plays an important role that extends beyond root wounding (Rotenberg et al. 2004; Saeed et al. 1999) (Fig. 7.2a, b). There is some evidence that *P. penetrans* may also interact synergistically with *Rhizoctonia solani* to reduce yield of potato under Wisconsin conditions (Kotcon et al. 1985).

Pratylenchus penetrans has also been demonstrated to play a role in the root rot of canning pea in Wisconsin (Oyekan and Mitchell 1972). The recovery of *P. penetrans* from fields damaged by *Aphanomyces euteiches* (Temp and Hagedorn 1967) led to greenhouse studies that showed that symptom expression was synergistic when both pathogens were present at low levels (Oyekan and Mitchell 1972). At higher population densities this nematode can cause severe damage to pea in the absence of *A. euteiches* (Fig. 7.2c).

Interactions for *Pratylenchus* spp., particularly *P. penetrans* and fungi resulting in yield reductions of other crops grown in Wisconsin, have been verified in other states. These associations follow the same pattern in that the combined effects of the pathogens are synergistic at low population densities and additive when one of the organisms is present at levels sufficient to cause disease as the sole pathogen. This was the case for *P. penetrans* and *Rhizoctonia fragariae* on strawberry in Connecticut (LaMondia and Cowles 2005) and *P. penetrans* and *Fusarium oxysporum* on alfalfa in Canada (Mauza and Webster 1982).

Current evidence argues against blanket generalizations suggesting that *Pratylenchus* spp. predisposes roots to fungal infection. Enhanced root infection by fungi in the presence of *P. penetrans* occurred for some fungi such as *R. fragariae* on strawberry (LaMondia and Cowles 2005), but not for others such as *R. solani* and *Colletotrichum coccodes* on potato (Kotcon et al. 1985). Bowers et al. (1996),

Fig. 7.1 Russet Burbank potato in a microplot experiment at the Hancock Research Station, Hancock, WI. The plants in the foreground are showing symptoms of the potato early dying disease

showed that the presence of *P. penetrans* increased infection and colonization of potato roots by *V. dahliae*, but not in the areas where nematodes were feeding. It is intuitive to assume that root wounding by root lesion nematodes attract and facilitate infection by other pathogens. This model may be specific to only certain root-infecting fungi since many are only successful in living tissue or in cases where they control the nature and timing of cell death (Kabbage et al. 2013). Educational tools used in Wisconsin to demonstrate *Pratylenchus* spp. are important pathogens in their own right include root explant cultures infected with only *P. penetrans* (Fig. 7.2d).

7.3.4 Management

There are three challenges to managing *Pratylenchus* spp. in Wisconsin. The first is the widespread distribution of this genus. Management practices for other nematode genera are sometimes conducive to reproduction by *Pratylenchus* spp., which can become problematic as populations of the primary pest decline. A second challenge is the lack of host resistance to *Pratylenchus* spp. Reports of tolerance or resistance are in the literature for alfalfa (Barnes et al. 1990) and wheat (Vanstone et al. 1998), but those cultivars are not grown in Wisconsin. It is well known that there are differences in nematode reproduction among genotypes of corn, soybean and other crops, but the market life of cultivars is too short for screening programs to be feasible for most crops. The final challenge is the lack of data on *Pratylenchus* species.

Fig. 7.2 (**a**) Injecting conidia of *Vertcillium dahliae* into the stem of Russet Burbank potato infected with *Pratylenchus penetrans*; (**b**) The fungal and nematode pathogens intecting different plant organs interacted to cause disease as expressed by cholorsis and wilting; (**c**) A pea field in central Wisconsin infested with *Pratylenchus penetrans*. The distribution of the nematode was patchy; (**d**) Root explant of I. O. Chief sweet corn that has been inoculated by placing a root infected with *Pratylenchus penetrans* retrieved from a mature in vitro culture onto healthy roots

Sustainable management recommendations, especially nuanced practices like crop rotation, will be greatly improved when diagnostics and field experiments include that level of detail. Presently, *P. penetrans* is the only species that is routinely singled out. The prevalence of this species in the central sandy vegetable production area of Wisconsin and the distinguishing high incidence of males allow diagnosticians to make a cursory identification of samples. Formal surveys have identified the majority of populations with males as *P. penetrans* (Kutsuwa and MacGuidwin unpublished).

7.3.4.1 Soil Disinfection

Fumigation of soil with biocidal chemicals is a cost effective and common practice in Wisconsin on high value crops such as ginseng, potato and carrot. In Wisconsin, soil is fumigated in the fall when soil temperature is above 10 °C and irrigation systems are still active so as to apply water after the material is injected into soil. Nematodes that come into contact with the fumigant are killed, but many are not, and residual root lesion nematode populations rebound over time (Morgan et al. 2002a). There is often a carry-over effect of fumigation for at least one additional year (Morgan et al. 2002a), but farmers are advised to manage root lesion nematode throughout the rotation to decrease the necessity of fumigating for the next potato crop.

Anaerobic soil disinfection (ASD), the practice of covering green manure with a plastic tarp, was demonstrated to be effective in Wisconsin (MacGuidwin et al. 2012). The study was conducted in potato fields with a high potential from damage due to *P. penetrans* and the Potato Early Dying disease. Inoculum levels of *P. penetrans* and *Verticillium dahliae* were reduced in 1 of 2 years and yields were increased both years in plots receiving ASD. Due to the logistics of applying tarps to large areas, ASD is not practiced in conventional production fields. A study showing that the distribution of *P. penetrans* was stable throughout a 3-year potato rotation in Central Wisconsin (Morgan et al. 2002a) indicates that using ASD as a site-specific approach is feasible.

7.3.4.2 Nematicides and Seed Treatments

Nematicides are used to control *Pratylenchus* spp. in potato and processing vegetables in Wisconsin. Product availability, and therefore use, declined over the past 30 years but has seen a recent upswing as new chemistries and biologicals have entered the market. Seed treatments for corn have generated a lot of interest based on early reports for *P. zeae* (Cabrera et al. 2009). A greenhouse study in Iowa found that nematode population densities of *P. penetrans* were reduced when corn seeds were treated with a nematicide in combination with a fungicide and insecticide (da Silva et al. 2016). The "early mitigation" strategy for seed treatments is supported by a yield loss model developed for corn in Wisconsin that explained yield loss using nematode population densities at planting (MacGuidwin and Bender 2016).

7.3.4.3 Crop Rotation

Crop rotation is difficult for *Pratylenchus* spp. due to the wide host range of multiple species, but there is sufficient variation in the host status of crops common to the state to plan rotations for nematode management in vegetable systems. Corn was a better host for *P. penetrans* (Dickerson et al. 1964; Morgan et al. 2002a) and *P. scribneri* (MacGuidwin and Forge 1991) than potato. Soybean was a better host than corn for *P. penetrans* in Canada (Bélair et al. 2002) and this seems to be the case in Wisconsin (Kutsuwa unpublished). The majority of fields planted with these long-season crops in the year prior to potato are fumigated to mitigate the potato early dying disease. Short season processing crops such as pea or green bean are good hosts to *P. penetrans* but offer opportunity to disrupt the nematode's life cycle with a period of fallow or cover crops which can increase the interval between fumigation events.

7.3.4.4 Cover Crops

In Wisconsin, cover crops are planted after harvest on the majority of potato and vegetable fields. Ninety-two percent of organic vegetable farmers in Wisconsin reported using cover crops (Moore et al. 2016). The most common cover crop grown is cereal rye because it can be planted throughout the state during the month of October. Rye is an excellent host for *P. penetrans* (Bélair et al. 2002), but the late planting and reduced reproduction by *P. penetrans* at low temperature (Thistlethwayte 1970) discourage the increase of nematodes until spring when the cover crop is destroyed. A rye cover crop provides valuable ecological services so farmers concerned about *Pratylenchus* spp. are advised to monitor their situation by soil sampling.

Cover crops reported to reduce root lesion nematodes have had mixed success in Wisconsin. Mustard, rapeseed, and other members of the Brassicaceae, excellent hosts for *P. penetrans*, are planted in the late summer when the nematodes are likely to increase. Sorghum-sudangrass and rapeseed maintained *Pratylenchus* spp. (MacGuidwin et al. 2012; MacGuidwin and Layne 1995). Population densities of *P. penetrans* were reduced at the time of planting potato in plots planted with African marigold (*Tagetes erecta*) and forage pearl millet (*Pennesetum glaucum*) the previous year (MacGuidwin et al. 2012). Subsequent research in Wisconsin supported results from earlier studies (Ball-Coelho et al. 2003; Bélair et al. 2005) showing forage pearl millet to be a good cover crop for managing *P. penetrans*. Millet is planted in August because it is sensitive to frost, so adoption has been highest for vegetable cropping systems.

7.4 Soybean Cyst Nematode, *Heterodera glycines*

7.4.1 Impact to Wisconsin

Heterodera glycines, the soybean cyst nematode, was intercepted from soil associated with cabbage transplants from Tennessee in 1980 and during the following year, a survey by the Wisconsin Department of Agriculture and Consumer Protection (DATCP) revealed an infested soybean field in Southeastern Wisconsin (Phibbs et al. 2016). As of 2016, 52 counties with 96% of the state's soybean acreage were known to be infested. The SCN also affects green bean production in the state. Research in Illinois (Melton et al. 1985) showed that green bean cultivars bred for root rot resistance in Wisconsin suffered less damage from *H. glycines* than other cultivars. Today, green bean cultivars grown for processing range in sensitivity to *H. glycines*, with some supporting even more nematode reproduction than soybean (MacGuidwin unpublished).

Due to its widespread distribution and potential for damage, the soybean cyst nematode is considered the greatest yield-reducing biotic factor affecting soybean production in the state. Studies in commercial fields in Wisconsin have consistently shown an advantage to planting soybean genotypes that limit reproduction by *H. glycines*. The yield advantage of planting a resistant variety was fairly consistent over a range of initial nematode population densities in a sandy soil, with an average gain of at least 30% (MacGuidwin et al. 1995). On silt loam soil the yield gain was at least 17% (Bradley et al. 2003). A regional study that included sites in Wisconsin, showed *H. glycines*-resistant varieties had greater yield than susceptible varieties with the magnitude of the difference depending on cultivar, location and initial population densities (Donald et al. 2006).

7.4.2 Life History in Wisconsin

The distribution of the soybean cyst nematode is aggregated in fields in Wisconsin (Kaszubowski and MacGuidwin 2000). Studies in commercial fields showed the field entrance to be the location most often infested with the nematode, but in some fields *H. glycines* was found only in one or two patches well away from vehicle access points (Kaszubowski unpublished), suggesting that birds or wind are important means of dispersal. One commercial field that was a site for multiple studies showed a remarkable range of pH that was positively associated with population densities of *H. glycines* (Pedersen et al. 2010). Information from these studies has been applied to sampling recommendations for farmers in Wisconsin.

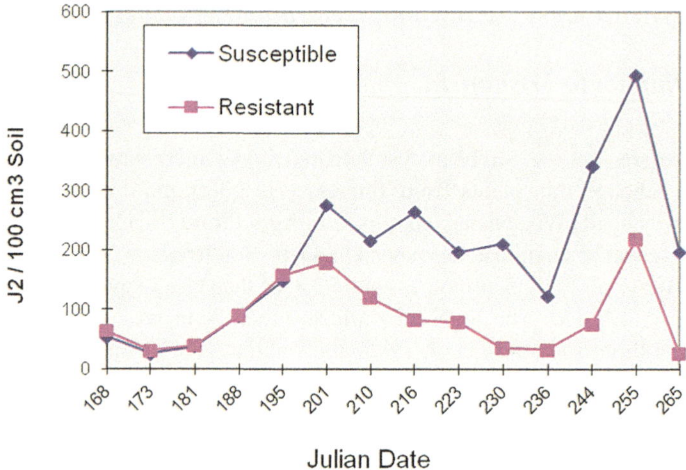

Fig. 7.3 Numbers of second-stage juveniles of *Heterodera glycines* recovered from 100 cm^3 soil samples on multiple dates from two soybean varieties: one susceptible and one resistant to *H. glycines*

The population dynamics of the soybean cyst nematode including temporal changes in the attrition of population densities, guide recommendations for managing this pest in Wisconsin. Soybeans are grown throughout the state, and in all but the northern counties it appears that there are at least two generations per year (Fig. 7.3). Second-stage juveniles can be detected in diagnostic and research samples collected at all times of the year but are particularly abundant in the fall. A collaborative study among northcentral states including Wisconsin, showed the rate of population growth during the season was inversely related to population densities in the spring (Wang et al. 2000). Wisconsin populations appear to be well adapted to winter, at least following the soybean year. In a study comparing *H. glycines* populations collected from five latitudes representing the U.S., populations from Wisconsin showed a high rate of overwinter survival even when they were moved to Florida (Riggs et al. 2001). Soil sampling confirms this finding; population densities are higher in the spring than fall in some years, suggesting that *H. glycines* can reproduce after soybeans senesce (MacGuidwin unpublished).

7.4.3 Interactions with Other Pathogens

Foliar symptoms of the sudden death syndrome caused by *Fusarium virguliforme* were exacerbated in *H. glycines*-infected plants in microplot studies in Mississippi (McLean and Lawrence 1993). Studies in Nebraska fields infested with both pathogens found the severity of foliar symptoms to be positively associated with initial population densities of the nematode (Brzostowski et al. 2014). The disease potential of the fungus was low in both studies and the presence of the soybean cyst nematode synergistically increased disease severity. Analysis of diagnostic samples in Wisconsin showed a negative association between the incidence of *F. virguliforme* and *H. glycines* (Marburger et al. 2014). Considering the many ways that the soybean cyst nematode can enter fields, farmers with a history of sudden death syndrome disease are advised to test their fields regularly for the nematode so that they can act quickly to amend their disease management plan if the soybean cyst nematode is also a factor.

Brown stem rot disease caused by *Phialophora gregata* was more severe in treatments with *H. glycines* in growth chamber studies in Iowa (Tabor et al. 2003). This fungus is common in Wisconsin soybean fields and often occurs together with the nematode. There is anecdotal evidence that the disease is more severe in joint infestions (Grau unpublished). Understanding the association of *P. gregata* and *H. glycines* in the field is difficult because soil pH has a strong and opposite effect on these pathogens (Pedersen et al. 2010) and many soybean cyst nematode-resistant cultivars derived from PI 88788 are also resistant to *P. gregata*. Crop rotation and host resistance, either alone or together, are the pillars of management plans for both pathogens in Wisconsin.

7.4.4 Management

Soybean fields in Wisconsin can experience yield loss without foliar symptoms, so infestations often escape notice until population densities are very high and difficult to manage. The soybean industry began offering free assays to Wisconsin farmers in the 1990s, yet to date most fields have never been tested. The majority of soybean varieties marketed in Wisconsin has PI 88788 in their pedigree and are labeled as soybean cyst nematode-resistant, so farmers are inadvertently managing the nematode. The challenge is to convince farmers to take a proactive approach by sampling soil to confirm and monitor soybean cyst nematode infestations on their farms.

7.4.4.1 Host Resistance

Most of the soybean varieties sold today in the Northcentral U.S. have resistance to the soybean cyst nematode derived from the source line PI 88788. Since there is significant genetic variation in *H. glycines*, it is not surprising that many populations are now adapted and reproducing on varieties with the PI 88788 source of resistance. One study, with sites ranging from Tennessee to Ontario, Canada, found that most *H. glycines* populations reproduced on PI 88788 (Faghihi et al. 2010). Analysis of 15 years of research data in Iowa verified an increase in the reproduction of *H. glycines* over time (McCarville et al. 2017). Populations of the nematode species from more than 325 farms in Wisconsin have been evaluated and 75% (10-year average) show some level of development (>10%) on PI 88788 (MacGuidwin unpublished). Adapted populations have been detected in Wisconsin on farms with only a brief or no history of planting soybean cyst nematode-resistant varieties, presumably because all populations are exotic to the state and may have been dispersed from a farm elsewhere that had relied on host resistance.

Populations of *H. glycines* adapted to the PI 548402 (Peking) source of resistance have also been detected in Wisconsin. About 25% (10-year average) of the populations tested over the past 10 years developed (>10%) on varieties with Peking-derived resistance (MacGuidwin unpublished). The level of adaptation to the Peking line reported for Tennessee and Ontario were higher and lower, respectively, than the estimate for Wisconsin (Faghihi et al. 2010). Adaptation of the soybean cyst nematode over a 15-year study period was not detected for varieties with the Peking pedigree in Iowa (McCarville et al. 2017).

Understanding the basis for adaptation and educating farmers to manage soybean germplasm resources wisely is a top priority for nematology in Wisconsin. Adaptation is characterized using the HG typing system (Niblack et al. 2002), awarding the "adapted" designation to any population with 10% or more of its members capable of developing on the germplasm being tested. While the majority of *H. glycines* populations in Wisconsin surpass the 10% level on PI 88788, most populations remain below 30%, the threshold commonly used to indicate the point at which growing a resistant variety is no longer economical (Schmitt and Shannon 1992). Data show the average level of adaptation is slowly increasing in the state, underscoring the importance of easing selection pressure on *H. glycines* populations.

7.4.4.2 Crop Rotation

Rotation, for the purpose of breaking the life cycle and minimizing nematode reproduction, is the most important cultural practice for soybean cyst nematode in Wisconsin. Crops of varying host status were planted for 1 or 2 years to create a range of *H. glycines* population densities in order to develop a damage model for soybean (MacGuidwin et al. 1995). The non-hosts, white clover and alfalfa, had the greatest impact on population attrition. Pea and hairy vetch treatments showed there is value to planting a host crop of another species as compared to a soybean monoculture. Rotating soybean varieties, even soybean cyst nematode-resistant varieties, is recommended in Wisconsin because many traits including root mass and architecture, affect the population dynamics of SCN.

7.4.4.3 Noxious Chemicals

Some of the practices to supplement host resistance involve noxious chemicals that must be taken up by *H. glycines* juveniles in soil. Examples include synthetic chemicals coated on to seeds, plant-derived compounds released during the breakdown of cover crop green manures such as glucosinolates, and toxins released by fungal antagonists. Schroeder and MacGuidwin (2007, 2010a) showed that juveniles, within eggs or hatched, accumulated fluorescein isothiocyanate in the intestine and the mode of entry was the stoma and other body openings. They demonstrated that the uptake of four plant-derived compounds varied according to the mobility of the nematode. Quiescent nematodes had decreased sensitivity compared to those that were actively moving (2010b). Based on these results, it is recommended that management practices based on noxious chemicals are deployed when conditions are warm and soil moisture is not limiting, thereby enhancing active movement of the nematodes.

7.5 Root Knot Nematode, *Meloidogyne hapla*

7.5.1 Impact to Wisconsin

The impact of *Meloidogyne hapla* in Wisconsin was first documented in a M.S. thesis in 1914 (McClintock 1914). Field and laboratory studies were conducted in Michigan and Wisconsin respectively, to control *M. hapla*, then called *Heterodera radicicola*, on ginseng. Soil treatments of carbon bisulfide, formalin, sulfuric acid, naphthaline, ammonia, nicotine, kerosene, gasoline and tobacco dust were unsuccessful in alleviating galling, and the author provided advice which is still followed today, "...the best solution of the problem is to take great pains to keep their soil free from this pest by planting only such seeds and roots as are known to be free from the parasite...".

Meloidogyne hapla, the northern root knot nematode, is distributed throughout Wisconsin today on a wide range of crops. Crops represented in diagnostic and research samples positive for the *M. hapla* include carrot, ginseng, potato, alfalfa, mint, onion, green bean, soybean, daisy and basil. Research in other locales has shown the host range to be extensive including crops and weeds that grow in Wisconsin (Bélair and Benoit 1996; Faulkner and McElroy 1964).

Reduced yields of potato have been demonstrated in the state (MacGuidwin and Rouse 1990a). The northern root knot nematode is often not diagnosed in potato fields because there is little to no galling and the primary impact is on tuber production rather than quality (MacGuidwin 2008). Crop quality is the major impact of *M. hapla* on carrot and nematode damage thresholds established for carrot in New York (Gugino et al. 2006) also apply to Wisconsin. Wisconsin produces about one third of the processing carrots in the U.S. (USDA NASS 2017), and farmers in Wisconsin consider the potential impact of the northern root knot nematode when managing the crop. Wang and Goldman (1996) identified resistance to Wisconsin populations of the nematode species in inbred carrot lines, however, commercial resistant cultivars are currently not available.

7.5.2 *Life History in Wisconsin*

Carrot, ginseng and other root crops are very sensitive to early season damage by the northern root knot nematode, so studies in Wisconsin were conducted to learn how the nematode survived winter conditions. Sampling during the winter months revealed second-stage juveniles as an abundant overwintering stage. Laboratory studies showed that exposure to low temperatures prior to freezing, as would occur in the field, increased survival of frozen conditions (Forge and MacGuidwin 1992a). Conditions of low water potential had the same effect (Forge and MacGuidwin 1992b). Manipulating temperature and water potential indicated that there were two mechanisms that allowed juveniles to escape freezing at subzero temperature: removal of water through desiccation (Fig. 7.4a) and the accumulation of cryoprotectant compounds. These studies suggest that the greatest mortality of *M. hapla* occurs during autumn with warm temperatures and high rainfall; conditions that might be manipulated, at least in part, in irrigated fields.

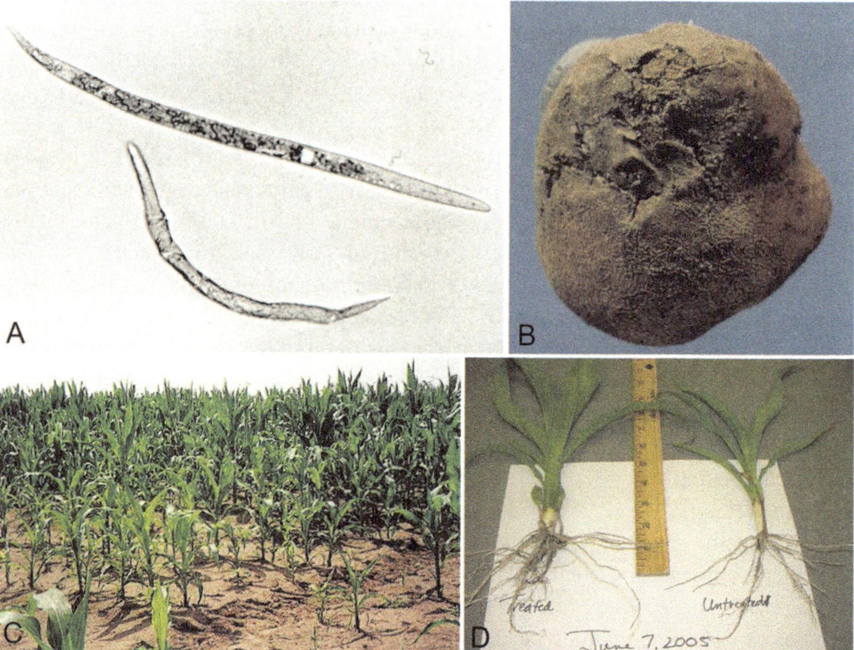

Fig. 7.4 (**a**) Second-stage juveniles of *Meloidogyne hapla*. The nematode below was preconditioned by exposure to 4 °C for 4 h and then the temperature was lowered to −10 °C. The nematode above was maintained at 24 °C for the entire period; (**b**) Potato tuber with dry rot symptoms following exposure to the potato rot nematode, *Ditylenchus destructor*; (**c**) A field near Spring Green Wisconsin showing stuting symptoms caused by the corn needle nematode, *Longidorus breviannulatus*; (**d**) Corn plants from field trials showing the benefit of a nonfumigant nematicide for mitigating damage caused by the corn needle nematode, *Longidorus breviannulatus*

7.5.3 *Interactions with Other Pathogens*

The northern root knot nematode is not known to have synergistic interaction with fungi in Wisconsin. Microplot experiments showed the effects of *M. hapla* and *V. dahliae* to be additive for potato (MacGuidwin and Rouse 1990a). Association between *M. hapla* and *F. oxysporum f. sp. medicaginis* on alfalfa have not been studied to determine if synergism between these two pathogens, reported for Utah (Griffin and Thyr 1988), also applies to populations in Wisconsin.

7.5.4 *Management*

Sampling before planting to determine nematode population densities is particularly important for *M. hapla* because many crops can't "outgrow" the damage inflicted by the nematode at the seedling stage. The challenge in making that assessment is that most *M. hapla* are deep in the soil profile in the spring. Sampling at the end of the preceding summer or in the fall to estimate population densities for the following spring is recommended for developing nematode management plans in Wisconsin.

7.5.4.1 Soil Disinfection

Many of the northern root knot nematode infestations in Wisconsin are in high organic soils that are difficult to fumigate because of their binding properties. Soil fumigants must penetrate deep within the soil profile in order to reach the nematode, and residual nematode populations are a concern. Biofumigation, the incorporation of *Brassica* spp. at flowering, is problematic because population densities of the *M. hapla* increase during the cover crop phase. Due to these issues, soil disinfection is either usually not practiced in Wisconsin for the northern root knot nematode or is done in combination with nematicides or seed treatments.

7.5.4.2 Nematicides and Seed Treatments

Nonfumigant nematicides are commonly used on carrot in Wisconsin. The product is targeted to the plant for protection against infection by nematodes. There is anecdotal evidence that systemic nematicides increase yield as documented for New York (Gugino et al. 2006). New seed treatments for nematodes have begun to enter the market and will likely become an important tool for managing the northern root knot nematode.

7.5.4.3 Crop Rotation and Cover Crops

Crop rotation is the cornerstone of northern root knot nematode management. The host range of *M. hapla* does not include corn or small grains, so planting these crops the year before carrot or other sensitive crops reduces nematodes below the detection level (MacGuidwin unpublished). Studies in New York (Viaene and Abawi 1998) concluded that rye, oat, and sudangrass grown for 7 weeks were not hosts for *M. hapla* and would, therefore, be good candidates to include as full season rotation crops or partial season cover crops. The authors also examined the benefit of incorporating sudangrass as a green manure, which they concluded, was superior to growing it as a cover crop. This practice has not been widely adopted in Wisconsin, because it has the unintended consequence of increasing root lesion nematode which does reproduce on sudangrass (MacGuidwin and Layne 1995).

7.6 Potato Rot Nematode, *Ditylenchus destructor*

7.6.1 Impact to Wisconsin

Wisconsin is one of seven U.S. states infested with *Ditylenchus destructor*, the potato rot nematode. The majority of the detections have been in one northern county and new infestations are relatively rare. The typical symptom on tubers is cracking and a dry rot that can spread among tubers in storage as the nematode multiplies (Fig. 7.4b). Since *D. destructor* is a nematode of regulatory concern worldwide, the WI DATCP (2017) places infested fields under quarantine for the production of seed potatoes and inspects new production fields, as well as fields with a prior history for the potato rot nematode. The seed potato farmers, DATCP, and the University of Wisconsin have worked together since 1953 to limit the impact of *D. destructor*, and no shipment of seed or commercial potatoes from the state has ever tested positive for the nematode.

7.6.2 Life History in Wisconsin

Much of what is known about the life cycle of *D. destructor* was based on populations from Wisconsin. Anderson and Darling (1964) determined that the gender of juveniles could be discerned by the third stage and that soon after the final moult, females mate with multiple males. These studies were facilitated by the studies of Faulkner and Darling (1961) that showed *D. destructor* can be cultured in vitro on more than 64 species of fungi. MacGuidwin and Slack (1991) expanded the host range of this highly polyphagus nematode to include corn and green bean, two crops commonly grown in Wisconsin, and detailed its proclivity to infect seed and to mature and survive in above-ground plant tissues (MacGuidwin et al. 1992).

One of the most important aspects of the life cycle is the low incidence of *D. destructor* in soil. This observation was made by Darling from field studies (Darling 1959; Darling et al. 1983), and MacGuidwin and Slack (1991) and MacGuidwin et al. (1992) found very few nematodes in soil in greenhouse studies. The probability of detecting *D. destructor* in soil is further thwarted by the large number of *Ditylenchus* species, all with very similar morphology in the juvenile stage. All discoveries of *D. destructor* in Wisconsin have only been from potato tubers.

7.6.3 Interactions with Other Pathogens

During the 1950s, fungi were detected from tubers infected with the potato rot nematode, therefore, it was important to verify that *D. destructor* was the primary cause of disease rather than a secondary invader. Faulkner and Darling (1961) were able to do so by rearing the nematodes in monoxenic callus culture. Today, fungi such as *Fusarium* spp. and *Rhizoctonia solani*, contribute to disease; their effects are considered to be additive to those of *D. destructor*.

7.6.4 Management

The challenge of managing *Ditylenchus destructor* is the inability to detect infestations by soil sampling prior to planting potato. Best management practices are aimed at seed potatoes in order to prevent the dispersal of potato rot nematode and regulatory procedures for Wisconsin are enforced (WI Administrative Code 2017). Seed potato fields new to certification and fields released from quarantine are inspected for potato rot nematode in Wisconsin. Commercial farmers are advised to plant only seed potatoes that are certified to be free from *D. destructor* and to be vigilant for symptoms on tubers during harvesting and washing activities.

7.6.4.1 Soil Disinfection

Experiments to eliminate potato rot nematode in the 1950s showed the soil fumigant ethylene di-bromide to be effective (Darling 1959). Soil fumigation was written into the state's regulatory standards and on the basis of case studies over 29 years, Darling et al. (1983) declared that potato rot nematode had been eradicated by fumigation in Wisconsin. However, soon afterward, some fumigated fields were found to be positive for potato rot nematode. Therefore, fumigation is now considered to be a mitigation measure that requires validation through crop inspection. Current regulations do not name a particular soil fumigant and fields are released from quarantine after fumigation and two successive potato crops with no detections of *D. destructor*. Fumigation is also recommended for fields in the 1st year of seed production until inspectors verify the field to be free of potato rot nematode at harvest. Additional recommended measures for 1st-year fields are to plant them last and disinfect equipment before exiting fields.

7.6.4.2 Crop Rotation

Fields adjacent to sites of potato rot nematode detection may also be placed under quarantine, but even if they are not, it is recommended to lengthen the interval between potato crops. Many crops can maintain *D. destructor* in the greenhouse, but oat was a poor host for a Wisconsin population even under greenhouse conditions (MacGuidwin and Slack 1991), thereby, making it a good candidate for fields at risk. Until more is known about the persistence of *D. destructor* in fields, crop rotation should be considered an important, but not the primary, practice for management.

7.7 Other Nematode Pests in Wisconsin

The corn needle nematode, *Longidorus breviannulatus*, causes severe stunting in sandy soils of Wisconsin (Fig. 7.4c). Infestations are highly aggregated as patches in fields, and in most cases, the patch size is small. Seasonal migration in the soil profile occurs as corn plants mature (MacGuidwin 1989), so detecting infestations beyond the seedling stage can be problematic. Nonfumigant nematicides are used to mitigate damage (Fig. 7.4d).

The stubby root nematode, *Paratrichodorus* spp., is commonly recovered from potato fields in Wisconsin. Crop loss due to nematodes alone has not been reported, but the corky ringspot disease, vectored by the nematode, was discovered in Wisconsin in 2007 (Phibbs and Leisso 2009). Detection of corky ringspot means total loss for infested fields since the potatoes are destroyed to prevent the spread of disease. Farmers are advised to plant only certified seed potatoes, as infected tubers are a known means of introducing the virus into potato fields.

The pine wood nematode (PWN), *Bursaphelenchus xylophilus*, was detected in Wisconsin in 1981 and 1982, associated primarily with insects in the family Cerambycidae (Wingfield and Blanchette 1983). Diagnostic samples positive for the pine wood nematode are detected periodically, accompanied by reports of tree death.

A number of other plant parasitic nematode species have been reported from Wisconsin (Table 7.2). The extent of damage and yield loss attributable to most of these species today has not been assessed. Some such as *Helicotylenchus digonicus* and *Pratylenchus penetrans*, are fairly common in agricultural fields across the state. Others such as *Meloidogyne ovalis* and *Nothocriconema sphagni*, have not been detected since the original report.

Table 7.2 Nematode species reported from Wisconsin

Species	Host	References
Bakernema inaequale	Maple	Hoffman (1974)
Bursaphelenchus xylophilus	Pine	Wingfield and Blanchette (1983)
Cactodera milleri	Lambsquarters	Schroeder et al. (2008)
C. rosae	Corn	Phibbs et al. (2017)
C. weissi	Unknown	Phibbs et al. (2017)
Criconema octangulare	Unknown	Hoffman (1974)
Crossonema menzeli	Unknown	Hoffman (1974)
Ditylenchus destructor	Potato	Darling (1959)
D. dipsaci	Phlox, garlic	WI DATCP (2017)
Helicotylenchus digiatus	Cranberry	Barker and Boone (1966)
H. digonicus	Blue grass, corn	Perry (1959) and Griffin (1964)
H. microlobus	Blue grass, corn	Perry (1959) and Griffin (1964)
H. platyurus	Blue grass, corn	Perry (1959) and Griffin (1964)
H. pseudorobustus	Cranberry	Barker and Boone (1966)
Hemicycliophora obtusa	Cranberry	Barker and Boone (1966)
H. typica	Maple	Riffle (1962)
Heterodera glycines	Soybean	MacGuidwin et al. (1995)
H. trifolii	Corn	WI PIB (2016)
Hirschmanniella gracilis	Unknown	Sher (1968)
Hoplolaimus galeatus	Pea	Temp and Hagedorn (1967)
Lobocriconema thornei	Unknown	Powers et al. (2017)
Longidorus breviannulatus	Corn	MacGuidwin (1989)
Meloidogyne hapla	Ginseng	McClintock (1914)
M. ottersoni	Canary grass	Thorne (1969)
M. ovalis	Maple, elm	Riffle (1963)
Nanidorus minor	Cranberry	Barker and Boone (1966)
Nothocriconema sphagni	Maple	Riffle (1962)
Ogma octangularis	Unknown	Powers et al. (2017)
Pratylenchus crenatus	Corn	Dickerson et al. (1964)
P. neglectus	Corn	Dickerson et al. (1964)
P. penetrans	Corn	Dickerson et al. (1964)
P. scribneri	Potato, corn	MacGuidwin and Stanger (1991)
P. vulnus	Corn	Griffin (1964)
P. thornei	Corn	Griffin (1964)
Rotylencus buxophilus	Corn	Griffin (1964)
R. pumilus	Blue grass	Perry (1959)
Tylenchorhynchus maximus	Corn	Griffin (1964)
Trichodorus californicus	Corn, cranberry	Griffin (1964) and Barker and Boone (1966)
Xiphinema americanum	Strawberry, corn	Perry (1958) and Griffin (1964)
X. chambersi	Strawberry	Perry (1958)

7.8 Sustainable Nematode Management in Wisconsin

Wisconsin farmers have been leaders in agricultural sustainability. Collaborations between the University of Wisconsin and commodity groups advance science-based programs to promote environmental stewardship and inform management decisions. The Healthy Grown Program, an industry-led initiative to use best-management practices, reduce pesticides, and to support native plants and animals (Zedler et al. 2009), developed standards and third-party certification for the sustainable production of potatoes and onions. The National Soybean Initiative, piloted in Wisconsin, developed an assessment process to help soybean farmers document practices and quantify progress in adopting sustainable approaches to soybean production (Dong et al. 2016). Farmers in both of these programs include nematodes in pest management plans, but much work remains in educating the agricultural industry at large about the importance of plant parasitic nematodes and the yield gains that can be realized when pest nematodes are maintained at low population densities. Collaborations between nematologists in the Northcentral region of the U.S. and nationally bolster Wisconsin efforts to evaluate and develop management practices for nematodes and to advance awareness of important nematode pests.

References

Anderson, R. V., & Darling, H. M. (1964). Embryology and reproduction of *Ditylenchus destructor* Thorne, with emphasis on gonad development. *Proceedings of the Helminthological Society of Washington, 31*, 240–256.

Ball-Coelho, B., Bruin, A. J., Roy, R. C., & Riga, E. (2003). Forage pearl millet and marigold as rotation crops for biological control of root lesion nematodes in potato. *Agronomy Journal, 95*, 282–292.

Barker, K. R., & Boone, D. M. (1966). Plant parasitic nematodes on cranberries in Wisconsin. *Plant Disease Report, 50*, 957–959.

Barnes, D. K., Thies, J. A., Rabas, D. L., Nelson, D. L., & Smith, D. M. (1990). Registration of two alfalfa germplasms with field resistance to the root lesion nematode. *Crop Science, 30*, 751–752.

Bélair, G., & Benoit, D. L. (1996). Host suitability of 32 common weeds to *Meloidogyne hapla* in organic soils of southwestern Quebec. *Supplement Journal of Nematology, 28*(4S), 643–647.

Bélair, G., Fournier, Y., Dauphinais, N., & Dangi, O. P. (2002). Reproduction of *Pratylenchus penetrans* on various rotation crops in Quebec. *Phytoprotection, 83*, 111–114.

Bélair, G., Dauphinais, N., Fournier, Y., Dangi, O. P., & Clement, M. F. (2005). Effect of forage and grain pearl millet on *Pratylenchus penetrans* and potato yields in Quebec. *Journal of Nematology, 37*, 78–82.

Bélair, G., Dauphinais, N., Benoit, D. L., & Fournier, Y. (2007). Reproduction of *Pratylenchus penetrans* on 24 common weeds in potato fields in Quebec. *Journal of Nematology, 39*, 321–326.

Bowers, J. H., Nameth, S. T., Riedel, R. M., & Rowe, R. C. (1996). Infection and colonization of potato roots by *Verticillium dahliae* as affected by *Pratylenchus penetrans* and *P. crenatus*. *Phytopathology, 86*, 614–621.

Bradley, C. A., Noel, G. R., Grau, C. R., Gaska, J. M., Kurtzweil, N. C., MacGuidwin, A. E., Wax, L. M., Hartman, G. L., & Pedersen, W. L. (2003). Impact of herbicides on *Heterodera glycines* susceptible and resistant soybean cultivars. *Journal of Nematology, 35*, 88–97.

Brzostowski, L. F., Schapaugh, W. T., Rzodkiewicz, P. A., Todd, T. D., & Little, C. R. (2014). Effect of host resistance to *Fusarium virguiliforme* and *Heterodera glycines* on sudden death syndrome disease severity and soybean yield. *Plant Health Progress, 15*, 1–8. https://doi.org/10.1094/PHP-RS-13-0100.

Cabrera, J. A., Kiewnick, S., Grimm, C., Dababat, A. A., Sikora, R. A., & R. A. (2009). Efficacy of abamectin seed treatment on *Pratylenchus zeae*, *Meloidogyne incognita*, and *Heterodera schachtii*. *Journal of Plant Diseases and Protection, 116*, 124–128.

Darling, H. M. (1959). Control of the potato rot nematode in Wisconsin. *Plant Disease Report, 43*, 239–242.

Darling, H. M., Adams, J., & Norgren, R. L. (1983). Field eradication of the potato rot nematode, *Ditylenchus destructor*: A 29-year history. *Plant Disease, 67*, 422–423.

Dickerson, O. J., Darling, H. M., & Griffin, G. D. (1964). Pathogenicity and population trends of *Pratylenchus penetrans* on potato and corn. *Phytopathology, 54*, 317–322.

Donald, P. A., Pierson, P. E., St. Martin, S. K., Sellers, P. R., Noel, G. R., MacGuidwin, A. E., Faghihi, J., Ferris, V. R., Grau, C. R., Jardine, D. J., Melakeberhan, H., Niblack, T. L., Steinstra, W. C., Tylka, G. L., Wheeler, T. A., & Wysong, D. S. (2006). Assessing *Heterodera glycines*-resistant and susceptible cultivars yield response. *Journal of Nematology, 38*, 76–82.

Dong, F., Mitchell, P. D., Knuteson, D., Wyman, J., Bussan, A. J., & Conley, S. (2016). Assessing sustainability and improvements in US Midwestern soybean production systems using a PCA-DEA approach. *Renewable Agriculture and Food Systems, 31*, 524–539.

Faghihi, J., Donald, P. A., Noel, G., Welacky, T. W., & Ferris, V. R. (2010). Soybean resistance to field populations of *Heterodera glycines* in selected geographic areas. Online. *Plant Health Progress*. https://doi.org/10.1094/PHP-2010-0426-01-RS.

Faulkner, L. R., & Darling, H. M. (1961). Pathological histology, hosts, and culture of the potato rot nematode. *Phytopathology, 51*, 778–786.

Faulkner, L. R., & McElroy, F. D. (1964). Host range of northern root knot nematode on irrigated crop plants and weeds in Washington. *Plant Disease Report, 48*, 190–193.

Forge, T. A., & MacGuidwin, A. E. (1992a). Impact of thermal history on freezing tolerance of second-stage *Meloidogyne hapla*. *Journal of Nematology, 24*, 262–268.

Forge, T. A., & MacGuidwin, A. E. (1992b). Effects of water potential and temperature on survival of the nematode *Meloidogyne hapla* in frozen soil. *Canadian Journal of Zoology, 70*, 1553–1560.

Griffin, G. D. (1964). Association of nematodes with corn in Wisconsin. *Plant Disease Report, 48*, 458–459.

Griffin, G. D., & Barker, K. R. (1966). Effects of soil temperature and moisture on survival and activity of *Xiphinema americanum*. *Proceedings of the Helminthological Society of Washington, 33*, 126–130.

Griffin, G. D., & Darling, H. M. (1964). An ecological study of *Xiphinema americanum* Cobb in an ornamental spruce nursery. *Nematologica, 10*, 471–479.

Griffin, G. D., & Thry, B. D. (1988). Interaction of *Meloidogyne hapla* and *Fusarium oxysporum* f. sp. *medicaginis* on alfalfa. *Phytopathology, 78*, 421–425.

Gugino, B. K., Abawi, G. S., & Ludwig, J. W. (2006). Damage and management of *Meloidogyne hapla* using oxamyl on carrot in New York. *Journal of Nematology, 38*, 483–490.

Hoffman, J. K. (1974). Morphological variation of *Bakernema*, *Criconema*, and *Criconemoides* (Criconematidae: Nematoda). *Iowa State Journal of Research, 49*, 137–153.

Kabbage, M., Williams, B., & Dickman, M. B. (2013). Cell death control: The interplay of apoptosis and autophagy in the pathogenicity of *Sclerotinia sclerotiorum*. *PLoS Pathogens, 9*(4), e1003287.

Kaszubowski, A., & MacGuidwin, A. (2000). Use of GPS/GIS technology for estimating distribution of the soybean cyst nematode. Proc. 5th Intl. Conf. Precision Agriculture, Bloomington, MN.

Kotcon, J. B., Rouse, D. I., & Mitchell, J. E. (1985). Interactions of *Verticillium dahliae*, *Colletotrichum coccodes*, *Rhizoctonia solani*, and *Pratylenchus penetrans* in the early dying syndrome of Russet Burbank potatoes. *Phytopathology, 75*, 68–74.

Krebill, R. G., Barker, K. R., & Patton, R. F. (1968). Plant parasitic nematodes of jack and red pine stands in Wisconsin. *Nematologica, 13*, 33–42.

LaMondia, J. A., & Cowles, R. S. (2005). Comparison of *Pratylenchus penetrans* infection and *Maladera castanea* feeding on strawberry root rot. *Journal of Nematology, 37*, 131–135.

MacGuidwin, A. E. (1989). Abundance and vertical distribution of *Longidorus breviannulatus* associated with corn and potato. *Journal of Nematology, 21*, 404–408.

MacGuidwin, A. E. (2008). Managing diseases caused by nematodes. In D. A. Johnson (Ed.), *Potato health management* (pp. 197–208). St. Paul: APS Press.

MacGuidwin, A. E., & Bender, B. E. (2012). Estimating population densities of root lesion nematodes, *Pratylenchus* spp., from soil samples using dual active and passive assays. *Plant Health Progress*. https://doi.org/10.1094/PHP-2012-1120-01-RS.

MacGuidwin, A. E., & Bender, B. E. (2016). Development of a damage function model for *Pratylenchus penetrans* on corn. *Plant Disease, 100*, 764–769.

MacGuidwin, A. E., & Forge, T. A. (1991). Winter survival of *Pratylenchus scribneri*. *Journal of Nematology, 23*, 198–204.

MacGuidwin, A. E., & Layne, T. L. (1995). Response of nematode communities to sudangrass and sorghum-sudangrass hybrids grown as green manure crops. *Journal of Nematology, 27*(4S), 606–616.

MacGuidwin, A. E., & Rouse, D. I. (1990a). Effect of *Meloidogyne hapla*, alone and in combination with subthreshold populations of *Verticillium dahliae*, on disease symptomology and yield of potato. *Phytopathology, 80*, 482–486.

MacGuidwin, A. E., & Rouse, D. I. (1990b). Role of *Pratylenchus penetrans* in the potato early dying disease of russet Burbank potato. *Phytopathology, 80*, 1077–1082.

MacGuidwin, A. E., & Slack, S. A. (1991). Suitability of alfalfa, corn, oat, red clover, and snapbean as hosts for the potato rot nematode, *Ditylenchus destructor*. *Plant Disease, 75*, 37–39.

MacGuidwin, A. E., & Stanger, B. A. (1991). Changes in the vertical distribution of *Pratylenchus scribneri* under potato and corn. *Journal of Nematology, 23*, 73–81.

MacGuidwin, A. E., Wixted, D. A., & Hudelson, B. D. (1992). Above-ground infection of snap bean by *Ditylenchus destructor*, the potato rot nematode. *Plant Disease, 76*, 1097–1102.

MacGuidwin, A. E., Grau, C. R., & Oplinger, E. S. (1995). Impact of planting "Bell", a soybean cultivar resistant to *Heterodera glycines* in Wisconsin. *Journal of Nematology, 27*, 78–85.

MacGuidwin, A. E., Knuteson, D. L., Connell, T., Bland, W. L., & Bartelt, K. D. (2012). Manipulating inoculum densities of *Verticillium dahliae* and *Pratylenchus penetrans* with green manure amendments and solarization influence potato yield. *Phytopathology, 102*, 519–527.

Marburger, D., Conley, S., Esker, P., MacGuidwin, A., & Smith, D. (2014). Relationship between *Fusarium virguliforme* and *Heterodera glycines* in commercial soybean fields in Wisconsin. *Plant Health Progress, 15*, 11–17. https://doi.org/10.1094/PHP-RS-13-0107.

Martin, M. J., Riedel, R. M., & Rowe, R. C. (1982). *Verticillium dahliae* and *Pratylenchus penetrans*: Interactions in the early dying complex of potato in Ohio. *Phytopathology, 72*, 640–644.

Mauza, B. F., & Webster, J. M. (1982). Suppression of alfalfa growth by concomitant populations of *Pratylenchus penetrans* and two *Fusarium* species. *Journal of Nematology, 14*, 364–367.

McCarville, M. T., Marett, C. C., Mullaney, M. P., Gebhart, G. D., & Tylka, G. L. (2017). Increase in soybean cyst nematode virulence and reproduction on resistant soybean varieties in Iowa from 2001 to 2015 and the effects on soybean yield. *Plant Health Progress, 18*, 146–155.

McClintock, J. (1914). Experiments on the control of the root knot nematode. Ms. Thesis, University of Wisconsin, Madison, WI.

McLean, K. S., & Lawrence, G. W. (1993). Interrelationship of *Heterodera glycines* and *Fusarium solani* in sudden death syndrome of soybean. *Journal of Nematology, 25*, 434–439.

Melton, T. A., Noel, G. R., Jacobsen, B. J., & Hagedorn, D. J. (1985). Comparative host suitabilities of snap beans to the soybean cyst nematode (*Heterodera glycines*). *Plant Disease, 69*, 119–122.

Moore, V. M., Mitchell, P. D., Silva, E. M., & Barham, B. L. (2016). Cover crop adoption and intensity on Wisconsin's organic vegetable farms. *Agroecology and Sustainable Food Systems, 40*, 693–713.

Morgan, G. D., MacGuidwin, A. E., Zhu, J., & Binning, L. (2002a). Population dynamics and distribution of root lesion nematode (*Pratylenchus penetrans*) over a three-year potato crop rotation. *Agronomy Journal, 94*, 1146–1155.

Morgan, G. D., Stevenson, W. R., MacGuidwin, A. E., Kelling, K. A., Binning, L. K., & Zhu, J. (2002b). Plant pathogen population dynamics in potato fields. *Journal of Nematology, 34*, 189–193.

Niblack, T. L., Arelli, P. R., Noel, G. R., Opperman, C. H., Orf, J. H., Schmitt, D., Shannon, J. G., & Tylka, G. L. (2002). A new classification scheme for genetically diverse populations of *Heterodera glycines*. *Journal of Nematology, 234*, 279–288.

Oyekan, P. O., & Mitchell, J. E. (1972). The role of *Pratylenchus penetrans* in the root rot complex of canning pea. *Phytopathology, 62*, 369–373.

Pedersen, P., Tylka, G. L., Mallarino, A., MacGuidwin, A. E., Koval, N. C., & Grau, C. R. (2010). Correlation between soil pH, *Heterodera glycines* population densities, and soybean yield. *Crop Science, 50*, 1458–1464.

Perry, V. G. (1958). Parasitism of two species of dagger nematode (*Xiphinema americanum* and *X. chambersi*) to strawberry. *Phytopathology, 48*, 420–423.

Perry, V. G. (1959). Anatomy, taxonomy, and control of certain spiral nematodes attacking bluegrass in Wisconsin. PhD thesis, University of Wisconsin.

Phibbs, A., & Leisso, R. (2009). Plant disease fact sheet: Corky ringspot of potato. http://pestsurvey.wi.gov.

Phibbs, A., Barta, A., Lueloff, S. (2016). Wisconsin pest survey report: Soybean cyst nematode survey. http://datcp.wi.gov/Documents/StatusSoybeanCystNematodeWisconsin.pdf.

Phibbs, A., Christianson, S., Barta, A. (2017). Wisconsin crop disease survey. https://datcp.wi.gov/Pages/Programs_Services/PestSurvey.aspx.

Powers, T., Harris, R., Higgins, R., Mullin, P., & Powers, K. (2017). An 18S rDNA perspective on the classification of Criconematoidea. *Journal of Nematology, 49*, 236–244.

Riffle, J. W. (1962). Nematodes associated with maple dieback and maple blight. *Phytopathology, 52*, 749.

Riffle, J. W. (1963). *Meloidogyne ovalis* (Nematode: Heteroderidae) a new species of root knot nematode. *Proceedings of the Helminthological Society of Washington, 30*, 287–292.

Riggs, R. D., Niblack, T. L., Kinloch, R. A., MacGuidwin, A. E., Mauromoustakos, A., & Rakes, L. (2001). Overwinter population dynamics of *Heterodera glycines*. *Journal of Nematology, 33*, 219–236.

Rotenberg, D., MacGuidwin, A. E., Saeed, I. A. M., & Rouse, D. I. (2004). Interaction of spatially separated *Pratylenchus penetrans* and *Verticillium dahliae* on potato measured by impaired photosynthesis. *Plant Pathology, 53*, 294–302.

Saeed, I. A. M., MacGuidwin, A. E., Rouse, D. I., & Sharkey, T. D. (1999). Stomatal limitation to photosynthesis in potato infected by *Verticillium dahliae* and *Pratylenchus penetrans*. *Crop Science, 39*, 1340–1346.

Schmitt, D. P., & Shannon, G. (1992). Differentiating soybean responses to *Heterodera glycines* races. *Crop Science, 32*, 275–277.

Schroeder, N. E., & MacGuidwin, A. E. (2007). Incorporation of a fluorescent compound by live *Heterodera glycines*. *Journal of Nematology, 39*, 43–49.

Schroeder, N. E., & MacGuidwin, A. E. (2010a). Behavioral quiescence reduces the penetration and toxicity of exogenous compounds in second-stage juveniles of *Heterodera glycines*. *Nematology, 12*, 277–287.

Schroeder, N. E., & MacGuidwin, A. E. (2010b). Mortality and behavior in *Heterodera glycines* juveniles following exposure to isothiocyanate compounds. *Journal of Nematology, 42*, 194–200.

Schroeder, N. E., Gerhardt, D. J., Phibbs, A., & MacGuidwin, A. (2008). First report of *Cactodera milleri* in Wisconsin. *Plant Disease, 92*, 656.

Sher, S. A. (1968). Revision of the genus *Hirschmanniella* Luc and Goodey, 1963 (Nematoda: Tylenchoidea). *Nematologica, 14*, 243–275.

da Silva, M. P., Tylka, G. L., & Munkvold, G. P. (2016). Seed treatment effects on maize seedlings coinfected with *Fusarium* spp. and *Pratylenchus penetrans*. *Plant Disease, 100*, 431–437.

Smart, G. C., Jr., & Darling, H. M. (1963). Pathogenic variation and nutritional requirements of *Ditylenchus destructor*. *Phytopathology, 53*, 374–381.

Tabor, G. M., Tylka, G. L., Behm, J. E., & Bronson, C. R. (2003). *Heterodera glycines* infection increases incidence and severity of brown stem rot in both resistant and susceptible soybean. *Plant Disease, 87*, 665–661.

Temp, M., & Hagedorn, D. J. (1967). Plant parasitic nematodes in soil samples from pea fields with *Aphanomyces* root rot potential. *Plant Disease Reporter, 52*, 190–192.

Thistlethwayte, B. (1970). Reproduction of *Pratylenchus penetrans* (Nematoda: Tylenchida). *Journal of Nematology, 2*, 101–105.

Thorne, G. (1969). *Hypsoperine ottersoni* sp. n. (Nemata, Heteroderidae) infesting Canary grass, *Phalaris arundinacea* (L.) Reed in Wisconsin. *Proceedings of the Helminthological Society of Washington, 36*, 98–102.

USDA NASS. (2017). United States Department of Agriculture, National Agriculture Statistics Service. https://www.nass.usda.gov.

Vanstone, V., Rathjen, A. J., Ware, A. H., & Wheeler, R. D. (1998). Relationship between root lesion nematodes (*Pratylenchus neglectus* and *P. thornei*) and performance of wheat varieties. *Australian Journal of Experimental Agriculture, 38*, 181–188.

Viaene, N. M., & Abawi, G. S. (1998). Management of *Meloidogyne hapla* on lettuce in organic soil with sudangrass as a cover crop. *Plant Disease, 82*, 945–952.

Wang, M., & Goldman, I. L. (1996). Resistance to root knot nematode (*Meloidogyne hapla* Chitwood) in carrot is controlled by two recessive genes. *The Journal of Heredity, 87*, 119–123.

Wang, J., Donald, P. A., Niblack, T. L., Bird, G. W., Faghihi, J., Ferris, J. M., Grau, C., Jardine, D. J., Lipps, P. E., MacGuidwin, A. E., Melakeberhan, H., Noel, G. R., Peirson, P., Riedel, R. M., Sellers, P. R., Stienstra, W. C., Todd, T. C., Tylka, G. L., Wheeler, T. A., & Wysong, D. S. (2000). Soybean cyst nematode reproduction in the North Central United States. *Plant Disease, 84*, 77–82.

WAS. (2017). Wisconsin Agricultural Statistics. https://www.nass.usda.gov/Statistics_by_State/Wisconsin/Publications/Annual_Statistical_Bulletin/2017AgStats_web.pdf.

Webber, A. J., Jr., & Barker, K. R. (1967). Biology of the pseudo root knot nematode *Hypsoperine ottersoni*. *Phytopathology, 57*, 723–728.

WI DATCP. (2017). Wisconsin Department of Agriculture, Trade, and Consumer Protection. http://pestsurvey.wi.gov/plantdisease/pdf/ornamentals/Stem_and_bulb_nematode-ALERT.pdf.

Wingfield, M. J., & Blanchette, R. A. (1983). The pine-wood nematode, *Bursaphelenchus xylophilus* in Minnesota and Wisconsin: Insect associates and transmission studies. *Canadian Journal of Forest Research, 13*, 1068–1076.

Wisconsin Plant Industry Bureau Annual Report. (2016). https://datcp.wi.gov/Documents/PIBAnnualReport2016.pdf.

Zedler, P. H., Anchor, T., Knuteson, D., Gratoon, C., & Barzen, J. (2009). Using an ecolabel to promote on-farm conservation: The Wisconsin Healthy Grown experience. *International Journal of Agricultural Sustainability, 7*, 61–74.

Chapter 8
Plant Parasitic Nematodes of North Dakota and South Dakota

Guiping Yan and Richard Baidoo

8.1 Introduction

Nematode studies began in the States of North Dakota and South Dakota only in the mid-twentieth century when nematologists were first employed by different state research institutions. Since then, a number of surveys and experiments have been conducted to annotate occurrence, abundance, economic importance and develop management strategies for different plant parasitic nematodes in the Dakotas. This chapter devotes to plant parasitic nematodes which limit or potentially threaten crop production in these states and their management strategies in sustainable agriculture.

8.2 Economically Important Crops in North Dakota and South Dakota

Production agriculture is the largest sector of the economies of both North and South Dakota making up to 25% of their economic bases (USDA-NASS 2015a, b). In North Dakota, the value of crop production in recent years has been estimated at \$7–10 billion, with an economic impact of \$20–30 billion (Anonymous 2016; USDA-NASS 2016b).

Major crops produced in North Dakota include soybean, wheat, sunflower, corn, dry edible beans, sugar beet and canola. Soybeans, corn, wheat, sugar beet and canola are the top revenue-producing cash crops for the state (USDA-NASS 2016b). The state maintained its position as the top U.S. producer of spring wheat, durum

G. Yan (✉) · R. Baidoo
Department of Plant Pathology, North Dakota State University, Fargo, ND, USA
e-mail: guiping.yan@ndsu.edu

S. A. Subbotin, J. J. Chitambar (eds.), *Plant Parasitic Nematodes in Sustainable Agriculture of North America*, Sustainability in Plant and Crop Protection,
https://doi.org/10.1007/978-3-319-99588-5_8

wheat, dry edible beans, pinto beans, canola and flaxseed in 2015 (Table 8.1). These crops are produced not only for their numerous food and industrial uses, but also for export, contributing immensely to the economy of the state.

About 94% of the soybeans produced in North Dakota is shipped to other states, whereas, approximately 70–75% of the soybeans are exported out of the country. Soybeans are used as food products, animal feed, and hundreds of industrial applications including productions of vegetable oil, margarine, inks, paints, biodiesel fuel, solvents and hydraulic fluids. The canola biodiesel facility at Velva, North Dakota is capable of producing 322 million liters of biodiesel annually. Corn ethanol is also a growing industry in North Dakota. Ethanol plants currently in operation

Table 8.1 Major crops produced in North and South Dakota (2015)

	Planted hectares ($\times 10^6$)	Harvested hectares ($\times 10^6$)	Production in kilogram ($\times 10^9$)	Sales in \$ ($\times 10^6$)	U.S. rank (2015)
Soybean	2.44	2.42	6.22	2,253.45	8
Corn	–	1.38	15.93	1,627.48	9
Wheat	3.07	2.99	8.99	1,544.51	1
Canola	0.59	0.59	1.64	436.04	1
Hay	–	1.01	2.57	312.41	9
Beans, dry edible	0.25	0.23	0.55	245.86	1
Potato	0.03	0.024	1.28	210.08	4
Sunflower	0.27	0.27	0.71	205.96	2
Barley	0.29	0.26	0.94	192.96	1
Pea, dry edible	0.23	0.22	0.76	131.21	1
Sugar beet	0.08	0.08	7.74	–	3
Lentil	0.12	0.12	0.23	104.86	2
Flaxseed	0.14	0.13	0.21	64.15	1
Oat	0.12	0.05	0.11	16.34	4
South Dakota					
Corn	2.3	2.22	28.67	2,642.98	6
Soybean	2.10	2.09	6.91	2,328.83	8
Wheat	0.91	0.87	3.0	439.63	6
Sunflower	0.23	0.22	0.66	178.67	1
Sorghum	0.10	0.09	0.99	42.47	7
Oats	0.12	0.04	0.13	17.59	1
Millet	0.02	0.01	0.04	4.76	3
Safflower	0.008	0.008	0.01	3.64	1
Flaxseed	0.004	0.004	0.004	1.15	3
Hay and alfalfa	–	0.03	0.60	–	3

'–' Means data is not available
Source: USDA-NASS (2016a, b)

produce nearly 1.5 billion liters of ethanol annually. North Dakota is number one in the production of two wheat classes: hard red spring and durum. Hard red spring is known for its gluten strength used for the production of high quality bread flours. Durum wheat is used for making spaghetti, lasagna and, at least, 350 other pasta shapes. North Dakota's production of spring wheat and durum wheat in 2014 accounted for 53% and 52% of the total U.S. production, respectively. Canola accounted for 87% and flaxseed accounted for 92% of what was produced in Minnesota. North Dakota produced nearly 45% of the nation's sugar beet crop. Monetary contribution from the sales of the sugar beet produce to the economies in the two states in 2014 were estimated at $2,066 per hectare or $544.6 million.

South Dakota's agriculture industry has more than 7 million ha of cropland and 9 million ha of pastureland and $25.6 billion of economic impact each year, constituting more than 30% of the state's total output (Anonymous 2014). Revenue generated from crop production and further processing alone is more than $13.3 billion annually and is responsible for 70,104 jobs (Anonymous 2014; USDA-NASS 2016a).

South Dakota consistently ranked amongst the top ten states for production of several crops including spring wheat, flaxseed, hay, oat, rye and sunflower seeds. Corn, soybean, oat and wheat are South Dakota's major cash crops; sunflowers, sorghum, flaxseed and barley are also grown. In 2015, total planted area of principal crops including hay, was 3.9 million ha (USDA-NASS 2015a, 2016a). The most economically important crops within the top ten in the US ranking and their production acreages, total production or total sales in 2015 for both states are summarized in Table 8.1.

8.3 Common Plant Parasitic Nematodes in North and South Dakota Fields

8.3.1 Historical Perspective

In 1949, Chitwood discovered the grass cyst nematode, *Punctodera punctata*, during routine soil inspections of potato fields for the golden cyst nematode in Pembina County, North Dakota (Chitwood 1949). This was the first record of this species in the United States. Following this, a *Cactodera* sp. (former *H. cacti* group) was discovered in a soil sample from North Dakota (Spears 1956). In 1958, *Heterodera schachtii* was reported in a soil sample from Cass County (Caveness 1958) but the occurrence of the nematode in the state was not confirmed at that time. From 1963 to 1968, several other nematode genera were detected during surveys in commercial fields of barley, wheat and forage grasses in North Dakota. The plant parasitic nematodes were identified from the genera *Tylenchorhynchus*, *Aphelenchoides*, *Xiphinema*, *Heterodera*, *Pratylenchus*, *Paratylenchus*, *Meloidogyne*, *Hoplolaimus*, *Tetylenchus*, *Helicotylenchus* and *Trichodorus* (Pepper 1963, 1968).

Tylenchorhynchus spp. were commonly associated with cereals and grasses showing marked root damage, but the cause of the root damage was not ascertained. The associated *Meloidogyne* sp. was identified as *M. incognita* and was detected in greenhouse flower beds adjacent to underground steam lines at the North Dakota State University (NDSU) campus at Fargo, probably since it could not survive North Dakota winter temperatures (Pepper 1963, 1968). Since then, several other plant parasitic nematodes surveys have been conducted and suggested that only selected nematode genera are frequently encountered in North and South Dakota fields (Thorne and Malek 1968; Donald and Hosford 1980; Krupinsky et al. 1983; Bradley et al. 2004; Nelson et al. 2012).

Recent nematode surveys conducted at the North Dakota State University (NDSU) on field crops such as corn, wheat, barley, potato and pea also resulted in detections of *Paratylenchus* spp., *Pratylenchus* spp., *Helicotylenchus* spp., *Tylenchorhynchus* spp. and *Xiphinema* spp. as the most common genera of plant parasitic nematodes in North Dakota agricultural fields (Plaisance and Yan 2015; Upadhaya et al. 2016; Yan et al. 2015b; Yan and Plaisance 2016). These findings corroborate the previous assertion that only specific adapted groups of plant parasitic nematodes are present in the state. In another survey, soybean fields or fields with history of soybean cyst nematode (SCN) were selected to ascertain the incidence and abundance of plant parasitic nematodes and their possible association with SCN. The nematodes identified per 200 g soil were *Helicotylenchus* spp. (incidence: 49%; highest density: 1800 specimens; average density: 174 specimens), *Tylenchorhynchus* spp. (41%; 340; 30), *Paratylenchus* spp. (37%; 2480; 151), *Pratylenchus* spp. (19%; 245; 9), *Xiphinema* spp. (7%; 180; 4), *Paratrichodorous* spp. (4%; 60; 1), *Hoplolaimus* spp. (3%; 140; 2), *Mesocriconema* spp. (1%; 300; 2), SCN juveniles from soil (24%; 1200; 46) and SCN eggs from cysts (56%; 21,540; 501). Interestingly, these nematodes had no or poor association with SCN in the 155 fields surveyed in 2015 (Yan and Plaisance 2016). A summary of plant parasitic nematodes identified in North and South Dakota and their associated crops are presented in Table 8.2.

8.3.2 *Soybean Cyst Nematode,* Heterodera glycines *in North Dakota*

8.3.2.1 Detection and Distribution

The soybean cyst nematode is considered the most damaging pathogen of soybeans in the USA and by far, the most economically important nematode in North Dakota. Heavily infested fields show patchy yellowing of the foliage (chlorosis), stunting of plants, and thin stands with swollen females and cysts attached to roots. The females first appear as lemon-shaped, cream-colored cysts, which later turn brown while still attached to plant roots (Fig. 8.1a, b). Losses of up to 30% have been reported

Table 8.2 Plant parasitic nematodes identified in North Dakota and South Dakota and their associated host plants

Nematode	Host and rhizosphere soil	Reference
Cactodera sp.	Potato	Spears (1956)
Criconema permistus	Grasses	Donald and Hosford (1980) and Donald (1978)
Geocenamus tenidens	Prairie sod	Thorne and Malek (1968)
Helicotylenchus digonicus	Red clover	Donald (1978) and Donald and Hosford (1980)
H. dihystera	Sugar beet	Caveness (1958)
H. erythrinae	Sugar beet	Caveness (1958)
H. exallus	Grasses, corn	Donald (1978) and Krupinsky et al. (1983)
H. glissus	Grasses	Krupinsky et al. (1983)
H. leiocephalus	Unknown	Krupinsky et al. (1983)
H. microlobus	Soybean	Yan et al. (2017c)
H. pseudorobustus	Grasses, red clover	Donald and Hosford (1980), Donald (1978), and Krupinsky et al. (1983)
Helicotylenchus sp.	Grasses, barley, wheat	Caveness (1958), Donald (1978), and Plaisance et al. (2016a, b)
Hemicycliophora sp.	Alfalfa	Caveness (1958) and Donald (1978)
Heterodera glycines	Soybean	Bradley et al. (2004), Smolik (1995), and Baidoo et al. (2017)
H. schachtii	Sugar beet	Caveness (1958) and Nelson et al. (2012)
Heterodera sp.	Grasses	Donald (1978), Pepper (1968), and Krupinsky et al. (1983)
Hoplolaimus galeatus	Grasses	Krupinsky et al. (1983)
H. stephanus	Soybean	Yan et al. (2016a)
Hoplolaimus sp.	Sugar beet	Caveness (1958) and Plaisance et al. (2016a)
Meloidogyne incognita	Flower bed, NDSU	Pepper (1968)
Merlinius lineatus	Barley	Pepper (1968)
Mesocriconema raskiensis	Grasses	Donald and Hosford (1980), Donald (1978), and Thorne and Malek (1968)
M. xenoplax	Grasses	Krupinsky et al. (1983)
Nagelus aberrans	Prairie sod	Thorne and Malek (1968)
Neodolichodorus pachys	Grasses	Thorne and Malek (1968) and Krupinsky et al. (1983)
Paratylenchus hamatus	Alfalfa, grasses	Donald and Hosford (1980) and Donald (1978)
Paratylenchus sp.	Barley, sugar beet	Caveness (1958) and Pepper (1968)
Paratrichodorus allius	Potato	Yan et al. (2016e) and Huang et al. (2017a, b)
Pratylenchus agilis	Prairie sod	Thorne and Malek (1968)
P. minyus	Sugar beet	Caveness (1958)
P. neglectus	Wheat	Yan et al. (2016d)

(continued)

Table 8.2 (continued)

Nematode	Host and rhizosphere soil	Reference
P. scribneri	Potato	Huang and Yan (2017) and Yan et al. (2016c)
P. vexans	Grasses	Donald and Hosford (1980), Donald (1978), and Tylka and Marett (2014)
Pratylenchus sp.	Grasses, soybean	Krupinsky et al. (1983) and Yan et al. (2017d, e)
Punctodera punctata	Potato, wheat	Chitwood (1949), Pepper (1968), and Spears (1956)
Quinisulcius acutus	Barley, sugar beet, wheat	Caveness (1958), Pepper (1968), and Thorne and Malek (1968)
Q. acutoides	Unknown	Donald (1978) and Pepper (1968)
Rotylenchus spp.	Sugar beet	Caveness (1958)
Trichodorus sp.	Barley	Pepper (1968)
Trophurus minnesotensis	Unknown	Thorne and Malek (1968)
Tylenchorhynchus canalis	Grasses	Krupinsky et al. (1983)
T. claytoni	Barley, sugar beet, wheat	Pepper (1968)
T. cylindricus	Barley, wheat	Pepper (1968)
T. latus	Barley	Pepper (1968)
T. macrurus	Barley	Pepper (1968)
T. maximus	Grasses	Donald (1978) and Krupinsky et al. (1983)
T. nudus	Barley, corn, grasses, sage	Donald (1978), Peper (1968), and Krupinsky et al. (1983)
T. robustus	Grasses	Krupinsky et al. (1983)
Tylenchorhynchus sp.	Grasses, barley, wheat, sugar beet	Donald (1978), Pepper (1968), and Plaisance et al. (2016a, b)
Xiphinema americanum	Barley, wheat, shelter belt trees, cottonwood	Caveness (1958), Donald (1978), and Plaisance et al. (2016a, b)
Xiphinema sp.	Barley	Pepper (1968)

even when there are no obvious above-ground symptoms (Nelson et al. 2012; Niblack et al. 2004).

Since its first detection in 1954 in North Carolina, USA (Winstead et al. 1955), the nematode has spread to other soybean producing areas in many states (Tylka and Marett 2014) and was reported in 2003 from Richland County in North Dakota (Bradley et al. 2004). By 2012, the nematode had been confirmed in 12 other counties of North Dakota (Berghuis 2016), and currently is present in 19 soybean-producing counties in the eastern half of the state (Berghuis 2016; Yan et al. 2015a, b) (Fig. 8.1d).

In 2013, a grower-based SCN sampling program, sponsored by the North Dakota Soybean Council, was established to increase SCN awareness and to monitor its occurrence and distribution in North Dakota. The participants receive prepaid sampling bags at their County Extension office, the North Dakota Soybean Council

Fig. 8.1 (**a**) Soybean field showing patchy distribution of chlorotic foliage as a result of soybean cyst nematode infestation. (Courtesy of Smolik J. D., SDSU); (**b**) Soybean roots showing soybean cyst nematode cysts. Cream-colored cysts (vertical arrow) and one nodule on soybean roots (horizontal arrow); (**c**) Brown cysts on soybean roots (vertical arrow). (Courtesy of Sam Markell, NDSU.); (**d**) Detection year and distribution of *Heterodera glycines* (SCN) in North Dakota. (Credit: Dr. Sam Markell, NDSU)

offices, field days and other events and submit the samples to Agvise Laboratories (Benson, MN, USA) for analysis. The number of samples submitted in 2013, 2014, and 2015 were 193, 579, and 943 respectively. Approximately, 30% of the samples submitted had, at least, 50 eggs/100 cm^3 soil of which approximately 50% had more than 200 eggs/100 cm^3 and 10% exceeded 10,000 eggs/100 cm^3. Between 2013 and 2015, sampling was done in 39 of the 53 North Dakota counties and resulted in 19 counties being positive for SCN. The highest SCN population densities (≥2,000 eggs/100 cm^3 soil) occurred in Cass, Richland and Trail Counties (Berghuis 2016). Previously, higher numbers of 550–20,000 eggs per 100 cm^3 soil were detected in Richland County, North Dakota (Bradley et al. 2004). A SCN distribution map for North Dakota was then generated based on the data. The spread of the nematode from the southeastern part across the mid and northeastern parts of the state strongly suggests that preemptive control measures against this species need to be implemented.

During surveys, samples with low level of egg densities (<50 eggs/100 cm^3) were excluded due to the possibility of false positives. This is due to the fact that it is always difficult to morphologically differentiate soybean cyst nematode eggs from those of other cyst-forming nematodes. Therefore, investigators report counties as positive only if multiple samples typically have over 50 eggs/100 cm^3 of soil. Undoubtedly, the morphological diagnostic approach used in such investigations has the potential to underestimate the number of counties with positive SCN in North Dakota. New technologies that sensitively and specifically detect SCN directly from soil with low densities, have been developed (Baidoo et al. 2017; Yan and Baidoo 2017). Such molecular-based detection techniques undoubtedly provide a viable alternative or compliment the traditional diagnostic methods. The spread of the nematode from the southeastern part across the mid and northeastern parts of the state indicates prophylactic control measures against this nematode are necessary.

8.3.2.2 Variation in Virulence Phenotypes

Soybean cyst nematode populations are either classified into different races or HG types. The race test was based on resistance or susceptible reaction to four SCN differential lines: Peking, Picket, PI 88788, and PI 90763, and standard susceptible check, Lee 74 (Golden et al. 1970; Riggs and Schmitt 1988; Riggs 1988). However, as more soybean differential lines were introduced, not only did race-based characterization become more complicated, but variability of SCN populations were not fully characterized by the race system. A new system of characterizing SCN populations was developed known as the *Heterodera glycines* (HG) type test (Niblack et al. 2002). With the HG type test, SCN populations are characterized by their ability to reproduce on soybean indicator lines with seven different sources of genetic resistance. HG typing considers phenotypic diversity and SCN reproduction differences on soybean lines PI 548402 (Peking), PI88788, PI 90763, PI 437654, PI 209332, PI 89772, and PI 548316 (Cloud) with respect to a standard susceptible check (Niblack et al. 2002). HG type determination not only reveals the diversity of the SCN populations, but can provide information of resistance sources that are effective against SCN. After a SCN population has been characterized using the HG type test, a grower can determine which sources of resistance to grow that would minimize the buildup of SCN in a particular field. Thus, the knowledge of the occurrence and distribution of virulent phenotypes (HG types) provides valuable information regarding sustainable and effective use of resistant cultivars.

The HG type 0, previously known as race 3, was the only SCN type reported in North Dakota until 2016 (Bradley et al. 2004). Soil samples collected in 2015 and 2016, and HG type tests conducted at the North Dakota State University suggested that other HG types are present in North Dakota (including HG type 0, 7, 2.7, 2.5, 5, and 2.5.7.) even though the HG type 0 and 7 are the most predominant populations. Interestingly, some North Dakota SCN populations were able to reproduce on the most widely used resistance, PI 88788 (Chowdhury et al. 2016 2017). The SCN populations in North Dakota are increasing in virulence diversity, as reported for other states (Niblack et al. 2002). As HG types diversify in North Dakota, the use of resistance for management of this nematode may no longer be sustainable.

8.3.2.3 Management of Soybean Cyst Nematode in Soybean Fields in North Dakota

Management of SCN in fields begins with soil sampling to determine egg levels. Once SCN is detected, the most common practices include the use of SCN resistant varieties and crop rotation. These two methods have been found to be most effective (Mathew et al. 2014).

8.3.2.3.1 Resistant Varieties

The use of resistant varieties is a major SCN management tool. SCN reproduction is inhibited on roots of SCN-resistant varieties. In North Dakota, early maturing varieties are being developed with SCN resistance and varieties containing the two common sources of resistance, PI88788 and Peking that are still effective against SCN. Each year, the North Dakota State University (NDSU) evaluates nearly 40 soybean varieties for SCN resistance under greenhouse and field conditions at three to four locations within the state. This program is funded by the North Dakota Soybean Council. Thereafter, information on SCN resistance is made available to growers through an annual bulletin of NDSU Extension Service publication A843, "*North Dakota Soybean Performance Testing*."

It is important to note that while varieties may have the same source of resistance, the degree of resistance in each variety varies. Thus, varieties marketed as SCN-resistant may be truly resistant or have only low to moderate levels of resistance. Therefore, selection of the most resistant variety possible and subsequent monitoring of the field for SCN are important. Previously, only HG type 0 (Race 3) was known in ND, but other HG types have recently been reported. Interestingly, the HG type 2.5.7 population of ND could reproduce on the most widely known source of resistance PI88788 which suggests that new sources of resistance are needed, in the future, for sustainable management of this nematode.

8.3.2.3.2 Crop Rotation

Crop rotation is another critical component of SCN management. Rotation of soybean varieties with different sources of resistance or non-host crops is imperative for long-term management. Common rotational crops such as wheat and corn, are used by growers to reduce population levels of SCN in North Dakota. Continuous reductions in SCN population levels can be achieved over years of planting non-host crops, but the greatest reduction in egg levels occurs the first year a non-host is planted, meaning that many years of crop rotation with non-hosts may be required to reduce high egg levels to low levels. On the other hand, when susceptible crops are grown sequentially, egg levels can become high enough so that growing soybeans may not be practical. Also, the pathogen may overcome resistance if soybean varieties with the same source of resistance are sequentially planted. A minimum of a 2-year rotation is critical for SCN management, although a rotation out of soybean for 2 years is beneficial.

Dry bean is an excellent host for soybean cyst nematode, but canola, dry edible peas, alfalfa, corn, forage grasses, sorghum and sugar beet are considered non- or poor- hosts. Soybean cyst nematode can reproduce on some weeds. Henbit and field pennycress, allow substantial reproduction of SCN. About 31 weed species are known to support SCN reproduction in North Dakota (Poromarto et al. 2015). These weed species and other crop hosts in North Dakota and Northern Minnesota that potentially support SCN reproduction can undermine the effectiveness of SCN management by crop rotation.

A recent study revealed that annual ryegrass (variety not stated: VNS), camelina (Bison), carinata (VNS), Ethiopian cabbage (VNS), faba bean (VNS), foxtail millet (Siberian), radish (Daikon), dwarf essex rape, red clover (Allington), sweet clover (VNS), triticale (Winter 336) and winter rye (Dylan) do not support SCN reproduction (Acharya et al. 2017). However, cowpea (VNS), crimson clover (Dixie) and turnips (Purple Top, Pointer), Austrian winter pea (VNS), field pea (Aragorn, Cooper), forage pea (Arvika) and hairy vetch (VNS) could support some levels of SCN reproduction. Cover crops that are non-host to SCN can be incorporated into a crop rotation system for a sustainable management of this pathogen.

8.3.2.3.3 Seed Treatment

A number of seed treatment products aimed at SCN control are being evaluated at NDSU (Mathew et al. 2014). Preliminary results suggest that some chemical products may reduce SCN numbers. Few seed treatment products aimed at SCN management are labeled and marketed as, (1) Avicta®500FS and (2) Avicta® Complete Beans 500 and (3) Poncho Votivo®. The Avicta products are a blend of different proportions of nematicide, insecticide and fungicide, while the Poncho Votivo product contains a *Bacillus firmus* bacterium which creates a living barrier that prevents nematodes from reaching the roots. The performance of these seed treatments is generally unpredictable, depends on specific soil and weather conditions and does not guarantee increased yields.

It is worthy of note that no single management approach provides an adequate control of SCN and hence, an integrated management scheme in which many other strategies including use of resistant varieties, crop rotation, cover crops, tillage practices, phytosanitary practices, chemical seed treatment, etc., are required for a sustainable management of this nematode.

8.3.2.4 Soybean Cyst Nematode Is a Threat to Dry Bean Production in North Dakota

Between 2007 and 2009, the effect of soybean cyst nematode (HG type 0) on dry bean was investigated. The cultivars GTS-900 (pinto bean), Montcalm (kidney bean) and Mayflower (navy bean) were evaluated in eight field experiments at four locations in North Dakota. The soybean cyst nematode reproduced on all three dry bean cultivars with reproduction factors ranging from 6.1 to 1.2. Plant growth and seed yield including pod number (PN), pod weight (PW), seed number (SN) and seed weight (SW), were significantly reduced by SCN (Poromarto and Nelson 2009; Poromarto et al. 2010). Recently, SCN was implicated in irregular patches of stunting and yellowing in a commercial dry bean field in the neighboring State of Minnesota (Yan et al. 2017a). These results indicate that SCN is a potential threat to the large dry bean industry in the North Dakota and Northern Minnesota region (Pormarto et al. 2010). Consequently, SCN resistance sources from plant introductions of *Phaseolus vulgaris* have been identified and SCN resistance is currently being introduced into breeding materials for the NDSU Dry Bean Breeding Program, while at the same time, the genetic basis for SCN resistance or susceptibility in dry bean is also being characterized (Nelson 2017; Shalu et al. 2017).

8.3.3 Sugar Beet Cyst Nematode, Heterodera schachtii

Sugar beet cyst nematode (SBCN) is a major problem for many sugar beet (*Beta vulgaris*) growing regions. The species was first described in 1859 in Germany and is now distributed worldwide. In the United States, SBCN was first reported in Utah in 1895 and is present in all sugar beet producing states except Minnesota and Eastern North Dakota.

The sugar beet cyst nematode was confirmed to be present in the Yellowstone Valley of Western North Dakota in 2011 (Nelson et al. 2012), even though it was first reported, although not confirmed, in the state in 1958 (Caveness 1958). Population densities ranged from 100 to 1,750 eggs/100 cm^3 soil in four fields in the Yellowstone Valley. Plants infected with SBCN show stunting and reduced leaf growth, with older outer leaves turning yellow and wilted during the hot period of the day. The taproot tends to be stunted with fibrous "bearded roots" (Fig. 8.2a). The most important confirmation of SBCN infection is the presence of white to yellow lemon-shaped females attached to feeder roots (Fig. 8.2a) or yellow-brown cysts (dead mature females) in soil (Fig. 8.2b) (Khan et al. 2016). Interestingly, the nematode has not been detected in Eastern North Dakota in the Red River Valley where sugar beet is mainly produced (Porter and Chen 2005).

Fig. 8.2 (**a**) White, lemon-shaped females feeding on root hairs of sugar beet. (Photo: courtesy of Steve Poindexter, Michigan State University). (**b**) Yellow-brown female cysts from a sugar beet cyst nematode-infested field in North Dakota. (Photo: courtesy of Guiping Yan, NDSU). (**c**) Detection and distribution of *Heterodera glycines* (SCN) in South Dakota. (Credit: Dr. Emmanuel Byamukama, SDSU)

8.3.3.1 Management of SBCN

The first step in management is sampling soil for the presence of SBCN cysts or juveniles or the presence of white females on root. Field symptoms such as patchy distribution of chlorotic leaves, stunted plants, profuse fibrous roots, *etc.*, may be similar to that caused by other stress factors. If soil sampling shows that SBCN is absent from a field, then prevention of SBCN introduction into the field will be the key strategy. This can be achieved by avoiding movement of machinery and equipment from areas infested with SBCN into non-infested fields, washing thoroughly machineries and equipment after use, especially those coming from nematode-infested regions with known SBCN problems, avoiding or limiting the use of host crops in rotation, good control of weed hosts, and taking proper sanitation measures between infested areas and non-infested areas (Khan et al. 2016; Anonymous 2017).

Various strategies are recommended to reduce cyst nematode populations below the economic threshold: use of tolerant cultivars, rotation with non-host crops, use of trap crops, early planting, weed control, phytosanitation, nematicide treatment, *etc.* Sugar beet cyst nematode-tolerant cultivars should be planted, if available, and rotated with non-host crops, including wheat, soybean, barley, corn, potato and alfalfa. Weeds that are hosts for SBCN such as shepherd's purse, common lambsquarters, chickweed, pigweed, dock, and purslane, must be controlled. Rotations with non-host crops may reduce initial SBCN population by 40–60% annually and a 3 to 4-year rotation is needed in heavily infested fields to reduce population density below damaging levels (Khan et al. 2016). Trap crops attract SBCN but prevent them from developing and reproducing, thus reducing population densities drastically. Some SBCN tolerant cultivars of oilseed radish including Defender, Image, and Colonel, and White mustard, are effective (Khan et al. 2016). Early planting is recommended, when soil temperatures are not favorable for infection and less than 15 °C. Chemical nematicides may be effective, but are typically difficult to apply and may be uneconomical. Biological seed treatment with *Pasteuria nishizawe* may help manage SBCN on tolerant sugar beet cultivars (Khan et al. 2016).

8.3.4 Lesion Nematodes, Pratylenchus *spp.*

Pratylenchus is a major nematode genus frequently found in North Dakota potato fields. These nematodes, apart from the damage they cause through their feeding activities, also interact with other organisms to increase disease incidence and severity. *Pratylenchus* spp. infect potato tubers causing a scabby appearance with sunken lesions or dark, wart-like bumps that turn purple on tubers in storage. Yield losses may be exacerbated by interaction with the fungus *Verticillium dahliae* causing a disease known as Potato Early Dying disease complex (PED) (MacGuidwin and Rouse 1990).

During 2015, 48 out of 54 soil samples collected from potato fields in Sargent County, North Dakota contained root lesion nematodes with population densities ranging from 125 to 1,900/kg of soil. Initial population density of 1,540 root lesion nematodes/kg soil increased to 9,163 specimens/kg soil and 48 specimens/g roots on potato cultivar, 'All Blue' after 10 weeks. In April 2016, the nematode was identified as *P. scribneri* (Huang and Yan 2017; Yan et al. 2016c), and found to be the most prevalent plant parasitic nematode infesting potato fields in Sargent County, North Dakota (Plaisance et al. 2016b). Preliminary greenhouse studies showed that potato and corn were good hosts of *P. scribneri* while wheat and soybeans were poor and intermediate hosts, respectively (Plaisance et al. 2016b).

Similarly, in 2015, soil samples collected from a wheat field in Walsh County, North Dakota were found to have root lesion nematodes from 125 to 1,044/kg soil. This nematode, with an initial density of 500 root lesion nematodes/kg soil, could reproduce on commercial and common wheat cultivars Glenn and Faller to an average of 24 or 20 root lesion nematodes per gram root after 10 weeks. The nematode was identified as *P. neglectus* (Yan et al. 2016d).

Two new, unnamed *Pratylenchus* species have been reported in two different fields in Richland County, North Dakota (Yan et al. 2017d, e). In 2015, two soil samples collected from a soybean field in Walcott, North Dakota contained 125 and 350 root lesion nematodes per kg of soil. In 2016, four soil samples were collected from the same field and all the samples had root lesion nematodes ranging from 300 to 2000. One soil sample with 350 root lesion nematodes per kg soil was planted to a commercial soybean cultivar, Barnes. After 15 weeks of growth in a greenhouse (22 °C, 16 h light), the final population density in soil was 1,518 ± 541 root lesion nematodes per kg soil and 25 ± 20 per g of fresh roots. Reproduction factor of the nematode was 5.02, indicating that this nematode infected and reproduced well on the soybean cultivar (Yan et al. 2017d). Again in 2015 and 2016, 10 of 11 soil samples collected from a soybean field in Hankinson, North Dakota, contained root lesion nematodes ranging from 150 to 875/kg of soil. One soil sample with 300 lesion nematodes/kg was used to inoculate soybean cultivar, Barnes. After 15 weeks of growth in the greenhouse, the population had increased to a final density of 460 ± 181 lesion nematodes/kg in soil and 34 ± 21 lesion nematodes/g of fresh roots. The reproduction factor of the nematode from both roots and soil was 3.76, indicating that this lesion nematode had reproduced well on the commercial soybean cultivar (Yan et al. 2017e).

8.3.5 Stubby Root Nematodes, Trichodorus *and* Paratrichodorus *spp.*

Stubby root nematodes are a major concern in potato production since they transmit *Tobacco rattle virus* (TRV) which causes the corky ringspot disease. *Paratrichodorus allius* has been identified in soil samples from potato fields in Sargent County, North Dakota (Huang et al. 2017a, b; Yan et al. 2016b, e). Previously, TRV associated with corky ringspot on potato in North Dakota was reported, but stubby root nematodes were not investigated (David et al. 2010). This virus is widespread in North Dakota as well as reported from the neighboring states of Minnesota and Wisconsin (Gudmestad et al. 2008). A research study on the association between the virus, nematode and occurrence of corky ringspot is underway at the North Dakota State University. Many potato processing companies have a zero-tolerance policy for potato tubers with the disease, and an entire shipment can be rejected if a single infected tuber is detected, thereby, making disease incidence a critical qualitative parameter (Plaisance et al. 2016a).

8.3.6 Other Plant Parasitic Nematodes

Other plant parasitic nematodes of concern include *Tylenchorhynchus* spp., *Paratylenchus* spp., *Hoplolaimus* spp., *Helicotylenchus* spp. and *Xiphinema* spp. (Upadhaya et al. 2017; Yan et al. 2015b, 2016a, 2017b, c). Some of these nematode species have been frequently detected at relatively high densities, however, the economic damage they cause in the North and South Dakota's agroecosystem is largely unknown.

8.3.6.1 Stunt Nematodes, *Tylenchorhynchus* spp.

North Dakota and South Dakota are part of the Great Plains region known for supporting extensive cattle ranching and dry farming. Western wheatgrass, *Agropyron smithii*, blue grama, *Bouteloua gracilis*, and warm-season short grass, *Buchloe dactyloides*, are predominant grasses in short and mixed-grass prairies of the Northern Great Plains (Sims et al. 1978; Smolik and Lewis 1982). *Tylenchorhynchus robustus* is reported to be the dominant member of Dolichodoridae in a mixed prairie (Smolik and Lewis 1982) reducing growth, clipping plant weight and root/crown weights (Smolik 1982). Nematicide treatments increased growth of native range grasses 28–59% in Western South Dakota (Smolik 1977a). Recently, an unknown *Tylenchorhynchus* sp. was reported from a soybean field in North Dakota (Yan et al. 2017b). The greenhouse bioassay showed that this new species was capable of infecting soybean plants. However, the impact of this nematode on soybean growth and yield need to be assessed.

8.3.6.2 Pin Nematodes, *Paratylenchus* spp.

Pin nematodes (PN) were found to be the major plant parasitic nematodes in pea fields in North Dakota (Upadhaya et al. 2016). In 2015, 91 soil samples were collected from 31 fields in 9 counties. Pin nematodes were present in 60% of the samples with a highest density of 21,500 per kg of soil, followed by spiral (22%), stunt (21%), dagger (8%), root lesion (2%) and stubby root (1%) nematodes. In a separate survey, a total of 135 soil samples were collected during 2015 and 2016. Pin nematodes were the dominant plant parasitic nematodes, detected in 72% of the soil samples with mean and highest population densities of 3,560 and 35,572 specimens per kg of soil, respectively. Interestingly, in this survey, more than 97% of the PN populations in the fields were fourth stage juveniles (J-4) without a distinct stylet, whereas less than 3% of the populations were stylet-bearing, plant-feeding adults. The nematode was identified as *Paratylenchus nanus* (Upadhaya et al. 2016; Thorne and Smolik 1971). Reproductions of the PN were evaluated at four initial population levels (3,000, 5,000, 6,000, and 13,000 nematodes/kg soil) and it reproduced on different cultivars of pea (Columbian, Aragorn and Cooper), in a greenhouse study. However, those without stylet had lower reproduction factor compared to the stylet-bearing ones. Moreover, the proportion of PN adults with stylet (15–33%) in all the final populations was significantly greater for each cultivar than in the initial populations (<3%). In a separate preliminary greenhouse study, *P. nanus* reduced the plant height of six field pea cultivars by 37% (Arcadia), 36% (Columbian), 29% (Bridger), 22% (Cruiser), 20% (Salamanca) and 19% (Aragorn) after 11 weeks of growth with an initial inoculum of 4,500 nematodes/kg of soil (Upadhaya et al. 2017). This study showed that significant populations of stylet-bearing, plant-feeding pin nematodes could parasitize these pea cultivars.

8.3.6.3 Spiral Nematodes, *Helicotylenchus* spp.

Spiral nematodes are common plant parasitic nematodes in many fields of North Dakota. In June 2015, two soil samples were collected from a soybean field in Richland County, North Dakota. Both samples contained spiral nematodes at 1,500–3,300 per kg of soil. In June and August 2016, ten soil samples were collected from the same field. Nine of the samples had spiral nematodes ranging in numbers from 125 to 3,065 per kg of soil. One soil sample containing *H. microlobus,* with 1,500 nematodes per kg soil, was used to inoculate two soybean cultivars, Sheyenne and Barnes, commercially cultivated in the state. After 15 weeks of growth at 22 °C in a greenhouse, the final population density was 9,300 ± 1,701 *H. microlobus* per kg soil for Sheyenne and 9,451 ± 2,751 for Barnes. The reproduction factor in Sheyenne and Barnes was 6.2 and 6.3, respectively, indicating that this spiral nematode invades and reproduces well on these soybean cultivars. Infected soybean roots had small brown lesions on the surface (Yan et al. 2017c).

8.3.6.4 Lance Nematodes, *Hoplolaimus* spp.

In August 2015, *Hoplolaimus* spp. were collected from a soybean field near Cogswell, Sargent County, North Dakota with density at 210 nematodes per 100 cm^3 of soil. Four soil samples collected in October 2015 from the same field had lance nematodes ranging in numbers from 30 to 100 per 100 cm^3 soil. One soil sample containing *H. stephanus*, with 60 nematodes per 100 cm^3 soil, was used to inoculate soybean cultivar, Lamour, in three replicates. After 12 weeks of growth in a greenhouse (22 °C, 16 h light), mean population numbers of lance nematodes had only increased slightly (68 ± 50 per 100 cm^3 soil). Stunted and shortened lateral roots branching from the main root were observed (Yan et al. 2016a).

8.3.7 *Soybean Cyst Nematode,* Heterodera glycines *in South Dakota*

8.3.7.1 Detection and Distribution

Soybean cyst nematode was first detected in 1995 in Union County in South Dakota (Smolik 1995). By 2007, the nematode was confirmed in 19 counties in South Dakota (Smolik and Draper 2007). From 2003 to 2012, the South Dakota State University (SDSU) received a total of 4,578 soybean soil samples that were voluntarily submitted by soybean growers as well as collected during annual soybean disease surveys from 43 counties in South Dakota by the SDSU Plant Diagnostic Clinic. Subsequently, 33% of soybean fields were found to have SCN. The top four counties with the highest number of positive samples for SCN were Turner, Clay, Union and Lincoln Counties. The years 2005 and 2012 had the highest SCN population densities averaging 3,124 and 2,245 eggs and second stage juveniles per 100 cm^3 of soil, respectively. Turner County had the highest incidence (50%) followed by Clay, Union and Lincoln Counties. Interestingly, as in North Dakota, the counties found to be infested with SCN span the eastern part of South Dakota. This shows the expanded risk of SCN from the south-eastern corner to the north-eastern corner in both states. Currently 28 counties in South Dakota have been found to be infested with SCN (Acharya et al. 2014, 2016) (Fig. 8.2c).

8.3.7.2 *Heterodera glycines* HG Types

HG refers to *Heterodera glycines* and the type indicates seven Plant Introduction lines with various sources of resistance. For example, HG type 2.5.7 refers to a SCN population that is capable of reproducing on the PI line numbers 2, 5, and 7. HG types that are prevalent in South Dakota include HG types 0, 1, 2, 7, 2.7, 5.7, 1.3.6 and 2.5.7, with HG type 7 being the most predominant (36%), followed by HG type 0 (29%) and HG type 2.5.7 (16%) (Acharya et al. 2016). These HG types collectively accounted for 80% of *H. glycines* populations in South Dakota. HG type 7 means, at least 10% female index (FI) on indicator line #7. The diversity of the *H. glycines* populations in HG types varied between and within the counties, with Brookings, Clay, Turner and Union Counties having more diverse SCN populations. HG types with greater than 10% reproduction on indicator lines PI 88788, PI 209332, and PI 548316 were prevalent in the soil samples tested, suggesting that the use of these sources of resistance for developing SCN-resistant cultivars in the state is no longer effective.

8.3.7.3 Management

Any approach to managing soybean cyst nematode in fields is aimed at reducing the nematode population below the level that may result in significant yield losses. Once the nematode is established in a field, there is no practical way to eliminate it. Therefore, early detection of the nematode, rotation with a non-host crop, and use of resistant soybean varieties are the critical components of SCN management in South Dakota (Smolik and Draper 2007). The SDSU Plant Diagnostic Clinic provides SCN diagnostic services to soybean growers in the state. Growers are provided with a Soybean Cyst Nematode Soil Sampling Information Sheet which contains field location, cropping history, grower's address, instructions for collecting the soil samples and other information. This practice has tremendously helped in obtaining the early detection and distribution of the nematode in South Dakota.

8.3.7.3.1 Crop Rotation

Crop rotation with non-host crops to reduce nematode populations is an essential component of SCN management. High SCN population densities (>1,000 eggs/100 cm^3 soil) are best managed by crop rotation with non-host crops such as corn, small grains, sunflowers, flax, canola or alfalfa followed by a SCN-resistant soybean variety. In the absence of locally adapted, SCN-resistant soybean varieties, growers opt for longer rotations with non-host crops between soybean crops. Dry beans are an excellent host for SCN and are not rotated with soybeans.

8.3.7.3.2 Resistant Varieties

Soybean cyst nematode-resistant soybean varieties, in combination with crop rotation, are a very important management tool. Planting SCN-resistant soybean varieties reduces yield loss and SCN population densities. In field plot tests conducted over an 11-year period, yields of SCN-resistant lines were 23–63% higher than those of susceptible lines (Smolik and Draper 2007). It is best to plant a SCN-resistant variety in fields where SCN has been detected, even with population densities as low as 150 eggs per 100 cm^3 soil or less. Fields with fairly high SCN populations (>5,000 eggs per 100 cm^3) are rotated to non-host crops to reduce SCN numbers before planting resistant soybean varieties.

8.3.7.3.3 Cultural Practices

Provision of optimal growing conditions of the crop will reduce plant stress and yield loss due to SCN. Good soil tillage practices and adequate soil fertility improve plant growth and development. Also, management of weeds, diseases and insects reduces plant stress and minimizes SCN damage. Efforts should be made to avoid spreading SCN from infested to un-infested fields by movement of infested soil on farm equipment and tools. Equipment and farm tools should be power-washed after working in infested fields. Tillage practices that reduce wind and water erosion also can slow the spread of SCN.

8.3.7.3.4 Seed Treatment

Use of nematicides for control of SCN has not been popular amongst growers in South Dakota. However, few nematicides or fungicides with nematicidal properties are being marketed. Soybean seeds are treated before planting. Avicta Complete Beans 500® (abamectin + thiomethoxam + mefenoxam + fludioxonil) applied at 6.2 fl oz/cwt (100 lb) seed, targets SCN, as well as, damping off, seedling rots, early-season Phytophthora root rot, *Fusarium* and *Rhizoctonia* root rot diseases. Clariva pn® (*Pasteuria nishizawae*-PN1) is also being labelled for control of SCN. Pocho/Votivo® (*Bacillus firmus* I-1582 + clothianidin) applied at 0.13 mg ai/seed targets SCN.

8.3.8 *Lesion Nematodes,* Pratylenchus *spp., in South Dakota*

Pratylenchus scribneri and *P. hexincisus* are commonly associated with corn in South Dakota (Smolik 1977b, 1978; Draper et al. 2009). Under high nematode population densities, infected plants are stunted with uneven plant height along rows. Infected plants also show yellowing of leaves, root necrosis, stubby roots and eventually, poor ear fill (Draper et al. 2009). Population density of *P. scribneri* at harvest was related to yield loss in irrigated corn in South Dakota (Smolik and Evenson 1987). In the absence of nematicide, the mean number of *P. scribneri* could be as higher as 8,000 nematodes/g dry root at midseason and 6,000 nematodes/g dry root at harvest, resulting in estimated yield losses of 246–361 kg/ha. Similarly, the mean number of *P. hexincisus* could be as high as 3,400/g dry root at midseason and 4,092/g root at harvest, resulting in an estimated yield loss of 599 kg/ha (Smolik and Evenson 1987). This indicates that *P. hexincisus* and *P. scribneri* may have significant impact on corn production in South Dakota.

8.3.9 Dagger Nematodes, Xiphinema americanum, *in South Dakota*

The American dagger nematode, *Xiphinema americanum*, is one of the most commonly encountered nematodes in South Dakota soils (Thorne and Malek 1968). Furthermore, it is one of the most common nematode species found in the Great Plains, and feeds ectoparasitically on roots of all kinds of plants, from native grass to cotton trees (Thorne 1974). Apart from the damage caused by direct feeding on plant roots, the nematode is also economically important due to its ability to transmit nepoviruses.

Symptoms of stunting and premature decline and dieback of shelterbelt trees have been associated with *X. americanum* infestation. It has also been suggested that the nematode serves as the primary parasite that makes openings through which fungi and bacteria can enter and join in the destruction of root systems. Generally, in severely infested trees, it is almost impossible to find a single live feeder root (Malek 1969). Thorne (1974) suggested that *X. americanum* caused more damage to crops, orchards and timber than any other single nematode species in the USA. *Xiphinema americanum* was pathogenic to cottonwood and green ash under greenhouse conditions. Experimental demonstrations of pathogenic capabilities have been infrequent and often inconclusive because of difficulties in maintaining *X. americanum* populations in laboratory or greenhouse conditions (Malek 1969).

8.3.10 General Nematode Management Tactics for Vermiform Nematodes

8.3.10.1 Disease Diagnosis

Nematodes from the genera *Pratylenchus, Helicotylenchus, Hoplolaimus, Longidorus*, *Trichodorus*, *Paratrichodorus*, *Paratylenchus* and *Tylenchorhynchus* have been reported on field crops including corn, soybean, dry edible peas, barley, potato, wheat, etc., in both North Dakota and South Dakota, but to date, their effects have been inconsequential or not extensively investigated (Smolik 1978; Draper et al. 2009). General nematode management strategies can be used to reduce their impacts, where necessary. In any management strategy, accurate detection and estimation of population density of the nematode species through soil analysis is paramount. Molecular methods have been used for detection, identification and quantification of *Pratylenchus scribneri* and *Paratrichodorus allius* in North Dakota (Huang and Yan 2017; Huang et al. 2017a, b). Practically, no single strategy should suffice, therefore, an integrated, sustainable management approach is recommended, contingent on a number of factors including field situation, cropping sequence, nematode species, nematode density, available resources, crop rotation, cover cropping, trap cropping, fallowing, removal of infested plants, weed control, resistant varieties, soil amendment, fertilization and tillage practices (Heald 1987; McKenry 1987; Young 1992; Westphal et al. 2006, 2009; Xing and Westphal 2009; Westphal 2011; Anonymous 2017).

8.3.10.2 Crop Rotation

Growing crops that are non-hosts or poor-hosts to a particular nematode species in rotation with a primary crop reduces the target nematode population. Thus, in the absence of a suitable host, nematode population density reduces over time. Successful nematode management through crop rotation depends on the species of the nematode present in a field, damage threshold level, host range of the species present, weed host, the expected rate of population decline or increase, the number of nematode species present, crop plants, availability of resistant varieties and time of planting (Anonymous 2017). Nematodes with a narrow host range such as soybean cyst nematode, which only reproduces on soybeans and its closely related legumes, are easily managed by crop rotation, unlike nematodes such as lesion nematodes and dagger nematodes, which have a wider host range. The three *Pratylenchus* species most commonly associated with field crops in North and South Dakota (*P. hexincisus*, *P. scirbneri* and *P. neglectus*) often occur in mixed populations. In a situation where, for instance, soybean cyst nematode with a narrow host range is present together with a nematode with a wide host range such as lesion nematode, a rotation ideal for soybean cyst nematode reduction may favor buildup of lesion nematode (Smolik 1978; Draper et al. 2009; Yan et al. 2016c).

Corn and potato are good hosts for both *Pratylenchus hexincisus* and *P. scribneri* whereas, alfalfa and red clover are non-hosts. *Pratylenchus hexincisus* reproduces more on wheat and rye than *P. scribneri* whereas sorghum, soybean, tomato and white clover are better hosts for *P. scribneri*. Other species of lesion nematodes damage both grasses and broad-leaf plants. The wide host range of lesion nematodes limits their effective management with crop rotation. Again, an important part of a crop rotation strategy is what is grown in the field during the offseason period when a cash crop is not grown. When the field is allowed to fallow, free of weeds and volunteer plants, nematode populations plummet. To minimize the problem of soil erosion during the period of fallow, weeds known to be non-host may be left on the field as barriers to erosion. Alternatively, cover crops which are non-hosts to the target nematode can be grown during the offseason period. Cover crops such as radish, mustards and other Brassicas may have nematicidal properties against the target nematode, if the shoots of these crops are incorporated into the soil (Heald 1987; Westphal et al. 2006; Xing and Westphal 2009).

8.3.10.3 Use of Resistant and Tolerant Varieties

The use of resistant varieties is another practical means of controlling nematodes (Young 1992; Roberts 2002). Crop plants may be described as non-host or immune, resistant, susceptible, tolerant or intolerant to a particular nematode species. Plants may be invaded by nematodes and may show damage, but chemical or physical barriers within the plant will prevent population increases (resistant varieties). However, when plants do not allow nematodes to attack including initial root invasion, these are called non-host or immune. Tolerant hosts allow nematode invasion and reproduction but are able to withstand nematode attack whereas intolerant hosts are more likely to be damaged by nematode attack (Anonymous 2017). Resistant or non-host varieties should be used whenever possible to reduce yield loss. Population development of *P. scribneri* and *P. hexincisus* varies among varieties of corn, soybean and wheat. Currently, there are no resistant varieties available against most vermiform nematodes.

8.3.10.4 Chemical Control

Nematicide application has been frequently used to control lesion nematodes on corn in South Dakota and North Dakota (Bergeson 1978; Draper et al. 2009). Foliar sprays with oxamyl drastically reduced nematode populations on ash seedlings and generally improved seedling growth. Soil fumigation with 1, 3 – dichloropropene and related chlorinated C3 hydrocarbons (1, 3 – D) increased growth of green ash and golden willow over a 4-year period on land infested with initially low populations of dagger nematodes, but did not affect growth of cottonwood, Siberian pea tree or honey locust (Malek and Smolik 1975). In North Dakota, field applications of oxamyl combined with clothianidin showed a potential efficacy against *Paratrichodorus* spp., but did not result in increased yield (Plaisance et al. 2016a).

Owing to environmental and health concerns, most fumigant nematicides have been phased out and are replaced by less toxic, environmentally friendly chemical products. Some of these emerging products include: abamectin or Avid® containing avermectin; Nimitz® containing fluensulfone; Multiguard Protect® containing furfural; and Kontos® or Movento® containing spirotetramat. The specific situation determines whether the expense of chemical application is warranted, however, unless a soil analysis reveals exceptionally high nematode populations, it is not economically viable to use nematicides for the control of nematodes (Draper et al. 2009). For effective management, these chemical products must be applied in conjunction with other long-term sustainable management practices (Westphal 2011).

8.3.11 Nematode Management in Sustainable Agriculture

Management of plant parasitic nematodes in sustainable agriculture aims at optimizing resources, skills and technology to achieve long-term control of nematodes without adverse effect on the environment or humans. Thus, methods of nematode control that curb the threats that nematodes pose to crop production without compromising other life forms and the environment now or in the future, form part of sustainable farming practices. It is in compliance with this background that many fumigant nematicides have been phased out and have been replaced with other innocuous cultural practices and technologies, to ensure sustained levels of control of plant pathogens and returns to growers, while minimizing adverse impacts to immediate and off-farm environments.

Consequently, the use of crop rotation practices, resistant crop cultivars, timing of planting, prevention and exclusion practices, tillage practices, fallowing, organic amendments, cover cropping, trap cropping, green manuring, etc. to mitigate nematode problems, form an integral part of sustainable agriculture. These practices, in addition to providing nematode control, may provide alternative sources of soil nitrogen, reduce soil erosion, improve soil structure, improve water retention and provide ecological niches to soil fauna and flora. They are not harmful to natural systems, farmers, their neighbors or consumers and pose no risk of environmental (water and atmospheric) pollution, yet may provide nematode control and generate the required crop productivity to growers.

References

Acharya, K., Byamukama, E., & Tande, C. (2014). The status of soybean cyst nematode, *Heterodera glycines*, in South Dakota. *Phytopathology, 104*, S3.3. https://doi.org/10.1094/PHYTO-104-11-S3.1.

Acharya, K., Tande, C., & Byamukama, E. (2016). Determination of *Heterodera glycines* virulence phenotypes occurring in South Dakota. *Plant Disease, 100*, 2281–2286.

Acharya, K., Plaisance, A., & Yan, G. P. (2017). Evaluation of cover crops for hosts and population reduction of soybean cyst nematode. *Phytopathology, 107*, S1.1. https://doi.org/10.1094/PHYTO-107-1-S1.1.

Anonymous. (2014). *South Dakota Ag economic contribution study*. South Dakota Department of Agriculture. https://www.google.com/webhp?sourceid=navclientandie=UTF-8andgws_rd=ssl#q=South+Dakota+Ag+Economic+contribution+study.+South+Dakota+Department+of+Agriculture+and*andspf=1

Anonymous. (2016). *North Dakota wheat statistics*. North Dakota Wheat Commission. http://www.ndwheat.com/buyers/default.asp?ID=539

Anonymous. (2017). *Biology and management of plant parasitic nematodes*. Davis: Department of Nematology, University of California http://www.google.com/url?sa=tandrct=jandq=andesrc=sandsource=webandcd=1andcad=rjaanduact=8andved=0ahUKEwjIjruO4q7TAhVMVWMKHVL4DQ0QFggnMAAandurl=http%3A%2F%2Fplpnemweb.ucdavis.edu%2Fnemaplex%2FCourseinfo%2FHandouts%2FWesterdahl.pdfandusg=AFQjCNHcL6YTjld8THFgoTZ1MlyiGbHVzA.

Baidoo, R., Yan, G. P., Nelson, B., Skantar, A. M., & Chen, S. (2017). Use of chemical flocculation and nested PCR for *Heterodera glycines* detection in DNA extracts from field soils with low population densities. *Plant Disease, 101*, 1153–1161.

Bergeson, G. B. (1978). Control of the lesion nematode *Pratylenchus* spp. in corn with carbofuran. *Plant Disease Reporter, 62*, 295–297.

Berghuis, B. (2016). *Increasing soybean cyst nematode awareness in North Dakota through a grower-based sampling program*. 2016 Soybean Cyst Nematode Conference. S1.2. Coral Gables, Florida, FL, December 13–15.

Bradley, C. A., Biller, C. R., & Nelson, B. D. (2004). First report of soybean cyst nematode (*Heterodera glycines*) on soybean in North Dakota. *Plant Disease, 88*, 1287.

Caveness, F. E. (1958). *A study of nematodes associated with sugarbeet production in selected northwest and north central states* (p. 157). Ft. Collins: Sugarbeet Development Foundation.

Chitwood, B. G. (1949). Cyst-forming *Heterodera* encountered in soil sampling. *Plant Disease Reporter, 33*, 130–131.

Chowdhury, I., Yan, G. P., Plaisance, A., Nelson, B., Markell, S., Helms, T. C., & Upadhaya, A. (2016). *Population diversity of soybean cyst nematode in North Dakota Fields*. In Abstracts of 55th annual meeting of the society of nematologists, Montreal Quebec, Canada, July 17–21 (pp. 68–69).

Chowdhury, I. A., Yan, G. P., & Plaisance, A. (2017). Characterizing virulence phenotypes of soybean cyst nematode (*Heterodera glycines*) in infested fields of North Dakota. *Phytopathology, 107*, S1.3. https://doi.org/10.1094/PHYTO-107-1-S1.1.

David, N., Mallik, I., & Gudmestad, N. C. (2010). First report of the *Tobacco rattle virus* associated with Corky Ringspot in potatoes grown in North Dakota. *Plant Disease, 94*, 130. https://doi.org/10.1094/PDIS-94-1-0130B.

Donald, P. A. (1978). *Plant nematodes in and relevant to North Dakota*. M.S. thesis, North Dakota State University, Fargo. p. 70.

Donald, P. A., & Hosford, R. M., Jr. (1980). Plant parasitic nematodes of North Dakota. *Plant Disease, 64*, 45–47.

Draper, M. A., Langham, M. A., Clay, S. A., & Bradley, R. E. (2009). *Best management practices for corn production in South Dakota: Corn diseases in South Dakota*. Extension Circulars. p. 499. http://openprairie.sdstate.edu/extension_circ/499

Golden, A. M., Epps, J. M., Riggs, R. D., Duclos, L. A., Fox, J. A., & Bernard, R. L. (1970). Terminology and identity of infraspecific forms of the soybean cyst nematode (*Heterodera glycines*). *Plant Disease Reporter, 54*, 544–546.

Gudmestad, N. C., Mallik, I., Pasche, J. S., & Crosslin, J. M. (2008). First report of *Tobacco rattle virus* causing corky ringspot in potatoes grown in Minnesota and Wisconsin. *Plant Disease, 92*, 1254. https://doi.org/10.1094/PDIS-92-8-1254C.

Heald, C. M. (1987). Classical nematode management practices. In J. A. Veech & D. W. Dickson (Eds.), *Vistas on nematology* (pp. 100–105). Hyatssville: Society of Nematologists.

Huang, D., & Yan, G. P. (2017). Specific detection of the root lesion nematode *Pratylenchus scribneri* using conventional and real-time PCR. *Plant Disease, 1010*, 359–365.

Huang, D., Yan, G. P., & Skantar, A. M. (2017a). Development of real-time and conventional PCR assays for identifying stubby root nematode *Paratrichodorus allius*. *Plant Disease, 1010*, 964–972.

Huang, D., Yan, G. P., Gudmestad, N., & Skantar, A. (2017b). Quantification of *Paratrichodorus allius* in DNA extracted from soil using TaqMan Probe and SYBR Green real-time PCR assays. *Nematology, 19*, 987–1001.

Khan, M., Arabiat, S., Chanda, A. K., & Yan, G. P. (2016). *Sugar beet cyst nematode* (North Dakota Extension Bulletin, p. 1788). Fargo: North Dakota State University. 2 p. https://www.ag.ndsu.edu/pubs/plantsci/pests/pp1788.pdf

Krupinsky, J. M., Barker, R. E., & Donald, P. A. (1983). Frequency of plant parasitic nematodes associated with blue gama and western wheatgrass in the Western Dakotas. *Plant Disease, 67*, 399–401.

MacGuiwin, A. E., & Rouse, D. I. (1990). Role of *Pratylenchus penetrans* in the potato early disease of Russet Burbank potato. *Phytopathology, 80*, 1077–1082.

Malek, R. B. (1969). Population fluctuations and observations of the life cycle of *Xiphinema americanum* associated with cottonwood (*Populus deltoids*) in South Dakota. *Proceedings of the Helminthological Society of Washington, 36*, 270–274.

Malek, R. B., & Smolik, J. D. (1975). Effects of *Xiphinema americanum* on growth of shelterbelt trees. *Plant Disease Reporter, 59*, 144–148.

Mathew, F., Markell, S., Jantzi, D., Yan, G. P., Nelson, B., & Helms, T. (2014). *Soybean cyst nematode, North Dakota Extension Bulletin* (p. 1732). Fargo: North Dakota State Univ, 4 p.

McKenry, M. V. (1987). Control strategies in high-value crops. In R. H. Brown & B. R. Kerry (Eds.), *Principles and practices of nematode control in crops* (pp. 330–349). Sydney: Academic.

Nelson, B. (2017). *The confrontation: Soybean cyst nematode and common bean*. 2017 Soybean cyst nematode conference. S1.7. Coral Gables. Florida. December 13–15.

Nelson, B. D., Bolton, M. D., Lopez-Nicora, H. D., & Niblack, T. L. (2012). First confirmed report of sugarbeet cyst nematode, *Heterodera schachtii*, in North Dakota. *Plant Disease, 96*, 772. https://doi.org/10.1094/PDIS-02-12-0112-PDN.

Niblack, T. L., Arelli, P. R., Noel, G. R., Opperman, C. H., Orf, J. H., Schmitt, D. P., Shannon, J. G., & Tylka, G. L. (2002). A revised classification scheme for genetically diverse populations of *Heterodera glycines*. *Journal of Nematology, 34*, 279–288.

Niblack, T. L., Tylka, G. L., & Riggs, R. D. (2004). Nematode pathogens of soybean. In J. R. Wilcox (Ed.), *Soybeans: Improvement, production, and uses* (3rd ed., pp. 821–851). Madison: American Society of Agronomy.

Pepper, E. H. (1963). Nematodes in North Dakota. *Plant Disease Report, 47*, 102–106.

Pepper, E. H. (1968). *Nematology in the North Central Region 1956–1966, NCR publication No. 187. Special Report 58*. Ames: Agric. Home Econ. Exp. Stn, Iowa State Univ. Sci. Tech..

Plaisance, A., & Yan, G. P. (2015). *Comparison of two nematode extraction techniques*. In Abstracts of 54th annual meeting of the society of nematologists, East Lansing, MI, July 19–24 (p. 120).

Plaisance, A., Yan, G. P., Peterson, D., Gudmestad, N. C., Thorsness, K. B. (2016a). *Experimental chemical applications for control of corky ring spot disease of potato vectored by Paratrichodorus allius*. In Abstracts of 55th annual meeting of the society of nematologists, Montreal Quebec, Canada, July 17–21 (p. 159).

Plaisance, A., Yan, G. P., Upadhaya, A. (2016b). *Selected northern-grown crops as hosts of Pratylenchus scribneri*. In Abstracts of 55th annual meeting of the society of nematologists, Montreal Quebec, Canada, July 17–21 (pp. 159–160).

Poromarto, S., & Nelson, B. (2009). Reproduction of soybean cyst nematode on dry bean cultivars adapted to North Dakota and northern Minnesota. *Plant Disease, 93*, 507–511.

Poromarto, S., Nelson, B., & Goswami, R. S. (2010). Effect of soybean cyst nematode on growth of dry bean in the field. *Plant Disease, 94*, 1299–1304.

Poromarto, S. H., Gramig, G. G., Nelson, B. D., Jr., & Jain, S. (2015). Evaluation of weed species from the Northern Great Plains as hosts of soybean cyst nematode. *Plant Health Progress*. https://doi.org/10.1094/PHP-RS-14-0024.

Porter, P. M., & Chen, S. (2005). Sugarbeet cyst nematode not detected in the Red River Valley of Minnesota and North Dakota. *Journal of Sugarbeet Research, 42*, 79–84.

Riggs, R. D. (1988). Races of *Heterodera glycines*. *Nematropica, 18*, 163–170.

Riggs, R. D., & Schmitt, D. P. (1988). Complete characterization of the race scheme for *Heterodera glycines*. *Journal of Nematology, 22*, 392–395.

Roberts, P. A. (2002). Concepts and consequences of resistance. In J. L. Starr, R. Cook, & J. Bridge (Eds.), *Plant resistance to parasitic nematodes* (pp. 23–41). London: CAB International.

Shalu, J., Kishore C., Brueggeman, R., Osorno, J., Richards, J., & Nelson, B. (2017). *Soybean cyst nematode in dry bean (Phaseolus vulgaris): Understanding and managing underground enemies in the era of genomics*. Soybean cyst nematode conference. S1.5. December 13–15. Coral Gables. Florida.

Sims, P. L., Singh, J. S., & Lauenroth, W. K. (1978). The structure and function of ten Western North American grasslands. I. Abiotic and vegetational characteristics. *Journal of Ecology, 66*, 251–258.

Smolik, J. D. (1977a). Effect of nematicide treatment on growth of range grasses in field and glasshouse studies. In J. K. Marshall (Ed.), *Proceedings of below-ground ecosystem symposium* (pp. 257–260). Colorado State University Range Science Series No. 26.

Smolik, J. D. (1977b). Nematodes associated with irrigated corn in South Dakota. *Proceedings of the South Dakota Academy of Science, 56*, 93–99.

Smolik, J. D. (1978). Influence of previous insecticidal use on ability of carbofuran to control nematode populations in corn and effect on corn yield. *Plant Disease Report, 62*, 95–99.

Smolik, J. D. (1982). Effect of *Tylenchorhynchus robustoides* on growth of buffalo grass and western wheatgrass. *Journal of Nematology, 14*, 585–588.

Smolik, J. D. (1995). First report of *Heterodera glycines* on soybean in South Dakota. *Plant Disease, 80*, 224.

Smolik, J. D., & Draper, M. A. (2007). *Soybean cyst nematode*. SDSU Extension Fact Sheets. Paper 81. http://openprairie.sdstate.edu/extension_fact/81

Smolik, J. D., & Evenson, P. D. (1987). Relationship of yields and *Pratylenchus* spp. population densities in dryland and irrigated corn. *Annals of Applied Nematology, 1*, 71–73.

Smolik, J. D., & Lewis, J. K. (1982). Effect of range condition on density and biomass of nematodes in a mixed ecosystem. *Journal of Range Management, 35*, 657–663.

Spears, J. F. (1956). Occurrence of the grass cyst nematode, *Heterodera punctata*, and *Heterodera cacti* group cysts in North Dakota and Minnesota. *Plant Disease Reporter, 40*, 583–584.

Thorne, G. (1974). Nematodes of the Northern Great Plains part II. *Technical Bulletins*, 5. 94 pp.

Thorne, G., & Malek, R. B. (1968). Nematodes of the Northern Great Plains. Part I. Tylenchida. *Technical Bulletin of South Dakota Agricultural Experiment Station, 31*, 1–111.

Thorne, G., & Smolik, J. D. (1971). The identity of *Paratylenchus nanus* Cobb, 1923. *Proceedings of the Helminthological Society of Washington, 38*, 90–92.

Tylka, G. L., & Marett, C. C. (2014). *Distribution of the soybean cyst nematode, Heterodera glycines, in the United States and Canada: 1954 to 2014*. https://doi.org/10.1094/PHP-BR-14-0006.

Upadhaya, A., Yan, G. P., Plaisance, A., Pasche, J. S., & McPhee, K. (2016). *Identification and reproduction of pin nematodes on field pea (Pisum sativum) in North Dakota*. In Abstracts of 55th annual meeting of the society of nematologists, Montreal Quebec, Canada, July 17–21 (pp. 185–186).

Upadhaya, A., Yan, G. P., Plaisance, A., Pasche, J., & McPhee, K. (2017). Pin nematode: a potential threat to pea production in North Dakota. *Proceedings of the North Dakota Academy of Science, 71*, 45.

USDA-NASS. (2015a). *State agriculture overview, South Dakota*. https://www.nass.usda.gov/Statistics_by_State/South_Dakota/

USDA-NASS. (2015b). *State agriculture overview, North Dakota*. https://www.nass.usda.gov/Statistics_by_State/North_Dakota/

USDA-NASS. (2016a). *News release. South Dakota's rank in U.S agriculture*. p. 1. https://www.google.com/webhp?sourceid=navclientandie=UTF8andgws_rd=ssl#q=South+Dakota's+rank+in+US+agricultureand*andspf=107

USDA-NASS. (2016b). News release. North Dakota's rank in U.S agriculture. p. 1. https://www.google.com/webhp?sourceid=navclientandie=UTF8andgws_rd=ssl#q=North+Dakota's+rank+in+US+agricultureand*andspf=1

Westphal, A. (2011). Sustainable approaches to the management of plant parasitic nematodes and disease complexes. *Journal of Nematology, 43*, 122–125.

Westphal, A., Xing, L. J., & Egel, D. S. (2006). *Use of cover crops for management of root knot nematodes in cucurbits*. Annual International Research Conference Methyl Bromide Alternatives Emission Reduction, Orlando, FL, 6–9, 2006 22-1-3.

Westphal, A., Xing, L. J., Pillsbury, R., & Vyn, T. J. (2009). Effects of tillage intensity on population densities of *Heterodera glycines* in intensive soybean production systems. *Field Crops Research, 113*, 218–226.

Winstead, N. N., Skotland, C. B., & Sasser, J. N. (1955). Soybean cyst nematode in North Carolina. *Plant Disease Report, 39*, 9–11.

Xing, L. J., & Westpal, A. (2009). Effects of crop rotation of soybean with corn on severity of sudden death syndrome and population densities of *H. glycines* in naturally infested soil. *Field Crops Research, 112*, 107–117.

Yan, G. P., & Baidoo, R. (2017). Molecular detection of soybean cyst nematode in North Dakota. *Phytopathology, 107*, S1.9. https://doi.org/10.1094/PHYTO-107-1-S1.1.

Yan, G. P., & Plaisance, A. (2016). Vermiform plant parasitic nematodes on soybean in North Dakota and their relationship with soybean cyst nematode. *Phytopathology, 106*, S4.104. https://doi.org/10.1094/PHYTO-106-12-S4.1.

Yan, G. P., Markell, S., Nelson, B. J., Helms, T. C., & Osorno, J. M. (2015a). *The status of soybean cyst nematode occurrence and management in North Dakota*. In Abstracts of 54th annual meeting of the society of nematologists, East Lansing, MI, July 19–24 (pp. 126–127).

Yan, G. P., Plaisance, A., & Ye, W. (2015b). Plant parasitic nematodes on field crops in Southeastern and Northeastern North Dakota. *Phytopathology, 105*(S4), 153.

Yan, G. P., Plaisance, A., Huang, D., & Handoo, Z. A. (2016a). First report of the lance nematode *Hoplolaimus stephanus* from a soybean field in North Dakota. *Plant Disease, 100*, 2536. https://doi.org/10.1094/PDIS-07-16-1012-PDN.

Yan, G. P., Plaisance, A., Huang, D., & Handoo, Z. A. (2016b). First detection of the stubby root nematode *Paratrichodorus allius* on potato in North Dakota and on sugarbeet in Minnesota. (Abstr.). *Phytopathology, 106*, S4.125. https://doi.org/10.1094/PHYTO-106-12-S4.1.

Yan, G. P., Plaisance, A., Huang, D., Gudmestad, N. C., & Handoo, Z. A. (2016c). First report of the root lesion nematode *Pratylenchus scribneri* infecting potato in North Dakota. *Plant Disease, 100*, 1023. https://doi.org/10.1094/PDIS-10-15-1227-PDN.

Yan, G. P., Plaisance, A., Huang, D., Liu, Z., Chapara, V., & Handoo, Z. A. (2016d). First report of the root lesion nematode *Pratylenchus neglectus* on wheat (*Triticum aestivum*) in North Dakota. *Plant Disease, 100*, 1794. https://doi.org/10.1094/PDIS-02-16-0260-PDN.

Yan, G. P., Plaisance, A., Huang, D., Upadhaya, A., Gudmestad, N. C., & Handoo, Z. A. (2016e). First report of the stubby root nematode *Paratrichodorus allius* on potato in North Dakota. *Plant Disease, 100*, 1247. https://doi.org/10.1094/PDIS-11-15-1350-PDN.

Yan, G. P., Plaisance, A., Chowdhury, I., Baidoo, R., Upadhaya, A., Pasche, J., Markell, S., Chen, S., & Nelson, B. (2017a). First report of the soybean cyst nematode *Heterodera glycines* infecting dry bean (*Phaseolus vulgaris* L.) in a commercial field in Minnesota. *Plant Disease, 101*, 391.

Yan, G. P., Plaisance, A., Huang, D., Handoo, Z. A. (2017b). First report of a new stunt nematode *Tylenchorhynchus* sp. from a soybean field in North Dakota. *Plant Disease, 102*, 453. https://doi.org/10.1094/PDIS-05-17-0616-PDN

Yan, G. P., Plaisance, A., Huang, D., & Handoo, Z. A. (2017c). First report of the spiral nematode *Helicotylenchus microlobus* infecting soybean in North Dakota. *Journal of Nematology, 49*, 1.

Yan, G. P., Plaisance, A., Huang, D., Chowdhury, I. A., & Handoo, Z. A. (2017d). First report of the new root lesion nematode *Pratylenchus* sp. on soybean in North Dakota. *Plant Disease, 101*, 1554.

Yan, G. P., Plaisance, A., Huang, D., Handoo, Z. A., & Chitwood, D. J. (2017e). First report of a new, unnamed lesion nematode *Pratylenchus* sp. infecting soybean in North Dakota. *Plant Disease, 101*, 1555.

Young, L. D. (1992). Problems and strategies associated with long-term use of nematode resistant cultivars. *Journal of Nematology, 24*, 228–233.

Chapter 9
Management of Plant Parasitic Nematode Pests in Florida

William Crow and Larry Duncan

9.1 Introduction

Florida is unique in the continental USA for its subtropical and tropical climates. The state is a major producer of fresh fruits, vegetables and ornamentals, in addition to its most renown crop, citrus. The value of Florida agricultural products in 2016 ($8.46 billion), ranked 20th among the states. Nationally, Florida ranks first in production of citrus ($1.24 B), sugarcane ($515 M), tomato ($382 M), cucurbits such as watermelon, cucumber and squash ($270 M), and second in the value of ornamental or floriculture products ($1.04 B), strawberry ($450 M), bell pepper ($210 M), sweet corn ($160 M) and avocado ($19 M) (FDACS 2016).

Turfgrasses grown on lawns, golf courses, athletic fields and sod farms are a major driver in Florida's economy. In 2007, turfgrass in Florida was grown on more than 1.6 million ha, provided 157,240 jobs and generated $6.26 billion (Hodges and Stevens 2010).

In addition to climate, Florida's edaphic conditions are unusual for the very high proportion of sand in most soils, combined frequently with shallow water tables. The warm, humid climate and porous soil conditions permit extended nematode population growth with short overwintering, often resulting in extensive crop damage by one or multiple nematode species including tropical nematodes not encountered elsewhere in North America.

W. Crow
IFAS, Department of Entomology and Nematology, University of Florida, Gainesville, FL, USA
e-mail: wtcr@ufl.edu

L. Duncan (✉)
IFAS, Department of Entomology and Nematology, CREC, University of Florida, Lake Alfred, FL, USA
e-mail: lwduncan@ufl.edu

S. A. Subbotin, J. J. Chitambar (eds.), *Plant Parasitic Nematodes in Sustainable Agriculture of North America*, Sustainability in Plant and Crop Protection,
https://doi.org/10.1007/978-3-319-99588-5_9

Growers of high value crops obtained a respite from nematode damage during several decades following the discovery and use of soil fumigants (Taylor 2003; Chitwood 2003). However, subtropical rainy seasons combined with porous soils resulted in widespread groundwater contamination by several fumigant and non-fumigant nematicides which are now deregistered or severely limited in their use. Loss of these materials, especially methyl bromide due to ozone depletion, has refocused attention on non-chemical nematode management tactics and sustainable methods of nematicide use (Noling 2016). Although Florida has been accurately described as a nematologist's paradise (Fig. 9.1), here we focus on just five genera that are economically important in Florida agriculture. Additional nematodes of importance in the state are given in Table 9.1.

9.2 Sting Nematode, *Belonolaimus longicaudatus*

Sting nematode is among the most damaging plant parasitic nematodes in the State of Florida. This ectoparasitic nematode has an extensive host range. In Florida, hosts include native plants, trees and grasses growing in natural areas as well as plants cultivated for agronomic, horticultural and ornamental purposes in farms and landscapes (Crow and Han 2005). Most crops grown in Florida can suffer damage from this nematode. *Belonolaimus longicaudatus* predominates in sandy soil (>80% sand content) and because most of Florida is very sandy, has a widespread distribution in that state. Florida is believed to be the point of origin for *B. longicaudatus* and hence, there is much variation in morphology, host-range and behavior among populations of *B. longicaudatus* from Florida (Gozel et al. 2006).

Belonolaimus longicaudatus feeds primarily near root tips and causes lesions at the feeding site. Feeding by multiple *B. longicaudatus* can kill the root meristem and stop root growth (Crow and Han 2005). This is particularly problematic for seedlings and transplants whose growth can be arrested and cease (Christie et al. 1952). While almost all crops grown in Florida can be damaged by *B. longicaudatus*, damage caused by the nematode is of primary concern on turfgrasses, strawberry, corn, potato and young citrus.

All of the turfgrass species grown in Florida are hosts to *B. longicaudatus*. A survey of golf courses in Florida found that 84% were infested by *B. longicaudatus* and population densities of the nematode were sufficient to be classified as moderate or high-risk on 60% of them (Crow 2005b). Similar to most other grasses, corn is an excellent host of *B. longicaudatus* and is highly susceptible to damage. Depending on population densities, *B. longicaudatus* causes stunting, chlorosis, premature inflorescence and death of corn (Christie et al. 1952). *Belonolaimus longicaudatus* is especially damaging to strawberry, causing stunting, decline and death (Christie et al. 1952; Noling 2016). Impacted areas can range in size from a few stunted areas to complete fields. On potato, *B. longicaudatus* causes stunting of potato plants, reduction in yield and malformation of tubers (Crow et al. 2000a).

Fig. 9.1 (**a**) Athletic field turf damaged by *Hoplolaimus galeatus*; (**b**) Stubby root symptoms on St. Augustinegrass caused by *Trichodorus obtusus*; (**c**) Strawberry leaves showing crimping due to *Aphelenchoides besseyi*, a recently detected pest (Courtesy of Johan Desaeger); (**d**) Peach tree on Flordaguard rootstock damaged by *Meloidogyne arenaria* (courtesy of Janete Brito) (**e**) Unknown peach rootstocks infected by *M. floridensis* (Courtesy of Jose Chaparro); (**f**) Citrus roots with stubby root symptoms from deep in soil infested with *Belonolaimus longicaudatus*

One of the attributes of *B. longicaudatus* that makes it difficult to manage on strawberry is its great vertical mobility. The nematode is detected as deep as 1-m below the bed surface, below the effective treatment zone of fumigants (Hamill 2006). Following fumigation these deep-occurring nematodes can migrate upward

Table 9.1 Plant parasitic nematodes of economic importance in Florida

Nematode species	Hosts	References
Aphelenchiodes besseyi	Various ornamentals; strawberry	Stokes (1979) and Desaeger and Noling (2017)
A. fragariae	Various ornamental plants; strawberry	Christie (1938) and Stokes (1979)
A. ritzamabosi	Various ornamental plants	Stokes (1979)
Belonolaimus longicaudatus	Turfgrasses; citrus; potato; strawberry; corn; cotton; various vegetables; many others	Duncan et al. (1996), Crow and Han (2005), and Gozel et al. (2006)
Bursaphelenchus xylophilus	Pine	Esser et al. (1983)
Dolichodorus heterocephalus	Celery; corn	Rhoades (1985)
Helicotylenchus pseudorobustus	Corn; various agronomic crops	O'Bannon and Inserra (1989)
H. paxilli	Turfgrass	Pang et al. (2011a, b)
H. multicinctus	Banana	McSorley and Parrado (1983)
Hemicriconemoides wessoni	Turfgrass	Inserra et al. (2014)
H. strictathecatus	Palms	Inserra et al. (2014)
Hemicycliophora parvana	Turfgrass	Tarjan and Frederick (1981)
Heterodera glycines	Soybean	Lehman and Dunn (1987)
H. leuceilyma	Turfgrass	Di Edwardo and Perry (1964)
Hoplolaimus galeatus	Turfgrass	Giblin-Davis et al. (1995)
H. concaudajuvencus	Turfgrass	Golden and Minton (1970)
Meloidogyne arenaria	Peanut; various vegetables; various ornamentals	Osman et al. (1985), Brito et al. (2008, 2010), and Grabau and Dickson (2018)
M. enterolobii	Various ornamentals; various vegetables, various fruits	Brito et al. (2008, 2010)
M. floradensis	Peach; various vegetables; various ornamentals	Brito et al. (2008, 2010, 2015)
M. graminis	Turfgrass; forage grasses	McGowan (1984) and Crow (2018)
M. hapla	Strawberry	Nyoike et al. (2012)
M. haplanaria	Tomato	Joseph et al. (2016)
M. incognita	Various vegetables; various ornamentals; various fruits; various field crops	Brito et al. (2008, 2010)
M. javanica	Various vegetables; various ornamentals; various fruits; various field crops	Brito et al. (2008, 2010) and Grabau and Dickson (2018)
M. partityla	Pecan	Brito et al. (2006)
Mesocriconema ornatum	Turfgrass	Ratanaworabhan and Smart (1970)

(continued)

Table 9.1 (continued)

Nematode species	Hosts	References
M. xenoplax	Peach	Malo (1963)
Nanidorus minor	Turfgrass; forage grasses; corn, various vegetables	Christie and Perry (1951), Perez et al. (2000), and Crow and Welch (2004)
Peltamigratus christiei	Turfgrasses	Crow and Walker (2003)
Pratylenchus brachyurus	Peanut	Dickson (1985)
P. coffeae	Citrus	MacGowan (1978) and Duncan et al. (1999)
P. hippiastri	Amaryllus	Inserra et al. (2006) and Crow (2012)
Radopholus similis	Citrus; banana	McSorley (1986) and Inserra et al. (2005)
Rotylenchulus reniformis	Cotton; various vegetables; various ornamentals	McSorley et al. (1982), Kinloch and Sprenkel (1994), and Inserra et al. (1994)
Trichodorus obtusus	Turfgrass; forage grasses; corn	Rhoades (1968) and Crow and Welch (2004)
Tylenchulus semipenetrans	Citrus	Tarjan (1967)
Xiphinema vulgare	Citrus	Tarjan (1964)

and attack newly transplanted strawberry plants. *Belonolaimus longicaudatus* has been shown to migrate 1 m vertically in 45 days (Hamill 2006).

Because *B. longicaudatus* has no long-term survival stage, it can be managed more effectively by cropping system and fallow schedules, than certain other nematodes (Crow et al. 2000b). It is an ectoparasite, so it responds well to contact nematicides and other soil treatments. Because of its requirement for sandy soil, it can be avoided by producing crops on soil with less sand content. Nevertheless, being an ectoparasite makes it a difficult target for resistance breeding. Because of its wide host-range, management of *B. longicaudatus* with rotation and cover crops is more problematic than it is for certain other plant parasitic nematodes. *Belonolaimus longicaudatus* is such a virulent nematode that for some hosts the damage threshold is below the detection level (Crow et al. 2000a, c). This is a difficult standard to achieve for any plant parasitic nematode.

9.2.1 Rotation and Replacement

Rotation with a non-host is not feasible in perennial crops such as turfgrasses. In some cases nematode-damaged turf can be replaced with a non-host turf substitute. For example, in some lawn, park and roadside situations, turf that is being damaged by *B. longicaudatus* can be replaced with the non-host ground cover, perennial peanut (*Arachis glabrata*).

Forage/silage corn is commonly double and triple-cropped with sorghum (*Sorghum* spp.) and other small grains and rotated with perennial grasses such as bermudagrass and limpograss (*Hemarthria* spp.). All of these grasses are excellent hosts of *B. longicaudatus*. Even alfalfa and clover, winter legumes that are often double-cropped with corn for forage/silage production, are hosts of *B. longicaudatus*. This cropping scheme makes problems because *B. longicaudatus* is a very common pest of forage/silage corn. Fortunately, several summer legumes are resistant to *B. longicaudatus* and are suited for rotation in forage/silage production (Vendramini et al. 2012). These include hairy indigo (*Indigofera hirsuta*), aeschynomene (*Aeschynomene americana*), and perennial peanut (Rhoades 1983, 1984; Macchia et al. 2003).

9.2.2 Cropping Systems

The practice of winter overseeding is performed in some situations in Florida. During the winter, cool temperatures in Northern Florida induce dormancy in warm-season turfgrasses. The turf will become brown and high-traffic areas will decline and die from wear. To counter this, a cool-season grass, typically ryegrass (*Lolium* spp.) or bluegrass (*Poa* spp.), is seeded directly onto the turf surface in the fall. This "overseed" grass will provide a green appearance and reduce turf-wear throughout the winter. As temperatures warm in the spring, the overseed grass will decline due to unfavorable environmental conditions, while the warm-season grass will break dormancy and grow actively. Unfortunately, these winter overseed grasses are also excellent hosts to *B. longicaudatus*, and soil temperatures during the winter in Florida generally stay in the optimum range for *B. longicaudatus* activity. Therefore, population densities of *B. longicaudatus* double during the winter on overseeded grass compared to non-overseeded (Crow et al. 2005), causing nematode damage to the warm-season grass to be more severe. For this reason, University of Florida recommends that turf areas infested with *B. longicaudatus* be overseeded only when absolutely necessary.

Strawberry is an annual winter crop in Florida, normally transplanted in the fall (September through November) and harvested from November into April. If strawberry plants are left in the ground at the end of the season they can support *B. longicaudatus* over the summer and lead to increased nematode problems during the next season. Therefore, it is recommended that strawberry plants be rogued by physical or chemical means soon after the final harvest. Once old strawberry plants are removed or killed, the field should be planted to a non-host cover crop or maintained clean fallow. Many common weeds found in Florida strawberry fields are hosts to *B. longicaudatus*, particularly Carolina geranium (*Geranium carolinianum*), black medic (*Medico* spp.), and nutsedge (*Cyperus* spp.). If these weeds are allowed to proliferate in the beds, or between the beds, they can serve as alternate hosts for *B. longicaudatus* and increase its damage on strawberry. Therefore, weed management is a very important consideration for nematode management in strawberry, particularly during the summer months. While most strawberry growers in Florida practice good field sanitation, those few who let their fields go unmaintained over the summer generally have increased nematode problems.

Because of the investment in bed formation, fumigation and other inputs, growers often practice double-cropping on the same beds after strawberry. In fields where *B. longicaudatus* is present, double-cropping with a poor or non-host is advised. Common double crops used for Florida strawberry are cantaloupe, watermelon, squash, cucumber, pepper, eggplant and tomato. Among these, only watermelon was found to be a poor to non-host of *B. longicaudatus* in Florida's strawberry fields (Hamill 2006). If a host of *B. longicaudatus* is used as a double-crop, the nematodes can cause considerable damage to the double-crop and increase their numbers for the subsequent strawberry crop. Therefore, the standard practice is to inject a soil fumigant, usually metam sodium, through drip lines to kill the strawberry plants and at the same time, decrease population densities of *B. longicaudatus* before transplanting the double-crop.

9.2.3 Cover Crops

Use of summer cover crops is a common practice in Florida strawberry production. Some of these cover crops such as sunn hemp (*Crotalaria juncea*) are resistant to *B. longicaudatus* and are recommended for fields where the nematode is present (Noling 2016). Other summer crops such as sorghum-sudangrass (*Sorghum bicolor* x *S. arundinaceum*) are excellent hosts of *B. longicaudatus* and should be avoided (Crow et al. 2001). Currently 90% of the cover-crop used in Florida strawberry fields is sunn hemp (J. Noling pers. comm.).

Following potato harvest, the standard practice is to grow sorghum-sudangrass as a summer cover crop. The sorghum-sudangrass crop is allowed to mature and die, then the stalks and residue are incorporated into the soil prior to potato bed construction. The incorporated residues provide bed structure and prevent bed erosion until the rows are covered by the potato canopy (Crow et al. 2001). The cover crop also reduces weeds that serve as alternate hosts of the bacterial wilt pathogen (*Ralstonia solanacearum*) (Weingartner et al. 1993). However, sorghum-sudangrass is an excellent host of *B. longicaudatus*, and most of the nematode yearly population increase occurs on this cover crop (Crow et al. 2001). Alternative summer cover crops have been identified that are resistant to *B. longicaudatus* including velvetbeen (*Mucuna pruriens*) and hairy indigo (Weingartner et al. 1993; Crow et al. 2001). However, these legumes decompose much more quickly than sorghum-sudangrass and do not provide the same bed structure benefits, nor the same degree of weed and bacterial wilt suppression. Florida potato growers perceive that the benefits of sorghum-sudangrass outweigh nematicides costs. Therefore, as long as affordable nematicides are available, the industry has not adopted alternative cover crops.

9.2.4 Tolerance

As already stated, developing resistance to an ectoparasite such as *B. longicaudatus*, is difficult and has not been a priority for breeding programs. However, some cultivars of Saint Augustinegrass (Busey et al. 1993; Quesenberry et al. 2015) and bermudagrass (Pang et al. 2011a, b; Aryal et al. 2015) have relative tolerance to *B. longicaudatus* compared to other cultivars. This tolerance has been attributed to two mechanisms; (a) genotypes that suffer minimal root loss from feeding by *B. longicaudatus*, and (b) genotypes that have a more vigorous root system that can perform well despite having substantial root loss caused by the nematode (Pang et al. 2011a, b; Aryal et al. 2015). Among Saint Augustinegrass cultivars, 'Floratam' is more tolerant of sting nematode than some diploid cultivars and the University of Florida recommends Floratam be used when replanting Saint Augustinegrass in lawns infested with *B. longicaudatus* (Crow 2017b). Similarly, the bermudagrass cultivar 'Celebration' is more vigorously-rooting and, therefore, more tolerant of *B. longicaudatus* than the older standard cultivar 'Tifway' (Pang et al. 2011a, b; Aryal et al. 2015) and is recommended for athletic fields infested with *B. longicaudatus* (Crow 2017a). Ongoing trials have identified other bermudagrass genotypes with improved tolerance to *B. longicaudatus* that should lead to increased use of tolerance as a management tool in the future.

9.2.5 Avoidance

Historically, the majority of sweet corn production in Florida has occurred on organic soils that are not conducive to *B. longicaudatus*. While the primary reasons for this is to reduce fertilizer and water inputs, at the same time crops grown on these organic soils seldom suffer from damage caused by *B. longicaudatus*. However, due to environmental concerns, agriculture production on these organic soils is increasingly restricted and the amount of organic land available for sweet corn production has been decreasing. This has increased the production of sweet corn on mineral soils where *B. longicaudatus* is more abundant. Forage/silage corn is generally produced on lower-value sandy areas and not high-value organic soil.

9.2.6 Biological Control and Biopesticides

Research at the University of Florida has explored many biological products for suppression of *B. longicaudatus* on turf including entomopathogenic nematodes (Crow et al. 2006), plant extracts and materials, live microbial products and killed microbial fermentation products, all of which have been without success (Crow 2005a; Crow et al. 2006). Despite lack of evidence for efficacy, many of these suppressive biological products are being used commercially because they are less expensive than conventional nematicides and because of the negative perception of conventional pesticides. However, several biological agents that are suppressive to *B. longicaudatus* on turf are discussed below. Additionally, ongoing University of Florida research with some other biologicals is yielding promising results.

Candidatus *Pasteuria usgae* is an obligate parasite of *B. longicaudatus* (Giblin-Davis et al. 2003). Endospores of this bacterium are commonly found attached to *B. longicaudatus* from turf. In fact, it is difficult to find *B. longicaudatus* from turf in Florida without 5–20% of the nematodes having *P. usgae* attached to them. In controlled experiments *P. usgae* has suppressed *B. longicaudatus* and has shown potential as a biological control agent (Luc et al. 2010). *Pasteuria usgae* undoubtably provides some degree of suppression of *B. longicaudatus* naturally (Giblin-Davis 2000), but apparently not to the degree that turf damage is prevented. A commercial biopesticide formulation of *P. usgae*, Econem®, (Pasteuria Bioscience, Alachua, FL) was launched for management of *B. longicaudatus* in 2010, but was ineffective (Crow et al. 2011) and was only on the market for a few years. A commercial formulation of the nematphagous fungus *Purpureocillium lilacinum* was found to be suppressive of *B. longicaudatus* on turf, but is currently not labeled for turf use in the United States (Crow 2013).

A formulation of *Bacillus firmus*, Nortica®, (Bayer CropScience, Research Triangle Park, NC) is labeled and continues to be used for the suppression of plant parasitic nematodes in turf since 2011. Nortica® is a wettable powder formulation that is sprayed onto the turf surface and then irrigated so to move the bacteria into the soil. Research in Florida has shown this formulation to be suppressive to *B. longicaudatus* on turf (Crow 2014). However, it is a preventative rather than a curative treatment and correct application timing is essential to achieve desired results. Crow (2014) reports that it is best used as part of an IPM program, rather than as a stand-alone treatment. From 2011 to 2016, Nortica® was one of the most commonly used treatments for nematodes on turf in Florida, mostly due to a lack of other effective treatment options. With the launch of several turfgrass nematicides in 2016, Nortica® sales have slowed, but it still has its place in the market, albeit, in a more limited role.

9.2.7 Broad-Spectrum Soil Fumigation

Almost all commercial strawberry fields in Florida are fumigated, prior to planting, for suppression of weeds, diseases and nematodes, primarily *B. longicaudatus*. The high cost of land in Florida, the high value of strawberry and the specialized nature of strawberry production, all contribute to strawberry being grown on the same land each year. Without soil fumigation strawberry cultivation without crop rotation would be impractical. In the past, methyl bromide was the fumigant of choice for most strawberry growers in Florida due to its broad-spectrum efficacy, excellent soil dispersal and consistency. With the loss of methyl bromide, combinations of fumigants, and combinations of fumigants with non-fumigant pesticides, are required to achieve the same spectrum of activity. Often alternative fumigants do not disperse as well as methyl bromide, contributing to inconsistency of results.

On turf, the primary purpose of preplant fumigation is to maintain genetic conformity. In Florida, turf is generally propagated vegetatively and planted as sod or sprigs. Sod certification standards do not allow perennial grasses or objectionable weeds in fields to be harvested. Fumigants are used to rid sod fields of grasses and weeds prior to establishing the desired species and cultivar. Similarly, fumigants are used in golf course and athletic field construction and renovation to remove old grasses and weeds in order to insure a uniform playing surface free of "off-type" grasses. While not typically the primary target of pre-plant fumigation, nematodes are also killed during fumigation. Infestation by *B. longicaudatus* in sod causes the sod mat to lack the root system necessary to hold the mat together, so that it falls apart on harvest. This can cause complete yield loss by making the sod unharvestable. When turf sprigs or new sod is planted, very few *B. longicaudatus* are required to cause slow establishment and sod installation failure. Therefore, failure to fumigate adequately can lead to major nematode issues for both sod harvest and installation. Since the loss of the methyl bromide critical-use exemption, nematode problems on newly planted turf in Florida are increasing (Crow per. comm.).

9.2.8 *Nematicides*

Nematicides remain the primary nematode management tool for turfgrasses in Florida, particularly for golf and sports turf, and on sod farms. The target treatment zone for *B. longicaudatus* in most turf systems is 5 to 15-cm-deep in the soil profile. Most turf has a thatch layer, a partially decomposed organic mat, ranging from 1 to 5-cm-thick. Many pesticides bind to thatch, and this can greatly reduce movement of nematicides into the soils infested with *B. longicaudatus*. From the 1970s the most commonly used turfgrass nematicide used in Florida was fenamiphos. Fenamiphos was labeled for use on golf courses and had a Section 24(c) Special Local Needs label for sod farms in Florida. Fenamiphos worked very well against *B. longicaudatus* and most of the other plant parasitic nematodes that cause damage to turf. However, sales of fenamiphos in the USA ceased in 2007 and it can no longer be used as of October 2017.

The soil fumigant 1,3-D is very effective on *B. longicaudatus*. While normally used as a pre-plant soil fumigant on other crops, two Section 24(c) Special Local Needs labels allow it to be used in Florida for nematode management on established turf: Telone® II for use on sod farms and Curfew® Soil Fumigant for use on golf and sports turf. Both products are injected below thatch, 13-cm deep in the soil profile using specialized slit-injection equipment. On sod, Telone II is applied 6 months or more prior to harvest to allow the production of a dense sod root mat which improves harvestability. On golf and sports turf Curfew is best used in the spring and fall to reduce population densities of *B. longicaudatus* when the turf is actively producing new roots (McGroary et al. 2013). In golf and sports turf infested with *B. longicaudatus* 1,3-D has consistently improved turf quality and root structure, and improved fertilizer uptake and drought tolerance (Crow et al. 2003; Trenholm et al. 2005; Luc et al. 2007). While 1,3-D is a potent turfgrass nematicide, it has reentry, buffer, and geological restrictions that limit its use.

From 2010 to 2017, several new turfgrass nematicides were labeled and began to be used on turf in Florida. A furfural nematicide was found to be effective against *B. longicaudatus* at 5 to 15-cm-deep, but was ineffective at depths of <5-cm-deep in the soil profile (Crow and Luc 2014). An abamectin turfgrass nematicide, while lethal to *B. longicaudatus* if it contacts the nematode, has limited functional utility against *B. longicaudatus* due to its poor movement below 5-cm-deep in the soil profile, where the majority of *B. longicaudatus* occur. Fluensulphone, formulated for use on turf, has shown good efficacy against *B. longicaudatus* in some turf trials but not others. A turfgrass formulation of fluopyram has shown excellent suppression of *B. longicaudatus*. Fluopyram provides longer-term suppression than the other turf nematicides, but it also can be slower-acting due to its slow movement through thatch (Crow unpublish.).

Until recently, no effective non-fumigant nematicides were labeled for use on strawberry. However, some new nematicides including fluensulfone and fluopyram, were recently labeled for strawberry and have shown promising results. These can be used both pre- and post-transplant. Because their use is so new, it is currently unknown how widespread adoption of non-fumigant nematicides will be by Florida strawberry growers. Research on developing best uses for these nematicides and their integration with current nematode management tactics is underway.

In Florida, nematicide-use on corn is generally restricted to sweet corn production on sandy soil. The most common nematicide used on sweet corn in Florida is 1,3-D, although small amounts of ethoprop and turbufos are also used. The low value of forage/silage corn does not make nematicides a cost-effective option in most cases. Use of nematicidal seed treatments on corn is not currently a common practice in Florida.

The predominant nematicide used by Florida potato growers for management of *B. longicaudatus* is 1,3-D (Weingarnter et al. 1993), which provides consistent and effective results at a relatively low cost. In the past, aldicarb was also used, primarily for suppression of Trichodorid nematodes that vector the *tobacco rattle virus*, but it is currently unavailable in Florida. Some new nematicides, fluensulphone and fluopyram, have shown good results in University of Florida field trials. Presently, some potato growers are evaluating their effectiveness as future management tools for *B. longicaudatus*, but for now, their use is limited.

9.3 Root Knot Nematodes, *Meloidogyne* spp.

Florida has a diversity of important *Meloidogyne* species that impact many different important crops. *Meloidogyne incognita*, *M. javanica*, *M. arenaria*, *M. enterolobii*, and *M. floridensis* are widespread and are problems to production of a variety of fruits, vegetables, ornamentals and agronomic crops produced in Florida. Other species cause problems to only a few kinds of crops; some examples are *M. partityla,* that is major concern to pecan and *M. graminis,* that is a problem on turfgrasses and forages. Management of these nematodes depends on the economic value of the crop, whether it is an annual or perennial, the nematode species present, and grower practices.

Commercial production of tomato, pepper, squash, melons and most other high-value horticultural annuals occurs on raised beds covered with plastic mulch. *Meloidogyne incognita, M. javanica, M. arenaria, M. enterolobii, M. floridensis* and recently *M. haplanaria* (Joseph et al. 2016), have all been found causing crop damage in these systems. In Southern Florida, production occurs in the winter as high temperature and humidity during the summer are not suited to plant health. In Northern Florida most production is in the spring, but often a fall double-crop is planted to generate additional income.

Many moderate-value annual crops such as potato, peanut, cotton and green beans are important crops in Florida. These are typically grown without use of plastic mulch and broad-spectrum fumigants. *Meloidogyne incognita* is the most common root knot nematode damaging potato and cotton in Florida. In Florida, the primary root knot nematode species on peanut *is M. arenaria,* although populations of *M. javanica* on peanut have also been reported (Cetinitas et al. 2003). To date, *M. haplanaria* has only been reported in Southern Florida where peanut is not grown. In addition to those mentioned, Florida produces a wide variety of other moderate-value annual crops that are impacted by *M. incognita, M. javanica, M. arenaria, M. enterolobii,* and *M. floridensis.*

Florida produces ornamentals in container and field nurseries and also produces cut foliage ornamentals used in flower arrangements and other decorative purposes. Root knot nematodes reported from ornamental nurseries in Florida include *M. incognita, M. javanica, M. enterolobii, M. floridensis* and *M. graminis* (Brito et al. 2008, 2010). A recent survey of cut-foliage farms in Florida found M. *incognita* and *M. javanica* to be the primary root knot nematode species infecting cut foliage (Baidoo et al. 2016).

Tropical and temperate fruit and nuts produced in Florida are damaged by root knot nematodes. Examples are *M. partityla* that infects pecan in nurseries and nut production and *M. incognita, M. javanica, M. arenaria* and *M. floridensis* that infect peach. Tropical fruit can be infected by many *Meloidogyne* spp. including *M. javanica, M. incognita,* and *M. enterolobii.*

Meloidogyne graminis is of increasing importance in lawn, sports, and golf course turf. All turfgrass species used in Florida are susceptible to *M. graminis.* On golf greens, *M. graminis* causes chlorosis and patchy growth. High infestations cause turf to wilt, thin and have increased weed and disease problems. Many sod farms are infested by *M. graminis* and since most turfgrass in Florida is planted as sod or sprigs this nematode is widely distributed in planting material.

9.3.1 Rotation

Rotation is seldom practiced with high-value crops since the soil is fumigated before planting for management of weeds and soilborne pathogens. However, rotation is utilized as a management tool for some mid-value crops. For example, in Northern Florida bahiagrass (*Paspalum notatum*), a forage grass that is a non-host to *M. incognita* and *M. arenaria*, is sometimes rotated with peanut and cotton for nematode suppression and other benefits (Wright et al. 2015).

9.3.2 Cropping Systems

If plants are left after final harvest they can maintain or increase the population density of root knot nematodes. Similar to sting nematode on strawberry, crop destruction after final harvest is important for management of root knot nematodes grown on raised beds with plastic mulch. In these systems the crop can be sprayed with an herbicide, or a fumigant like metam sodium can be applied by chemigation through drip tape. Weeds also should be managed to avoid providing alternate hosts for the nematodes between double-crops (Crow et al. 2001).

9.3.3 Cover Crops

Sorghum-sudangrass is a non-host for *M. incognita* and its use as a cover crop for potato helps prevent this nematode from being a major potato production concern. Sunn hemp and other leguminous cover-crops that are non-hosts for common *Meloidogyne* spp. are sometimes used during the summer in vegetable production in Southern Florida. In Northern Florida, rye (*Secale cereal*) is commonly used as a winter cover crop for suppression of *M. incognita* and *M. arenaria* on cotton and peanut, respectively.

9.3.4 Resistance

While resistance to several species of root knot nematodes using the Mi-gene is available in commercial tomato and pepper cultivars, use of resistant cultivars has not been widely adopted in Florida. One problem is that the Mi-gene does not impart resistance to all *Meloidogyne* spp. present in the state. The Mi-gene imparts resistance to *M. incognita*, *M. arenaria*, and *M. javanica*, but not to *M. enterolobii*,

M. floridensis, or *M. haplanaria*. There also are concerns about heat instability of Mi-genes, as temperatures in Florida are often high in the early season for fall crops and late-season for spring crops. Another issue is that growers practice soil fumigation for management of weeds and pathogens whether or not nematodes are a problem. When grown in fumigated soil, resistant cultivars do not produce yields as high as those of high-yielding susceptible cultivars (Vau 2017).

For certain perennial crops, grafting onto resistant root stock is used for management of root knot nematodes. 'Flordaguard' peach rootstock was developed by the University of Florida for adaptability to Florida conditions and for resistance to root knot nematodes. In addition to *M. incognita* and *M. javanica*, Flordaguard peach rootstock confers resistance to some, but not all populations of *M. floridensis*, a trait that is lacking in other rootstocks. Consequently, Flordaguard is the only peach rootstock recommended by the University of Florida for use in that state (Olmstead et al. 2012). In Southern Florida, the common landscape flowering ornamental gardenia (*Gardenia jasminoides*) is grafted onto the root knot nematode resistant rootstock, *G. thunbergia* (Brown and Bradshaw 2013).

9.3.5 Avoidance

Certain agricultural areas of Florida have organic soil that is not conducive to *Meloidogyne* spp. Problems with root knot nematodes, and other nematodes, are rare on these soils. Potato is grown in the winter in Florida when temperatures, and *M. incognita* activity, are low. Therefore, damage from *M. incognita* is rare and typically occurs only if weather is uncommonly warm or if harvest is excessively delayed.

9.3.6 Biological Control and Biopesticides

Suppression of root knot nematodes by *Pasteuria penetrans* has been the subject of a great deal of research in Florida (Chen and Dickson 1998). By continuously planting hosts to the target nematode, naturally occurring *P. penetrans* have increased to and maintained, suppressive levels. Ground roots or peanut hulls containing endospores of *P. pentrans* have been used to inoculate small fields and induce suppressiveness to *Meloidogyne* spp. However, adoption of *P. penetrans* as a management tool for root knot nematodes has occurred on only a limited scale. Most commercial growers are not willing to accept the multiple years of crop loss required to build a suppressive soil. Also, the selectivity of *P. penetrans* makes it difficult to match an effective *Pasteuria* isolate to a target nematode.

Purpureocillium lilacinum is suppressive to *Meloidogyne* spp. in some situations in Florida. A commercial formulation of *P. lilacinum* is being used on a limited scale in Florida agriculture. It is best used as a component of an integrated pest management program and not as a stand-alone treatment for *Meloidogyne* spp. (Baidoo et al. 2017).

9.3.7 *Broad-Spectrum Soil Fumigation*

Soil fumigation is used for most high-value annual crops grown in Florida. It is important to remember that fumigation is used for management of weeds and pathogens in addition to nematodes. Methyl bromide was the fumigant of choice used for this purpose in Florida due to its broad-spectrum of activity and excellent soil movement. Since the cancellation of methyl bromide came into effect it has been replaced with other fumigants, combinations of fumigants, and combinations of fumigants with herbicides.

Sod fields growing high-value grass for planting on golf courses and athletic fields are fumigated before planting to maintain genetic purity of the grass. Formerly methyl bromide was used for this purpose and provided excellent incidental control of *M. graminis*. The fumigants currently being used in sod are not as effective on nematodes as methyl bromide. Consequently, infestation of newly-planted golf and sports turf by *M. graminis* in infested planting material has been increasing in recent years (Crow pers. comm.).

9.3.8 *Nematicides*

In past decades organophoshate and carbamates were commonly used in Florida for management of root knot nematodes on many crops. For example, aldicarb was used for *M. arenaria* on peanut and *M. incognita* on cotton, and fenamiphos for *M. incognita* and *M. javanica* on cut foliage crops. However, these nematicides are no longer available in Florida. Currently, the most common nematicide used for management of root knot nematodes on most crops is 1,3-D. New nematicides such as fluopyram and fluensulfone are only beginning to be used, so it is unknown how widespread their use will be in the coming years.

Fenamiphos was commonly used for management of *M. graminis* on sod farms and golf courses until its recent cancellation. On turf, *M. graminis* predominates in roots growing in the thatch and upper few cm of soil. Since abamectin binds in this upper zone, it provides excellent results on *M. graminis*, especially on golf greens. Hence, abamectin is currently recommended for management of *M. graminis* on golf greens in Florida (Crow 2017c). Other new nematicides, like fluopyram and fluensulfone, also are effective on *M. graminis* and can be rotated with abamectin in a management program (Crow 2017c).

9.4 Citrus Nematode, *Tylenchulus semipenetrans*

Citrus in Florida supports several economically important nematode species that are not associated with citrus grown in other states. Worldwide, most nematodes damaging to citrus tend to be regional or local problems, with the notable exception of the citrus nematode, which occurs widely in every citrus growing region (Duncan 1999; Shokoohi and Duncan 2018). The narrow host range of the citrus nematode is largely responsible for its current distribution in Florida. When Florida's citrus industry expanded southward following the freezes in the 1980s, nurseries were required by law to sell only trees certified as nematode-free. As a result, most southerly orchards have remained free of citrus nematodes for decades. Moreover, in older citrus-growing areas where growers replanted trees on new rootstocks that are resistant to most biotypes of *T. semipenetrans*, infestation is less frequent than in the past (Duncan et al. 1994b). In a survey for sting nematode that resulted in collections of more than 200 samples from 84 orchards located in traditional citrus-growing regions of Central Florida (Duncan et al. 1996), *T. semipenetrans* was detected in only 15% of the total number of orchards (unpublished).

9.4.1 Biology and Ecology

The life cycle of the citrus nematode, *Tylenchulus semipenetrans,* from egg to egg at 25 °C, is about 6 weeks (O'Bannon et al. 1966). Female juveniles feed for up to 2 weeks on epidermal cells before resuming development (Van Gundy 1958). The female anterior body penetrates the fibrous root cortex to induce formation of several "nurse cells" from which it feeds (Cohn 1964; Kallel et al. 2006). Soon the female swells and its posterior body remains exposed on the root surface where a gelatinous egg mass is secreted from the excretory pore (Van Gundy 1958; Cohn 1964).

The disease caused by *T. semipenetrans* is termed 'slow decline' because the damage to trees can take years to become obvious. The nematode does not kill trees, however, it is an excellent parasite that causes few symptoms on its own other than those resulting from the appropriation of carbon by the nematode: smaller fruit and leaves and less fibrous root biomass despite more frequent root flushes (Hamid et al. 1988; Duncan and Eissenstat 1993; Duncan et al. 1995). In combination with other diseases or suboptimal conditions (salinity, *Phytophthora* infection, drought, flooding, *etc.*) effects of citrus nematode parasitism are more apparent (Van Gundy et al. 1964; Heald and O'Bannon 1987). The condition of nematode-infected trees depends heavily on the nematode infection rate (females per root weight), which varies with season, climate and soil conditions. Identifying variables that affect nematode population size (below) has revealed some that are amenable to intervention by Florida growers.

Unlike more virulent nematodes, the ecology of *T. semipenetrans* reflects its coevolution with citrus and other deep-rooted woody trees and vines. The nematode reproduces on just a few plant species. It induces nurse cell formation with little damage to the host, which then supports large numbers of the parasite for years. Indeed, the nematode defends the infection site from invasion by other pathogens (El-Borai et al. 2003). Citrus nematode develops fastest at moderate temperatures, typical of those beneath tree canopies. Unlike many nematodes, *T. semipenetrans* is unable to survive at a low soil water potential, which occurs infrequently in the surface rhizosphere of deep-rooted trees (see below) compared to shallow-rooted herbaceous plants.

The citrus nematode displays distinct, predictable patterns of annual population growth. Depending on the climate, one (Prasad and Chawla 1966; Bello et al. 1986; Sorribas et al. 2000; Maafi and Damadzadeh 2008) or two (Vilardebo 1964; O'Bannon et al. 1972; Salem 1980; Baghel and Bhatti 1982; Duncan et al. 1993; Al Hinai and Mani 1998; Sorribas et al. 2000; Galeano 2002) periods of active population development typically occur annually. Low winter temperatures arrest population growth (Duncan et al. 1993; Maafi and Damadzadeh 2008) and high summer soil temperatures are associated with seasonally low populations in Egypt, Texas, Oman, and Spain (Salem 1980; Davis 1984; Al Hinai and Mani 1998; Sorribas et al. 2000; Korayem and Hasabo 2005). Many nematodes are well adapted to survive periodic dry soil conditions, whereas *T. semipenetrans* has little capacity for desiccation survival (Tsai and Van Gundy 1988). Although citrus nematode population densities decline rapidly if drought causes citrus trees to wilt (Van Gundy and Martin 1961; Van Gundy et al. 1964), surprisingly, in the absence of wilt, soil moisture is frequently inversely related to population growth of *T. semipenetrans* in the field (Tuong 1963; Duncan et al. 1993; Sorribas et al. 2000; Galeano 2002). In Florida and elsewhere, citrus nematode population densities increase most rapidly in the dry season and decline abruptly as the rainy season begins (Duncan et al. 1993; Sorribas et al. 2000). Passive downward movement of nematodes due to precipitation is a potential cause of the relationship (Sorribas et al. 2000), although the vertical distribution of some nematodes is unaffected by heavy rainfall events (Chabrier et al. 2008). Alternatively, split-root pot studies showed that populations of *T. semipenetrans* in extremely dry parts of the rhizosphere can either grow very rapidly or decline precipitously, depending on whether part or all the root system is affected by drought (Duncan and El-Morshedy 1996). Hydraulic lift of water deep in soil to drier surface soil horizons via the root xylem (Caldwell et al. 1991) creates a humid zone at the rhizoplane that may not be measurable in the soil. This environment of hydrated roots in dry soil favors population growth of *T. semipenetrans* compared to more humid soil conditions. Similarly, Sorribas et al. (2000) observed that population densities were higher under drip than under flood irrigation. As a parasite of deep-rooted perennials, *T. semipenetrans* likely experienced less selection pressure than many nematodes for anhydrobiotic survival. The response of *T. semipenetrans* to hydraulic lift is one factor explaining regional patterns of population density which tend to be higher in arid (Cohn 1966; Macaron 1972; Willers 1979; Davis 1985; Sorribas et al. 2000, 2008) than in sub-tropical regions such as Florida (Davide 1971; O'Bannon et al. 1972; Duncan et al. 1993).

The population dynamics of *T. semipenetrans* are also regulated by seasonal availability of nutrients in roots (Duncan et al. 1993). Young roots are suitable for penetration and development of *T. semipenetrans* and new cohorts of developing nematodes are created during root flushes (Cohn 1964; O'Bannon et al. 1972). However, the nematode can alter the normal pattern of carbon allocation. Trees heavily infected by the nematode have less root mass than lightly infected trees, but root growth is initiated more frequently to replace those damaged by the nematode (Hamid et al. 1988). Starch is a major nutrient requirement of *T. semipentrans* (Cohn 1965a, b), whereas lignin and phenolic compounds inhibit root infections (Kaplan and O'Bannon 1981). The seasonal concentrations of these compounds in fibrous roots have been shown to be correlated in the expected ways with *T. semipentrans* population growth in field surveys (Van Gundy and Kirkpatrick 1964; Duncan et al. 1993) and in field experiments in which root carbohydrates were manipulated (Duncan and Eissenstat 1993).

Salinity and soil pH modulate crop loss to *T. semipenetrans*. In Florida's coastal orchards with saline irrigation water, sodium accumulates at much higher levels in leaves of nematode-infected compared to non-infected trees (Mashela et al. 1992a). The nematode develops poorly in saline soils (Kirkpatrick and Van Gundy 1966) and yet, salinity is associated with high nematode numbers and increased crop loss in the field (Machmer 1958; Willers and Holmden 1980). An explanation for this phenomenon is that soil salinity is seasonal, being least during rainy seasons when salt from saline irrigation is leached from surface soils. Citrus previously exposed to salinity supports greater than normal population growth of *T. semipenetrans* only after salts are washed from the rhizosphere (Mashela et al. 1992a, b). Reduced production of phenylalanine ammonia lyase in salt-stressed citrus may limit production of phenolic-based defensive compounds in roots (Dunn et al. 1998). Nematode-infected roots accumulate less sodium and chloride ions, whereas these elements increase in leaves (Willers and Holmden 1980; Mashela et al. 1992a, b), perhaps due to increased osmotic pressure from carbohydrate transfer to nematode-infected roots (Mashela and Nthangeni 2002).

Equilibrium density of *T. semipenetrans* is greater as pH increases (Van Gundy and Martin 1962; Bello et al. 1986; El-Borai et al. 2003). Citrus grown in calcareous soils, typical of dry climates, is likely to experience larger nematode numbers and greater damage than citrus grown in more acidic, subtropical soils of Florida. Although population growth in pots was greater in moderately fine-textured soil than in sand (Van Gundy et al. 1964; Davide 1971; Bello et al. 1986), high pH and low moisture can cause large population densities of the nematode in coarse-textured field soils (Duncan et al. 1993). Organic matter favors population growth of the nematode (Van Gundy 1958).

There are no yield loss studies in the field in which *T. semipenetrans* is the only independent variable. Instead experiments have manipulated nematode density with nematicides, or surveys have compared natural patterns of nematode density to the fruit yield or tree properties. Despite recognized limitations of both approaches (Shookohi and Duncan 2017), a large body of research consistently shows that *T. semipenetrans* requires high population density to cause measurable crop loss, but at high densities can seriously affect profitability. Nematicide treatments in Florida increased citrus yield by about 15% on average (O'Bannon and Tarjan 1973; Childers et al. 1987; Duncan 1989). A Florida orchard was identified in which randomly distributed trees were infested or not infested by *T. semipenetrans*. Trees were also damaged by *Phytophthora nicotianae* and salinity, but levels of those variables and others such as soil pH, texture and nutrients, did not differ between infested or non-infested trees. Visual tree condition was unrelated to presence of

T. semipenetrans, but leaf area, fibrous root mass density, and fruit yield of infested trees were 32%, 8%, and 22% lower, respectively, than on non-infested trees (Duncan et al. 1995). The economic impact of the nematode is greater when fruit are marketed fresh rather than for juice, because *T. semipenetrans* reduces fruit size and value (Philis 1989; McClure and Schmitt 1996).

9.4.2 Management

The tactics used in Florida to manage crop damage from citrus nematode, in order of economic importance include sanitation and use of resistance to exclude the pest. Growers minimize the economic loss to infected trees through crop management and use of nematicides. The exclusion of nematode pests from orchards through a mandatory nursery certification program, operational since 1955, is one of the most successful nematode management programs devised anywhere.

9.4.3 Sanitation

Most citrus-growing regions have few serious nematode pests and few naturally occurring hosts of citrus nematode. Therefore, exclusion of *T. semipenetrans* from orchards is a realistic goal to preclude the perennial expense of nematode management. *T. semipenetrans* moves very little on its own from tree to tree, so the occasional introduction into non-infested orchards does not negate the value of a conscientious sanitation program (Meagher 1967; Tarjan 1971; Baines 1974). In the absence of flooding, and particularly, with the use of low volume irrigation, trees may remain uninfected for long periods, despite the existence of nematodes on adjacent trees (Duncan et al. 1995). Infestation usually results from movement of infected planting stock (Van Gundy and Meagher 1977) or from contaminated equipment (Tarjan 1956). Programs to approve and monitor nursery sites so to ensure that nursery stock is nematode-free, have been highly effective. The Florida nursery certification program saved growers US$33 million in 1994 by reducing yield losses from *T. semipenetrans* that would have otherwise occurred from the spread of this nematode (Lehman 1996). Such programs must be thorough to be effective. Florida requires continuous monitoring through soil sampling, isolating nursery locations to avoid runoff water from infested orchards, and security to exclude contaminated planting media or equipment. If separate equipment for use in infested and non-infested orchards is not possible, equipment must be disinfested continually before entering non-infested orchards (Esser 1984). Irrigation from canals and rivers has been found to represent a serious source of rapid contamination by *T. semipenetrans* and *Phytophthora nicotianae* (Cohn 1976). Irrigation water can be decontaminated by the conscientious use of settling ponds and filtration systems (Cohn 1976).

9.4.4 Biotypes and Rootstock Resistance

All rutaceous species support the nematode, but few non-rutaceous hosts are known, the most important of which are grape, olive, persimmon and pomegranate. Three biotypes of *T. semipenetrans*, based on host suitability include a 'Citrus' biotype that reproduces poorly on *Poncirus trifoliata* (a citrus relative) but will reproduce on *Citrus* spp., olive, grape and persimmon. The 'Poncirus' biotype reproduces on most citrus including *P. trifoliata*, and on grape, but not olive. It is reported from USA including Florida, South Africa, Japan and Spain (Murguía et al. 2005; Kwaye et al. 2008; Mathabatha et al. 2015). A 'Mediterranean' biotype is similar to the 'Citrus' biotype, except that it does not reproduce on olive (Stokes 1969; Inserra et al. 1980, 2009; Mathabatha et al. 2015). It is found throughout the Mediterranean region and perhaps India.

Cultivars of *P. trifoliata* and some hybrids of *P. trifoliata* provide acceptable rootstocks in some regions including Florida. Swingle citrumelo (*C. paradisi* × *P. trifoliata*) and C-35 (*C. sinensis* × *P. trifoliata*) are rootstocks widely used in Florida with a high degree of resistance to most populations of *T. semipenetrans* as well as *citrus tristeza virus* and *P. nicotianae*. However, they are intolerant of calcareous soils. Several hybrids of *P. trifoliata* × various mandarin (*C. reticulata*) rootstocks have inherited high resistance to *T. semipenetrans*, and grow well in calcareous soils (Verdejo-Lucas et al. 2003). Factors identified as responsible for resistance of citrus to *T. semipenetrans* development include host cell hypersensitivity, wound periderm formation, and allelopathy (Van Gundy and Kirkpatrick 1964; Kaplan and O'Bannon 1981; Kallel et al. 2006). Several random amplified polymorphic DNA markers associated with resistance have been identified to facilitate selection for resistance (Ling et al. 2000; Xiang et al. 2010).

Resistance management is important because the Poncirus biotype occurs in Florida and elsewhere where trifoliate-derived rootstocks are extensively used (Baines et al. 1969; Duncan et al. 1994b). When resistant rootstocks are used to replant an entire orchard, they are challenged only by nematodes that remain from the previous trees. However, if resistant rootstocks are used to replace individual trees in orchards with susceptible rootstocks, the infection pressure provided by adjacent susceptible rootstocks is continuous (Duncan et al. 1994b; Verdejo-Lucas et al. 2003).

9.4.5 Cultural Practices

A key concept for successful management of *T. semipenetrans* is that of the 'limiting factor', in which all rhizosphere limitations must be resolved in order to obtain a response to nematode management (Thomason and Caswell 1987). Citrus trees damaged by *Phytophthora* spp., poor drainage, salinity, frequent drought or other problems are unlikely to respond consistently to management of only *T. semipenetrans*. Indeed, yield response to nematicide use against *T. semipenetrans* in Florida is typically most apparent in orchards with vigorous, healthy appearing trees. It is important to ensure that orchards are managed properly in all respects before investing in nematode management tactics. Although Florida soils are not typically calcareous, growers actively reduce soil pH to avoid bicarbonate toxicity to citrus roots in order to mitigate disorders caused by the recently introduced bacterial disease, huanglongbing (Graham et al. 2014). This practice appears to be reducing population levels of citrus nematodes and increasing a beneficial species of entomopathogenic nematode, thereby reducing soilborne arthropod pests (Campos-Herrera et al. 2014).

9.4.6 Chemical Control

Soil sampling, diagnosis of nematode species and, in the case of citrus nematode, population size, are fundamental pre-requisites for the rational use of nematicides. The aggregation of nematodes in space, time and along roots, means that sample size can be reduced by sampling during seasons of peak population size and in locations of highest feeder root and nematode concentration (Duncan 1986; Duncan and Phillips 2009). Seasonal variation in soil and root nematode densities are as much as tenfold; thus, for comparative purposes and accuracy, samples should be taken during the same season each year when peak densities occur during spring in Florida (Duncan et al. 1993; Sorribas et al. 2000). Similarly, feeder roots and nematodes are more abundant beneath the tree canopy than at the dripline or in rows between trees (Davis 1985; Duncan 1986). Low volume irrigation systems concentrate root and nematode densities even further in the wetted zones. Accurate estimation of *T. semipenetrans* population density is costly, requiring from 30 to 75 soil cores in 2 ha to estimate population levels within 20–40% of the true mean (McSorley and Parrado 1982a, b; Davis 1984; Duncan et al. 1989, 1994a). Despite its low precision, sampling is valuable since most density estimates are well above or below management threshold levels. A damage function derived in Spain revealed a tolerance limit (below which no loss is measurable) of fewer than three hundred females per gram of root, with economic thresholds ranging between 330 and 710 females per gram of root depending on the cost of the nematicide used and the value of the fruit (Sorribas et al. 2008). Similar estimates were reported in California where greater than 400 or 700 females per gram of root in early spring or early summer, respectively, are considered to merit management in orchards with a history of responding to management (Garabedian et al. 1984).

Soil fumigation to control citrus nematode is not practiced in Florida due to cost and local restrictions, based on soil texture and shallow water table. Aldicarb and fenamiphos were widely used until both were deregistered (Childers et al. 1987; Duncan 1989). Oxamyl was also widely used until it became unavailable in 2014 following an industrial accident. Fluopyram is a promising new nematicide that was recently registered for use in citrus. Nematicides in large commercial citrus orchards are applied in bands along the tree rows or through low volume irrigation systems.

Since the abundance of nematodes and feeder roots in the upper soil horizons declines quickly with distance from the trunk, nematicide bands, even for systemic products, are most effective when they are applied as much as possible beneath the tree canopy (Nigh 1981; Duncan 1986, 1989).

9.5 Burrowing Nematode, *Radopholus similis*

First described in 1928, the disease "spreading decline" remains one of the most damaging to Florida citrus. *Radopholus similis* and a lesion nematode, *Pratylenchus coffeae*, are the most virulent nematode parasites of citrus worldwide (O'Bannon et al. 1976). In infested orchards, the losses due to *R. similis* were estimated at 40–70% for oranges and slightly higher for grapefruit (DuCharme 1968). The damage by spreading decline within orchards has been mitigated in recent years by improved management practices described below. Unfortunately, the discontinuation of programs to prevent migration of these nematodes from infested to uninfested orchards has increased the rate of spread of this pest.

The discovery of *R. similis* as the causal agent of spreading decline (Suit and DuCharme 1953) set in motion research and regulatory efforts that were among the most intensive and instructive nematode management programs undertaken anywhere. At their peak in the 1960s and 1970s, these programs required growers to attempt to either eradicate the nematode by orchard removal, followed by soil fumigation and mandatory fallow, or to confine the nematode by surrounding known areas of infestation with buffer zones of plant-free land, which were periodically fumigated with ethylene dibromide (EDB) to kill roots that might carry the nematode across the buffer to un-infested orchards (Poucher et al. 1967). Undergirding these efforts were government offices throughout the central citrus growing areas that supported crews of workers who systematically inspected all orchards and sampled tens of thousands of trees each year to demarcate the borders of nematode infestation (Lehman 1996). Widespread detection of ethylene dibromide in groundwater led to the suspension of EDB in 1983 and the programs were eventually discontinued. However, the Florida Department of Agriculture nursery certification program initiated in 1955 to proscribe sale of planting stock infected by *R. similis, T. semipenetrans,* or *P. coffeae*, remains as one of the most cost-effective and important nematode management programs in existence (Lehman 1996).

The citrus race of *R. similis* occurs only in Florida and is the only race known to infect citrus, although *R. citri* also infects and damages citrus in Indonesia (Bridge et al. 1990; Hahn et al. 1994; Machon and Bridge 1996). The banana race of *R. similis* occurs throughout the tropics (DuCharme and Birchfield 1956). Both races infect banana and have similar host ranges of more than 250 plant species (O'Bannon 1977). In 1984, the citrus race of *R. similis* was designated as *R. citrophilus*, a sibling species to *R. similis* based on putative differences in chromosome number, isozyme patterns, mating behavior, host preference and morphology (Huettel et al. 1984; Huettel and Yaegashi 1988). Subsequent research refuted the designation based on karyotype identity, morphological and genetic congruence and reproductive compatibility (Kaplan and Opperman 1997, 2000; Valette et al. 1998). Remarkable genetic similarity among *R. similis* worldwide likely resulted from a modern dissemination worldwide on banana from its center of origin somewhere in Australasia (Kaplan 1994; Fallas et al. 1996; Machon and Bridge 1996).

Infected trees have very sparse foliage, leaves and fruit are small, and experience greater fruit drop than normal. Often, entire branches die. Large, contiguous groups of trees are affected, and expansion of the diseased area is rapid, as much as 15 m/year. Forced water uptake in the trunk of the tree is indistinguishable from normal trees, in contrast to trees suffering from citrus blight (Graham et al. 1983). Trees wilt more readily, particularly during Florida's long dry season (mid-autumn till late spring) when tree decline is most evident. The nematode penetrates the region of elongation and root tips can become swollen due to hyperplasia and stubby if terminals are penetrated (DuCharme 1959, 1968). Nematodes burrow in a section of root for several weeks, completely destroying the phloem and much of the cortex (DuCharme 1959). The normally abundant root system in the surface soil horizon (0–30 cm) is not damaged by burrowing nematode. However, at depths of 25–50 cm, just 75% of the root system may remain, and below this level the root system is almost totally lacking (Ford 1952, 1953). Mature citrus growing in the deep sands of Florida's central ridge establish as much as half of the feeder roots between 1 and 6 m. Thus, burrowing nematode is only an economic pest to citrus on the central ridge. Destruction of the deep roots accounts for the drought-related decline during the dry season. Moreover, the nematode moved farther (Tarjan 1971) and was more virulent to citrus (O'Bannon and Tomerlin 1971) in pot studies in sandy soils typical of the central ridge than loamy soils from other regions.

Compared to citrus nematode, *R. similis* on citrus has a short life cycle of 18–20 days under optimum conditions (DuCharme and Price 1966), growing rapidly when conditions are favorable (DuCharme and Suit 1967). Following root penetration, mature females begin to lay eggs at an average rate of nearly two per day and eggs hatch in 2–3 days. The nematodes normally reproduce sexually; however, females that do not mate, after a period of time reproduce as hermaphrodites (Brooks and Perry 1962; Kaplan and Opperman 2000). Mature males do not feed. The nematode remains within the root until forced by overcrowding and decay to migrate. The cardinal temperature for *R. similis* is 24 °C. Optimum temperatures occur each year in the deeper soil horizons where the highest level of reproduction is known to occur. Highest absolute soil densities are found in the late summer–early autumn period when optimum temperatures combine with an annual cycle of root growth to support population increase (DuCharme 1967, 1969).

9.5.1 Management

Following the deregistration of ethylene dibromide for buffer zone maintenance, the use of methyl bromide, mechanical trenching and physical barriers were evaluated as tactics to maintain the buffers root-free (Duncan et al. 1990). None of those methods have proven to be feasible or cost-effective. Management of spreading decline currently emphasizes diagnosis, restricting the spread of the nematode through planting-stock certification, sanitation, use of resistant rootstocks, orchard management and use of nematicides. Because most root damage occurs deep in the soil, pest managers traditionally sampled to depths of 120 cm to obtain roots most likely to contain high densities of the nematode (Poucher et al. 1967). The Florida Department of Agriculture and Consumer Services maintained crews with fleets of vehicles equipped with rear-mounted augers to survey orchard infestations. It was later shown that processing larger amounts of roots from near the soil surface (that are easily and inexpensively obtained by growers with a shovel) can more accurately detect nematode infected trees than processing smaller amounts of roots from deeper in the soil. Visual stratification of orchards based on tree decline symptoms is important. Current recommendations are for a single ‘shovel-deep’ sample from beneath 12 trees composited into a 4-l plastic bag. Samples should be immediately washed and incubated for 72 h before rinsing to recover nematodes.

A cost-benefit analysis of the value of Florida’s nursery certification program in reducing potential losses to burrowing nematode estimated a 14:1 return on investment resulting in increased yield worth US$40 million/year (Lehman 1996). As noted previously, nurseries are regularly sampled and inspected to remain certified. In addition, commercial movement of soil within and into citrus-producing areas of the state requires certification that the site of origin is pest free. Equipment used in infested orchards should be reserved for that purpose when possible or disinfested between operations (Esser 1984). There are also regulatory considerations associated with the burrowing nematode (and other restricted species) beyond citrus. The citrus race of the nematode can infect roots and sometimes, epigeal tissue of many ornamental crops destined for places beyond Florida. From the mid-1950s to the late 1990s nearly half a million floriculture shipments were inspected for burrowing nematode to prevent spread of the citrus race to other citrus growing regions (Lehman et al. 2000).

Within citrus and closely related genera, more than 1200 species, cultivars and hybrids have been screened for resistance or tolerance to *R. similis* (Ford and Feder 1961; O’Bannon and Ford 1976). Several cultivars of citrus, ‘Ridge Pineapple’ sweet orange, ‘Estes’ rough lemon and ‘Milam’ lemon have been released as rootstocks since 1958, and all of the rootstocks have been subsequently shown to support biotypes of *R. similis* capable of breaking resistance (Kaplan and O’Bannon 1985). In the case of ‘Carrizo’ citrange, considerable variability exists within the progeny for susceptibility to burrowing nematodes (Kaplan 1986); however, a now-widely used breeding line known as ‘Kuharski Carrizo’ was identified in which resistance is stable (Kaplan 1994). The occurrence of resistance-breaking populations of the burrowing nematode indicates a need for rootstocks with additional resistance genes (Kaplan and O’Bannon 1985; Kaplan 1994). Ongoing research is based on reports that transformation of potato and *Musa* spp. germplasm to express a maize protein, cystatin, against nematode digestive proteinases and a peptide that interferes with chemoreception, produced plants resistant to cyst and burrowing nematodes respectively, as well as several other endoparasitic and ectoparasitic nematodes that attack plantain (Lilley et al. 2011; Roderick et al. 2012).

Although citrus on Florida's central ridge produces very deep root systems, trees in the flatwoods ecoregions are grown on shallow root systems in bedded soil due to shallow groundwater. The discovery that *R. similis* damages primarily the deeper (below 45 cm) portion of the citrus root system revealed the potential for managing spreading decline with cultural practices designed to support a healthy, shallow root system. Growers now use herbicides and mowing rather than cultivation to avoid cutting surface roots (Tarjan and Simmons 1966). They adopted microsprinkler irrigation to provide adequate soil moisture in the dry season (Bryan 1969) and the ability to frequently fertigate in order to maintain nutrients in the shallow rhizosphere. Systemic nematicides such as oxamyl, have been used by growers to reduce *R. similis* in deeper roots (O'Bannon and Tomerlin 1977). Infested orchards in which sound practices are employed, have remained economically viable (Tarjan and O'Bannon 1977) and may even out-produce annual state production averages (Bryan 1969).

9.6 Lesion Nematodes *Pratylenchus* spp.

Lesion nematodes are widespread in Florida crops, but few are managed except in citrus, peanut and certain ornamental crops, primarily those requiring pest-free certification for movement. While there remains considerable phylogenetic variability in some species, the common agronomically important ones in Florida include *Pratylenchus coffeae*, *P. brachyurus*, *P. zeae*, *P. bolivianus* and *P. hippeastri*. *Pratylenchus vulnus* was found recently on peach and *P. scribneri* on grasses.

There are few recommendations for management of lesion nematodes on crops other than citrus, and the ornamental amaryllis. Lesion nematodes have for long been recognized as damaging to amaryllis, reducing bulb size, flower number and plant vigor. The species was described and renamed *P. hippeastri* (Inserra et al. 2006). It is also widespread, although not yet studied in bromeliads and turfgrasses in Florida. The nematode is managed by sanitation and by disinfesting bulbs by removing roots and treating with hot water for 20 min at 50 °C. *Pratylenchus brachyurus* can cause pod lesions and rot in peanut and reduced yields. 1,3-Dichloropropene, used to manage root knot nematode on peanut, is also effective against lesion nematode and, less frequently, is used specifically for *P. brachyurus* management.

Pratylenchus coffeae is easily the most virulent of any nematode on citrus in Florida. Although reported in numerous citrus industries worldwide, *P. coffeae* was shown to be a large species complex (Golden et al. 1992; Duncan et al. 1998, 1999). For example, *P. coffeae* on citrus in Sao Paulo State, Brazil was renamed *P. jaehni* (Inserra et al. 2001). Putative *P. coffeae* on native vegetation in Florida prevented the nematode-free certification of some citrus nurseries, until the nematodes were re-described as *P. loosi* and *P. pseudocoffeae*, neither of which reproduces on citrus (Inserra et al. 1996, 1998; Duncan et al. 1999).

Symptoms of damage by *P. coffeae* to citrus in Florida are similar to those of burrowing nematode. Infection occurs by all motile stages of the nematode and there is an abundance of males in the fibrous root cortex, while the vascular tissues remain intact until invaded by secondary organisms (Radewald et al. 1971b; Inserra et al. 2001). Reproduction is highest at 26–30 °C where the life cycle is completed in less than 1 month producing as many as 10,000 nematodes/g root (O'Bannon and Tomerlin 1969; Radewald et al. 1971a). In the greenhouse, *P. coffeae* reduced citrus root weights by as much as half (O'Bannon and Tomerlin 1969; Radewald et al. 1971a) and young tree growth and fruit production in the field were reduced by up to 80% and 20-fold, respectively (O'Bannon and Tomerlin 1973). Soil types ranging from sands to sandy loams did not affect the virulence of *P. coffeae* (O'Bannon et al. 1976). Although migration of the nematode was measured at 1 m/year (Tarjan 1971; O'Bannon and Tomerlin 1973; O'Bannon 1980), decline symptoms in orchards spread as quickly as those of spreading decline.

The once limited distribution of *P. coffeae* in Florida citrus is becoming more widespread due to discontinuation of buffer zones. It is a regulated species in the nursery certification program which has undoubtedly slowed the spread. No commercial rootstocks resistant to the nematode are available, so growers currently have few options for managing the nematode. There is urgent need for new systemic nematicides and especially resistant rootstocks.

Pratylenchus brachyurus is a pathogen of seedlings in greenhouse trials and on young trees in the field (Tarjan and O'Bannon 1969; Radewald et al. 1971a; Tomerlin and O'Bannon 1974; O'Bannon et al. 1974). However, it is not especially damaging to mature citrus (O'Bannon et al. 1974). Nevertheless, when *P. brachyurus* in mature trees was managed with aldicarb, trees better tolerated winter frost and yields increased (Wheaton et al. 1985).

9.7 Perspective on Future Management

Crop production systems in Florida vary importantly with region, and future improvements to nematode management tactics are likely to reflect those differences. While agriculture in the northern part of the state is typical of that in the Southern USA – field crops including corn, cotton, soybean, peanut, small grains, forage and oil seed crops – the central and southern industries tend to specialize in specific fruit, vegetable and ornamental industries. Ongoing research in Northern Florida seeks to exploit new methods to understand the role of nematode communities as they interact with the rhizosphere microbiome, the physical properties of soils, and the responses of plants in cropping systems. Increased knowledge of nematode ecology can provide insights for developing cultural practices that minimize their damage to crops. In contrast, it will be difficult for growers of high value crops in Florida to manage nematodes without broad spectrum fumigants in the near term.

The many reasons include the subtropical climate and coarse textured soils that favor population growth of diverse nematode species and subsequent, heavy crop damage, inadequate efficacy of other nematicides because of short persistence and inability to kill nematodes deep in the soil, and the need for multiple additional pesticides that often fail to control weeds and diseases consistently for myriad reasons. Crop rotation, a critical management tactic in more temperate regions with fewer damaging species and shorter growing seasons, is of limited use because many of the crops that could be grown in rotation are of too little economic value to support high value cropping systems reliant on special equipment and infrastructure.

The recent devastation of Florida's largest and signature crop, citrus, by the introduction of the bacterial disease huanglongbing (production of citrus fruit declined by 70% in the past decade), will potentially cause citrus growers to be the first in Florida to embrace genetically modified germplasm that may become available. This eventuality would increase the likelihood that genetically modified, nematode resistant rootstocks with durable resistance to several of the most economically important nematode species could be introduced into the industry (Roderick et al. 2012). Combined with the nursery certification program, the use of such rootstocks could eliminate the need to use expensive, less sustainable methods of reducing crop loss to nematodes in hundreds of thousands of acres in the state. A similar scenario would seem feasible for the large ornamental industry in Southern Florida.

There is a rich history of nematological research in Florida dating to before the mid-twentieth century. Following the discovery of *R. similis* as the causal agent of spreading decline, more than a dozen state, federal and university research and regulatory programs dealing with nematodes have continuously been supported. Especially important has been collaborations with taxonomic experts who continue to reveal the state's species diversity and facilitating management programs based on regulation, resistance/tolerance and economic thresholds. The increasing need to feed more people with fewer inputs and less detriment to Florida's vulnerable soil and water resources argues for the continued broad support of diverse regulatory and research programs to improve the means by which growers manage plant parasitic nematodes.

References

Al Hinai, M. S., & Mani, A. (1998). Seasonal population changes and management of *Tylenchulus semipenetrans* using organic amendments and fenamiphos. *Nematologia Mediterranea, 26*, 179–184.

Aryal, S. A., Crow, W. T., McSorley, R., Giblin-Davis, R. M., Rowland, D. L., Poudel, B., & Kenworthy, K. E. (2015). Effects of *Belonolaimus longicaudatus* on rooting dynamics among St. Augustinegrass and bermudagrass genotypes. *Journal of Nematology, 47*, 322–331.

Baghel, P. P. S., & Bhatti, D. S. (1982). Vertical and horizontal distribution of phytonematodes associated with citrus. *Indian Journal of Nematology, 12*, 339–344.

Baidoo, R., Joseph, S., Mengistu, T. M., Brito, J. A., McSorley, R., Stamps, R. H., & Crow, W. T. (2016). Mitochondrial haplotype-based identification of root knot nematodes (*Meloidogyne* spp.) on cut foliage crops in Florida. *Journal of Nematology, 48*, 193–202.

Baidoo, R., Mengistu, T., McSorley, R., Stamps, R. H., Brito, J., & Crow, W. T. (2017). Management of root knot nematode (*Meloidogyne incognita*) on *Pittosporum tobira* under greenhouse, field, and on-farm conditions in Florida. *Journal of Nematology, 49*, 133–139.

Baines, R. C. (1974). The effect of soil type on movement and infection rate of larvae of *Tylenchulus semipenetrans*. *Journal of Nematology, 6*, 60–62.

Baines, R. C., DeWolfe, T. A., Klotz, L. J., Bitters, W. P., Small, R. H., & Garber, M. J. (1969). Susceptibility of six *Poncirus trifoliata* selections and Troyer citrange to a biotype of the citrus nematode and growth response on fumigated soil. *Phytopathology, 59*, 1016–1017.

Bello, A., Navas, A., & Belart, C. (1986). Nematodes of citrus-groves in the Spanish Levante Ecological study focused to their control. In R. Cavalloro & E. Di Martino (Eds.), *Integrated pest control in citrus-groves: Proceedings of the experts' meeting*, Acireale, 26–29 March 1985. Published for the Commission of the European Communities by AA Balkema, Rotterdam, pp. 217–226.

Bridge, J., Machon, D., & Djatmiadi, D. (1990). A new nematode problem of citrus caused by *Radopholus* sp. in Java. *Nematologica, 36*, 336.

Brito, J. A., Kaur, R., Dickson, D. W., Rich, J. R., & Halsey, L. A. (2006). *The pecan root knot nematode, Meloidogyne partityla Kleynhans, 1986, Nematology Circular 222*. Gainesville: Florida Department of Agriculture and Consumer Services.

Brito, J. A., Kaur, R., Cetintas, R., Stanley, J. D., Mendes, M. L., McAvoy, E. J., Powers, T. O., & Dickson, D. W. (2008). Identification and isozyme characterisation of *Meloidogyne* spp. infecting horticultural and agronomic crops, and weed plants in Florida. *Nematology, 5*, 757–766.

Brito, J. A., Kaur, R., Centintas, R., Stanley, J. D., Mendes, M. L., Powers, T. O., & Dickson, D. W. (2010). *Meloidogyne* spp. infecting ornamental plants in Florida. *Nematropica, 40*, 87–103.

Brito, J. A., Dickson, D. W., Kaur, R., Vau, S., & Stanley, J. D. (2015). *The peach root knot nematode, Meloidogyne floridensis, and its potential impact for the peach industry, Nematology Circular 224*. Gainesville: Florida Department of Agriculture and Consumer Services.

Brooks, T. L., & Perry, V. G. (1962). Apparent parthenogenetic reproduction of the burrowing nematode *Radopholus similis*. *Soil Crop Science Society of Florida Proceedings, 22*, 160–162.

Brown, S. P., & Bradshaw, J. (2013). *Gardenias at a Glance, Circular No. 1098*. Gainesville: Environmental Horticulture Department, Florida Cooperative Extension Service, Institute of Food and Agricultural Sciences, University of Florida.

Bryan, O. C. (1969). Living with the burrowing nematode. *Citrus and Vegetable Magazine, 33*, 29–38.

Busey, P., Giblin-Davis, R. M., & Center, B. J. (1993). Resistance in *Stenotaphrum* to the sting nematode. *Crop Science, 33*, 1066–1070.

Caldwell, M. M., Richards, J. H., & Beyschlag, W. (1991). Hydraulic lift: Ecological implications of water efflux from roots. In D. Atkinson (Ed.), *Plant root growth: An ecological perspective* (pp. 423–436). Oxford: Blackwell Scientific.

Campos-Herrera, R., El-Borai, F. E., Ebert, T. E., Schumann, A., & Duncan, L. W. (2014). Management to control citrus greening alters the soil food web and severity of a pest–disease complex. *Biological Control, 76*, 41–51.

Cetinitas, R., Lima, R. D., Mendes, M. L., Brito, J. A., & Dickson, D. W. (2003). *Meloidogyne javanica* on peanut in Florida. *Journal of Nematology, 35*, 433–436.

Chabrier, C., Carles, C., Queneherve, P., & Cabidoche, Y. M. (2008). Nematode dissemination by water leached in soil: Case study of *Radopholus similis* (Cobb) Thorne on nitisol under simulated rainfall. *Applied Soil Ecology, 40*, 299–308.

Chen, Z. X., & Dickson, D. W. (1998). Review of *Pasteuria penetrans*: Biology, ecology, and biological control potential. *Journal of Nematology, 30*, 313–340.

Childers, C. C., Duncan, L. W., Wheaton, T. A., & Timmer, L. W. (1987). Arthropod and nematode control with Aldicarb on Florida citrus. *Journal of Economic Entomology, 80*, 1064–1071.

Chitwood DJ (2003) *Nematicides*. New York: Wiley. Published Online: 15 Apr 2003. https://doi.org/10.1002/047126363X.agr171.

Christie, J. R. (1938). Two distinct strains of the nematode *Aphelenchoides fragariae* occurring on strawberry plants in the United States. *Journal of Agricultural Research, 57*, 73–80.

Christie, J. R., & Perry, V. G. (1951). A root disease of plants caused by a nematode of the genus *Trichodorus*. *American Association for the Advancement of Science, 113*, 491–493.

Christie, J. R., Brooks, A. N., & Perry, V. G. (1952). The sting nematode, *Belonolaimus gracilus*, a parasite of major importance on strawberries, celery, and sweet corn in Florida. *Phytopathology, 42*, 173–176.

Cohn, E. (1964). Penetration of the citrus nematode in relation to root development. *Nematologica, 10*, 594–600.

Cohn, E. (1965a). On the feeding and histopathology of the citrus nematode. *Nematologica, 11*, 47–54.

Cohn, E. (1965b). The development of the citrus nematode on some of its hosts. *Nematologica, 11*, 593–600.

Cohn, E. (1966). Observations on the survival of free-living stages of the citrus nematode. *Nematologica, 12*, 321–327.

Cohn, E. (1976). *Report of investigations on nematodes of citrus and subtropical fruit crops in South Africa*. Nelspruit: Citrus and Subtropical Fruit Research Institute.

Crow, W. T. (2005a). Alternatives to fenamiphos for management of plant parasitic nematodes on Bermudagrass. *Journal of Nematology, 37*, 477–482.

Crow, W. T. (2005b). How bad are nematode problems on Florida's golf courses? *Florida Turf Digest, 22*(1), 10–12.

Crow, W. T. (2012). *Pratylenchus hippiastri*. Gainesville: University of Florida.

Crow, W. T. (2013). Effects of a commercial formulation of *Paecilomyces lilacinus* strain 251 on overseeded bermudagrass infested with *Belonolaimus longicaudatus*. *Journal of Nematology, 45*, 223–227.

Crow, W. T. (2014). Effects of a commercial formulation of *Bacillus firmus* I-1582 on golf course bermudagrass infested with *Belonolaimus longicaudatus*. *Journal of Nematology, 46*, 331–335.

Crow, W. T. (2017a). *Nematode management on athletic fields*. Gainesville: University of Florida.

Crow, W. T. (2017b). *Nematode management in residential lawns*. Gainesville: University of Florida.

Crow, W. T. (2017c). *Nematode management for golf courses in Florida*. Gainesville: University of Florida.

Crow, W. T. (2018). *Nematode management on golf courses in Florida*. Gainesville: University of Florida.

Crow, W. T., & Han, H. (2005). Sting nematode. Plant Health Instructor. https://doi.org/10.1094/PHI-I-2005-1208-01. http://www.apsnet.org/edcenter/intropp/lessons/Nematodes/Pages/StingNematode.aspx

Crow, W. T., & Luc, J. E. (2014). Field efficacy of furfural as a nematicide on turf. *Journal of Nematology, 46*, 8–11.

Crow, W. T., & Walker, N. R. (2003). Diagnosis of *Peltamigratus christiei*, a plant parasitic nematode associated with warm-season turfgrasses in the Southern United States. *Plant Health Progress*. https://doi.org/10.1094/PHP-2003-0513-01-DG http://www.plantmanagementnetwork.org/pub/php/diagnosticguide/2003/turf/.

Crow, W. T., & Welch, J. K. (2004). Root reductions of St. Augustinegrass (*Stenotaphrum secundatum*) and hybrid bermudagrass (*Cynodon dactylon* × *C. transvaalensis*) induced by *Trichodorus obtusus* and *Paratrichodorus minor*. *Nematropica, 34*, 31–37.

Crow, W. T., Weingartner, D. P., McSorley, R., & Dickson, D. W. (2000a). Damage function and economic threshold for *Belonolaimus longicaudatus* on potato. *Journal of Nematology, 32*, 318–322.

Crow, W. T., Weingartner, D. P., McSorley, R., & Dickson, D. W. (2000b). Population dynamics of *Belonolaimus longicaudatus* in a cotton production system. *Journal of Nematology, 32*, 210–214.

Crow, W. T., Dickson, D. W., Weingartner, D. P., McSorley, R., & Miller, G. L. (2000c). Yield reduction and root damage to cotton induced by *Belonolaimus longicaudatus*. *Journal of Nematology, 32*, 205–209.

Crow, W. T., Weingartner, D. P., Dickson, D. W., & McSorley, R. (2001). Effect of sorghum-sudangrass and velvetbean covercrops on plant parasitic nematodes associated with potato production in Florida. *Supplement to the Journal of Nematology, 33*, 285–288.

Crow, W. T., Giblin-Davis, R. M., & Lickfeldt, D. W. (2003). Slit injection of 1,3-dichloropropene for management of *Belonolaimus longicaudatus* on established bermudagrass. *Journal of Nematology, 35*, 302–305.

Crow, W. T., Lowe, T., & Lickfeldt, D. W. (2005). Effects of fall overseeding and nematicide applications on populations of sting nematode. *USGA Green Section Record, 43*(6), 8–11.

Crow, W. T., Porazinska, D. L., Giblin-Davis, R. M., & Grewal, P. S. (2006). Entomopathogenic nematodes are not an alternative to fenamiphos for management of plant parasitic nematodes on golf courses in Florida. *Journal of Nematology, 38*, 52–58.

Crow, W. T., Luc, J. E., & Giblin-Davis, R. M. (2011). Evaluation of Econem™, a formulated *Pasteuria sp. bionematicide*, for management of *Belonolaimus longicaudatus* on golf course turf. *Journal of Nematology, 43*, 101–109.

Davide, R. G. (1971). *Survey of the distribution of different plant parasitic nematodes associated with the citrus decline in the Philippines, A report of NSDB project No 2203*. Laguna: University of Philippines, College of Agriculture.

Davis, R. M. (1984). Distribution of *Tylenchulus semipenetrans* in a Texas grapefruit orchard. *Journal of Nematology, 16*, 313–317.

Davis, R. M. (1985). Citrus nematode control in Texas. *Citrograph, 70*, 212–213.

Desaeger, J., & Noling, J. (2017). *Foliar or bud nematodes in Florida strawberries*. Gainesville: University of Florida.

Di Edwardo, A. A., & Perry, G. (1964). *Heterodera leuceilyma n. sp. (Nemata: Heteroderidae), a severe pathogen of St. Augustine grass in Florida, Bulletin 637*. Gainesville: Florida Agricultural Experiment Station, University of Florida.

Dickson, D. W. (1985). *Nematode diseases of peanut, Nematology Circular 121*. Gainesville: Florida Department of Agriculture and Consumer Services.

DuCharme, E. P. (1959). Morphogenesis and histopathology of lesions induced on citrus roots by *Radopholus similis*. *Phytopathology, 49*, 388–395.

DuCharme, E. P. (1967). Annual population periodicity of *Radopholus similis* in Florida citrus groves. *Plant Disease Reporter, 51*, 1031–1034.

DuCharme, E. P. (1968). Burrowing nematode decline of citrus: A review. In G. C. Smart & V. G. Perry (Eds.), *Tropical nematology* (pp. 20–37). Gainesville: University of Florida Press.

DuCharme, E. P. (1969). Temperature in relation to *Radopholus similis* (Nematoda) spreading decline of citrus. In *Proceedings of the 1st International citrus symposium* (Vol. 2, pp. 979–983).

DuCharme, E. P., & Birchfield, W. (1956). Physiologic races of the burrowing nematode. *Phytopathology, 46*, 615–616.

DuCharme, E. P., & Price, W. C. (1966). Dynamics of multiplication of *Radopholus similis*. *Nematologica, 12*, 113–121.

DuCharme, E. P., & Suit, R. F. (1967). Population fluctuations of burrowing nematodes in Florida citrus groves. *Proceedings of the Florida State Horticultural Society, 80*, 63–67.

Duncan, L. W. (1986). The spatial distribution of citrus feeder roots and of the citrus nematode, *Tylenchulus semipenetrans*. *Revue de Nématologie, 9*, 233–240.

Duncan, L. W. (1989). Effect of Fenamiphos placement on *Tylenchulus semipenetrans* and yield in a Florida citrus orchard. *Supplement to the Journal of Nematology, 21*(4S), 703–706.

Duncan, L. W. (1999). Nematode diseases of citrus. In L. W. Timmer & L. W. Duncan (Eds.), *Citrus health management* (pp. 136–148). St. Paul: APS Press.

Duncan, L. W., & Eissenstat, D. M. (1993). Responses of *Tylenchulus semipenetrans* to citrus fruit removal: Implications for carbohydrate competition. *Journal of Nematology, 25*, 7–14.

Duncan, L. W., & El-Morshedy, M. M. (1996). Population changes of *Tylenchulus semipenetrans* under localized versus uniform drought in the citrus root zone. *Journal of Nematology, 28*, 360–368.

Duncan, L. W., & Phillips, M. S. (2009). Sampling root knot nematodes. In R. N. Perry, M. Moens, & J. L. Starr (Eds.), *Root knot nematodes* (pp. 275–300). St. Albans: CAB International.

Duncan, L. W., Ferguson, J. J., Dunn, R. A., & Noling, J. W. (1989). Application of Taylor's Power Law to sample statistics of *Tylenchulus semipenetrans* in Florida citrus. *Supplement to the Journal of Nematology, 21*(4S), 707–711.

Duncan, L. W., Kaplan, D. T., & Noling, J. W. (1990). Maintaining barriers to the spread of *Radopholus citrophilus* in Florida citrus orchards. *Nematropica, 20*, 71–88.

Duncan, L. W., Graham, J. H., & Timmer, L. W. (1993). Seasonal patterns associated with *Tylenchulus semipenetrans* and *Phytophthora parasitica* in the citrus rhizosphere. *Phytopathology, 83*, 573–581.

Duncan, L. W., El-Morshedy, M. M., & McSorley, R. (1994a). Sampling citrus fibrous roots and *Tylenchulus semipenetrans*. *Journal of Nematology, 26*, 442–451.

Duncan, L. W., Inserra, R. N., O'Bannon, J. H., & El-Morshedy, M. M. (1994b). Reproduction of a Florida population of *Tylenchulus semipenetrans* on resistant citrus rootstocks. *Plant Disease, 78*, 1067–1071.

Duncan, L. W., Mashela, P., Ferguson, J., Graham, J., Abou-Setta, M. M., & El-Morshedy, M. M. (1995). Estimating crop loss in orchards with patches of mature citrus trees infected by *Tylenchulus semipenetrans*. *Nematropica, 25*, 43–51.

Duncan, L. W., Noling, J. W., Inserra, R. N., & Dunn, D. (1996). Spatial patterns of *Belonolaimus* spp. among and within citrus orchards on Florida's central ridge. *Journal of Nematology, 28*, 352–359.

Duncan, L. W., Inserra, R. N., & Dunn, D. (1998). Seasonal changes in citrus fibrous root starch concentration and body length of female *Pratylenchus coffeae*. *Nematropica, 28*, 263–266.

Duncan, L. W., Inserra, R. N., Thomas, W. K., Dunn, D., Mustika, I., Frisse, L. M., Mendes, M. L., Morris, K., & Kaplan, D. T. (1999). Molecular and morphological analysis of isolates of *Pratylenchus coffeae* and closely related species. *Nematropica, 29*, 61–80.

Dunn, D. C., Duncan, L. W., & Romeo, J. T. (1998). Changes in arginine, PAL activity, and nematode behavior in salinity-stressed citrus. *Phytochemistry, 49*, 413–417.

El-Borai, F. E., Duncan, L. W., & Graham, J. H. (2003). Infection of citrus roots reduces root infection by *Phytophthora nicotianae*. *Journal of Nematology, 34*, 384–389.

Esser, R. P. (1984). *How nematodes enter and disperse in Florida nurseries via vehicles*. Florida Department of Agriculture and Consumer Services, Division of Plant Industry. Nematology Circular No 109.

Esser, R. P., Wilinson, R. C., Jr., & Harkcom, J. (1983). Pinewood nematode (*Bursaphelenchus xylophilus*) survey in Florida. *Proceedings of the Soil and Crop Science Society of Florida, 42*, 127–132.

Fallas, G. A., Hahn, M. L., Fargette, M., Burrows, P. R., & Sarah, J. L. (1996). Molecular and biochemical diversity among isolates of *Radopholus* spp. from different areas of the world. *Journal of Nematology, 28*, 422–430.

FDACS. (2016). http://www.freshfromflorida.com/Divisions-Offices/Marketing-and-Development/Education/For-Researchers/Florida-Agriculture-Overview-and-Statistics; https://www.nass.usda.gov/Quick_Stats/Ag_Overview/stateOverview.php?state=FLORIDA.

Ford, H. W. (1952). The effect of spreading decline on the root distribution of citrus. *Proceedings of the Florida State Horticultural Society, 65*, 47–50.

Ford, H. W. (1953). Effect of spreading decline disease on the distribution of feeder roots of orange and grapefruit trees on rough lemon rootstocks. *Journal of the American Society for Horticultural Science, 61*, 68–72.

Ford, H. W., & Feder, W. A. (1961). Additional citrus rootstock selections that tolerate the burrowing nematode. *Proceedings of the Florida State Horticultural Society, 74*, 50–53.

Galeano, M. (2002). Dinamica di popolazione di *Tylenchulus semipenetrans* e della nematofauna di vita libera nella rizosfera di agrumi in Spagna. *Supplemento Nematologia Mediterranea, 30*, 49–53.

Garabedian, S., Van Gundy, S. D., Mankau, R., Radewald, J. D. (1984). Nematodes. In *Integrated pest management for citrus* (pp. 129–131). Berkeley: Division of Agriculture and Natural Resources Publications, University of California.

Giblin-Davis, R. M. (2000). *Pasteuria* sp. for biological control of the sting nematode, *Belonolaimus longicaudatus*, in turfgrass. In J. M. Clark & M. P. Kenna (Eds.), *American chemical society symposium series no. 743: Fate and management of turfgrass chemicals* (pp. 408–426). New York: Oxford Press.

Giblin-Davis, R. M., Busey, P., & Center, B. J. (1995). Parasitism of *Hoplolaimus galeatus* on diploid and polyploid St. Augustinegrass. *Journal of Nematology, 27*, 472–477.

Giblin-Davis, R. M., Williams, D. S., Bekal, S., Dickson, D. W., Brito, J. A., Becker, J. O., & Preston, J. F. (2003). '*Candidatus Pasteuria usgae*' sp. nov. an obligate endoparasite of the phtyoparasitic nematode *Belonolaimus longicaudatus*. *International Journal of Systemic and Evolutionary Microbiology, 53*, 197–200.

Golden, A. M., & Minton, N. A. (1970). Description and larval heteromorphism of *Hoplolaimus concaudajuvencus*, n. sp. (Nematoda: Hoplolaimidae). *Journal of Nematology, 2*, 161–166.

Golden, M. A., Lopez, C. H. R., & Vilchez, R. H. (1992). Description of *Pratylenchus gutierrezi* n.sp. (Nematoda: Pratylenchidae) from coffee in Costa Rica. *Journal of Nematology, 24*, 298–304.

Gozel, U., Adams, B. J., Nguyen, K. B., Inserra, R. N., Giblin-Davis, R. M., & Duncan, L. W. (2006). A phylogeny of *Belonolaimus* populations in Florida inferred from DNA sequences. *Nematropica, 36*, 155–171.

Grabau, Z. J., & Dickson, D. W. (2018). *Management of plant parasitic nematodes in Florida peanut*. Gainesville: University of Florida.

Graham, J. H., Timmer, L. W., & Lee, R. F. (1983). Comparison of zinc, water uptake by gravity infusion and syringe injection tests for diagnosis of citrus blight. *Proceedings of the Florida State Horticultural Society, 96*, 45–47.

Graham, J. H., Gerberich, K. M., Bright, D. B., & Johnson, E. G. (2014). Excess bicarbonate in soil and irrigation water increases fibrous root loss and decline of Huanglongbing-affected citrus trees in Florida. *Phytopathology, 104*(S3), 47.

Hahn, M. L., Burrows, P. R., Gnanapragasam, N. C., Bridge, J., Vines, N. J., & Wright, D. J. (1994). Molecular diversity amongst *Radopholus similis* populations from Sri Lanka detected by RAPD analysis. *Fundamental and Applied Nematology, 17*, 275–281.

Hamid, G. A., Van Gundy, S. D., & Lovatt, C. J. (1988). Phenologies of the citrus nematode and citrus roots treated with oxamyl. *Proceedings of the International Society of Citriculture, 2*, 993–1004.

Hamill, J. E. (2006). *Population dynamics and management of Belonolaimus longicaudatus on strawberry in Florida.* Ph.D. Dissertation. University of Florida, Gainesville.

Heald, C. M., & O'Bannon, J. H. (1987). *Citrus declines caused by nematodes. V. slow decline.* Florida Department of Agriculture and Consumer Services, Division of Plant Industry. Nematology Circular No 143.

Hodges, A. W., & Stevens, T. J. (2010). *Economic contributions of the turfgrass industry in Florida*. Gainesville: University of Florida.

Huettel, R. N., & Yaegashi, T. (1988). Morphological differences between *Radopholus citrophilus* and *R. similis*. *Journal of Nematology, 20*, 150–157.

Huettel, R. N., Dickson, D. W., & Kaplan, D. T. (1984). *Radopholus citrophilus* n.sp. (Nematoda): A sibling species of *Radopholus similis*. *Proceedings of the Helminthological Society of Washington, 51*, 32–35.

Inserra, R. N., Vovlas, N., & O'Bannon, J. H. (1980). A classification of *Tylenchulus semipenetrans* biotypes. *Journal of Nematology, 12*, 283–287.

Inserra, R. N., Lehman, P. S., & Overstreet, C. (1994). *Ornamental hosts of the reniform nematode, Rotylenchulus reniformis, Nematology Circular No. 209*. Gainesville: Florida Department of Agriculture and Consumer Services, Division of Plant Industry.

Inserra, R. N., Duncan, L. W., Vovlas, N., & Loof, P. A. A. (1996). *Pratylenchus loosi* from pasture grasses in Central Florida. *Nematologica, 42*, 159–172.

Inserra, R. N., Duncan, L. W., Dunn, D. C., Kaplan, D. T., & Porazinska, D. (1998). *Pratylenchus pseudocoffeae* from Florida and its relationship with *P. gutierrezi* and *P. coffeae*. *Nematologica, 44*, 683–712.

Inserra, R. N., Duncan, L. W., Troccoli, A., Dunn, D., Maia Sos Santos, J., & Vovlas, N. (2001). *Pratylenchus jaehni* n.sp. from citrus in Brazil and a redescription of *P. coffeae*. *Nematology, 3*, 653–665.

Inserra, R. N., Stanley, J. D., O'Bannon, J. H., & Esser, R. P. (2005). Nematode quarantine and certification programmes implemented in Florida. *Nematologia Mediterranea, 1*, 113–123.

Inserra, R. N., Trocolli, R., Gozel, U., Bernard, E., Dunn, D., & Duncan, L. W. (2006). *Pratylenchus hippeastri* n.sp. from Amaryllis in Florida with notes on *Pratylenchus scribneri* and *P. hexincisus*. *Nematology, 8*, 1–18.

Inserra, R. N., Duncan, L. W., O'Bannon, J. H., & Fuller, S. A. (2009). *Citrus nematode biotypes and resistant citrus rootstocks in Florida, Circular 205/CH115*. Gainesville: University of Florida, IFAS Extension.

Inserra, R. N., Stanley, J. D., Ochoa, A., Schubert, T. S., Subbotin, S. A., Crow, W. T., & McSorley, R. (2014). *Hemicriconemoides species as crop damaging parasitic nematodes in Florida, Nematology Circular 223*. Gainesville: Florida Department of Agriculture and Consumer Services.

Joseph, S., Mekete, T., Danquah, W. B., & Noling, J. (2016). First report of *Meloidogyne haplanaria* infecting Mi-resistant tomato plants in Florida and its molecular diagnosis based on mitochondrial haplotype. *Plant Disease, 100*, 1438–1445.

Kallel, S., Louhichi, A., & B'Chir, M. M. (2006). Resistance of *Citrus aurantium* induced by *Poncirus trifoliata* vis-à-vis the *Tylenchulus semipenetrans* Cobb. *Nematology, 8*(5), 671–679.

Kaplan, D. T. (1986). Variation in *Radopholus citrophilus* population densities in the citrus rootstock Carrizo citrange. *Journal of Nematology, 18*, 31–34.

Kaplan, D. T. (1994). An assay to estimate citrus rootstock resistance to burrowing nematodes. *Proceedings of the Florida State Horticulture Society, 107*, 85–89.

Kaplan, D. T., & O'Bannon, J. H. (1981). Evaluation and nature of citrus nematode resistance in Swingle citrumelo. *Proceedings of the Florida State Horticultural Society, 94*, 33–36.

Kaplan, D. T., & O'Bannon, J. H. (1985). Occurrence of biotypes in *Radopholus citrophilus*. *Journal of Nematology, 17*, 158–162.

Kaplan, D. T., & Opperman, C. H. (1997). Genome similarity implies that citrus-parasitic burrowing nematodes do not represent a unique species. *Journal of Nematology, 29*, 430–440.

Kaplan, D. T., & Opperman, C. H. (2000). Reproductive strategies and karyotype of the burrowing nematode, *Radopholus similis*. *Journal of Nematology, 32*, 126–133.

Kinloch, R. A., & Sprenkel, R. K. (1994). Plant parasitic nematodes associated with cotton in Florida. *Journal of Nematology, 26*, 749–752.

Kirkpatrick, J. C., & Van Gundy, S. D. (1966). Scion and rootstock as factors in the development of citrus nematode populations. *Phytopathology, 56*, 438–441.

Korayem, M., & Hassabo, S. A. A. (2005). Citrus yield in relation to *Tylenchulus semipenetrans* in silty loam soil. *International Journal of Nematology, 15*, 179–182.

Kwaye, R. G., Mashela, P. W., Shimelis, H., & Mapope, N. (2008). Determination of *Tylenchulus semipenetrans* biotype in Zebediela and Champagne, Republic of South Africa. *Plant Disease, 92*, 639–641.

Lehman, P. S. (1996). *Role of plant protection organizations in nematode management, XIX Congress of Brazilian Society of Nematology* (pp. 137–148). Rio Quente: Brazilian Society of Nematology.

Lehman, P. S., & Dunn, R. A. (1987). Distribution of Florida populations of the soybean cyst nematode with previously undescribed genetic variation. *Plant Disease, 71*, 68–70.

Lehman, P. S., Vovlas, N., Inserra, R. N., Duncan, L. W., & Kaplan, D. T. (2000). Colonization of foliar tissues of an aquatic plant *Anubias barteri* Schott by *Radopholus similis*. *Nematropica, 30*, 63–75.

Lilley, C. J., Wang, D., Atkinson, H. J., & Urwin, P. E. (2011). Effective delivery of a nematode-repellent peptide using a root-cap-specific promoter. *Plant Biotechnology Journal, 9*, 151–161.

Ling, P., Duncan, L. W., Deng, Z., Dunn, D., Hu, X., Huang, S., & Gmitter, F. G., Jr. (2000). Inheritance of citrus nematode resistance and its linkage with molecular markers. *Theoretical and Applied Genetics, 100*, 1010–1017.

Luc, J. E., Crow, W. T., Stimac, J. L., Sartain, J. B., & Giblin-Davis, R. M. (2007). Effects of *Belonolaimus longicaudatus* management and nitrogen fertility on turf quality of golf course fairways. *Journal of Nematology, 39*, 62–66.

Luc, J. E., Crow, W. T., McSorley, R., & Giblin-Davis, R. M. (2010). Suppression of *Belonolaimus longicaudatus* with in vitro-produced *Pasteuria* sp. endospores. *Nematropica, 40*, 217–225.

Maafi, Z. T., & Damadzadeh, M. (2008). Incidence and control of the citrus nematode, *Tylenchulus semipenetrans* Cobb, in the north of Iran. *Nematology, 10*, 113–122.

Macaron, J. (1972). Contribution h l'6tude du nematode phytophase *Tylenchulus semipenetrans* Cobb 1913—(Nematode-Tylenchida), Ph.D. thesis, Universit des Sciences et Techniques du Languedoc, Montpellier, France.

Macchia, E. T., McSorley, R., Duncan, L. W., & Syvertsen, J. S. (2003). Effects of perennial peanut (*Arachis glabrata*) ground cover on nematode communities in citrus. *Journal of Nematology, 35*, 450–457.

MacGowan, J. B. (1978). *The lesion nematode, Pratylenchus coffeae, affecting citrus in Florida, Nematology Circular 37*. Gainesville: Florida Department of Agriculture and Consumer Services.

MacGowan, J. B. (1984). *Meloidogyne graminis, a root knot nematode of grass, Nematology Circular 107*. Gainesville: Florida Department of Agriculture and Consumer Services.

Machmer, J. H. (1958). Effect of soil salinity on nematodes in citrus and papaya plantings. *Journal of the Rio Grande Valley Horticultural Society, 12*, 57–60.

Machon, J. E., & Bridge, J. (1996). *Radopholus citri* n. sp. (Tylenchida: Pratylenchidae) and its pathogenicity on citrus. *Fundamental and Applied Nematology, 19*, 127–133.

Malo, S. E. (1963). Pathogenic nematodes of peach trees in Florida and others of potential importance. *Proceedings of the Florida State Horticultural Society, 76*, 377–379.

Mashela, P. W., & Nthangeni, M. E. (2002). Osmolyte allocation in response to *Tylenchulus semipenetrans* infection, stem girdling, and root pruning in citrus. *Journal of Nematology, 34*, 273–277.

Mashela, P., Duncan, L. W., Graham, J. H., & McSorley, R. (1992a). Leaching soluble salts increases population densities of *Tylenchulus semipenetrans*. *Journal of Nematology, 24*, 103–109.

Mashela, P., Duncan, L. W., & McSorley, R. (1992b). Salinity reduces resistance to *Tylenchulus semipenetrans* in citrus rootstocks. *Nematropica, 22*, 7–12.

Mathabatha, R. V., Mashela, P. W., Mokgalong, N. M., & Dube, Z. P. (2015). Potential existence of multiple *Tylenchulus semipenetrans* biotypes in South Africa. *Acta Agriculturae Scandinavica, 65*(7), 673–679.

McClure, M. A., & Schmitt, M. E. (1996). Control of citrus nematode, *Tylenchulus semipenetrans*, with cadusafos. *Supplement to the Journal of Nematology, 28*(4S), 624–628.

McGroary, P. C., Crow, W. T., Giblin-Davis, R. M., & McSorley, R. (2013). Timing of nematicide applications on turfgrass to reduce damage caused by *Belonolaimus longicaudatus*. *International Turfgrass Society Research Journal, 12*, 363–367.

McSorley, R. (1986). *Nematode problems on bananas and plantains in Florida, Nematology Circular 133*. Gainesville: Florida Department of Agriculture and Consumer Services.

McSorley, R., & Parrado, J. L. (1982a). Plans for the collection of nematode soil samples from fruit groves. *Nematropica, 12*, 257–267.

McSorley, R., & Parrado, J. L. (1982b). Relationship between two nematode extraction techniques on two Florida soils. *Soil Crop Science Society of Florida Proceedings, 41*, 30–36.

McSorley, R., & Parrado, J. L. (1983). The spiral nematode *Helicotylenchus multicinctus* on bananas in Florida and its control. *Proceedings of the Florida State Horticultural Society, 96*, 201–207.

McSorley, R., Campbell, C. W., & Parrado, J. L. (1982). Nematodes associated with tropical and subtropical fruit trees in south Florida. *Proceedings of Florida State Horticultural Society, 95*, 132–135.

Meagher, J. W. (1967). Observations on the transport of nematodes in subsoil drainage and irrigation water. *Australian Journal of Experimental Agricultural and Animal Husbandry, 7*, 577–579.

Murguía, C., Abad, P., Jordá, C., & Bello, A. (2005). Identification of the Poncirus biotype of *Tylenchulus semipenetrans* in Valencia, Spain. *Spanish Journal of Agricultural Research, 3*(1), 130–133.

Nigh, E. L., Jr. (1981). Relation of citrus nematode to root distribution in flood irrigated citrus. *Journal of Nematology, 13*, 451–452.

Noling JW (2016) *Nematode management in strawberries*, ENY-031 (12 pp). Entomology and Nematology Department, UF/IFAS Extension, University of Florida, Gainesville.

Nyoike, T. W., Mekete, T., McSorley, R., Weibelzahl-Karigi, E., & Liburd, O. E. (2012). Confirmation of *Meloidogyne hapla* on strawberry in Florida using molecular and morphological techniques. *Nematropica, 42*, 253–259.

O'Bannon, J. H. (1977). Worldwide dissemination of *Radopholus similis* and its importance in crop production. *Journal of Nematology, 9*, 16–25.

O'Bannon, J. H. (1980). Migration of *Pratylenchus coffeae* and *P. brachyurus* on citrus and in soil. *Nematropica, 10*, 70.

O'Bannon, J. H., & Ford, H. W. (1976). An evaluation of several *Radopholus similis*-resistant or -tolerant citrus rootstocks. *Plant Disease Reporter, 60*, 620–624.

O'Bannon, J. H., & Inserra, R. N. (1989). *Helicotylenchus species as crop damaging parasitic nematodes. Nematology Circular 165*. Gainesville, FL: Florida Department of Agriculture and Consumer Services.

O'Bannon, J. H., & Tarjan, A. C. (1973). Preplant fumigation for citrus nematode control in Florida. *Journal of Nematology, 5*, 88–95.

O'Bannon, J. H., & Tomerlin, A. T. (1969). Population studies on two species of *Pratylenchus* on citrus. *Journal of Nematology, 1*, 299–300.

O'Bannon, J. H., & Tomerlin, A. T. (1971). Response of citrus seedlings to *Radopholus similis* in two soils. *Journal of Nematology, 3*, 255–259.

O'Bannon, J. H., & Tomerlin, A. T. (1973). Citrus tree decline caused by *Pratylenchus coffeae*. *Journal of Nematology, 5*, 311–316.

O'Bannon, J. H., & Tomerlin, A. T. (1977). Control of the burrowing nematode, *Radopholus similis*, with DBCP and oxamyl. *Plant Disease Reporter, 61*, 450–454.

O'Bannon, J. H., Reynolds, H. W., & Leathers, C. R. (1966). Effects of temperature on penetration, development, and reproduction of *Tylenchulus semipenetrans*. *Nematologica, 12*, 483–487.

O'Bannon, J. H., Radewald, J. D., & Tomerlin, A. T. (1972). Population fluctuation of three parasitic nematodes in Florida citrus. *Journal of Nematology, 4*, 194–199.

O'Bannon, J. H., Tarjan, A. C., & Bistline, F. W. (1974). Control of *Pratylenchus brachyurus* on citrus and tree response to chemical treatment. *Soil Crop Science Society of Florida Proceedings, 33*, 65–67.

O'Bannon, J. H., Radewald, J. D., Tomerlin, A. T., & Inserra, R. N. (1976). Comparative influence of *Radopholus similis* and *Pratylenchus coffeae* on citrus. *Journal of Nematology, 8*, 58–63.

Olmstead, M., Chaparro, J., & Ferguson, J. (2012). *Rootstocks for Florida stone fruit*. Gainesville: University of Florida.

Osman, H. A., Dickson, D. W., & Smart, G. C., Jr. (1985). Morphological comparisons of host races 1 and 2 of *Meloidogyne arenaria* from Florida. *Journal of Nematology, 17*, 379–385.

Pang, W., Luc, J. E., Crow, W. T., Kenworthy, K. E., Giblin-Davis, R. M., McSorley, R., & Kruse, J. K. (2011a). Field responses of bermudagrass and seashore paspalum cultivars to sting and spiral nematodes. *Journal of Nematology, 43*, 195–202.

Pang, W., Luc, J. E., Crow, W. T., Kenworthy, K. E., Giblin-Davis, R. M., McSorley, R., & Kruse, J. K. (2011b). Bermudagrass cultivar responses to sting nematode. *Crop Science, 51*, 2199–2203.

Perez, E. E., Weingartner, D. P., & McSorley, R. (2000). Correlation between *Paratrichodorus minor* population levels and corky ringspot symptoms on potato. *Nematropica, 30*, 247–251.

Philis, I. (1989). Yield loss assessment caused by the citrus nematode *Tylenchulus semipenetrans* on Valencia oranges in Cyprus. *Nematologia Mediterranea, 17*, 5–6.

Poucher, C., Ford, H. W., Suit, R. F., & DuCharme, E. P. (1967). *Burrowing nematode in citrus*. Florida Department of Agriculture, Division of Plant Industry. Bulletin No 7.

Prasad, S. K., & Chawla, M. L. (1966). Observations on the population fluctuations of citrus nematode, *Tylenchulus semipenetrans* Cobb 1913. *Indian Journal of Entomology, 27*, 450–454.

Quesenberry, K. H., Crow, W. T., & Kenworthy, K. E. (2015). Effect of *Belonolaimus longicaudatus* on root parameters of St. Augustinegrass cultivars. *Nematropica, 45*, 96–101.

Radewald, J. D., O'Bannon, J. H., & Tomerlin, A. T. (1971a). Temperature effects on reproduction and pathogenicity of *Pratylenchus coffeae* and *P. brachyurus* and survival of *P. coffeae* in roots of *Citrus jambhiri*. *Journal of Nematology, 3*, 390–394.

Radewald, J. D., O'Bannon, J. H., & Tomerlin, A. T. (1971b). Anatomical studies of *Citrus jambhiri* roots infected by *Pratylenchus coffeae*. *Journal of Nematology, 3*, 409–416.

Ratanaworabhan, S., & Smart, G. C., Jr. (1970). The ring nematode, *Criconemoides ornatus*, on peach and centipede grass. *Journal of Nematology, 2*, 204–208.

Rhoades, H. L. (1968). Effect of nematicides on yield of field corn in central Florida. *Proceedings of the Soil and Crop Science Society of Florida, 28*, 262–265.

Rhoades, H. L. (1983). Effects of cover crops and fallowing on population densities of *Belonolaimus longicaudatus* and *Meloidogyne incognita* and subsequent crop yields. *Nematropica, 13*, 6–16.

Rhoades, H. L. (1984). Effects of fallowing, summer cover crops, and fenamiphos on nematode populations and yields in a cabbage-field corn rotation in Florida. *Nematropica, 14*, 131–138.

Rhoades, H. L. (1985). Effects of separate and concomitant populations of *Belonolaimus longicaudatus* and *Dolichodorus heterocephalus* on *Zea mays*. *Nematropica, 15*, 171–174.

Roderick, H., Tripathi, L., Babirye, A., Wang, D., Tripathi, J., Urwin, P. L., & Atkinson, H. (2012). Generation of transgenic plantain with resistance to plant pathogenic nematodes. *Molecular Plant Pathology, 13*, 842–851.

Salem, A. A.-M. (1980). Observations on the population dynamics of the citrus nematode, *Tylenchulus semipenetrans* in Sharkia Governorate. *Egypt Journal of Phytopathology, 12*, 31–34.

Shokoohi, E., & Duncan, L. W. (2017). Nematode parasites of citrus. In R. Sikora, P. Timper, & D. Coyne (Eds.), *Plant parasitic nematodes in tropical and subtropical agriculture* (3rd ed.). St Albans: CAB International.

Shokoohi, E., & Duncan, L. W. (2018). Nematode parasites of citrus. In R. Sikora, P. Timper, & D. Coyne (Eds.), *Plant-Parasitic Nematodes in Tropical & Subtropical Agriculture* (Third ed., pp. 446–476). St. Albans, UK: CAB International.

Sorribas, F. J., Verdejo-Lucas, S., Forner, J. B., Alcaide, A., Pons, J., & Ornat, C. (2000). Seasonality of *Tylenchulus semipenetrans* Cobb and *Pasteuria* sp. in citrus orchards in Spain. *Supplement to the Journal of Nematology, 32*(4S), 622–632.

Sorribas, F. J., Verdejo-Lucas, S., Pastor, J., Ornat, C., Pons, J., & Valero, J. (2008). Population densities of *Tylenchulus semipenetrans* related to physicochemical properties of soil and yield of clementine mandarin in Spain. *Plant Disease, 92*, 3.

Stokes, D. E. (1969). *Andropogon rhizomatus* parasitized by a strain of *Tylenchulus semipenetrans* not parasitic to four citrus rootstocks. *Plant Disease Reporter, 53*, 882–885.

Stokes, D. E. (1979). *Some plant symptoms associated with Aphelenchoides spp. in Florida, Nematology Circular 49*. Gainesville: Florida Department of Agriculture and Consumer Services.

Suit, R. F., & DuCharme, E. P. (1953). The burrowing nematode and other parasitic nematodes in relation to spreading decline. *Citrus Leaves, 33*(8–9), 32–33.

Tarjan, A. C. (1956). The possibility of mechanical transmission of nematodes in citrus groves. *Proceedings of the Florida State Horticultural Society, 69*, 34–37.

Tarjan, A. C. (1964). Two new American dagger nematodes (*Xiphinema*: Dorylaimida) associated with citrus, with comments on the variability of *X. bakeri* Williams, 1961. *Proceedings of the Helminthological Society of Washington, 31*, 65–76.

Tarjan, A. C. (1967). Citrus nematode found widespread in Florida. *Plant Disease Reporter, 51*, 317.

Tarjan, A. C. (1971). Migration of three pathogenic citrus nematodes through two Florida soils. *Soil Crop Science Society of Florida Proceedings, 31*, 253–255.

Tarjan, A. C., & Frederick, J. J. (1981). Reaction of nematode-infected centipedegrass turf to pesticidal and non-pesticidal treatments. *Proceedings of the Florida State Horticultural Society, 94*, 225–227.

Tarjan, A. C., & O'Bannon, J. H. (1969). Observations on meadow nematodes (*Pratylenchus* spp.) and their relation to decline of citrus in Florida. *Plant Disease Reporter, 53*, 683–686.

Tarjan, A. C., & O'Bannon, J. H. (1977). Nonpesticidal approaches to nematode control. *Proceedings of the International Society of Citriculture, 3*, 848–853.

Tarjan, A. C., & Simmons, P. N. (1966). The effect of interacting cultural practices of citrus trees with spreading decline. *Soil Crop Science Society of Florida Proceedings, 26*, 22–31.

Taylor, A. L. (2003). Nematocides and nematicides – a history. *Nematropica, 33*, 225–232.

Thomason, I. J., & Caswell, E. P. (1987). Principles of nematode control. In R. H. Brown & B. R. Kerry (Eds.), *Principles and practice of nematode control in crops* (pp. 87–130). Australia: Academic.

Tomerlin, A. T., & O'Bannon, J. H. (1974). Effect of *Radopholus similis* and *Pratylenchus brachyurus* on citrus seedlings in three soils. *Soil Crop Science Society of Florida Proceedings, 33*, 95–97.

Toung, M.-C. (1963). A study on seasonal influence in quantitative variation of the citrus nema, *Tylenchulus semipenetrans* Cobb. *National Taiwan University Plant Protection Bulletin, 5*, 323–327.

Trenholm, L. E., Lickfeldt, D. W., & Crow, W. T. (2005). Use of 1,3-dichloropropene to reduce irrigation requirements of sting nematode infested bermudagrass. *Hortscience, 40*(5), 1543–1548.

Tsai, B. Y., & Van Gundy, S. D. (1988). Comparison of anhydrobiotic ability of the citrus nematode with other plant parasitic nematodes. *Proceedings of the International Society of Citriculture, 2*, 983–992.

Valette, C., Mounport, D., Nicole, M., Sarah, J. L., & Baujard, P. (1998). Scanning electron microscope studies of two African populations of *Radopholus similis* (Nematoda: Pratylenchidae) and proposal of *R. citrophilus* as a junior synonym of *R. similis*. *Fundamental and Applied Nematology, 21*, 139–146.

Van Gundy, S. D. (1958). The life history of the citrus nematode *Tylenchulus semipenetrans* Cobb. *Nematologica, 3*, 283–294.

Van Gundy, S. D., & Kirkpatrick, J. D. (1964). Nature of resistance in certain citrus rootstocks to citrus nematodes. *Phytopathology, 54*, 419–427.

Van Gundy, S. D., & Martin, J. P. (1961). Influence of *Tylenchulus semipenetrans* on the growth and chemical composition of sweet orange seedlings in soils of various exchangeable cation ratios. *Phytopathology, 51*, 146–151.

Van Gundy, S. D., & Martin, J. P. (1962). Soil texture, pH and moisture effects on the development of citrus nematode (*Tylenchulus semipenetrans*). *Phytopathology, 52*, 31.

Van Gundy, S. D., & Meagher, J. W. (1977). *Citrus nematode (Tylenchulus semipenetrans) problems world-wide*. Orlando: International Citrus Congress.

Van Gundy, S. D., Martin, J. P., & Taso, P. H. (1964). Some soil factors influencing reproduction of the citrus nematode and growth reduction of sweet orange seedlings. *Phytopathology, 54*, 294–299.

Vau, S. (2017). *Mi-gene tomato for management of root knot disease in Florida*. Ph.D. Dissertation. University of Florida, Gainesville.

Vendramini, J., Adesogan, A., & Wasdin, J. (2012). *Silage crops for dairy and beef cattle*. Gainesville: University of Florida.

Verdejo-Lucas, S., Galeano, M., Sorribas, F. J., Forner, J. B., & Alcaide, A. (2003). Effect on resistance to *Tylenchulus semipenetrans* of hybrid citrus rootstocks subjected to continuous exposure to high population densities of the nematode. *European Journal of Plant Pathology, 109*, 427–433.

Vilardebo, A. (1964). Etude sur *Tylenchulus semipenetrans* Cobb au Maroc. II. Al-Awamia. *Rabat, 11*, 31–49.

Weingartner, D. P., McSorley, R., & Goth, R. W. (1993). Management strategies in potato for nematodes and soil-borne diseases in subtropical Florida. *Nematropica, 23*, 233–245.

Wheaton, T. A., Childers, C. C., Timmer, L. W., Duncan, L. W., & Nikdel, S. (1985). Effects of aldicarb on yield, fruit quality, and tree condition on Florida citrus. *Proceedings of the Florida State Horticultural Society, 98*, 6–10.

Willers, P. (1979). *Influence of citrus nematode, Tylenchulus semipenetrans, on Navel yield in Sundays River Valley Orchard* (pp. 9–10). Nelspruit: Citrus and Subtropical Fruit Research Institute.

Willers, P., & Holmden, E. (1980). The influence of citrus nematode, *Tylenchulus semipenetrans,* on the performance of trees growing under saline conditions. *Information Bulletin, Citrus and Subtropical Fruit Research Institute, 99*, 13–16.

Wright, D. L., Marois, J. J., George, S., & Katsvairo, T. W. (2015). *Use of perennial grasses in peanut/cotton rotations: Effect on pests*. Gainesville: University of Florida.

Xiang, X., Deng, Z., Zheng, Q., Chen, C., & Gmitter, F. G., Jr. (2010). Developing specific markers and improving genetic mapping for a major locus *Tyr1* of Citrus nematode resistance. *Molecular Plant Breeding, 7*, 497–504.

Chapter 10
Nematodes of Agricultural Importance in North and South Carolina

Weimin Ye

10.1 Introduction

North Carolina's agricultural industry including food, fiber, ornamentals and forestry, contributes $84 billion to the state's annual economy, accounts for more than 17% of the state's income, and employs 17% of the work force. North Carolina is one of the most diversified agricultural states in the nation. Approximately, 50,000 farmers grow over 80 different commodities in North Carolina utilizing 8.2 million of the state 12.5 million hectares to furnish consumers a dependable and affordable supply of food and fiber. North Carolina produces more tobacco and sweet potatoes than any other state, ranks second in Christmas tree and third in tomato production. The state ranks nineth nationally in farm cash receipts of over $10.8 billion (NCDA Agricultural Statistics 2017).

Plant parasitic nematodes are recognized as one of the greatest threat to crops throughout the world. Nematodes alone or in combination with other soil microorganisms have been found to attack almost every part of the plant including roots, stems, leaves, fruits and seeds. Crop damage caused worldwide by plant nematodes has been estimated at $US80 billion per year (Nicol et al. 2011). All crops are damaged by at least one species of nematode. Most plant parasitic nematodes live in soil and damage plants by feeding in large numbers on roots impairing the plant's ability to take up water and nutrients. Severe root damage caused by nematodes typically results in aboveground symptoms that may include stunting, yellowing of leaves, incipient wilt, loss of plant vigor and/or an overall general decline in plant performance. Damage is often more pronounced when plants are under stress from lack of

W. Ye (✉)
Nematode Assay Section, Agronomic Division, North Carolina Department of Agriculture and Consumer Services, Raleigh, NC, USA
e-mail: weimin.ye@ncagr.gov

S. A. Subbotin, J. J. Chitambar (eds.), *Plant Parasitic Nematodes in Sustainable Agriculture of North America*, Sustainability in Plant and Crop Protection,
https://doi.org/10.1007/978-3-319-99588-5_10

water or nutrients or when damaged by other diseases or insects. Although nematodes rarely kill plants, they can drastically limit plant growth and yields. Plant parasitic nematodes are usually confined to localized areas in soil and spread very slowly by their own power. However, they may be dispersed more rapidly by movement of infested soil through cultivation, on soil clinging to farming tools and tillers, in water, wind or on roots of transplants.

The Nematology Research Program in North Carolina began with C. J. Nusbaum who joined the Plant Pathology Department of North Carolina State University (NCSU) in 1948. In 1951, Nusbaum observed that pre-plant soil fumigation was "growing in popularity" around the nation, while in North Carolina "comparatively few growers have used these treatments extensively". He began testing the efficacy of nematicides for local conditions and commodities, particularly tobacco and, by 1956, at least half of the tobacco land (approximately 81,000 ha) was fumigated and losses from root disease were decreasing rapidly. Due to the increased awareness of nematode damage to the state's other primary crops such as peanuts, soybeans, cotton, ornamentals, fruits and vegetables, Nusbaum advocated for a nematode assay project to examine population dynamics and epidemiology. J. N. Sasser joined NCSU Plant Pathology Department in 1953 and brought widespread recognition and public acclaim to the Nematology program. Sasser was at one time referred to as "the most widely known nematologist around the world". He became a pioneer in the work of identifying host races within nematode species, particularly *Meloidogyne* spp., by using differential host specificity. Perhaps most significant, though, was his efforts to expand and organize the discipline of Nematology on a regional, national and international basis. In 1975, he obtained a multimillion-dollar grant from the U.S. Agency for International Development to launch the International *Meloidogyne* Project (IMP) to focus on the biology, ecology, genetics, pathogenicity and control of root knot nematodes affecting economic food crops in the developing world.

In the 1960s and 1970s, many scientists helped guide and develop the Nematology program at NCSU during its height of productivity. Among the most unique was the husband and wife research and teaching team of H. H. Triantaphyllou and A. C. Triantaphyllou. H. H. Triantaphyllou joined NCSU in 1954 and A. C. Triantaphyllou in 1960. Their primary contributions were in the areas of taxonomy, morphology, ultrastructure, developmental biology, cytogenetics and biochemistry, particularly in regards to *Meloidogyne* and *Heterodera*. A. C. Triantaphyllou and his student, P. R. Esbenshade, developed new biochemical methods to assist in reliable identification of *Meloidogyne* species (Esbenshade and Triantaphyllou 1985, 1987, 1988). He also developed and refined a technique for fixing and staining the chromosomes of nematodes.

C. J. Nusbaum first conceived of a nematode assay service for North Carolina in 1953. Four years later, Nusbaum and W. M. Powell started basic research and application towards this goal. In 1965, a 5-year plan for development of an advisory service was funded by the state of North Carolina. In 1966, K. R. Barker was hired by the NCSU Plant Pathology Department to research and develop a pilot nematode assay program. His research focused on charting the complex interactions of plant parasitic nematodes with their hosts and environment. He established the nematode assay program with funds provided by the state legislature. Barker and associates developed the semi-automatic elutriator, a new tool for extracting nematodes from soil. Barker's long career also was focused on conceptual breakthroughs regarding quantitative population dynamics and related application to crop performance. He modified techniques for estimating crop damage based on the population and distribution of nematodes in fields. Studies on the role of the environment on nematode populations, particularly seasonal changes in nematode densities, allowed the formulation of practical nematode damage thresholds and hazard indices that helped put diagnostic work and nematode disease management for a wide spectrum of economically valuable commodities on a new scientific basis (Barker and Noe 1987). The research was expanded by extensive excellent work by colleagues and graduate students in the NCSU Plant Pathology Department. By 1974, the Nematode Advisory Service was on solid conceptual and practical footing, and was transferred to the North Carolina Department of Agriculture and Consumer Services (NCDA & CS), where it continues today (Barker and Imbriani 1984).

By the turn of the twenty-first century, the peak of classical Nematology had passed. North Carolina State University altered its focus from classical Nematology to molecular research in order to remain progressive and competitive in a shifting science. C. Opperman was hired in 1987, E. L. Davis in 1993 and D. M. Bird in 1995, all of whom have made considerable contributions in nematode genomics, host-parasite relationships, transgenic host resistance and gene function. In 1986, S. R. Koenning was hired by NCSU to focus on nematode management and ecology on most field crops grown in North Carolina, including soybean, corn, cotton, and small grains.

The Nematode Assay Laboratory in the Agronomic Division of the NCDA & CS has been successively led by Nematologists D. A. Richard (1972–1979), J. L. Starr (1979–1981), Jack Imbriani (1981–2005) and Weimin Ye (2005 to present). It is now the largest nematode assay lab in the United States, processing 51,223 samples and issuing 6,498 reports in fiscal year 2018. Among those, 14,050 samples (1,378 reports) from many different states were specifically tested for the presence of pinewood nematode (*Bursaphelenchus xylophilus*) so that shipments of lumber and wood products could be cleared by United States Department of Agriculture – Animal and Plant Health Inspection Service – Plant Protection and Quarantine for export from the United States. In addition, a molecular diagnostic service was developed and implemented from 2011 to identify nematodes to the species level when this level of taxonomic resolution is needed.

In the State of North Carolina, nematodes are a major threat to most crops. About 82 plant parasitic nematode species have been recorded. Among them, root knot, soybean cyst, reniform, sting, lesion, lance, stubby root, tobacco cyst, ring, foliar, and stem and bulb nematodes are considered the most important and are the subjects in this chapter. The list of plant parasitic nematodes in Table 10.1 is based in part on the nematode section of the North Carolina Plant Disease Index by K. R. Barker (Grand 1985) and includes all recent research findings (Ye et al. 2012, 2013, 2015a, b; Ye 2017; Zeng et al. 2012b, 2015; Holguin et al. 2015a, 2016) and nematode

Table 10.1 Plant parasitic nematodes in North Carolina and South Carolina

Nematode species	State*	Crop or plant	References
Anguina tritici	NC, SC	Rye, wheat	Grand (1985)
Aorolaimus leipogrammus	SC	Bamboo	Sher (1963)
Aorolaimus spp.	NC	Camellia	Grand (1985)
Aphelenchoides fragariae	NC, SC	Ornamentals, strawberry	Williamson et al. (2000), McCuiston et al. (2007), Kohl et al. (2010), and Fu et al. (2012)
A. myceliophagus	NC	Mushrooms, turfgrass	Zeng et al. (2012b)
A. parietinus	NC, SC	Cotton	Steiner (1938) and Society of Nematologists (1984)
A. ritzemabosi	NC	Florist's daisy	Strider (1979) and Grand (1985)
A. subtenuis	NC, SC	Unknown	Society of Nematologists (1984)
Belonolaimus euthychilus	SC	Longleaf pine	Rau (1963)
B. gracilis	NC, SC	Corn, cotton	Graham and Holdeman (1953) and Ruehle and Sasser (1962)
B. longicaudatus	NC, SC	Bermuda grass, common bean, corn, cotton, creeping grasses, peanut, soybean, tomato	Lewis et al. (1993) and Zeng et al. (2012b)
B. maritimus	NC	American beachgrass	Grand (1985)
Cactodera weissi	NC	Bermuda grass, knotweed, tobacco	Grand (1985) and Ye (2012)
Criconema demani	SC	Unknown	Taylor (1936)
C. lamellatum	SC	Ferns, grass, trees	Raski and Golden (1966)
C. permistum	SC	Swamp soil	Raski and Golden (1966)
C. sphagni	NC	Trees	Ruehle (1968)
Criconemoides annulatus	SC	Unknown	Society of Nematologists (1984)
Crossonema fimbriatus	SC	Unknown	Mehta and Raski (1971)
Ditylenchus dipsaci	NC	Alfalfa, ornamentals, white clover, wild onion	Barker and Sasser (1959) and Grand (1985)
D. triformis	NC	Iris	Hirschmann and Sasser (1955)
Dolichodorus heterocephalus	NC, SC	Bermuda grass, camellia	Zeng et al. (2012b)
D. marylandicus	NC	Pine	Lewis and Golden (1981)
Globodera tabacum solanacearum	NC	Carolina horsenettle, tobacco	Melton et al. (1991)

(continued)

Table 10.1 (continued)

Nematode species	State*	Crop or plant	References
Helicotylenchus caroliniensis	SC	Swamp soil	Sher (1966)
H. dihystera	NC, SC	Bermuda grass, common bean, corn, cotton, creeping grasses, loblolly pine, peach, peanut, slash pine, soybean	Ruehle and Sasser (1962), Aycock et al. (1976), Schmitt and Barker (1988), Lewis et al. (1993), and Zeng et al. (2012b)
H. erythrinae	NC	Grass	Ruehle and Sasser (1962)
H. exallus	SC	Soybean	Nyczepir and Lewis (1979)
H. hydrophilus	SC	Swamp soil	Sher (1966)
H. microlobus	SC	Grass, strawberry	Perry et al. (1959)
Hemicaloosia graminis	NC, SC	Turfgrass	Zeng et al. (2012a)
Hemicriconemoides chitwoodi	NC, SC	Camellia, turfgrass	Ye and Robbins (2000), López et al. (2012a), and Zeng et al. (2012b)
H. wessoni	NC, SC	Bermuda grass	Ye and Robbins (2000) and Zeng et al. (2012b)
Hemicycliophora conida	NC, SC	Creeping bentgrass	Zeng et al. (2012b)
H. gigas	NC	Forest soil	Thorne (1955)
H. gracilis	NC	Grass	Thorne (1955)
H. mettleri	NC	Trees	Brzeski (1974)
H. parvana	NC	Turfgrass	López et al. (2013) and Van den Berg et al. (2018)
H. robbinsi	NC	Turfgrass	López et al. (2013), Subbotin et al. (2014), and Van den Berg et al. (2018)
H. sheri	NC	Grass	Society of Nematologists (1984)
H. thienemanni	SC	Turfgrass	Zeng et al. (2012b)
H. vaccinium	SC	Swamp soil	Brzeski (1974)
H. vidua	SC	Camellia	López et al. (2013)
Heterodera cyperi	NC	Ornamentals	Golden et al. (1962)
H. glycines	NC, SC	Common bean, soybean	Schmitt and Barker (1988) and Lewis et al. (1993)
H. lespedezae	NC	Bush clover	Golden and Cobb (1963)
H. trifolii	NC	Daylilly, white clover	Grand (1985) and Ye (2012)

(continued)

Table 10.1 (continued)

Nematode species	State*	Crop or plant	References
Hoplolaimus columbus	NC, SC	Cotton, slash pine, soybean	Ruehle and Sasser (1962), Sher (1963), Fassuliotis et al. (1968), Schmitt and Barker (1988), Lewis et al. (1993), and Holguin et al. (2015a, c, 2016)
H. galeatus	NC, SC	Bermuda grass, boxwood, Chinese holly, corn, cotton, creeping bentgrass, creeping grasses, slash pine, soybean, tall fescue, white clover	Schmitt and Barker (1988), Lewis et al. (1993), Martin et al. (1994), Zeng et al. (2012b), and Holguin et al. (2015a)
H. stephanus	NC	Soybean	Sher (1963) and Holguin et al. (2015a)
Longidorus breviannulatus	NC	Turfgrass	Society of Nematologists (1984)
L. crassus	SC	Turfgrass	Ye and Robbins (2005)
L. elongatus	SC	Unknown	Society of Nematologists (1984)
L. longicaudatus	SC	Unknown	Siddiqi (1962)
L. paralongicaudatus	SC	Turfgrass	Zeng et al. (2012b)
Meloidodera floridensis	NC	Azalea, grass, loblolly pine, slash pine	Ruehle and Sasser (1962), Triantaphyllou and Hirschmann (1973), and Ye (2012)
Meloidogyne arenaria	NC, SC	Asparagus, azalea, carrot, daylily, peanut, corn, sweet potato, tobacco, soybean, tomato	Tedford and Fortnum (1988), Lewis et al. (1993), and Agudelo et al. (2011)
M. carolinensis	NC	Blueberry	Eisenback (1982)
M. enterolobii	NC, SC	Cotton, horseweed, morning glory, sicklepod, soybean, sweet potato, tobacco	Ye et al. (2013) and Rutter et al. (2018)
M. graminis	NC, SC	Bermuda grass, blue oat grass, centipedegrass, creeping bentgrass, creeping grasses, meadow fescue, St. Augustine grass	Ye et al. (2015b) and Zeng et al. (2012b)

(continued)

Table 10.1 (continued)

Nematode species	State*	Crop or plant	References
M. hapla	NC, SC	Boxwood, cabbage, cantaloupe, common bean, cotton, gingseng, Irish potato, pea, peanut, peony, sage, soybean, strawberry, sweet potato, tobacco, tomato, watermelon, wheat	Schmitt and Barker (1988)
M. incognita	NC, SC	Alfalfa, asparagus, azalea, bean, begonia, bentgrass, bermuda grass, boxwood, Buddleja, butterbean, cabbage, camellia, cantaloupe, carrot, collards, common bean, corn, cotton, creeping bentgrass, creeping grasses, cucumber, daylily, dogwood, eggplant, English oak, fig, forsythia, gardenia, green bean, holly, hyacinth bean, hydrangea, kiwifruit, Korean boxwood, lantana, liriope, lima bean, mondo grass, milo, muskmelon, oak, okra, ornamentals, peach, pepper, peony, potato, pumpkin, sage, snap bean, soybean, spinach, squash, St. Augustine grass, strawberry, sweet potato, tobacco, tomato, tube rose, watermelon, white clover, zoysiagrass	Sitterly and Fassuliotis (1965), Schmitt and Barker (1988), Tedford and Fortnum (1988), Haygood et al. (1990), Lewis et al. (1993), and Agudelo et al. (2011)
M. javanica	NC, SC	Boxwood, southern peas, soybean, sweet potato, tobacco, tomato	Schmitt and Barker (1988)
M. marylandi	NC, SC	Bermuda grass, creeping bentgrass, creeping grasses	Ye et al. (2015b)
M. megatyla	NC	Loblolly pine	Baldwin and Sasser (1979)
M. naasi	NC, SC	Bermuda grass, creeping bentgrass, creeping grasses	Zeng et al. (2012b) and Ye et al. (2015b)
M. partityla	SC	Laurel oak	Eisenback et al. (2015)
M. spatinae	NC, SC	Smooth cordgrass	Rau and Fassuliotis (1965)
Mesoanguina plantaginis	NC, SC	Bracted plantain	Vargas and Sasser (1976)
Mesocriconema curvatum	NC, SC	Bermuda grass, creeping bentgrass, creeping grasses, soybean	Lewis et al. (1977, 1993) and Zeng et al. (2012b)
M. ornatum	NC, SC	Corn, blueberry, grasses, ornamentals, peanut	Ratanaworabhan and Smart Jr. (1970), Schmitt and Barker (1988), Powers et al. (2014), and Jagdale et al. (2013)

(continued)

Table 10.1 (continued)

Nematode species	State*	Crop or plant	References
M. rusticum	SC	Unknown	Society of Nematologists (1984)
M. sphaerocephalum	NC, SC	Corn, peanut	López et al. (2012b) and Zeng et al. (2012b)
M. xenoplax	NC, SC	Bermuda grass, Chinese holly, creeping bentgrass, ornamentals, peach	Aycock et al. (1976), Nyczepir et al. (1985), Zeng et al. (2012b), and Powers et al. (2014)
Nanidorus minor	NC, SC	American holly, azalea, bentgrass, bermuda grass, boxwood, camellia, centipedegrass, Chinese holly, corn, cotton, creeping bentgrass, holly, loblolly pine, longleaf pine, ornamentals, peach, peanut, slash pine, soybean, tall fescue, tobacco, white clover	Schmitt and Barker (1988), Lewis et al. (1993), Boutsika et al. (2004), Zeng et al. (2012b), and Huang et al. (2018)
Ogma decalineatum	SC	Ferns	Mehta and Raski (1971)
O. floridense	NC, SC	Bermuda grass	Zeng et al. (2012b)
Paratrichodorus allius	NC, SC	Creeping bentgrass	Zeng et al. (2012b)
P. porosus	NC, SC	Camellia, corn, sorghum, soybean	Huang et al. (2018)
Paratylenchus goldeni	NC	Boxwood, centipedegrass	Zeng et al. (2012b)
Pratylenchus brachyurus	NC, SC	Cotton, peach, peanut, soybean, tobacco	Schmitt and Barker (1988) and Lewis et al. (1977, 1993)
P. coffeae	NC, SC	Peach	Grand (1985)
P. macrostylus	NC	Fraser fir, red spruce	Hartman and Eisenback (1991)
P. penetrans	NC, SC	Bermuda grass, boxwood, peach, potato, soybean, tobacco	Grand (1985) and Zeng et al. (2012b)
P. pratensis	NC	Cotton, ornamentals, potato, tobacco	Steiner (1938) and Grand (1985)
P. scribneri	NC, SC	Common bean, peach, soybean, tobacco	Grand (1985) and Lewis et al. (1993)
P. vulnus	NC	Boxwood, Chinese holly, peach	Grand (1985)
P. zeae	NC	Corn, peach, tall fescue	Grand (1985) and Schmitt and Barker (1988)
Quinisulcius capitatus	NC	Unknown	Society of Nematologists (1984)

(continued)

Table 10.1 (continued)

Nematode species	State*	Crop or plant	References
Rotylenchulus reniformis	NC, SC	Cotton, sweet potato	Grand (1985), Lewis et al. (1993), Koenning et al. (2004), Leach et al. (2012), and Holguin et al. (2015b)
Rotylenchus buxophilus	NC	Azalea, boxwood	Grand (1985)
R. pumilus	NC	Trees	Ruehle (1968)
Scutellonema brachyurus	NC, SC	Bermuda grass, corn, cotton	Kraus-Schmidt and Lewis (1979), Lewis et al. (1993), Agudelo and Harshman (2011), and Zeng et al. (2012b)
Trichodorus elefjohnsoni	NC	Tulip poplar	Bernard (1992)
T. obtusus	NC, SC	Bermuda grass, St. Augustine grass, zoysiagrass	Shaver et al. (2013, 2015, 2016), Ye et al. (2015a, b), and Huang et al. (2018)
Trophurus sculptus	SC	Ferns, grass, hardwood trees	Society of Nematologists (1984)
Trophurus spp.	NC	Azalea	Grand (1985)
Tylenchorhynchus claytoni	NC, SC	American holly, azalea, bermuda grass, camellia, centipedegrass, Chinese holly, common bean, corn, cotton, creeping bentgrass, lima bean, loblolly pine, longleaf pine, ornamentals, peach, peanut, potato, slash pine, soybean, strawberry, tall fescue, tobacco, tomato, white clover	Steiner (1937), Ruehle and Sasser (1962), and Aycock et al. (1976), Schmitt and Barker (1988), Lewis et al. (1993), and Zeng et al. (2012b)
T. maximus	NC	Grasses, tomato	Grand (1985)
Tylenchulus sp.	SC	Peach	Wehunt et al. (1987)
Xenocriconemella macrodora	NC	Boxelder maple	López et al. (2012a)
Xiphinema americanum sensu lato	NC, SC	American holly, azalea, bermuda grass, boxwood, centipedegrass, Chinese holly, corn, cotton, loblolly pine, ornamentals, peach, peanut, slash pine, soybean, strawberry, sweet potato, tall fescue, tobacco, tomato	Ruehle and Sasser (1962), and Zeng et al. (2012b)
Xiphinema bakeri	NC	Bermuda grass	Zeng et al. (2012b)
Xiphinema chambersi	NC, SC	Azalea, bermuda grass, centipedegrass, ornamentals, rose, soybean	Ye (2012) and Zeng et al. (2012b)
Xiphinema krugi	NC	Bermuda grass, boxwood, cotton, ornamentals, soybean, tomato	Ye and Robbins (2010)

*Names of the states are represented by two letter abbreviations: NC North Carolina, SC South Carolina assay services provided by the NCDA & CS. Nematode entries in this list do not necessarily indicate that all nematodes associated with a given plant species are pathogenic to that plant species. Although all these nematodes are obligate parasites, some of them often cause little or no damage. Furthermore, nematodes are primarily soil inhabitants and may feed on associated weeds and grasses. In a few instances, the nematodes included under a given plant species probably were feeding primarily on these associated plants/weeds or had fed on previous crops.

South Carolina is bordered to the north by North Carolina with similar climate and agricultural crops and thus should harbor similar nematode species (Alexander 1963; Lewis et al. 1993; Dickerson et al. 2000; Zeng et al. 2012a, b; Shaver et al. 2013; Holguin et al. 2015a, c, 2016; Mueller and Agudelo 2015; Ye et al. 2015a). There are many reports of plant nematodes in South Carolina published by Prof. S. A. Lewis and Prof. P. Agudelo from the Clemson University. So far, 105 plant nematodes were recorded in the Carolinas (Table 10.1), but only 41 species were recorded in both states. Some of the species were never recorded in either North Carolina or South Carolina but may very likely have been omitted due to lack of study and are still present in another state.

10.2 Root Knot Nematodes, *Meloidogyne* spp.

The most common plant parasitic nematodes identified in North Carolina are the root knot nematodes (*Meloidogyne* spp.) (Meadows et al. 2018). These pests can be a serious problem for most field crops, vegetables, fruit trees, turfgrasses and ornamental plants especially in sandy soil. Infected plants are stunted and pale, drop flowers and fruits, wilt often, and decline even when plants are generously watered and fertilized. Growers most often realize they have root knot nematode at the end of the season, when they are pulling up spent crops and notice multiple bumpy, knot-like swellings on the roots of plants. Concomitant infection of roots galled by *Meloidogyne* spp. by other soilborne pathogens occurs frequently and increases the decline of host plant vigor and productivity. Currently, there are 11 species recorded in North Carolina, including *Meloidogyne arenaria, M. carolinensis, M. enterolobii, M. graminis, M. hapla, M. incognita, M. javanica, M. marylandi, M. megatyla, M. naasi* and *M. spatiniae.* Three tropical species (*M. incognita, M. javanica* and *M. arenaria*) and temperate species (*M. hapla*) are the predominant species and cause major damage to plants in this state.

10.2.1 Meloidogyne enterolobii

Meloidogyne enterolobii is a recently detected and emerging species causing severe damage to sweet potato, soybean, cotton and tobacco (Ye et al. 2013). It is believed to be an introduced species only confirmed in Columbus, Greene, Harnett, Johnston,

Fig. 10.1 Distribution of three plant parasitic nematodes in North Carolina as of September 2018 (yellow color counties): (**a**) *Meloidogyne enterolobii;* (**b**) *Rotylenchulus reniformis;* (**c**) *Globodera tabacum*

Nash, Sampson, Wayne and Wilson Counties in North Carolina as of September 2018 (Fig. 10.1a). It has a wide host range, including field crops and weeds. This species is a major concern to sweet potato growers because it affects not only the yield, but also the quality of sweet potato (Fig. 10.2). Once infested, the nematode

Fig. 10.2 Symptoms of sweet potato caused by *Meloidogyne enterolobii* in North Carolina. (**a**, **c**) Infected sweet potato; (**b**) Female and an egg-mass; (**d**) Galls

often causes total crop loss and sweet potatoes are not marketable. *Meloidogyne enterolobii* is a tropical species that was recorded only from Florida in the United States (Brito et al. 2004) until recently from field crops in North Carolina (Ye et al. 2013). *Meloidogyne enterolobii* was originally described from a population collected from the pacara earpod tree (*Enterolobium contortisiliquum*) in China in 1983. In 2001, it was detected for the first time in the continental United States in South Florida from regulatory samples of ornamental plants. *Meloidogyne enterolobii* is now considered one of the most important root knot nematode species because of its ability to reproduce on root-knot-nematode-resistant (*Mi-1, Mh, Mir1, N, tabasco*, and *Rk* gene carrying genotypes) crops. The greenhouse bioassay revealed this species could cause galls on the North Carolina differential hosts of tomato, bell pepper, tobacco, water melon, cotton, but not on peanuts, and has the same differential plant hosts as southern root knot nematode (*Meloidogyne incognita*) race 4 (Ye et al. 2013). There are no resistant varieties available against this root knot nematode species and the use of soil fumigation before planting and use nematode-free transplants are recommended. This introduced species is having a significant impact on North Carolina agriculture and poses a threat to sweet potato and other crop hosts if it is not controlled.

10.2.2 Other Meloidogyne *Species*

Meloidogyne marylandi, M. graminis and *M. naasi* are limited to turfgrass in North Carolina (Ye et al. 2015b; Zeng et al. 2012b, 2015). *Meloidogyne marylandi* and *M. graminis* are widely distributed in the state and cause damage to turfgrass, whereas *M. naasi* has limited distribution and damage in turfgrasses. Blueberry root knot nematode, *M. carolinensis,* was described from cultivated highbush blueberry (cultivars derived from hybrids of *Vaccinium corymbosum* and *V. lamarckii*) in Rose Hill, Duplin County, North Carolina (Eisenback 1982). Host range studies showed that only blueberry (*Vaccinium* sp.) and azalea (*Rhododendron* sp.) were good hosts. *Meloidogyne megatyla* (Baldwin and Sasser 1979) was described from loblolly pine (*Pinus taeda*) in Bladen County, North Carolina and has a host range that is different from other *Meloidogyne* species. *Meloidogyne spartinae* was originally described as *Hypsoperine spartinae* (Rau and Fassuliotis 1965). However, the genus *Hypsoperine* was synonymized and two of its species were renamed as *M. graminis* and *M. spartinae* (Whitehead 1968; Plantard et al. 2007).

Root knot nematodes are very difficult to manage because they are soilborne pathogens with a wide host range. The most reliable control of root knot nematodes can be achieved by integrating two or more control tactics. Combining an effective non-host rotation, resistant varieties, and selected cultural practices can give excellent control with little added cost. Crop rotation is one of the oldest and most economical methods of controlling nematodes. Rotation is simply the practice of not growing a susceptible host in the same site for more than 1 year. Nonhost plants that are especially suitable for rotation in root knot-infested fields include fescue, small grains, marigolds, sweet corn, asparagus and cool season crops in the cabbage family such as broccoli, kale, collards and mustard. Cultural methods may minimize root knot nematode damage. Practices such as removing the roots of each crop as soon as harvest is completed, followed by tilling the soil two to three times is very effective in reducing nematode levels, followed by a winter cover crop of annual rye grass, rye or wheat. Maintaining optimum conditions for plant growth in terms of soil pH, fertility, and soil moisture will increase the tolerance of low to moderate nematode pressure and will make the plants less susceptible to other stresses as well. Frequent incorporation of organic matter, especially high rates of composted leaves and manure into the soil is also beneficial for improving soil structure and moisture retention. It also encourages natural enemies in the soil for biological control of the nematodes. In some situations, nematode severity is sufficiently high so that chemical control is the only effective option. Fumigants are commonly applied as pre-plant treatments to reduce nematode numbers, but they must thoroughly penetrate large soil volumes to be effective. Nonfumigant and systemic compounds are less effective, but they have some advantage in ease of application and handling.

10.3 Soybean Cyst Nematode, *Heterodera glycines*

Soybean cyst nematode (SCN), *Heterodera glycines*, is an obligate and sedentary plant parasitic species that is the number one pathogen of soybean, causing more than twice as much yield loss than any other disease (Allen et al. 2017). SCN was first discovered in the United States in New Hanover County, North Carolina, in 1954 (Winstead et al. 1955) and is believed to have been introduced from Asia (Riggs 2004). SCN distribution in the United States has spread rapidly, although the underlying cause is debated. By 2017, SCN was found in every soybean-producing state in the United States except for West Virginia (Tylka and Marett 2017).

The Agronomic Division of NCDA & CS has analyzed numerous soil samples to monitor the spread and distribution of SCN. From July 1, 2014 to June 30, 2017, 100,118 soil samples were submitted for routine nematode assay by growers of various crops in 97 North Carolina counties. Only Alleghany, Clay, and McDowell Counties were not represented in this sample population. SCN was detected in 21,922 of the soil samples (21.9%) (Ye 2017). The overall mean population density of the second-stage juveniles and females was 110 ± 266 (10–14,600) per 500 cm^3 of soil. The total of SCN-positive counties included 57 counties in North Carolina (Fig. 10.3). Catawba is the only county not shown on the recent North American SCN distribution map published by Tylka and Marett (2017). Johnston (3462 SCN-positive samples), Wayne (3274), Nash (2960), Wilson (2039), and Pasquotank (1513) counties had the most SCN-positive samples. Population density (the second-stage juveniles and females of SCN/500 cm^3 of soil in average) were highest in Montgomery (831), Bladen (790), Washington (610), Carteret (607), and Harnett (368) Counties. According to the most recent NCDA & CS data of soybean-planted fields and yield statistics by county in North Carolina (NCDA Agricultural Statistics 2017), soybean is mainly grown in the eastern half of the state. In general, the SCN-negative counties are those with no soybean acreage or less than 200 planted hectares of soybeans in the western part of the state. The high number of samples in this

Fig. 10.3 Known distribution of SCN in North Carolina as of September 2018 (green counties) and average SCN population density (number of the second-stage juveniles and females of SCN/500 cm^3 soil) from each county

work gave a clear picture of where SCN is occurring in North Carolina and its population density in each county since its first detection in the United States in 1954. Given the yield losses that SCN is capable of causing, SCN continues to be a yield-limiting factor in the state, and growers should be actively managing this obligate parasite to mitigate yield loss.

To reduce the crop losses caused by SCN, an integrated management approach using multiple practices is recommended. Managing reproduction by using soybeans with an earlier maturity group can help reduce population numbers the following year. Crop rotation with nonhost crops such as corn, cotton, tobacco, sweet potato, peanuts or sorghum is very effective in decreasing SCN populations. The use of resistant varieties gives excellent control. Resistant varieties can only be used effectively when matched with the correct SCN race. The continued use of the same resistant variety can lead to a race shift in the population (Koenning 2004). Field populations of SCN have historically been characterized as races 1–16 based on four soybean hosts (Riggs and Schmitt 1988); however, these are now characterized as HG types based upon source of SCN resistance (Niblack et al. 2002). A study by NCDA & CS and North Carolina Cooperative Extension Agents of soybean cyst populations was conducted to determine the most abundant races in North Carolina using the old SCN race system (Riggs and Schmitt 1988). Of the 18 counties sampled in 2017, races 2 (87%), 4 (10%), and 5 (3%) were identified. Few varieties have resistance to the races recently identified in North Carolina. Some of the varieties that have resistance to SCN races 2, 4 or 5 include: Fowler (Race 1,2,3,5,14), Jake (1,2,3,5,14), JTN-5503 (2,3,5,14), N7003CN (1,2,3,4,5,14), Osage (2,3,14), and P52T86R (1,2,3,5,14) (Joyce and Thiessen 2017). Chemical controls for SCN are often expensive and do not guarantee a positive yield response, and few nematicides are registered for use against SCN. The fumigant 1,3-dichloropropene (Telone II) can be used for SCN control (Thiessen 2018). Seed treatments using *Pasteuria nishizawae* (Clariva), abamectin (Avicta), *Bacillus fermis* (Poncho/Votivo), fluopyram (Ilevo) are recommended in managing yield losses when used in fields with low to moderate populations of SCN (Thiessen 2018). Seed treatments may help soybean plants establish larger root volume but provide a short window of efficacy since they treat a small volume of soil/roots. These should be used in conjunction with other management practices described above.

10.4 Reniform Nematode, *Rotylenchulus reniformis*

Reniform nematode, *Rotylenchulus reniformis,* has been recorded only from Cleveland, Cumberland, Harnett, Hoke, Johnston, Richmond, Robeson, Sampson, Scotland and Wayne Counties in North Carolina (Fig. 10.1b). This species is semi-endoparasitic in which the females penetrate the root cortex, establish a permanent feeding site in the stele region of a root and become sedentary or immobile. The anterior portion of the female body remains embedded in the root, whereas the posterior portion protrudes from the root surface and swells during maturation. It is only considered hazardous to cotton, soybean and sweet potato in North Carolina. Corn and peanuts are very poor hosts and small grains, sorghum and common bermudagrass are nonhosts. Management includes fallow, rotation with nonhost or poor host crops, nematicides and weed control. Crop rotation with resistant or immune plant species is recommended. These include mustard (*Brassica nigra*), oats, rhodesgrass (*Chloris gayana*), onion and sun hemp (*Crotalaria juncea*) (Robinson et al. 1997; Caswell et al. 1991). Sorghum, corn and reniform nematode resistant soybeans are recommended as rotation crops for cotton (Starr and Page 1990; Starr et al. 2007).

10.5 Sting Nematode, *Belonolaimus longicaudatus*

Belonolaimus longicaudatus was originally described from soil around the roots of corn in Sanford, Florida (Rau 1958). It is a major plant parasite in the sandy soils in Southeastern United States, with widespread distribution throughout the Atlantic Coastal Plain from Virginia to Florida (Lucas et al. 1974; Orton Williams 1974). It is considered the most important pest of turf and pasture grasses (Heald and Perry 1970; Crow 2005). While this species has been documented in association with many grass species, in a survey, *B. longicaudatus* was found in 131 turfgrass samples in three turf management zones (green, fairway and tee) in both North Carolina and South Carolina and three grass species (bermudagrass, creeping bentgrass, zoysiagrass) from 24 counties (Zeng et al. 2012b). In addition to turfgrasses and home owner lawns, this nematode causes considerable damages to cotton, corn, peanuts, soybean and strawberry in North Carolina, but it is restricted to sandy soil.

Sting nematodes can be effectively managed with nematicides. Unlike many of the endoparasitic nematodes that spend a majority of their life within roots, contact nematicides often work well on sting nematode. 1,3-dichloropropene (Curfew), abamectin (Avid) 0.15 EC, abamectin (Divanem) 0.7 SC, *Bacillus firmus* (Nortica), fluensulfone (Nimitz Pro G), fluopyram (Indemnify) and furfural (Multiguard Protect) are currently registered and can be effectively used to reduce sting nematode populations (Kerns and Butler 2018). Nortica is a bacterial biological control agent, *Bacillus firmus,* used for the protection of plant roots against plant parasitic nematodes in several crop species. On turfgrasses, relieving additional stresses by raising mowing height, increasing irrigation frequency, improving aeration to roots, and reducing traffic can improve tolerance to sting nematodes. The addition of organic, and some inorganic, amendments to soil also can improve tolerance to sting nematodes by improving the water and nutrient-holding capacity of the soil (Crow and Han 2005).

10.6 Lesion Nematodes, *Pratylenchus* spp.

The lesion nematodes of the genus *Pratylenchus* are recognised worldwide to have a great economic impact in crop production. This is not only due to their wide host range, but also their distribution in almost every temperate and tropical environment. At present, more than 80 species of *Pratylenchus* have been described, with a combined host range of greater than 400 crop plant species (Loof 1991; Castillo and Vovlas 2007). The species recorded in North Carolina include *P. brachyurus, P. coffeae, P. macrostylus, P. penetrans, P. pratensis, P. scribneri, P. vulnus* and *P. zeae*. These migratory endoparasites of plants mainly feed and move within plant roots. Crops of primary economic importance that are attacked by lesion nematodes include potato, corn, cotton, soybean, tobacco, peanuts, forage legumes, ornamental plants and many fruit trees. Symptoms of lesion nematode disease often go unrecognized initially because the nematodes are microscopic pathogens of belowground plant parts (mainly roots), and the aboveground symptoms are often general symptoms of plant-root stress. Lesion nematodes induce characteristic necrotic lesions (darkened areas of dead tissue) on the surface and throughout the cortex of infected roots. The lesions turn from reddish-brown to black and are initially spotty along the root surface. As the nematodes continue to migrate and feed within the roots, the lesions can coalesce to become large necrotic areas of tissue that may eventually girdle the root. The wounds inflicted on plant roots and other below ground plant parts by lesion nematodes can serve as infection courts for pathogenic soil microbes, primarily fungi. This causes disease complexes that involve lesion nematodes and wilt fungi such as *Fusarium* and *Verticillium* (Rowe et al. 1987; MacGuidwin and Rouse 1990).

Lesion nematodes are difficult to control. Cultivars bred for resistance to lesion nematodes are not currently commercially available. Rotations to nonhost crops offer limited opportunities to manage lesion nematode field populations since most *Pratylenchus* species have wide host ranges. If the species of *Pratylenchus* is accurately diagnosed, and a suitable economic nonhost can be grown, rotations offer some promise as a management practice. The two most effective tactics for lesion nematode management remain sanitation and the use of nematicides.

10.7 Lance Nematodes, *Hoplolaimus* spp.

Columbia lance nematode, *Hoplolaimus columbus*, was first described from soybean in Richland County, South Carolina (Sher 1963). Since that time, *H. columbus* has been found mainly in South Carolina, North Carolina, and Georgia (Lewis and Fassuliotis 1982; Koenning et al. 1999; Holguin et al. 2015a, 2016). It is primarily associated with cotton and soybean and has limited distribution in North Carolina (Fig. 10.4). Losses to *H. columbus* in cotton are typically 10–25% and on soybean as high as 70% (Mueller and Sanders 1987; Noe 1993). *Hoplolaimus columbus* is

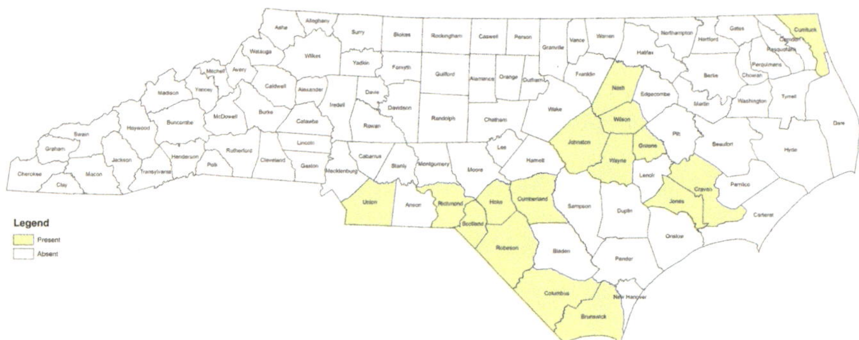

Fig. 10.4 Known distribution of *Hoplolaimus columbus* in North Carolina as of September 2018 (yellow color counties)

the predominant species associated with soybean in South Carolina, whereas another lance species *H. stephanus* was the prevalent species in North Carolina (Holguin et al. 2015a, 2016). The Columbia lance nematode feeds both externally and internally on soybean roots. Lesions may develop on roots which can coalesce and give the appearance of root rot. The amount of damage to soybean and subsequent yield loss will be directly proportional to the soil population density of this nematode at soybean planting. Corn, cotton and soybean are good hosts for this nematode. Nematode densities are generally in the moderate to high range following these crops. Peanuts, tobacco and small grains are poor or nonhosts for Columbia lance nematode.

Hoplolaimus galeatus was first described from soil in Virginia (Cobb 1913). In a survey, *H. galeatus* was found from 22 counties in North Carolina and South Carolina (Zeng et al. 2012b). It feeds and reproduces on a wide range of plant hosts and can cause serious damage to cotton (Krusberg and Sasser 1956; Wrather et al. 1992; Martin et al. 1994; Gazaway and Mclean 2003), soybean (Lewis et al. 1993), and corn (Norton and Hinz 1976). It is also considered an economically important pest of turfgrasses such as St. Augustine grass (*Sternotaphrum secundatum*) and bermudagrass (*Cynodon dactylon*) (Henn and Dunn 1989; Giblin-Davis et al. 1995), but is considered to be a moderate damaging species in North Carolina.

10.8 Stubby Root Nematodes, *Nanidorus minor* and *Trichodorus obtusus*

Stubby root nematode, *Nanidorus minor*, is a common and widespread nematode in North Carolina. They feed primarily on meristem cells of root tips causing a stunted or "stubby" appearing root system (Fig. 10.5). *Nanidorus minor* is important because of the direct damage to plant roots and also transmitting specific plant

Fig. 10.5 (**a**) Stubby-root nematode, *Nanidorus minor*; (**b**) Stubby root symptom caused by *Nanidorus minor* on corn from Bladen County in North Carolina

tobraviruses (Decraemer 1995). Stubby root nematodes are migratory, ectoparasitic and obligate plant parasites that feed on plants while their bodies remain in the soil. *Nanidorus minor* is a parthenogenic species and reproduces without sexual activity; males are rare. The life cycle of *P. minor* is fairly short for a plant parasite, being as short as 16 days at 29 °C. Stubby root nematode is considered a serious pest on corn, cotton, azalea and turfgrass in North Carolina. In a survey conducted in North Carolina and South Carolina, *N. minor* was found in 121 turfgrass samples taken in 33 counties (Zeng et al. 2012b).

Another stubby-root nematode, *Trichodorus obtusus*, was recently identified from South Carolina from 'Tifway' Bermudagrass (*Cynodon dactylon* × *C. transvaalensis*), 'Emerald' Zoysia (*Zoysia japonica*), 'Empire' Zoysia (Shaver et al. 2013) and from North Carolina from bermudagrass, St. Augustinegrass (*Stenotaphrum secundatum*) and Zoysiagrass (Ye et al. 2015a). This species is clearly different from the parthenogenic *Nanidorus minor* because of the presence of males, larger body sizes and DNA sequences of ribosomal DNA near-full-length small subunit (18S) and large subunit domain 2 and 3 (28S D2/D3). *Trichodorus obtusus* is known to occur only in the United States and damages turfgrasses. It is reported in the states of Virginia, Florida, South Carolina, Texas, Iowa, Kansas, Michigan, New York, North Carolina and South Dakota (Crow and Welch 2004; Shaver et al. 2013; Ye et al. 2015a).

10.9 Tobacco Cyst Nematode, *Globodera tabacum*

The tobacco cyst nematode, *Globodera tabacum*, is a serious and important soil-borne parasite of the tobacco roots. The species comprises three subspecies: *G. t.* subsp. *tabacum*, *G. t.* subsp. *solanacearum* and *G. t.* subsp. *virginiae*. These subspecies are differentiated by host preference. *G. t.* subsp. *tabacum* parasitizes shade-grown cigar wrapper and fieldgrown broadleaf cigar tobacco (*Nicotiana tabacum*) (Lownsbery and Peters 1955); *G. t.* subsp. *virginiae*, the horsenettle nematode, does not reproduce well in *Nicotiana* species (Miller 1977); and *G. t.* subsp. *solanacearum* attacks flue-cured tobacco cultivars (Komm et al. 1983). Only *G. t.* subsp. *virginiae* and *G. t.* subsp. *solanacearum* were found in North Carolina (Shepherd and Barker 1990; Melton et al. 1991). *Globodera tabacum solanacearum* was reported in North Carolina for the first time in 1991 in flue-cured tobacco fields in Warren County (Melton et al. 1991), but it is now distributed in Caswell, Granville, Person, Rockingham, Stokes, Vance and Warren Counties adjoining the Virginia border (Fig. 10.1c). Infection of the tobacco root system by tobacco cyst nematodes causes dramatic stunting, yield loss and decreases leaf quality. The use of resistant crops is useful in an integrated pest management program for controlling tobacco cyst nematode (Herrero et al. 1996).

Long-term control of any cyst nematode is difficult as viable eggs within cysts can survive in the soil for many years. Methods used include trap cropping, early destruction of roots and stalks, chemical control, crop rotation and the use of resistant tobacco genotypes.

10.10 Ring Nematodes, *Mesocriconema* spp.

Ring nematodes are very common, widespread and often occur at high population densities. However, symptoms of injury are not consistently associated with high numbers of this nematode and they are not often pathogenic. Plants which often support high populations of ring nematodes include peach, peanut and turfgrass in North Carolina.

The ring nematode, *Mesocriconema xenoplax*, is one of the most important nematode pathogens on peach due to its association with the disease complex known as Peach-Tree-Short-Life (PTSL) disease complex (Barker and Clayton 1973; Brittain and Miller 1978; Nyczepir et al. 1985; Parker 2000). Infection of ring nematodes can cause peach trees to be more susceptible to bacterial canker and cold damage, thus vastly reducing peach yields. PTSL is apparent in the spring when trees, or portions of trees, fail to grow. Trees, especially ones in their third to sixth season, may be killed back to the soil line. This problem most often occurs where trees are replanted in recent peach tree sites. Although nematodes rarely kill trees, they can predispose them to PTSL, especially ring nematodes. Cultivars tolerant to PTSL include 'Guardian' (root knot resistant), 'Lovell' (root knot susceptible) and 'Halford'. The root knot resistant cultivar 'Nemaguard' is very susceptible to

PTSL. The root knot resistant cultivar 'Guaradian' is a good choice for management of PTSL, but pretreatment of infested planting sites with a nematicide is needed to control nematodes and limit PTSL. Long-term cropping systems should also be considered in peach tree establishment and PTSL management. Successful management strategies may include use of cover crops such as bermudagrass or wheat.

Ring nematodes commonly infest golf course turfgrasses throughout the United States. High numbers of ring nematodes can cause visual chlorosis and decline of turfgrasses, particularly on putting greens. The above-ground symptoms of nematode feeding are slow growth, thinning of the turfgrass, poor response to adequate fertilization and irrigation, rapid wilting during dry weather, and weed invasion. These symptoms typically appear as irregular patterns across the turfgrass stand, not in circular patches or other distinct patterns. By the time above-ground symptoms of ring nematode injury appear, significant damage to the root system may have already occurred. There are three species of ring nematodes, *M. xenoplax, M. curvatum* and *M. sphaerocephalum,* that are known to damage turfgrasses in North Carolina (Ye et al. 2012; Zeng et al. 2012b).

Peanut ring nematode, *Mesocrionema ornatum*, commonly infects roots of peanut plants but has low damage potential. It is known to occur in a large percentage of the peanut production regions of the United States (Wheeler and Starr 1987; Dickson and De Waele 2005). It caused "yellows disease' symptom in microplots studies of several crops that used freshly-extracted, greenhouse-grown inoculum of *M. ornatum* (Barker et al. 1982). Greenhouse studies in North Carolina reavealed an interaction between *M. ornatum* and the black rot fungus *Cylindrocladium crotalariae* (Diomande and Beute 1981a, b).

10.11 Foliar Nematode, *Aphelenchoides fragariae*

Foliar nematodes (*Aphelenchoides* spp.) are an emerging problem on a number of landscape plant species in North Carolina, with *Aphelenchoides fragariae* as the predominant species detected. Unlike many other plant shoot pathogens that have narrow host ranges, foliar nematodes have broad host ranges and are capable of infecting hundreds of species of agronomic and ornamental plants including numerous ferns, foliage and flowering plants, and herbaceous and woody perennials (Decker 1989; Daughtrey et al. 1995; Knight et al. 1997, 2002). They live in the aboveground portions of plants, often without causing any obvious symptoms. The nematodes can remain on the outside of the plant, but most penetrate into leaf and stem tissue. Wetness on the stems and leaves provides an excellent environment for their movement. Splashing water during irrigation readily spreads the nematodes from leaf to leaf and plant to plant. Optimum temperatures for foliar nematode development are between 21 and 24 °C. The entire life cycle can be completed in 2–4 weeks, even sooner if the temperatures are higher. The most practical and effective control strategy is early detection and exclusion of this pest from growing facilities, including the use of nematode-free propagative stock.

10.12 Stem and Bulb Nematode, *Ditylenchus dipsaci*

Stem and bulb nematode, *Ditylenchus dipsaci*, is an endoparasitic migratory nematode that attacks aerial parts, bulbs and tubers of plants, causing the breakdown of the middle lamellae of cell walls. Feeding often causes swellings and distortion of aerial plant parts (stems, leaves, flowers) and necrosis or rotting of stem bases, bulbs, tubers and rhizomes. During cold storage of bulbs and tubers, *D. dipsaci* and related rotting may continue to develop. *Ditylenchus dipsaci* occurs locally in most temperate areas of the world and is occasionally found in ornamental plants in North Carolina (Fig. 10.6). A recent severe infestation on onion in Craven County in North Carolina revealed the presence of a species of *Ditylenchus*, but not *D. dipsaci* according to DNA sequencing data (Ye unpublished).

In most countries, regulatory measures such as certification schemes are applied to minimize further spread of *D. dipsaci*. However, North Carolina does not have a seed certification program. Bulbs and seeds can be disinfected by hot-water treatments. Races of *D. dipsaci* are highly host-specific, so employing a 3-year crop rotation can deprive the nematodes of a suitable host and starve the population. *Ditylenchus dipsaci* is known to attack over 450 different plant species, including many weeds (Hooper 1972). The use of tolerant or resistant cultivars can also reduce the damage. Soil fumigation in fields during fall can control nematodes on a susceptible crop in the spring.

Fig. 10.6 Photographs of symptoms caused by *Ditylenchus dipsaci* on *Hydrangea macrophylla* from Johnston County in North Carolina (**a**); *D. dipsaci* dults and juveniles (**b**) (lab ID: 14–38639)

10.13 Conclusion and Future Perspectives

Agriculture will continue to be the number one industry in the State of North Carolina and farmers will continue to fight against plant parasitic nematodes to increase crop production. Root knot nematodes are most common and destructive in this state due to their wide host range and wide distribution. In the past few years, *Meloidogyne enterolobii* has become an emerging species damaging several field crops, especially to sweet potato due to its severe damage on the storage roots resulting in unmarketable products and sometimes total loss. Soybean cyst nematode is still expanding in distribution posing a very serious threat to soybean production in North Carolina. Nematodes are microscopic hidden enemies of plants that are often difficult to detect, therefore, using nematode advisory service through state lab has been encouraged to determine the presence of nematode species, population density, hazard level and management action. New molecular diagnosis tool can be used to detect new emerging and or existing nematode species. Novel nematode management strategies should be developed, including more effective nematicides and resistant varieties to facilitate the reduction of crop losses to nematode damage.

References

Agudelo, P., & Harshman, D. (2011). First report of the spiral nematode *Scutellonema brachyurum* on lilyturf in the United States. *Plant Disease, 95*, 74–75.

Agudelo, P., Lewis, S. A., & Fortnum, B. A. (2011). Validation of a real-time polymerase chain reaction assay for the identification of *Meloidogyne arenaria*. *Plant Disease, 95*, 835–838.

Alexander, P. M. (1963). Stylet-bearing nematodes associated with various plants in South Carolina, 1959–1962. *Plant Disease Report, 47*, 978–982.

Allen, T. W., Bradley, C. A., Sisson, A. J., et al. (2017). Soybean yield loss estimates due to diseases in the United States and Ontario, Canada, from 2010 to 2014. *Plant Health Progress, 18*, 19–27. https://doi.org/10.1094/PHP-RS-16-0066.

Aycock, R., Barker, K. R., & Benson, D. M. (1976). Susceptibility of Japanes holly to *Criconemoides xenoplax*, *Tylenchorhynchus claytoni*, and certain other plant parasitic nematodes. *Journal of Nematology, 8*, 26–31.

Baldwin, J. G., & Sasser, J. N. (1979). *Meloidogyne megatyla* n. sp. a root-knot nematode from loblolly pine. *Journal of Nematology, 11*(1), 47–56.

Barker, K. R., & Clayton, C. N. (1973). Nematodes attacking cultivars of peach in North Carolina. *Journal of Nematology, 5*, 265–271.

Barker, K. R., & Imbriani, J. (1984). Nematode advisory programs-status and prospects. *Plant Disease, 68*, 735–741.

Barker, K. R., & Noe, J. P. (1987). Establishing and using threshold population levels. In J. A. Veech & D. W. Dickson (Eds.), *Vistas on nematlogy. A commoration of the twenty-fifth anniversary of the society of nematologists* (pp. 75–81). Hyattsville: Society of Nematologists.

Barker, K. R., & Sasser, J. N. (1959). Biology and control of the stem nematode, *Ditylenchus dipsaci*. *Phytopathology, 49*(10), 664–670.

Barker, K. R., Schmitt, D. P., & Campos, V. P. (1982). Response of peanut, corn, tobacco, and soybean to *Criconemella ornata*. *Journal of Nematology, 14*, 576–581.

Bernard, E. C. (1992). *Trichodorus elefjohnsoni* n. sp. (Nemata: Trichodoridae) from undisturbed Appalachian forest. *Journal of Nematology, 24*, 78–83.

Boutsika, K., Blok, V. C., Phillips, M. S., Lewis, S. A., Robbins, R. T., Ferraz, L. C. C. B., & Brown, D. J. F. (2004). Confirmation of the synonymy of *Paratrichodorus christiei* (Allen, 1957) Siddiqi, 1974 with *P. minor* (Colbran, 1956) Siddiqi, 1974 (Nematoda: Triplonchida) based on sequence data obtained for the ribosomal DNA 18S gene. *Nematology, 6*, 145–151.

Brito, J. A., Powers, T. O., Mullin, P. G., Inserra, R. N., & Dickson, D. W. (2004). Morphological and molecular characterization of *Meloidogyne mayaguensis* isolates from Florida. *Journal of Nematology, 36*, 232–240.

Brittain, J. A., & Miller, R. W. (1978). *Managing peach tree short life in the Southeast*. South Carolina Extention Circular 585.

Brzeski, M. W. (1974). Taxonomy of Hemicycliophorinae (Nematoda, Tylenchida). *Zeszyty Problemowe Postepów Nauk Rolniczych, 154*, 237–330.

Castillo, P., & Vovlas, N. (2007). Pratylenchus (Nematoda: Pratylenchidae): Diagnosis, biology, pathogenicity and management (Nematology monographs and perspectives. 6, D. J. Hunt, R. N. Perry, Eds.). Brill, Leiden-Boston, 530 pp.

Caswell, E. P., deFrank, J., Apt, W. J., & Tang, C. S. (1991). Influence of nonhost plants on population decline of *Rotylenchulus reniformis*. *Journal of Nematology, 23*, 91–98.

Cobb, N. A. (1913). New nematode genera found inhabiting fresh water and non-brackish soils. *Journal of the Washington Academy of Sciences, 3*, 432–444.

Crow, W. T. (2005). *Plant parasitic nematodes on golf course turf*. Outlooks on pest management. February, 10–15.

Crow, W. T., & Han, H. (2005). Sting nematode. *The Plant Health Instructor*. https://doi.org/10.1094/PHI-I-2005-1208-01.

Crow, W. T., & Welch, J. K. (2004). Root reductions of St. Augustinegrass (*Stenotaphrum secundatum*) and hybrid bermudagrass (*Cynodon dactylon* × *C. Transvaalensis*) induced by *Trichodorus obtusus* and *Paratrichodorus minor*. *Nematropica, 34*, 31–37.

Daughtrey, M. L., Wick, R. L., & Peterson, J. L. (1995). *Compendium of flowering potted plant diseases*. St. Paul: American Phytopathological Society.

Decker, H. (1989). Leaf-parasitic nematodes. In N. M. Sveshnikova (Ed.), *Plant nematodes and their control (Phytonematology)* (pp. 354–368). New York: E.J. Brill.

Decraemer, W. (1995). *The family Trichodoridae: Stubby root and virus vector nematodes*. Dordrecht: Kluwer Academic Publishers 360pp.

Dickerson O. J., Blake, J.H., & Lewis, S.A. (2000). *Nematode guidelines for South Carolina*. Clemson University Extension Bulletin EC 703.

Dickson, D. W., & De Waele, D. (2005). Nematode parasites of peanut. In M. Luc, R. A. Sikora, & J. Bridge (Eds.), *Plant parasitic nematodes in subtropical and tropical agriculture* (2nd ed.). Wallingford: CAB International.

Diomande, M., & Beute, M. K. (1981a). Effects of *Meloidogyne hapla* and *Macroposthonia ornata* on Cylindrocladium balck rot on peanut. *Phytopathology, 71*, 491–496.

Diomande, M., & Beute, M. K. (1981b). Relations of *Meloidogyne hapla* and *Macroposthonia ornata* popultions to *Cylindrocladium* balck rot on peanut. *Plant Disease, 65*, 339–342.

Eisenback, J. D. (1982). Description of the blueberry root-knot nematode. *Meloidogyne carolinensis* n. sp. *Journal of Nematology, 14*, 303–317.

Eisenback, J. D., Paes-Takahashi, V., Dos, S., & Graney, L. S. (2015). First report of the pecan root-knot nematode, *Meloidogyne partityla*, causing dieback to laurel oak in South Carolina. *Plant Disease, 99*(7), 1041.

Esbenshade, P. R., & Triantaphyllou, A. C. (1985). Identification of major *Meloidogyne* species employing enzyme phenotypes as differentiating characters. In J. N. Sasser & C. C. Carter (Eds.), *An advanced treatise on Meloidogyne, Biology and ccontrol* (Vol. I, pp. 135–140). Raleigh: North Carolina State University Graphics.

Esbenshade, P. R., & Triantaphyllou, A. C. (1987). Enzymatic relationships and evolution in the genus *Meloidogyne* (Nematoda: Tylenchida). *Journal of Nematology, 19*, 8–18.

Esbenshade, P. R., & Triantaphyllou, A. C. (1988). Genetic analysis of esterase polymorphism in the soybean-cyst nematode, *Heterodera glycines*. *Journal of Nematology, 20*, 486–492.

Fassuliotis, G., Rau, G. H., & Smith, F. H. (1968). *Hoplolaimus columbus*, a nematode parasite associated with cotton and soybeans in South Carolina. *Plant Disease Report, 52*, 571–572.

Fu, Z., Wells, C., & Agudelo, P. (2012). Induction of glutaredoxin expression in response to desiccation stress in the foliar nematode *Aphelenchoides fragariae*. *Journal of Nematology, 44*, 370–376.

Gazaway, W. S., & Mclean, K. S. (2003). A survey of plant parasitic nematodes associated with cotton in Alabama. *The Journal of Cotton Science, 7*, 1–7.

Giblin-Davis, R. M., Busey, P., & Center, B. J. (1995). Parasitism of *Hoplolaimus galeatus* on diploid and polyploidy St. Augustine grasses. *Journal of Nematology, 27*, 472–477.

Golden, A. M., & Cobb, G. S. (1963). *Heterodera lespedezae* (Heteroderidae), a new species of cyst-forming nematode. *Proceedings of the Helminthological Society of Washington, 30*, 281–286.

Golden, A. M., Rau, G. J., & Cobb, G. S. (1962). *Heterodera cyperi* (Heteroderidae), a new species of cyst-forming nematode. *Proceedings of the Helminthological Society of Washington, 29*, 168–173.

Graham, T. W., & Holdeman, Q. L. (1953). The sting nematode, *Belonolaimus gracilis* on cotton and other crops in South Carolina. *Phytopathology, 43*, 431–439.

Grand, L. F. (1985). *North Carolina plant disease index, Issue 240 (Revised) of Technical Bulletin*. Raleigh: North Carolina Agricultural Research Service. North Carolina State University 157pp.

Hartman, K. M., & Eisenback, J. D. (1991). Amended description of *Pratylenchus macrostylus* Wu, 1971 with SEM observations. *Journal of Nematology, 23*, 104–109.

Haygood, R. A., Saunders, J. A., & Miller, R. W. (1990). Widespread occurrence of *Meloidogyne incognita* on kiwifruit in the coastal areas of South Carolina. *Plant Disease, 74*, 81.

Heald, C. M., & Perry, V. G. (1970). Nematodes and other pests. *Turfgrass Science, Agronomy, 14*, 358–369.

Henn, R. A., & Dunn, R. A. (1989). Reproduction of *Hoplolaimus galeatus* and growth of seven St. Augustinegrass (*Stenotaphrum secundatum*) cultivars. *Nematropica, 19*, 81–87.

Herrero, S., Rufty, R. C., & Barker, K. R. (1996). Evaluation of tobacco germ plasm for resistance to the tobacco-cyst nematode, *Globodera tabacum solanacearum*. *Plant Disease, 80*, 61–65.

Hirschmann, H., & Sasser, J. N. (1955). On the occurrence of an intersexual form in *Ditylenchus triformis* n. sp. (Nematoda: Tylenchida). *Proceedings of the Helminthological Society of Washington, 22*, 115–123.

Holguin, C. M., Baeza, J. A., Mueller, J. D., & Agudelo, P. (2015a). High genetic diversity and geographic subdivision of three lance nematode species (*Hoplolaimus* spp.) in the United States. *Ecology and Evolution, 5*(14), 2929–2944. https://doi.org/10.1002/ece3.1568.

Holguin, C. M., Gerard, P., Mueller, J. D., Khalilian, A., & Agudelo, P. (2015b). Spatial distribution of reniform nematode in cotton as influenced by soil texture and crop rotations. *Phytopathology, 105*, 674–683.

Holguin, C. M., Mueller, J., Khalilian, A., & Agudelo, P. (2015c). Population dynamics and spatial distribution of Columbia lance nematode in cotton. *Applied Soil Ecology, 95*, 107–114.

Holguin, C. M., Ma, X., Mueller, J. D., & Agudelo, P. (2016). Distribution of *Hoplolaimus* species in soybean fields in South Carolina and North Carolina. *Plant Disease, 100*(1), 149–153.

Hooper, D. J. (1972). *Ditylenchus dipsaci*. Commonwealth Institute of Helminthology descriptions of plant parasitic nematodes, Set 1, no. 14. Farnham Royal: Commonwealth Agricultural Bureaux.

Huang, D., Yan, G., Gudmestad, N., Whitworth, J. L., Frost, K., Brown, C., Ye, W., Agudelo, P., & Crow, W. T. (2018). Molecular characterization and identification of stubby-root nematode species from multiple states in the United States. *Plant Disease*. https://doi.org/10.1094/PDIS-10-17-1668-RE.

Jagdale, G., Holladay, T., Brannen, P., Cline, W., Agudelo, P., Nyczepir, A., & Noe, J. (2013). Incidence and pathogenicity of plant parasitic nematodes associated with blueberry (*Vaccinium* spp.) replant disease in Georgia and North Carolina. *Journal of Nematology, 45*, 92–98.

Joyce, A., & Thiessen, L. (2017). *Soybean cyst nematode*. NC State Extension Publications. https://content.ces.ncsu.edu/management-of-soybean-cyst-nematode

Kerns, J.P., & Butler, E.L. (2018). *Turfgrass disease control*. North Carolina Agricultural Chemicals Manual. College of Agriculture and Life Sciences, North Carolina State University. NC State Extension Publications. AG-1. pp. 466–485.

Knight, K. W. L., Barber, C. J., & Page, G. D. (1997). Plant parasitic nematodes of New Zealand recorded by host association. *Journal of Nematology (Supplement), 29*, 640–656.

Knight, K. W. L., Hill, C. F., & Sturhan, D. (2002). Further records of *Aphelenchoides fragariae* and *A. ritzemabosi* (Nematoda: Aphelenchida) from New Zealand. *Australasian Plant Pathology, 31*, 93–94.

Koenning, S. R. (2004). Resistance of soybean cultivars to field populations of *Heterodera glycines* in North Carolina. *Plant Disease, 88*, 942–950.

Koenning, S. R., Overstreet, C., Noling, J. W., Donald, P. A., Becker, J. O., & Fortnum, B. A. (1999). Survey of crop losses in response to phytoparasitic nematodes in the United States for 1994. *Journal of Nematology (Supplement), 31*, 587–618.

Koenning, S. R., Kirkpatrick, T. L., Starr, J. L., Wrather, J. A., Walker, N. R., & Mueller, J. D. (2004). Plant parasitic nematodes attacking cotton in the United States – old and emerging production challenges. *Plant Disease, 88*, 100–113.

Kohl, L. M., Warfield, C. L., & Benson, D. M. (2010). Population dynamics and dispersal of *Aphelenchoides fragaria* in nursery-grown lantana. *Jornal of Nematology, 42*, 332–341.

Komm, D. A., Reilly, J. J., & Elliott, A. P. (1983). Epidemiology of a tobacco-cyst nematode (*Globodera tabacum*) in Virginia. *Plant Disease, 67*, 1249–1251.

Kraus-Schmidt, H., & Lewis, S. A. (1979). Seasonal fluctuations of various nematode populations in cotton fields in South Carolina. *Plant Disease Report, 63*, 859–863.

Krusberg, L. R., & Sasser, J. N. (1956). Host-parasitic relationship of the lance nematode on cotton roots. *Phytopathology, 46*, 505–510.

Leach, M. M., Agudelo, P., & Lawton-Rauh, A. L. (2012). Genetic variability of *Rotylenchulus reniformis*. *Plant Disease, 96*, 30–36.

Lewis, S. A., & Fassuliotis, G. (1982). Lance nematodes, Hoplolaimus spp., in the Southern United States. In R. D. Riggs (Ed.), *Nematology in the southern region of the United States*. Southern Co-op Ser. Bull. 276. Arkansas Agric. Exp. Stn., University of Arkansas, Fayetteville, AR. pp. 127–138.

Lewis, S. A., & Golden, A. M. (1981). Description and SEM observations of *Dolichodorus marylandicus* n. sp. with a key to species of *Dolichodorus*. *Journal of Nematology, 13*, 128–135.

Lewis, S. A., Skipper, H. D., & Musen, H. L. (1977). Nematode and nodulation studies in coastal plain soybean fields of South Carolina. *Journal of Nematology, 9*(4), 275.

Lewis, S. A., Drye, C. E., Saunders, J. A., Shipe, E. R., & Halbrendt, J. M. (1993). Plant-parasite nematodes on soybean in South Carolina. *Journal of Nematology (Supplement), 25*, 890–894.

Loof, P. A. A. (1991). The family Pratylenchidae Thorne 1949. In W. R. Nickle (Ed.), *Manual of agricultural nematology* (pp. 363–422). New York: Marcel Dekker.

López, M. A. C., Robbins, R. T., & Szalanski, A. L. (2012a). Taxonomic and molecular identification of *Bakernema*, *Criconema*, *Hemicriconemoides*, *Ogma* and *Xenocriconemella* species (Nematoda: Criconematidae). *Journal of Nematology, 44*, 427–446.

López, M. A. C., Robbins, R. T., & Szalanski, A. L. (2012b). Taxonomic and molecular identification of *Mesocriconema* and *Criconemoides* species (Nematoda: Criconematidae). *Journal of Nematology, 44*, 399–426.

López, M. A. C., Robbins, R. T., & Szalanski, A. L. (2013). Taxonomic and molecular identification of *Hemicaloosia*, *Hemicycliophora*, *Gracilacus* and *Paratylenchus* species (Nematoda: Criconematidae). *Journal of Nematology, 45*, 145–171.

Lownsbery, B. F., & Peters, B. G. (1955). The relation of the tobacco cyst nematode to tobacco growth. *Phytopathology, 45*, 163–167.
Lucas, L. T., Blake, C. T., & Barker, K. R. (1974). Nematodes associated with bentgrass and bermudagrass golf greens in North Carolina. *Plant Disease Report, 58*, 822–824.
MacGuidwin, A. E., & Rouse, D. I. (1990). Role of *Pratylenchus penetrans* in the potato early dying disease of Russet Burbank potato. *Phytopathology, 80*, 1077–1082.
Martin, S. B., Mueller, J. D., Saunders, J. A., & Jones, W. I. (1994). A survey of South Carolina cotton fields for plant parasitic nematodes. *Plant Disease, 78*, 717–719.
McCuiston, J. L., Hudson, L. C., Subbotin, S. A., Davis, E. L., & Warfield, C. Y. (2007). Conventional and PCR detection of *Aphelenchoides fragariae* in diverse ornamental host plant species. *Journal of Nematology, 39*(4), 343–355.
Meadows, I., Averre, C., Duncan, H., & Barker, K. (2018). *Control of root knot nematodes in the home vegetable garden*. NC State Extension Publications. AG-420.
Mehta, U. K., & Raski, D. J. (1971). Revision of the genus *Criconema* Hofmanner & Menzel, 1914 and other related genera (Criconematidae: Nematoda). *Indian Journal of Nematology, 1*, 145–198.
Melton, T. A., Phillips, J. A., Imbriani, J. L., & Barker, K. R. (1991). First report of *Globodera tabacum solanacearum* on flue-cured tobacco outside Virginia. *Plant Disease, 75*, 1074.
Miller, L. I. (1977). Pathogenicity of *Globodera virginiae* to Kentucky 16 burley tobacco. *Proceedings of the American Phytopathology Society, 4*, 217.
Mueller, J. D., & Agudelo, P. (2015). Lance nematodes. In: *Compendium of soybean diseases* (5th ed.). APS Press.
Mueller, J. D., & Sanders, G. B. (1987). Control of *Hoplolaimus columbus* on late-planted soybean with aldicarb. *Journal of Nematology (Supplement), 19*, 123–126.
NCDA Agricultural Statistics. (2017). *Annual statistics book*. http://www.ncagr.gov/stats/AgStat/NCAgStatBook.pdf
Niblack, T. L., Arelli, P. R., Noel, G. R., Opperman, C. H., Orf, J. H., Schmitt, D. P., Shannon, J. G., & Tylka, G. L. (2002). A new classification scheme for genetically diverse populations of *Heterodera glycines*. *Journal of Nematology, 34*, 279–288.
Nicol, J. M., Turner, S. J., Coyne, D. L., den Nijs, L., Hockland, S., & Maafi, Z. T. (2011). Current nematode threats to world agriculture. In J. T. Jones, G. Gheysen, & C. Fenoll (Eds.), *Genomics and molecular genetics of plant-nematode interactions* (pp. 21–44). Heidelberg: Springer.
Noe, J. P. (1993). Damage functions and population changes of *Hoplolaimus columbus* on cotton and soybean. *Journal of Nematology, 25*, 440–445.
Norton, D. C., & Hinz, P. (1976). Relationships of *Hoplolaimus galeatus* and *Pratylenchus hexincisus* to reduction of corn yields in sandy soils in Iowa. *Plant Disease Report, 60*, 197–200.
Nyczepir, A. P., & Lewis, S. A. (1979). Relative tolerance of selected soybean cultivars to *Hoplolaimus columbus* and possible effects of soil temperature. *Journal of Nematology, 11*, 27–31.
Nyczepir, A. P., Bertrand, P. F., Miller, R. W., & Motsinger, R. E. (1985). Incidence of *Criconemella* spp. and peach orchard histories in short-life and non-short-life sites in Georgia and South Carolina. *Plant Disease, 69*, 874–877.
Orton Williams, K. J. (1974). *Belonolaimus longicaudatus*. Commonwealth Institute of Helminthology. Descriptions of plant parasitic nematodes, Set 3, no. 40. St. Albans: Commonwealth Agricultural Bureaux.
Parker, M. (2000). *Growing peaches in North Carolina*. Raleigh: NC State Extension Publications.
Perry, V. G., Darling, H. M., & Thorne, G. (1959). *Anatomy, taxonomy and control of certain spiral nematodes attacking blue grass in Wisconsin*. University of Wisconsin Research Bulletin 207. 24 pp.
Plantard, O., Valette, S., & Gross, M. F. (2007). The root-knot nematode producing galls on *Spartina alterniflora* belongs to the genus *Meloidogyne*: Rejection of *Hypsoperine* and *Spartonema* spp. *Journal of Nematology, 39*, 127–132.

Powers, T. O., Bernard, E. C., Harris, T., Higgins, R., Olson, M., Lodema, M., Mullin, P., Sutton, L., & Powers, K. S. (2014). COI haplotype groups in *Mesocriconema* (Nematoda: Criconematidae) and their morphospecies associations. *Zootaxa, 3827*, 101–146.

Raski, D. J., & Golden, A. M. (1966). Studies on the genus *Criconemoides* Taylor, 1936 with descriptions of eleven new species and *Bakernema variabile* n. sp. (Criconematidae: Nematoda). *Nematologica, 11*, 501–565.

Ratanaworabhan, S., & Smart, G. C., Jr. (1970). The ring nematode, *Criconemoides ornatus*, on peach and centipede grass. *Journal of Nematology, 2*, 204–208.

Rau, G. J. (1958). A new species of sting nematode. *Proceedings of the Helminthological Society of Washington, 25*, 95–98.

Rau, G. J. (1963). Three new species of *Belonolaimus* (Nematoda: Tylenchida) with additional data on *B. longicaudatus* and *B. gracilis*. *Proceedings of the Helminthological Society of Washington, 30*, 119–128.

Rau, G. J., & Fassuliotis, G. (1965). *Hypsoperine spartinae* n. sp., a gall-forming nematode on the roots of smooth cordgrass. *Proceedings of the Helminthological Society of Washington, 32*, 159–162.

Riggs, R. D. (2004). History and distribution. In D. P. Schmitt, J. A. Wrather, & R. D. Riggs (Eds.), *Biology and management of soybean-cyst nematode: Second edition* (pp. 9–39). Marceline: Walsworth Publishing Company.

Riggs, R. D., & Schmitt, D. P. (1988). Complete characterization of the race scheme for *Heterodera glycines*. *Journal of Nematology, 20*, 392–395.

Robinson, A. F., Inserra, R. N., Caswell-Chen, E. P., Vovlas, N., & Troccoli, A. (1997). *Rotylenchulus* species: Identification, distribution, host ranges, and crop plant resistance. *Nematropica, 27*, 127–180.

Rowe, R. C., Davis, J. R., Powelson, M. L., & Rouse, D. I. (1987). Potato early dying: Causal agents and management strategies. *Plant Disease, 71*, 482–489.

Ruehle, J. L. (1968). *Distribution of plant-parasitic nematodes associated with forest trees of the world*. Ashville: U.S. Forestry Services. Southeast Forest Experimental Station.

Ruehle, J. L., & Sasser, J. N. (1962). The role of plant parasitic nematodes in stunting of pines in southern plantations. *Phytopathology, 52*(1), 56–68.

Rutter, W., Skantar, A. M., Handoo, Z. A., Mueller, J. D., Autma, S. P., & Agudelo, P. (2018). *Meloidogyne enterolobii* found infecting root-knot nematode resistant sweet potato in South Carolina, United States. *Plant Disease* (in press).

Schmitt, D. P., & Barker, K. R. (1988). Incidence of plant parasitic nematodes in the coastal plain of North Carolina. *Plant Disease, 72*, 107–110.

Shaver, J. B., Agudelo, P., & Martin, S. B. (2013). First report of stubby root caused by *Trichodorus obtusus* on zoysiagrass and bermudagrass in South Carolina. *Plant Disease, 97*(6), 852. https://doi.org/10.1094/PDIS-10-12-0932-PDN.

Shaver, J. B., Martin, S. B., Bridges, W. C., & Agudelo, P. (2015). Effects of *Trichodorus obtusus* on zoysiagrass and bermudagrass root weight and turfgrass quality. *Nematology, 17*, 671–678.

Shaver, J. B., Marchant, S., Martin, S. B., & Agudelo, P. (2016). 18S rRNA and COI haplotype diversity of *Trichodorus obtusus* from turfgrass in South Carolina. *Nematology, 18*, 53–65.

Shepherd, J. A., & Barker, K. R. (1990). Nematode parasites of tobacco. In M. Luc, R. A. Sikora, & J. Bridge (Eds.), *Plant parasitic plant parasitic nematodes in subtropical and tropical agriculture* (pp. 493–517). Wallingford: CAB International.

Sher, S. A. (1963). Revision of the Hoplolaiminae (Nematoda). II. *Hoplolaimus* Daday, 1905 and *Aorolaimus* n. gen. *Nematologica, 9*, 267–295.

Sher, S. A. (1966). Revision of the Hoplolaiminae (Nematoda). VI. *Helicotylenchus* Steiner, 1945. *Nematologica, 12*, 1–56.

Siddiqi, M. R. (1962). Studies on the genus *Longidorus* Micoletzky, 1922 (Nematoda: Dorylaimoidea), with descriptions of three new species. *Proceedings of the Helminthological Society of Washington, 29*, 177–188.

Sitterly, W. R., & Fassuliotis, G. (1965). Potato losses in South Carolina due to cotton root-knot nematode, *Meloidogyne incognita acrita*. *Plant Disease Report, 49*, 723.

Society of Nematologists. Nematode Geographical Distribution Committee. (1984). Distribution of plant-parasitic nematode species in North America: A project of the Nematode Geographical Distribution Committee of the Society of Nematologists. 205pp.

Starr, J. L., & Page, S. L. (1990). Nematode parasites of cotton and other tropical fiber crops. In M. Luc, R. A. Sikora, & J. Bridge (Eds.), *Plant parasitic nematodes in subtropical and tropical agriculture* (pp. 539–556). Oxon: CAB International.

Starr, J. L., Koenning, S. R., Kirkpatrick, T. L., Robinson, A. F., Roberts, P. A., & Nichols, R. L. (2007). The future of nematode management in cotton. *Journal of Nematology, 39*, 283–294.

Steiner, G. (1937). Opuscula miscellanca nematologica, V. *Proceedings of the Helminthological Society of Washington, 4*, 33–38.

Steiner, G. (1938). Opuscula miscellanca nematologica, VII. *Proceedings of the Helminthological Society of Washington, 5*, 35–40.

Strider, D. L. (1979). Control of *Aphelenchoides ritzemabosi* in African violet. *Plant Disease Report, 63*, 378–382.

Subbotin, S. A., Chitambar, J. J., Chizhov, V. N., Stanley, J. D., Inserra, R. N., Doucet, M. E., McClure, M., Ye, W., Yeates, G. W., Mollov, D. S., van den Berg, E., & Castillo, P. (2014). Molecular phylogeny, diagnostics and diversity of plant parasitic nematodes of the genus *Hemicycliophora* (Nematoda: Hemicycliophoridae). *Zoological Journal of the Linnean Society, 171*, 475–506.

Taylor, A. L. (1936). The genera and species of the Criconematinae, a sub-family of the Anguillulinidae (Nematoda). *Transactions of the American Microscopical Society, 55*, 391–421.

Tedford, E. C., & Fortnum, B. A. (1988). Weed hosts of *Meloidogyne arenaria* and *Meloidogyne incognita* in tobacco fields in South Carolina. *Annals of Applied Nematology, 2*, 102–105.

Thiessen, L. D. (2018). Soybean disease control. In: *North Carolina Agricultural Chemicals Manual*. College of Agriculture and Life Sciences, North Carolina State University. NC State Extension Publications. AG-1. 459–460.

Thorne, G. (1955). Fifteen new species of the genus *Hemicycliophora* with an emended description of *H. typica* de Man (Tylenchida: Criconematidae). *Proceedings of the Helminthological Society of Washington, 22*, 1–16.

Triantaphyllou, A. C., & Hirschmann, H. (1973). Environmentally controlled sex expression in *Meloidodera floridensis*. *Journal of Nematology, 5*, 181–185.

Tylka, G. L., & Marett, C. C. (2017). Known distribution of the soybean-cyst nematode, *Heterodera glycines*, in the United States and Canada, 1954 to 2017. *Plant Health Progress, 18*, 167–168. https://doi.org/10.1094/PHP-05-17-0031-BR.

Van den Berg, E., Tiedt, L. R., Liébanas, G., Chitambar, J. J., Stanley, J. D., Inserra, R. N., Castillo, P., & Subbotin, S. A. (2018). Morphological and molecular characterisation of two new *Hemicycliophora* species (Tylenchida: Hemicycliophoridae) with a revision of the taxonomic status of some known species and a phylogeny of the genus. *Nematology, 20*(4), 319–354.

Vargas, O. F., & Sasser, J. N. (1976). Biology of *Anguina plantaginis* parasitic on *Plantago aristata*. *Journal of Nematology, 8*(1), 64–68.

Wehunt, E. J., Golden, A. M., Weaver, C. F., & Rodriguez-Kabana, R. (1987). First report of a *Tylenchulus* sp. on peach in Alabama, Arkansas, Georgia, and South Carolina. *Annals of Applied Nematology, 1*, 127–128.

Wheeler, T. A., & Starr, J. L. (1987). Incidence and economic importance of plant parasitic nematodes on peanut in Texas. *Peanut Science, 14*, 94–96.

Whitehead, A. G. (1968). Taxonomy of *Meloidogyne* (Nematodea: Heteroderidae) with descriptions of four new species. *The Transactions of the Zoological Society of London, 31*, 263–401. https://doi.org/10.1111/j.1096-3642.1968.tb00368.x.

Williamson, M. R., Blake, J. H., Jeffers, S. N., & Lewis, S. A. (2000). First reports of *Aphelenchoides fragariae* on royal fern and on *Hosta* and other hosts in South Carolina. *Plant Disease, 84*, 593.

Winstead, N. N., Skotland, C. B., & Sasser, J. N. (1955). Soybean-cyst nematode in North Carolina. *Plant Disease Report, 39*, 9–11.

Wrather, J. A., Niblack, T. L., & Milam, M. R. (1992). Survey of plant parasitic nematodes in Missouri cotton fields. *Journal of Nematology, 24*, 779–782.

Ye, W. (2012). Development of PrimeTime-Real-Time PCR for species identification of soybean cyst nematode (*Heterodera glycines* Ichinohe, 1952) in North Carolina. *Journal of Nematology, 44*(3), 284–290.

Ye, W. (2017). Soybean cyst nematode (*Heterodera glycines*) distribution in North Carolina, U.S.A. *Plant Health Progress, 18*, 230–232. https://doi.org/10.1094/PHP-08-17-0050-BR.

Ye, W., & Robbins, R. T. (2000). Morphology of four species of *Hemicriconemoides* (Nematoda: Criconematidae) in the USA with the synonym of *H. annulatus*. *International Journal of Nematology, 10*, 101–111.

Ye, W., & Robbins, R. T. (2005). Morphological observation on *Longidorus crassus* Thorne, 1974 (Nematoda: Longidoridae) and its intraspecies variation. *Journal of Nematology, 37*, 83–93.

Ye, W., & Robbins, R. T. (2010). Morphology and taxonomy of *Xiphinema* (Nematoda: Longidoridae) occurring in Arkansas, USA. *Acta Agriculturae Universitatis Jiangxiensis, 32*(5), 928–945.

Ye, W., Zeng, Y., Tredway, L., Martin, S, Martin, M. (2012). Plant parasitic nematodes in Carolina turfgrass. *Carolina Green*. March/April 26–28. http://spectrumcreativegraphics.com/carolinasgreen_marapr12/

Ye, W., Koenning, S. R., Zhuo, K., & Liao, J. (2013). First report of *Meloidogyne enterolobii* on cotton and soybean in North Carolina, USA. *Plant Disease, 97*, 1262.

Ye, W., Zeng, Y., & Kerns, J. (2015a). First report of *Trichodorus obtusus* on turfgrass in North Carolina, USA. *Plant Disease, 99*(2), 291. https://doi.org/10.1094/PDIS-08-14-0830-PDN.

Ye, W., Zeng, Y., & Kerns, J. (2015b). Molecular characterization and diagnosis of root-knot nematodes (*Meloidogyne* spp.) from turfgrasses in North Carolina, USA. *PLoS One, 10*(11), e0143556. https://doi.org/10.1371/journal.pone.0143556.

Zeng, Y., Ye, W., Tredway, L., Martin, S., & Martin, M. (2012a). Description of *Hemicaloosia graminis* n. sp. (Nematoda: Caloosiidae) associated with turfgrasses in North and South Carolina, USA. *Journal of Nematology, 44*(2), 134–141.

Zeng, Y., Ye, W., Tredway, L., Martin, S., & Martin, M. (2012b). Taxonomy and morphology of plant parasitic nematodes associated with turf grasses in North and South Carolinas, USA. *Zootaxa, 3452*, 1–46.

Zeng, Y., Ye, W., Kerns, J., Martin, S., Tredway, L., & Martin, M. (2015). Molecular characterization and phylogenetic relationships of plant parasitic nematodes associated with turfgrasses in North Carolina and South Carolina, USA. *Plant Disease, 99*(7), 982–993.

Chapter 11
Plant Parasitic Nematodes of Virginia and West Virginia

Jonathan D. Eisenback

11.1 Introduction

The first nematode to cause alarm in Virginia and West Virginia was the wheat gall nematode, *Anguina tritici,* which was found in Rockingham County, Virginia in 1917 where it had been present for 10 years or more, causing a loss of 25–50% (Fromme 1919). A quarantine was threatened for wheat in Virginia, West Virginia and Georgia, but was not realized. The wheat gall was eventually eradicated by using crop rotation and clean seed (Leukel 1931). Root knot nematodes became the second recognized plant parasitic nematode (Zimmerley 1919).

Heterodera schachtii was reported parasitizing *Polygonum pennsylvanicum* in Fairfax County, Virginia (Steiner 1931), but was later described as a new species and now is known under the Latin name *Cactodera weissii.* Steiner also discovered root knot nematode on carrot. In Virginia, root knot was found on sweet potato, tomato and clover (Fenne 1940). The idea that these nematodes were being imported from other southeastern states on cuttings supported State Quarantine No. 6 that was enacted on April 1, 1940 to prevent further imports of plants unless they were shown to be "nematode-free" (VDAI 1940). Nevertheless, root knot nematodes continued to increase in their importance because of the injury that they caused. Parris (1943) estimated a 48% reduction in the yield of trimmed celery, and Parris and Jehle (1943) reported severe galling on lima bean which was causing a 20% loss to the grower.

Damage caused by lesion nematodes started to be noticed on boxwood in Virginia (Taylor 1944), and in that same year Jenkins (1944) showed that tobacco root disease complexes were caused by lesion nematode and root fungi. He went on to

J. D. Eisenback (✉)
Department of Plant Pathology, Physiology and Weed Science, Virginia Tech, Blacksburg, VA, USA
e-mail: jon@vt.edu

S. A. Subbotin, J. J. Chitambar (eds.), *Plant Parasitic Nematodes in Sustainable Agriculture of North America*, Sustainability in Plant and Crop Protection,
https://doi.org/10.1007/978-3-319-99588-5_11

demonstrate that small grains were also predisposed to infection to soil borne fungi by these nematodes (Jenkins 1948).

Bulb and stem nematode, *Ditylenchus dipsaci*, was found in Virginia in 1948 and selection of resistant cultivars was initiated immediately (Fenne et al. 1950); a few were found to be effective (Henderson 1950). Miller (1952) demonstrated the damage caused by the ectoparasitic nematode, *Belonolaimus longicaudatus,* and a new area of investigation was begun. He attended several nematode workshops, held at various universities in the southeast, to become more proficient in nematology. However, root knot nematodes continued to be the most important nematodes, and the use of DD and Dowfume was routinely practiced in tobacco fields and gardens (Fenne 1952).

The Virginia Department of Agriculture and Immigration published a bulletin warning about the occurrence of soybean cyst nematode in North Carolina, and strict measures were put in place to prevent its spread to Virginia (VDAI 1956), but it was probably too late; this nematode was soon found in Nansemond County.

Gruenhagen showed that decline of boxwood was caused by several plant parasitic nematodes and their health recovered with application of certain nematicides (Anonymous 1959). Miller continued to work with sting nematode and root knot on peanut (Anonymous 1959). He also discovered the horsenettle cyst nematode during this time and later described it as a new species, now known as *Globodera tabacum virginiae* (Miller and Gray 1968). Grover Smart, who joined Miller in 1960 to work on nematode problems around Holland, was the first professionally trained plant nematologist on the faculty, and he developed a technique to culture the soybean cyst nematode (Smart 1961). Soybean cyst nematode refused to be contained and rapidly spread to numerous farms despite the use of crop rotations, pesticides and resistant varieties. Grover Smart went on to show that soybean yields increased 46% in infested, fumigated soil compared to infested, unfumigated soil (Smart 1964). Miller became interested in the variability of the soybean cyst nematodes and repeatedly crossed interspecific and intergeneric species to show variation in morphology and host specificity (Miller 1983).

Al Williams became interested in nematodes when he was assigned to forage crops pathology. He worked with the alfalfa nematodes and also on *Meloidogyne graminis* (= *Hyposperine graminis*), but moved to the University of Kentucky where he eventually became Department Head of Horticulture.

The second professionally trained nematologist, Wyatt Osborne, was hired in 1961 to work with diseases of field crops, but his interest quickly switched to chemical control of nematodes. He and a county agent, Mr. H. M. Holmes, found a round cyst nematode in Amelia County parasitizing tobacco, similar to *Heterodera tabacum* (Osborne 1961), that was later described by Miller and Gray (1972) and now known as *Globodera tabacum solanacearum*. Wyatt demonstrated to growers that crop losses resulted from nematode damage and raised awareness with the development of a mobile nematode assay clinic that he towed to numerous field days.

Joseph Fox was the third professionally trained nematologist in Virginia, hired in 1965. He and Miller cooperated with Baalaway and Spasoff to find sources of resistance to the Virginia tobacco cyst nematode (Baalawy and Fox 1971; Spasoff et al. 1971).

Alma Elliot become the fourth trained nematologist. She worked on nematodes of small fruits, tobacco, soybean, peanut, and nematode parasitism (Elliot et al. 1982; Komm et al. 1983) and assisted in running the nematode assay lab (Elliot et al. 1986). She collaborated with Reilly to direct Earl Grant in his dissertation on the effects of plant parasitic nematodes on tobacco (Grant et al. 1982).

Elliot resigned in 1984 and was replaced by the fifth trained nematologist, J. D. Eisenback, in 1985 who continued his career refining descriptions of root knot nematodes (Eisenback and Gnanapragasam 1992; Eisenback and Hirschmann 2001) and developing techniques useful for studying nematode morphology (Eisenback 1988, 2010a, b, 2012; Eisenback and Rammah 1987). Several new species of nematodes were described by Eisenback and his colleagues (Eisenback 1982; Yang and Eisenback 1983; Eisenback and Hartman 1985; Eisenback et al. 1985, 1994; Bernard and Eisenback 1997; Charchar et al. 1999; Charchar and Eisenback2002; Eisenback et al. 2003; Charchar et al. 2008a, b, 2009). Eisenback published several first reports of nematodes for Virginia or the U.S. including *Meloidogyne chitwoodi* (Eisenback et al. 1986) which was later shown to be absent by survey, *Heterodera zeae* (Eisenback et al. 1993), *Tylenchulus palustris* (Eisenback et al. 2007) and *Rotylenchulus reniformis* (Eisenback et al. 2004); and a first report for New York, *Meloidogyne mali* (Eisenback et al. 2017). Eisenback also worked with Pat Phipps, evaluating the effect of nematodes on peanut, soybean and cotton (Phipps and Eisenback 2012a, b, Mehl et al. 2013); and Charles Johnson, investigating resistance in tobacco to root knot and the Virginia tobacco cyst nematode (Wang et al. 1997; Rideout et al. 2000; Parkunan et al. 2010; Pollock et al. 2016). In addition to his work on understanding the mechanisms of resistance, Johnson also provided the basis for the practical control on nematodes on tobacco (Johnson 1998, 2017; Johnson et al. 1989, 1994, 2005).

In West Virginia, nematology emphasized the nematodes of fruit trees, forest trees, strawberries and tomatoes (Barnet 1985; Nesiius 1988). Apple trees, with high populations of *Xiphinema americanum*, showed an increase in growth as the nematode population decreased; an integrated control system with the use of a systemic nematicide, application of organic matter (wood chips) on the soil surface and maintenance of adequate soil moisture provided reduced plant parasitic nematode populations. Increased plant vigor was suggested by the first trained nematologist in the state, R. E. Adams and his student S. E. Tamburo (Adams 1955; Tamburo and Adams 1962).

On forest trees, 26 genera of nematodes were found to be parasitic, but not pathogenic, on nursery seedlings of red pine (Sutherland and Adams 1964). The most common nematodes found were *Paratylenchus* sp., *Hoplolaimus galeatus* and *Trichodorus* sp., and their populations were reduced with the application of methyl bromide as demonstrated by Adam's student, Sutherland (Sutherland and Adams 1964, 1965, 1966). *Dolichodorus silvestris*, an awl nematode, was described as a pathogen of seedlings and a parasite of mature white pines (*Pinus strobus*) (Gillespie and Adams 1962; Lapp 1967). Likewise, another new species, *Gracilacus capitatus*, was described as a parasite on scarlet oak (*Quercus coccinea*) (Adams and

Eichenmuller 1962), and *Criconema grassator* (= *Criconemoides grassator*) was described from the roots of a native stand of yellow poplar (*Leriodendron tulipifera*) (Adams and Lapp 1967).

Interaction studies on tomato showed that nematodes increase the severity and incidence of bacterial wilt (Libman et al. 1964). Screening for resistance to root knot nematodes discovered five resistant lines from a single resistant plant; two were resistant, two were moderately resistant, and one was extremely susceptible and thought to demonstrate the occurrence of a new biotype of root knot nematode (Kish 1973). The biological control of root knot on tomato was tested with *Purpureocillium lilacinus* showing more than 50% reduction in gall production and nematode reproduction in both the greenhouse and in field production (Rodriques 1983).

The second trained nematologist in West Virginia, James Kotcon, made numerous contributions to plant nematology including fall fumigation with potato to manage lesion nematode, *Pratylenchus crenatus* (Kotcon 1987) the effect of *P. penetrans* on water relations and plant growth in potato (Kotcon and Loria 1987)and the interaction of *P. penetrans* with three fungi in the Early Dying Syndrome of potato (Kotcon 1984, 1985; Kotcon and Rouse 1984). He then switched his attention to peach orchards, where he looked at the distribution, frequency and density of nematodes (Kotcon 1990). Recently, he examined the value of intercropping with resistant varieties of tomato for organic farms (Kotcon 2018) and devoted much of his time to organic production including biological control (Biggs et al. 1994; King and Kotcon 2010; Kotcon et al. 1985, Nelson and Kotcon 2005; Panaccione et al. 2006; Salinas and Kotcon 2007; Salinas et al. 2007; Bull et al. 2018).

11.2 Agriculture in Virginia

Agriculture in Virginia remains strong and has increased significantly in the last decade. Currently, it is the most important sector in the economy with a value of $70 billion, plus an additional $36.2 billion in value-added products (Rephann 2017) (Table 11.1). More than 334,000 people are employed in agriculturally-related jobs. Although crops make up only one-third of the value of farm sales, they are also responsible for the nutrition and well-being of the remaining two-thirds of the farm receipts that are generated from livestock. The commonwealth ranked fourth in the production of tobacco (*Nicotiana tabacum*), seventh for apples (*Malus pumila*), seventh for cut Christmas trees (various fir and pine species), eighth for grapes (*Vitis vinifera*), eighth for peanuts (*Arachis hypogaea*), tenth for tomatoes (*Solanum lycopersicum* and fourteenth for cotton (*Gossypium hirsutum*) (Rephann 2017; USDA Census of Agriculture, 2016). Many of these crops are susceptible to damage by plant parasitic nematodes (Table 11.2).

Table 11.1 Dollar value of the major crop plants grown in Virginia and West Virginia (USDA 2016)

Field crops	Value in dollars	Crops supporting livestock	Value in dollars	Crop improving the quality of life	Value in dollars
Virginia					
Tobacco	$109,705,000	Hay	$357,075,000	Nursery, greenhouse, floriculture, sod	$251,871,000
Cotton	$29,616,000	Soybean	$198,720,000	Fruit, nut, berries	$65,820,000
Wheat	$44,056,000	Corn	$188,700,000	Vegetables	$78,323,000
Tomatoes	$23,184,000	Alfalfa	$41,410,000	Apples	$35,854,000
Potatoes	$18,192.000			Grapes	$16,830,000
Peanuts	$15,152,000			Christmas trees	$7,873,000
Pumpkins	$11,160,000			Peaches	$5,226,000
Barley	$2,332,000				
West Virginia					
Wheat	$1,171,000	Hay	$117,990,000	Nursery, greenhouse, floriculture, sod	$26,772,000
		Corn	$18,270,000	Christmas trees	$33,136,000
		Soybean	$12,199,000	Fruit, nut, berries	$31,338,000
		Alfalfa	$10,335,000	Vegetables	$5,600,000
				Apples	$4,823,000
				Peaches	$3,815,000
				Ginseng	$2,000,000

Table 11.2 Plant-parasitic nematodes that are widely distributed in Virginia and West Virginia

Nematode	State	Crop	References
Anguina agrostis	VA	Bentgrass	Eisenback and Roane (2006)
Aphenlenchoides besseyi	VA	Strawbeerry	Cairns et al. (1960
A. fragariae	VA	Ferns, hosta, peony, strawberry	Eisenback (unpublished)
A. ritzemabosi	VA	Ferns, strawberries	Eisenback (unpublished)
Aphenlenchus avenae	WV	Wheat	https://www.prevalentnematodes.org/state.cfm?id=us_wv
Bakernema inaequale	VA	Sugar maple	Taylor (1936)
Belononolaimus longicaudatus	VA	Beans, bentgrass, bermuda grass, boxwoods, corn, cotton, crucifers, curcurbits, vegetables, peach, peanut, sorghum, soybean, strawberries, sudan grass, sweet corn, sweet potato, yew	Cairns et al. (1960)

(continued)

Table 11.2 (continued)

Nematode	State	Crop	References
Bitylenchus dubius	WV		https://www.prevalentnematodes.org/state.cfm?id=us_wv
Cactodera cactii	VA, WV	Cactus	https://www.prevalentnematodes.org/state.cfm?id=us_wv
C. weissi	WV		https://www.prevalentnematodes.org/state.cfm?id=us_wv
Criconema sphagni	WV		https://www.prevalentnematodes.org/state.cfm?id=us_wv
Crossonema fimbriatum	VA	Rhododendron,	Eisenback (unpublished)
Ditylenchus dipsaci	VA	Alfalfa, ladino clover, red clover, white clover, daffodills, garlic, narcissus	Cairns et al. (1960)
Dolichodorus heterocephalus	VA	Beans, crucifers	
D. silvestris	WV	Eastern white pine	Gillespie and Adams (1962)
Globodera tabacum solanacearum	VA	Pepper, tomato, tobacco	Miller and Gray. (1972) and Osborne (1961)
Globodera tabacum virgniae	VA	Horsenettle	Miller and Gray (1968)
Gracilicus capitatus	WV	Scarlet oak	Adams and Eichenmuller (1962)
G. marylandicus	WV		https://www.prevalentnematodes.org/state.cfm?id=us_wv
Helicotylenchus crenacauda	WV		https://www.prevalentnematodes.org/state.cfm?id=us_wv
H. digonicus	WV	Peach	Kotcon (1990)
H. dihystera	VA, WV	Alfalfa, apple, boxwoods, red clover, gardenia, Japanese holly,lima bean, orchard grass, peach, peanut, privet, strawberries	Cairns et al. (1960) and Kotcon (1990)
H. nannis	VA	Boxwood, yew	Cairns et al. (1960)
H. platyurus	WV	Peach	Kotcon (1990)
H. pseudorobustus	WV	Peach	Kotcon (1990)
Hemicycliophora gigas	WV		https://www.prevalentnematodes.org/state.cfm?id=us_wv
H. typica	VA	Short leaf pine	Cairns et al. (1960)
H. vidua	WV	Peach	Kotcon (1990)
H. vivida	WV	Peach	Kotcon (1990)
Heterodera glycines	VA	Soybean	Cairns et al. (1960)
H. schachtii	WV	Sugarbeet	https://www.prevalentnematodes.org/state.cfm?id=us_wv

(continued)

Table 11.2 (continued)

Nematode	State	Crop	References
H. trifolii	VA, WV	Clover	Eisenback (unpublished)
H. weissi	VA	Smartweed	Cairns et al. (1960)
H. zeae	VA	Corn	Eisenback et al. (1993)
Hoplolaimus galeatus	VA, WV	Alfalfa, apple, arbor vitae, bentgrass, blueberries, boxwood, camellia, chrysanthemum, clover, corn, cotton, gardenia, grape, hemlock, Iiris, Kentucky bluegrass, lima bean, oak, peach, peanut, soybean, sweet corn, tomato	Cairns et al. (1960) and Kotcon (1990)
H. tylenchiformis	VA, WV	Abelia grandiflora, Arbor vitae, boxwood, legumes, ligustrum, violet	Cairns et al. (1960)
Longidorus breviannulatus	VA	Boxwood	Cairns et al. (1960)
Meloidogyne arenaria	VA, WV	Alfalfa, beet, cabbage, celery, crimson clover, cucumber, gardenia, ladino clover, white clover, gardenia, iris, lespedeza, lettuce, peach, peony, snapbean, soybean, St. Augustine grass, tobacco, tomato	Cairns et al. (1960) and Walters and Barker (1994)
M. hapla	VA, WV	Alfalfa, peach, peanut, soybean, tobacco, tomato	Kotcon (1990) and Walters and Barker (1994)
M. incognita	VA, WV	Alfalfa, soybean, tobacco, tomato	Walters and Barker (1994)
M. javanica	VA, WV	Alfalfa,Japanes holly, soybean, tobacco, tomato	Cairns et al. (1960) and Walters and Barker (1994)
M. graminis	VA	Bermuda grass	Williams (1968)
M. platani	VA	Sycamore	Hirschmann (1982)
M. querciana	VA	Pin oak	Golden (1979)
M. spartinae	VA	Beachgrass	Eisenback (unpublished)
Mesocriconema curvatum	WV	Peach	Kotcon (1990)
M. ornatum	VA, WV	Boxwood, corn, peanut, rose, strawberries	Cairns et al. (1960
M. rusticum	VA, WV	Boxwood	Cairns et al. (1960
M. sphaerocephala	WV		https://www.prevalentnematodes.org/state.cfm?id=us_wv
M. xenoplax	VA, WV	Blueberries, corn, English boxwood, grape, peach, peanut, soybean, bentgrass	Kotcon (1990)

(continued)

Table 11.2 (continued)

Nematode	State	Crop	References
Paratrichodorus minor	VA	Beans, blueberries, corn, cotton crucifers, vegetables, peanut, soybean, sweet corn, bentgrass	Eisenback (unpublished)
Paratylenchus bukowinensis	VA, WV		https://www.prevalentnematodes.org/state.cfm?id=us_wv
P. ciccaronei	WV	Peach	Kotcon (1990)
P. hamatus	WV	Peach	Kotcon (1990)
P. nanus	WV	Peach	Kotcon (1990)
P. projectus	WV	Peach	Kotcon (1990)
P. tenuicaudatus	WV	Peach	Kotcon (1990)
Pratylenchus brachyurus	VA	Corn	Eisenback (unpublished)
P. crenatus	VA, WV	Peach	Kotcon (1990)
P. neglectus	WV	Peach	Kotcon (1990)
P. penetrans	VA, WV	Alfalfa, peach, strawberries	Cairns et al. (1960) and Kotcon (1990)
P. pratensis	WV	Peach	Kotcon (1990)
P. scribneri	VA, WV	Peach	Kotcon (1990)
P. thornei	WV	Peach	Kotcon (1990)
P. vulnus	WV	Peach	Kotcon (1990)
P. zeae	VA	Corn, Bermuda grass	Cairns et al. (1960)
Quinisulcius capitatus	WV	Peach	Kotcon (1990)
Rotylechulus reniformis	VA	Cotton, soybean	Eisenback et al. (2004)
Rotylechus buxophilus	VA, WV	English boxwood, American boxwood, peach	Kotcon (1990)
Scutellonema brachyurum	VA	Gardenia	Cairns et al. (1960)
Tetylenchus sp.	VA	Strawberries, Zoysia grass	Cairns et al. (1960)
Trichodorus primitivus	VA, WV		https://www.prevalentnematodes.org/state.cfm?id=us_wv
Tylenchorhynchus claytoni	VA, WV	Arbor vitae, Japanese holly, English ivy, orchard grass, soybean, sweetpotato,tobacco,	Cairns et al. (1960) https://www.prevalentnematodes.org/state.cfm?id=us_wv
T. agri	WV	Peach	Kotcon (1990)
T. dubius	WV		https://www.prevalentnematodes.org/state.cfm?id=us_wv
T. maximus	WV		https://www.prevalentnematodes.org/state.cfm?id=us_wv
T. striatus	VA	Corn, soybean	Cairns et al. (1960)
Tylenchulus palustris	VA	Peach	Eisenback et al. (2007)

(continued)

Table 11.2 (continued)

Nematode	State	Crop	References
Xenocriconema macrodora	VA, WV	Oak	Eisenback (unpublished)
Xiphinema americanum	VA, WV	Apple, blueberries, boxwood, grape, peanut, Japanese holly, peach, soybean, strawberries, tobacco	Cairns et al. (1960) and Kotcon (1990)
X. californicum	WV	Peach	Kotcon (1990)
X. chambersi	VA, WV	Rhododendron	Eisenback (unpublished) https://www.prevalentnematodes.org/state.cfm?id=us_wv
X. rivesi	WV	Apple	Kotcon (1990)

11.3 Plant Parasitic Nematodes of Virginia

11.3.1 Root Knot Nematodes, Meloidogyne *spp.*

Root knot nematodes are the most prevalent and widespread plant parasitic nematodes in the commonwealth, affecting most of the major field crops such as tobacco, cotton, wheat, tomato, potato, peanut and pumpkin (Walters and Barker 1994). They also damage crops that support livestock such as alfalfa, soybean and corn and cause injury to crops that improve the quality of life including vegetables, grapes, fruit, nursery, greenhouse, floriculture and turfgrass.

Meloidogyne spp. are the most economically important nematode pests for many of these plants and minimizing their effects on crop production is essential for sustainable agriculture. If nematodes are not properly managed, production may drop below the level of profitability. Although sustainable techniques such as no-till or minimal till have been used in the production of some of these plants (Duiker et al. 2016), their high value justifies high value inputs. Therefore, in Virginia, because root knot nematodes are so widespread and damaging, chemicals remain a valuable management tactic; however, crop rotation and host plant resistance, when economically justified, are widely utilized. Crop rotation and use of genetic resistance often require the accurate identification of the predominant species in the field, since one or two or more of the most common species, *M. incognita, M. arenaria, M. javanica* and *M. hapla* may be present. Furthermore, *M. incognita* can be characterized as four host races and *M. arenaria* has two (Eisenback et al. 1885). Precise identification of species is time consuming and expensive and often requires the use of PCR and gene sequencing and greenhouse testing (Adams et al. 2009; Eisenback and Triantaphyllou 1991).

Other species of root knot that are known to occur in Virginia include the oak root knot nematode, *Meloidogyne querciana,* on pin oak (*Quercus palustris*) and sycamore root knot nematode, *M. platani*, on American sycamore (*Platanus occidentalis*) (Golden 1979; Hirschmann 1982). Nearby species that are potentially important pests in Virginia include *M. enterolobii*. This species occurs in North Carolina where it was found parasitizing cotton and soybean (Ye et al. 2013). It has the potential of causing much economic harm to crops in Virginia since it is very aggressive and readily reproduces on resistant tobacco, cotton, soybean and many other plants (Yang and Eisenback 1983). The problem of *Meloidogyne* spp. on major crops in Virginia will be discussed further.

11.3.1.1 Tobacco

Meloidogyne incognita was the most important root knot nematode on tobacco until resistant cultivars were introduced in the middle of the twentieth century. During the last 60 years, however, *M. arenaria* has become the most common species found attacking tobacco, with *M. javanica* and *M. hapla* also causing economic losses. Likewise, races 2 and 4 of *M. incognita* have also become more common (Eisenback, Johnson and Reed, unpublished data).

Tobacco cultivation in Virginia is limited by the occurrence of the four most common species of root knot, including *Meloidogyne incognita, M. arenaria, M. javanica* and *M. hapla* (Johnson et al. 2005). Because organically produced tobacco is worth 200% more than non-organic, nearly one-half of the crop is organically produced in Virginia (Johnson, personal comm., Kuepper and Thomas 2008). Cultural control of these nematodes includes destruction of the root system soon after harvest, deep plowing into high and wide ridges before transplanting, early planting, utilization of effective cover crops and crop rotation (Johnson 1998; Johnson et al. 2005). Of these tactics, the use of cover crops and crop rotation are difficult to implement because the four species of root knot have different but very diverse host ranges. Bare fallow and weedy fallow are not suitable for reducing nematode populations because bare fallow may cause significant soil erosion, and weedy fallow may allow nematodes to survive or actually increase in number on numerous plant species that may serve as host to the root knot nematodes (Johnson et al. 2005).

Most tobacco cultivars planted in Virginia are resistant to races 1 and 3 of *M. incognita* (Johnson et al. 1989). Unfortunately, this widespread, persistent use of a single source of resistance has selected populations of *M. arenaria, M. javanica, M. hapla* and *M. incognita* race 4 to become more common in flue-cured tobacco during the last 60 years since resistance was first used (Eisenback et al. unpublish.).

11.3.1.2 Cotton

Meloidogyne incognita race 1 is the only root knot nematode that reproduces on cotton in Virginia, therefore, if root knot nematode is present, the species and race is known. *Meloidogyne enterolobii* can also reproduce on cotton (Yang and Eisenback 1983) and has been found nearby in North Carolina causing significant injury to cotton (Ye et al. 2013).

For management, fields need to be sampled to determine the presence of nematodes and to measure their population levels. If none are found, preventing their introduction into additional fields should be a top management priority. If some are found, preventing their spread to other fields and reducing their population densities should be prioritized (Phipps 2013). Cleaning all equipment to remove soil and plant tissues infested with nematodes will help to prevent their spread from field to field. Infested fields should be worked last. When plants in the field are under stress from a lack of moisture, feeding by plant parasitic nematodes enhances that stress. If a hard pan occurs in the field, subsoiling may be necessary to break up that layer so that roots of the cotton plants will be able to penetrate deeper into the soil. The application of additional water through irrigation will help these plants to survive. At the end of the growing season the roots should be exposed to the drying air.

Planting a non-host crop for one or more years is an effective rotation scheme to reduce the root knot population to below damaging thresholds (Phipps 2013). This strategy is currently possible without identifying the root knot species since *M. incognita* race 1 is the only root knot in Virginia that parasitizes cotton. However, if *M. enterolobii* is found in Virginia, identification of the species in cotton will be become necessary.

Tolerant cotton plants are available in several commercial varieties (Phipps 2013; Cook and Robinson 2005; Starr et al. 2007). Seed treatments with nematicides or other chemicals have been shown to be effective for controlling root knot nematodes on cotton (Faske and Starr 2007). These tactics are gradually becoming adopted by growers in Virginia. Seed treatments used by growers in Virginia include abamectin (Avicta Complete Pak) and thiodicarb (Aeris Seed-applied insecticide/nematicide) (Mehl 2017).

11.3.1.3 Tomato

Tomato is most commonly attacked by the southern root knot nematode, *Meloidogyne incognita*; however, the other three common species, *M. arenaria, M. javanica* and *M. hapla*, occur within the commonwealth. *Meloidogyne enterolobii* is a potential threat since it has been found in neighboring North Carolina. Crop rotations with non-hosts of the root knot nematodes are the most important and recommended sustainable practices for use in tomato production (Gatton et al. 2007); however, the species of root knot has to be identified for effective management because these species have enormous and varied susceptible hosts.

Although many of the common varieties of tomatoes have genetic resistance to at least some of the species of root knot nematodes, Better Boy, Celebrity, Park's Whopper and Goliath are resistant to *M. incognita*, *M. arenaria* and *M. javanica*, but not *M. hapla* or *M. enterolobii* (Gardenweb 2017). *Trichoderma* spp. have been shown to reduce the number of galls produced by root knot nematode on tomato, (Sahebani and Hadavi 2008), but is not used by growers in Virginia.

11.3.1.4 Potato

Southern and northern root knot nematodes, *M. incognita* and *M. hapla*, respectively, are the most common species parasitizing potato in Virginia (Schooley et al. 2003). Crop rotation with non-host crops is the most commonly practiced sustainable tactic used to manage root knot nematodes on potato; however, because they are high-value crops, most growers rely on chemicals to control these pests. Plant resistance to root knot nematodes is not available for managing these nematodes.

11.3.1.5 Peanut

In Virginia, the northern root knot nematode, *M. hapla*, is the most common species of root knot nematode found damaging peanut; however, the peanut root knot nematode, *M. arenaria,* is also prevalent (Mehl 2017). Crop rotations of 3–4 years with non-hosts reduces the impact of both *M. arenaria* and *M. hapla* on peanut. Since these two species have different host ranges, species identification is necessary to select the most useful rotation crops (Alexander et al. 2002). Corn may be a good rotation with peanut if the problem nematode is *M. hapla*, but not if it is *M. arenaria* (Baldwin and Barker 1970). Resistant varieties of Virginia-type peanuts are not available for managing either of the two root knot species that parasitize them (Alexander et al. 2002).

11.3.1.6 Soybean

Root knot nematodes are very common on soybean in Virginia, and all four common species can reproduce on most varieties of this plant (Mehl 2017). The occurrence of *M. enterolobii* on soybean in North Carolina makes this nematode a potential threat to Virginia (Ye et al. 2013). In Virginia, the most common root knot nematode parasitizing soybean is *M. hapla*. Unfortunately, soybean cultivars are not tested for resistance to this species. However, different cultivars vary in their ability as hosts to the different species of *Meloidogyne.* Likewise, they also differ in their resistance to soybean cyst nematode and reniform nematode. Therefore, if rotation is to be effective in managing root knot nematodes on soybean, the species must be determined as well as the occurrence of cyst and reniform nematodes. Root knot resistant soybean cultivars are useful for minimizing the effect of three of the four common species of this nematode on soybean yields. Unfortunately, soybean cultivars are not screened for resistance to the northern root knot nematode, *M. hapla*. Abamectin (Activa Complete Beans)®, Fluopyram (ILeVO)® and *Bacillus firmus* (Poncho/VOTIVO)® are seed treatments that are commonly used in Virginia to reduce the effect of root knot nematodes on soybean yields (Mehl 2017).

11.3.1.7 Corn

Southern and other root knot nematodes may cause small losses in on corn in Virginia; however, more importantly, these nematodes may increase in number on corn so that they will be damaging to next year's crop (Mehl 2017). Managing root knot on corn with rotation is a difficult proposition because corn can be quite tolerant to root knot, allowing high populations to increase during the growing season without noticeable damage. Only *M. hapla* cannot reproduce on corn (Baldwin and Barker 1970). Crop rotation with non-hosts or fallow are the most commonly used sustainable tactics for managing nematodes on corn in Virginia (Mehl 2017). *Meloidogyne incognita, M. arenaria* and *M. javanica* all reproduce well on corn, a result that limits their effectiveness in crop rotations (Baldwin and Barker 1970), especially for crops that also serve as hosts of these species. Since *M. hapla* cannot reproduce on corn and it may be an effective rotation crop with crops that are good hosts for this species, such as peanut.

Different cultivars of corn have various responses to root knot nematodes. All cultivars are non-hosts to *M. hapla*. Unfortunately, for the other three common species, damage is not necessarily correlated with reproduction. Cultivars that are least favorable hosts may suffer the most damage, whereas those that are excellent hosts may be tolerant (Baldwin and Barker 1970). As seed treatments, both clothianidin + *Bacillis firmus* (Poncho/Votivo) and abamectin + thiamethoxam (Avicta Duo) give early season protection to young, vulnerable seedlings. These treatments also give some suppression of soil borne insects (Mehl 2017).

11.3.1.8 Alfalfa

Root knot nematodes of *M. incognita, M. arenaria, M. javanica* and *M. hapla* can reproduce on alfalfa and cause serious damage. If the nematode population is high when the seedlings are becoming established, the plants can be severely stunted and young seedlings can be killed. The stand may fail to thrive, and the full potential of the crop may not be realized. Furthermore, plants that are infected with root knot nematodes are more susceptible to other diseases such as *Fusarium, Phytophthora* and bacterial wilt. Crop rotation for 2–3 years with non-host crops, including some grasses, may reduce the population of root knot nematodes to below-economic levels. All four most common species parasitize alfalfa, sometimes making the identification of the species necessary. Alfalfa cultivars vary in their resistance to root knot nematodes (Potenza et al. 1996; Dhandaydham et al. 2008); however, the use of resistance is the best sustainable practice for managing these nematodes in Virginia (Schooley 2004).

11.3.1.9 Pumpkin

Root knot nematodes are very important on pumpkin where *M. incognita* and *M. arenaria* can be devastating and cause significant losses (Schooley 2013). The only sustainable tactic to manage these nematodes is to assay potential fields for them and select fields that are root knot nematode-free. Crop rotation with non-host plants may be useful for reducing their population, but resistance is not available.

11.3.1.10 Grape

During the last 35 years, vineyards in Virginia have not been surveyed for nematodes; however, root knot and other genera are commonly found whenever they have been assayed. In a small survey of ten vineyards throughout the commonwealth, completed in 2015, root knot nematodes were found in 50% of the fields and one had levels of second-stage juveniles that were high enough to expect significant economic loss (Noah Adamo, pers. comm.). Therefore, another larger survey is currently underway to more closely evaluate the presence of nematodes including root knot nematodes.

Growers are encouraged to take nematode assays before establishing new vineyards. If potentially harmful nematodes are found, the option of growing a cover crop may help reduce the nematode population, provided that a beneficial plant species is selected. Improving vine vigor by proper fertilization, irrigation during dry periods, addition of manures and soil amendments, prevention of soil compaction and other practices that reduce stress on the vines may reduce the impact of root knot and other nematodes (Verdegaal 2015). Periods of fallow will reduce the population of root knot nematode, provided that weeds are eliminated, since they may be hosts and allow the nematode population to be maintained or increased (Thomas et al. 2005). If root knot is a major concern in a vineyard, resistant rootstocks may play an important role in reducing the damage. Nematode resistant rootstocks include Freedom, Harmony, Ramsey and Teleki; however, they are not effective against all kinds of nematodes. Therefore, nematode assays are necessary to determine which root knot species is present to select the suitable resistant rootstock (Verdegaal 2015).

11.3.1.11 Vegetables

Root knot nematodes attack many plants utilized as vegetables. All four most common species can cause problems on a host of vegetables including asparagus, beets, broccoli, cabbage, cucumbers, eggplants, green beans, onions, peppers, snap beans, spinach, squash, sweet potatoes and others. Nematode assays of fields used for vegetable production are encouraged so that lands infested with root knot nematodes can be avoided. If nematodes are present, cover crops and fallow may be useful to reduce the population to non-damaging levels. Crop rotations may also be useful to retard the development of high levels of root knot nematodes. Likewise, good weed management during periods of fallow may also be helpful (Thomas et al. 2005).

11.3.1.12 Small Fruits

Cantaloupes, strawberries and watermelons are good hosts of root knot nematodes. Cantaloupes and watermelon are particularly sensitive to *M. incognita*, but not *M. hapla*. Likewise, strawberries are severely attacked by *M. hapla* but are resistant to *M. incognita*. Assays are useful in managing these pests, especially if the species of root knot nematode is identified. Cover crops, fallow and crop rotations are also helpful in reducing the population of root knot nematodes.

11.3.1.13 Sod

Turfgrass, especially greens used for golf, is commonly infected with root knot nematodes. Although the exact effect of these nematodes on turf is poorly understood, they probably cause significant injury if the population is high. In Virginia, the most common root knot nematode species on turfgrass is an undescribed species (Eisenback 2010a). No sustainable tactics for managing this species have been utilized.

11.3.1.14 Nursery Plants and Flowers

Root knot nematodes of all four most common species can be pathogens on many nursery plants and flowers. Since most of these plants are grown under controlled conditions in nurseries, high tunnels and greenhouses, root knot develops into a problem only when proper sanitation and other cultural practices have been ignored or improperly executed. In cases where root knot develops, it is necessary to practice good technique to prevent it from happening again. To prevent infection by root knot nematodes, growers may select certified planting material, use soilless media in greenhouses, wash all equipment with water after use or before moving to another location, allow excess irrigation water to settle into a holding pond and pump irrigation water from the pond near the surface and monitor plant roots for signs of root knot nematodes.

11.3.2 Soybean Cyst Nematode, Heterodera glycines

Soybean cyst nematode is a major pest of soybean in Virginia and occurs wherever the plant is grown in the commonwealth. Several host races have been found in Virginia and host race 4 is predominant. This is unfortunate because sources of resistance that are effective for managing this race, Hartwig, do not have the yield potential equal to that of other varieties (Koenning 2000). Crop rotation with corn, cotton, tobacco and peanut are useful options in Virginia, especially when combined with the use of resistant varieties. No-till and double cropping with winter wheat suppressed the cyst nematodes after 3 years' practice and increased yield by 5% as long as weeds was properly managed. Early planting causes more injury to soybean since cyst juveniles are highest and most active in early spring. Late planting reduces the amount of injury. Therefore, growers can avoid some injury in fields with the highest cyst populations by planting them last (Koenning 2000). Soybean cyst resistant soybean cultivars are considered to be useful for managing the soybean cyst nematodes. Where resistant variety are not suitable, tolerant ones may be suitable replacements. Abamectin (Activa Complete Beans), Fluopyram (ILeVO), *Pasteuria* spp. (Clariva) and *Bacillus firmus* (Poncho/VOTIVO) are seed treatments that are commonly used in Virginia to reduce the effect of soybean cyst nematodes on soybean yields (Mehl 2017).

11.3.3 Tobacco Cyst Nematode, **Globodera tabacum solanacearum**

Tobacco cyst nematode (TCN) occurs wherever flue-cured tobacco is produced in the Commonwealth of Virginia (Johnson 2017). Varieties of flue-cured tobacco that possess the *Php* gene reduce TCN populations to non-damaging levels after several years of use (Johnson 2017); however, without this resistance, populations rapidly increase to damaging levels if fields are planted continuously in susceptible tobacco. Crop rotation with a non-host plant such as corn, sorghum, barley, soybean, peanut, wheat or pasture grasses is very useful for managing this nematode, especially when resistant flue-cured tobacco varieties are utilized (Johnson 2017). Tomato, sweet pepper and eggplant are moderate hosts of TCN and are not useful in crop rotations to manage TCN.

11.3.4 Lesion Nematodes, **Pratylenchus** *spp.*

Lesion nematodes parasitize and may cause significant economic injury to alfalfa, apple, blueberries, corn, cotton, boxwood, vegetables, grape, peach, peanut, soybean, strawberries, sweet corn, tobacco and turfgrass. For tobacco, early root and stalk destruction and crop rotation can significantly reduce populations of these nematodes (Johnson 2017). Rotation crops that are useful to reduce root knot nematode or TCN in tobacco may not be effective for lesion nematodes.

11.3.5 Stem and Bulb Nematode, Ditylenchus dipsaci

Alfalfa and garlic, as well as other bulb crops, can be affected by stem and bulb nematode. Avoiding this nematode through certified nematode-free bulbs has reduced the occurrence of this nematode in Virginia to a rare event. Alfalfa, on the other hand is often infected with low levels of stem bulb nematode. Most growers are aware of this nematode and use resistant varieties. However, because of the complex genetics of alfalfa, even plants that are rated highly resistant may allow reproduction in 50% of the individual plants (Samac et al. 2015). Certified nematode-free seed prevents introduction of stem and bulb nematode to uninfested areas. Plant tissues can easily spread this nematode, including infested hay that is placed in uninfested areas. Crop rotation of 2–3 years with corn, sorghum, small grains and beans will significantly reduce the population. Also, fall burning of alfalfa stubble will reduce the stem and bulb nematode population as well as reducing the number of weeds (Samac et al. 2015). Adding soil amendments may also provide reduction in nematode numbers. Resistant varieties are available; however, because of the complex nature of the genetics of alfalfa, a portion (50%) of the plants remain susceptible (Graham et al. 1979; Samac et al. 2015).

11.3.6 Other Nematodes

Lance nematode, *Hoplolaimus galeatus*. Lance nematodes are important pathogens to alfalfa, apple, blueberries, corn, cotton, grapes, peaches, soybean, sweet corn and turfgrass. Cultural practices that reduce the populations of these nematodes to below-damaging levels include crop rotation, fallow and bio-fumigation.

Ring nematodes, *Mesocriconema* spp. Ring nematodes damage blueberries, corn, English boxwood, grapes, peaches, peanuts, soybeans and turfgrass. Crop rotation, fallow and bio-fumigants are useful tactics for managing these pests. Seed treatments may also be useful on soybean.

Sting nematode, *Belonolaimus longicaudatus*. Although sting nematodes are restricted to soils containing more than 95% sand, they can be extremely damaging to many different plant species growing in these soils. They are especially pathogenic on beans, corn, cotton, crucifers, cucurbits, vegetables, peanuts, soybeans, sweet corn and turfgrass. Managing sting nematodes with cultural means is difficult because their host range is very broad. Seed treatments for corn, soybean and cotton are available such as COPeO® Prime from Bayer and Nemastrike® Acceleron from Monsanto.

***Tylenchulus palustris*.** This nematode was found parasitizing peach in Virginia, but little is known about its effect on peach tree health (Eisenback et al. 2007).

Boxwood spiral nematode, *Rotylenchus buxophilus*. Spiral nematode causes a slow decline to English boxwood that may continue for many years before the plant must be replaced. Minimizing other stresses is useful for extending the life of boxwood that is declining from the spiral boxwood nematode. These practices include controlling insect pests, application of appropriate nutrients, deep watering during periods of prolonged drought, use of an anti-transpirants during the windy months of late fall and winter and placement of 2–5 cm of mulch to retain moisture and to encourage pathogenic fungi and other organisms.

Dagger nematodes, *Xiphinema americanum* and *X. rivesi*. Dagger nematode causes injury to apple, blueberries, brambles, grapes, peaches, peanuts, soybeans and strawberries. The dagger nematode is especially important on apple, peach, grape and blueberry because it transmits plant viruses. These viruses viruses which in turn cause stem pitting on peach, bud-union necrosis on apple and ringspot on grapes and blueberries. Cultural practices to manage dagger nematode include crop rotation, fallow and bio-fumigation.

Stubby root nematodes, *Paratichodorus* and *Trichodorus* spp. Stubby root nematodes affect beans, blueberries, corn, cotton, crucifers, vegetables, peanuts, soybeans, sweet corn and turfgrass. Cultural practices include crop rotation and fallow.

Spiral nematode, *Helicotylenchus dihystera*. Peanuts and soybean may be damaged by high populations of spiral nematode. Crop rotation and fallow are cultural practices to manage these nematodes.

Stunt nematodes, *Tylenchorhynchus* spp. Stunt nematodes may suppress the yields of soybean and sweet potato.

Reniform nematode, *Rotylenchulus reniformis*. Reniform has been found attacking cotton and soybean in Virginia, but the cold temperatures during winter months prevent them from developing high populations that cause crop losses (Eisenback et al. 2004).

Foliar nematodes, *Aphelenchoides* spp. Foliar nematodes on floral and ornamental crops can be important on many different hosts in Virginia. Most instances where these nematodes are a problem are from overhead irrigation that is collected as runoff and recycled. Good sanitation practices and destruction of infested material are used to prevent these nematodes from causing disease.

Sheath nematodes, *Hemicycliphora* spp. Blueberries and woody ornamentals may be damaged by the sheath nematode.

Awl nematode, *Dolichodorus heterocephalus*. Beans and crucifers are subject to damage caused by the awl nematode. Because this nematode is limited to very wet areas, they are rarely found damaging crops.

Corn cyst nematode, *Heterodera zeae*. Corn cyst nematode is found only on one farm in one county in Virginia (Eisenback et al. 1993). Apparently, soil temperatures keep the reproductive level of this nematode very low so that damaging populations do not develop in the Commonwealth of Virginia.

Bentgrass seed gall nematode, *Anguina agrostis*. Bentgrass seed gall nematode occurs on bentgrass in Virginia, but has it little effect on plant growth and management is not necessary (Eisenback and Roane 2006). The usual practice of mowing prevents the formation of seed heads and eliminates the nematodes in a short period of time.

11.4 Agriculture in West Virginia

Although agriculture in West Virginia is not one of the most important sectors in the economy, it contributed more than 1 billion dollars in 2014 (Table 11.1) (USDA Census of Agriculture 2016). Because of the mountainous terrain, most of the income from agriculture was generated from poultry and livestock sales. Only 1 out of 7 dollars was derived from crops, much of which was used for animal feed. However, on the bright side, the state ranked 11th for apples and 17th for peaches and American ginseng collected from wild areas and planted under shade cloth, generated more than $2 million dollars (USDA Census of Agriculture. 2016).

West Virginia, the mountain state, has a very rugged terrain with an average elevation of 460 m above sea level (Britanica.com 2017). The rocky, acidic soils limit agricultural production in the state; however, wherever plants are placed in the soil they are subject to damage caused by plant parasitic nematodes (Table 11.2).

11.5 Plant Parasitic Nematodes of West Virginia

11.5.1 Root Knot Nematodes, Meloidogyne *spp.*

11.5.1.1 Alfalfa

Root knot (*M. incognita*) nematode resistant varieties are available include and cultivar 'Achieva' and moderately resistant 'Vernal,' 'WL 225,' 'WL 317,' 'WL 320', 'Royalty', 'Allstar' and 'Chief' (Anonymous).

11.5.1.2 Peach

Root knot nematodes attacking peach in West Virginia is most commonly the northern root knot nematode, *Meloidogyne hapla* (Kotcon 1990). Most resistance root stocks are evaluated for *M. incognita*, *M. arenaria* and *M. javanica*, but not *M. hapla*. Anecdotal evidence suggests that the root stock Siberian C is more susceptible to *M. hapla* than are Halford and Lovell (Kotcon 1990).

11.5.1.3 Tomato

The most common nematode found in 20–30% of tomato fields in West Virginia is the northern root knot nematode, *Meloidogyne hapla*, although *M. incognita* may be found as well (Baniecki and Dabaan 2002). Root knot nematode is more common in sandy soils. Cover crops including rye and tillage radish and crop rotation with non-host vegetable crops is the most effective and utilized management tool; however, the lack of suitable land and marketing requirements make these options less than ideal. The use of nematode-free transplants, the addition of organic matter into the soil and practicing suitable weed management also help reduce the root knot populations. Bio-fumigation with brassica species including wild mustard, canola, tillage radish etc., may be effective as well. Finally, cleaning farm equipment with water to remove infested soil is useful for minimizing the spread of this nematode to non-infested fields (Baniecki and Dabaan 2002). Nematode resistant tomatoes are susceptible to *M. hapla*, the most common species in West Virginia. Also, the resistant varieties tend to have lower yields and less quality and are not competitive with other more suitable susceptible varieties (Baniecki and Dabaan 2002).

11.5.1.4 Potato

Crop rotation with corn or small grains reduces southern root knot nematode (*M. incognita*) populations (Baniecki and Dabaan 2003a).

11.5.1.5 Ginseng

American ginseng is extremely valuable in the Asian markets. Most ginseng that is collected in West Virginia is wild; however, a few growers plant it as a crop. Although northern root knot (*M. hapla*) can retard the growth and kill small seedlings, a slight infestation in this crop may actually increase the value because these nematodes cause the root to produce additional secondary roots which cause it to resemble the human body (Harrison et al. 2017). Since wild ginseng is collected in the forests, management of the root knot nematodes is not likely. In cultivated ginseng, however, sight selection and sampling for nematodes before planting is important to minimize the damage that could be caused by root knot nematodes.

11.5.2 Lesion Nematodes, **Pratylenchus** *spp.*

11.5.2.1 Apple

Lesion nematodes (*P. penetrans*) are very common in apple orchards, but their populations are frequently insufficient to cause significant economic loss. However, it is important to start the management of new orchards before they are planted by assaying for the occurrence and population levels of these potentially important pests. In West Virginia, lesion nematodes are known to be present in 50–70% or the orchards, but only 10–20% have population levels that are economically important (Baniecki and Dabaan 2003b). Lesion nematodes are often associated with replant problems. It is best to remove old stumps and large roots and leave the site in fallow, followed by a green manure cover crop of bio-fumigant plant like wild mustard or rapeseed.

11.5.2.2 Peach

Replanting of peach orchards in West Virginia is troubled by lesion nematodes. Several species were identified in a survey (*P. crenatus, P. neglectus, P. penetrans, P. pratensis, P. scribneri, P. thornei* and *P. vulnus* (Kotcon 1990). When peach seedlings are transplanted into soil with moderately high populations of lesion nematodes, they are stunted and never achieve their yield potential (Kotcon 1990). In old orchards that are to be replanted, the old stumps and roots are removed so that the site can be fallowed and then planted with green manure cover crops or bio-fumigant producing plants like wild mustard and rapeseed (Hogmire and Biggs 2005).

11.5.3 Dagger Nematodes, **Xiphinema americanum** *and* **X. rivesi**

Dagger nematode affects the growth of apples, peaches and grapes, but is more important because it transmits plant viruses. In apples, these viruses cause bud-union necrosis, in peaches they cause stem pitting and in grapes they transmit ring viruses which reduce the quality and amount of grape production. Cultural practices include crop rotation, fallow and bio-fumigation.

11.5.4 Stem and Bulb Nematode, **Ditylenchus dipsaci**

Stem and bulb nematodes are commonly found on alfalfa; however, most growers plant resistant cultivars that greatly minimize their impact. The best practice is to use only nematode-free seed and rotate corn, sorghum, small grains and beans for 2 or 3 years. Burning in the fall to control weeds will decrease stem and bulb nematodes in the following spring (Baniecki and Dabaan 1999). Resistant varieties are available; however, because of the complex nature of the genetics of alfalfa, a portion of the plants remain susceptible (Graham et al. 1979).

11.5.5 Ring Nematodes, **Mesocriconema** *spp.*

Ring nematodes may reach damaging levels on apple, potato, grape, peaches and turfgrass. Cultural practices include crop rotation, fallow and bio-fumigation.

11.6 Conclusion and Future Perspective

Agriculture in Virginia and West Virginia has adapted very well to the demands placed on it by growers, government and society; and, in all likelihood, it will continue to do so in the future. However, both states continue to lose vast areas of farmland to urban sprawl. As less available fertile agricultural land remains for farming, more pressure is placed on utilizing sustainable practices; therefore, crop rotation becomes more difficult because land is a limiting factor. Other concerns managing nematodes in a sustainable way have been expressed. For example, in Virginia, the use of resistance in tobacco against *M. incognita* race 1 has selected for more aggressive races (2 and 4) and other species (*M. arenaria, M. javanica* and *M. hapla*). In addition, *M. enterolobii* has been found in adjacent North Carolina and threatens Virginia with its more pathogenic to cotton, tobacco and soybean. Additional concerns occur in soybean, because all field populations of soybean cyst nematodes, tested in 2016, were shown to be race 4. This race does not have any known sources of resistance in commercially available cultivars.

Although nematode assays are freely available for growers in the commonwealth, few take advantage of this service. Last year more than 2,000 assays were processed by the Virginia Tech Nematode Assay Lab, but in adjacent North Carolina more than 40,000 assays were taken. Whereas in neighboring Maryland, Delaware and Pennsylvania, less than 100 fields in all three states were assayed because their states do not have the capability of running nematode program and they must be done out-of-state. A key component for the application of sustainable tactics for the management of plant parasitic nematodes is a good nematode assay.

References

Adams, R. E. (1955). Evidence of injury to fruit trees by an ectoparasitic nematodes (*Xiphinema* sp.) and a promising control measure. *Phytopathology, 45*, 477–479.

Adams, R. E., & Eichenmuller, J. J., Jr. (1962). *Gracilacis capitatus* n. sp. from scarlet oak in West Virginia. *Nematologica, 8*, 87–92.

Adams, R. E., & Lapp, N. A. (1967). *Criconemoides grassator* n. sp. from Yellow Poplar (*Liriodendron tulipifera*) in West Virginia. *Nematologica, 13*, 63–66.

Adams, B. J., Dillman, A. R., & C. Finlinson C. (2009). Molecular taxonomy and phylogeny. In R. N. Perry, M. Moens, & J. Starr (Eds.), *Root knot nematodes*. Wallingford: CABI Publishing.

Alexander, W., Allen, J., Bailey, J., Brandenburg, R., Dunn, J. M., Ellison, C., Flippen, B. L., Herbert, A., Jorda, D., Madre, D., Malone, S., Mizell, R., Phipps, P., Poarch, W., Rogers, B., Rogers, T., Royals, B., Schooley, T., Schools, R., Smith, L., Swann, C., Toth, S., Weaver, M., Wells, K. Williams, B.. (2002). *Pest management strategic plan for North Carolina/Virginia peanuts*. Summary of a workshop held on April 4, 2002 at the Tidewater Agricultural Research and Extension Center in Suffolk, Virginia sponsored by the: North Carolina Pest Management Information Program (North Carolina State University) Virginia Pest Management Information Program (Virginia Polytechnic Institute and State University) Southern Region Pest Management Center (University of Florida) Office of Pest Management Policy, U. S. Department of Agriculture. https://ipmdata.ipmcenters.org/documents/pmsps/NCVApeanutpmsp.pdf

Anonymous. (1959). Research report of the Virginia agricultural experiment station for the period July 1, 1957 to June 30, 1959.

Anonymous. *West Virginia University scout manual*. Integrated Pest Management Scouting Report, ALFALFA. West Virginia University, Extension Service.

Baalawy, H. A., & Fox, J. A. (1971). Resistance to Osborne's cyst nematode in selected *Nicotiana* species. *Journal of Nematology, 3*, 395–398.

Baldwin, J. G., & Barker, K. R. (1970). Host suitability of selected hybrids, varieties and inbreds of corn to populations of *Meloidogyne* spp. *Journal of Nematology, 2*, 345–350.

Baniecki, J. F., & Dabaan, M. E. (1999). *Crop profile for alfalfa in West Virginia*. https://ipmdata.ipmcenters.org/documents/cropprofiles/WValfalfa.pdf

Baniecki, J. F., & Dabaan, M. E. (2002). *Crop perofile for tomatoes in West Virginia*. https://ipmdata.ipmcenters.org/documents/cropprofiles/WVtomatoes.pdf

Baniecki, J. F., & Dabaan, M. E. (2003a). *Crop profile for potatoes in West Virginia*. http://www.ipmcenters.org/cropprofiles/docs/wvpotatoes.pdf

Baniecki, J. F., & Dabaan, M. E. (2003b). *Mid-Atlantic apple pest management strategic plan – 2003*. https://ipmdata.ipmcenters.org/documents/pmsps/MID_ATLANTIC_ApplePMSP.pdf

Barnet, H. L. (1985). Plant pathology research in the West Virginia agricultural and forestry experiment station 1892–1982.

Bernard, E. C., & Eisenback, J. D. (1997). *Meloidogyne trifoliophila* n.sp. (Nemata: Meloidogynidae), a parasite of clover from Tennessee. *Journal of Nematology, 29*, 43–54.

Biggs, A. R., Kotcon, J., Baugher, T. A., & Lightner, G. W. (1994). Comparison of corn and fescue rotations on pathogenic – nematodes, nematode biocontrol agents, and soil structure and fertility on an apple replant site. *Journal of Sustainable Agriculture, 4*, 39–56.

Bull, C., Greene, C., Kotcon, J., and Oberholtzer, L. (2018). *Organic agriculture: Innovations in organic marketing, technology, and research – introduction to the proceedings*.

Charchar, J. M., & Eisenback, J. D. (2002). *Meloidogyne brasilensis* n. sp. (Nematoda: Meloidogynidae), a root-knot nematode parasitizing 'Rossol' tomato in Brazil. *Nematology, 4*, 629–643.

Charchar, J. M., Eisenback, J. D., & Hirschmann, H. (1999). Description of *Meloidogyne petuniae* n. sp., a root-knot nematode parasitic on Petunia (*Petunia hybrida*) in Brazil. *Journal of Nematology, 31*, 352–362.

Charchar, J. M., Eisenback, J. D., Charchar, M. J., & Boiteux, M. E. (2008a). *Meloidogyne pisi* n. sp. (Nematoda: Meloidogynidae), a root-knot nematode parasitizing pea in Brazil. *Nematology, 10*, 479–493.

Charchar, J. M., Eisenback, J. D., Charchar, M. J., & Boiteux, M. E. (2008b). *Meloidogyne phaseoli* n. sp. (Nematoda: Meloidogynidae), a root-knot nematode parasitizing bean in Brazil. *Nematology, 10*, 525–538.

Charchar, J. M., Eisenback, J. D., Vieira, J. V., Foncesca-Boiteux, M. E., & Boiteux, L. S. (2009). *Meloidogyne polycephannulata* n. sp. (Nematoda: Meloidogynidae), a root-knot nematode parasitizing carrot in Brazil. *Journal of Nematology, 41*, 174–186.

Cook, C. G., & Robinson, A. F. (2005). Registration of RN96425, RN96527, and RN96625-1 nematode-resistant cotton germplasm lines. *Crop Science, 45*, 1667.

Dhandaydham, M., Charles, L., Zhu, H., Starr, J. L., Huguet, T., Cook, D. R., Prosperi, J.-M., & Opperman, C. S. (2008). Characterization of root-knot nematode resistance in *Medicago truncatula*. *Journal of Nematology, 40*, 46–54.

Duiker, S. W., Myers, J. C., Blazure, L. C. (2016). *Better soils with the no-till system*. USDA natural resources conservation service, Penn State University Extension, Capital resource conservation and development, and Clinton county conservation district: Molly McDonough, USDA NRCS. https://www.no-tillfarmer.com/articles/5824-better-soils-with-the-no-till-system

Eisenback, J. D. (1982). Description of the blueberry root-knot nematode, *Meloidogyne carolinensis* n. sp. *Journal of Nematology, 14*, 303–317.

Eisenback, J. D. (1988). Multiple focus and exposure photomicroscopy of nematodes for increased depth of field. *Journal of Nematology, 20*, 333–334.

Eisenback, J. D. (2010a). *A new species of root knot nematode parasitizing bentgrass in Virginia, Maryland, and Pennsylvania*. Society of Nematologists Annual Meeting, Boise.

Eisenback, J. D. (2010b). A new technique for photographing perineal patterns of root-knot nematodes. *Journal of Nematology, 42*, 33–34.

Eisenback, J. D. (2012). A technique for making high-resolution megapixel mosaic photomicrographs of nematodes. *Journal of Nematology, 44*, 260–263.

Eisenback, J. D., & Gnanapragasam, N. C. (1992). Additional notes on the morphology of *Meloidogyne brevicauda*. *Fundamental and Applied Nematology, 15*, 347–353.

Eisenback, J. D., & Hartman, K. (1985). *Sphaeronema sasseri* n. sp., a nematode parasitic on Fraser fir and red spruce. *Journal of Nematology, 17*, 346–354.

Eisenback, J. D., & Hirschmann, H. (2001). Additional notes on the morphology of *Meloidogyne spartinae*. *Nematology, 3*, 303–312.

Eisenback, J. D., & Rammah, A. (1987). Evaluation of the utility of a stylet extraction technique for understanding morphological diversity of several genera of plant parasitic nematodes. *Journal of Nematology, 19*, 116–122.

Eisenback, J. D., & Roane, C. W. (2006). First report of bentgrass seed gall nematode, *Anguina agrostis*, in Virginia and Minnesota. *Plant Disease, 90*, 1110.

Eisenback, J. D., & Triantaphyllou, H. H. (1991). Root-knot nematodes: *Meloidogyne* species and races. In W. R. Nickle (Ed.), *Manual of agricultural nematology* (pp. 191–274). New York: Marcel Dekker.

Eisenback, J. D., Yang, B., & Hartman, K. M. (1985). Description of *Meloidogyne pini* n. sp., a root-knot nematode parasitic on sand pine (*Pinus clausa*), with additional notes on the morphology of the *M. megatyla*. *Journal of Nematology, 7*, 206–219.

Eisenback, J. D., Stromberg, E. L., & McCoy, M. S. (1986). First report of the Columbia root-knot nematode (*Meloidogyne chitwoodi*) in Virginia. *Plant Disease, 70*, 801.

Eisenback, J. D., Reaver, D. M., & Stromberg, E. L. (1993). First report of corn cyst nematode (*Heterodera zeae*) in Virginia. *Plant Disease, 77*, 647.

Eisenback, J. D., Bernard, E. C., & Schmidt, D. P. (1994). Description of the Kona root-knot nematode, *Meloidogyne konaensis* n. sp. *Journal of Nematology, 26*, 363–374.

Eisenback, J. D., Bernard, E. C., Starr, J. L., Lee, T. A., Jr., & Tomazewski, E. K. (2003). *Meloidogyne haplanaria* n. sp. (Nematoda: Meloidogynidae), a root knot nematode parasitizing peanut in Texas. *Journal of Nematology, 35*, 395–403.

Eisenback, J. D., Hopkins, N., & Phipps, P. M. (2004). First report of the reniform nematode *Rotylenchulus reniformis* on cotton in Virginia. *Plant Disease, 88*, 683.

Eisenback, J. D., Reaver, D., & Ashley, J. E., Jr. (2007). First report of the nematode, *Tylenchulus palustris*, parasitizing peach in Virginia. *Plant Disease, 91*, 1683.

Eisenback, J. D., Graney, L. S., & Vieira, P. (2017). First report of the apple root-knot nematode, *Meloidogyne mali*, in North America found parasitizing *Euonymus* in New York. *Plant Disease, 101*, 510.

Elliott A. P., Phipps, P. M., Komm, D. A., Babinea, D., Yoder, K., Reilly, J., Harris, C., Meredith, S., Ravlin, R., Walker, N. (1982). Predictive nematode assay program at Virginia Tech. 1982. Ext. Publ. 450–70. 28 pp.

Elliott, A. P., Phipps, P. M., & Terrill, R. (1986). Effects of continuous cropping of resistant and susceptible cultivars on reproduction potentials of *Heterodera glycines* and *Globodera tabacum solanacearum*. *Journal of Nematology, 18*, 375–379.

Encyclopedia Britiannica. https://www.britannica.com/place/West-Virginia

Faske, T. R., & Starr, J. L. (2007). Cotton root protection from plant parasitic nematodes by Abamectin-treated seeds. *Journal of Nematology, 39*, 27–30.

Fenne, S. B. (1940). Some observations on the development of root knot nematode diseases in Virginia. *Phytopathology, 30*, 708.

Fenne, S. B. (1952). Unpublished annual report, Department of Plant Pathology. Virginia Polytechnic Institute. Blacksburg.

Fenne, S. B., Henderson, R. G., Smith, T. J., & White, W. C. (1950). Alfalfa-clover disease survey in Virginia. *Plant Disease Reporter, 34*, 204–205.

Fromme, F. D. (1919). *The nematode disease of wheat in Virginia*. Virginia Agricultural Experiment Station Bulletin 222.

Gardenweb. (2017). http://forums.gardenweb.com/discussions/2101097/list-of-some-tomato-varieties-resistant-to-nematodes

Gatton, H., Nessler, S., Kuhar, T., Jennings, K., King, S., Monks, D., Rideout, S., Troth, S., Waldenmeire, C., Weaver, M., Wison, H. (2007). *Pest management strategic plan for tomato in Virginia, North Carolina and Delaware*. Southern Region IPM Center, Virginia Tech, North Carolina State University, and University of Delaware. (https://docs.google.com/viewer?url=http%3A%2F%2Fwww.ipmcenters.org%2Fpmsp%2Fpdf%2FSRTomato.pdf).

Gillespie, W. H., & Adams, R. E. (1962). An awl nematode, *Dolichodorus silvestris* n. sp. from West Virginia. *Nematologica, 8*, 93–98.

Golden, A. M. (1979). Descriptions of *Meloidogyne camelliae* n. sp. and *M. querciana* n. sp. (Nematoda: Meloidogynidae), with SEM and host-range observations. *Journal of Nematology, 11*, 175–189.

Graham, J. H., Frosheiser, F. I., Stuteville, D. L., & Erwin, D. C. (1979). *A compendium of alfalfa diseases*. St. Paul: The American Phytopathological Society 65pp.

Grant, C. E., Reilly, J. J., & Elliot, A. P. (1982). Reproduction of *Globodera solanacearum* and its effect on growth and development of three flue-cured tobacco cultivars under greenhouse conditions. *Journal of Nematology, 14*, 443.

Harrison, H. C., Parke, J. L., Oelke, E. A., Kaminski, A. R., Hudelson, B. D., Martin, L. J., Kelling, K. A., & Binning, L. K. (2017). *Ginseng. Alternative field crops manual*. Madison: Univ. of Wisconsin Extension, Cooperative Extension University of Minnesota: Center for Alternative Plant and Animal Products and the Minnesota Extension Service https://hort.purdue.edu/newcrop/afcm/ginseng.html.

Henderson, R. G. (1950). Stem nematode, the cause of a new alfalfa disease in Virginia. *Virginia Journal of Science, 1*, 332.

Hirschmann, H. (1982). *Meloidogyne platani* n. sp. (Meloidgynidae), a root-knot nematode parasitizing American sycamore. *Journal of Nematology, 14*, 84–95.

Hogmire, H. W., & Biggs, A. R. (2005). *Crop profile for peaches in West Virginia*. Pest Management Alternatives Program-USDA, Cooperative State, Research, Education and Extension Service. https://ipmdata.ipmcenters.org/documents/cropprofiles/WVpeaches.pdf

Jenkins, W. A. (1944). Root rot disease-complexes of tobacco with reference to the meadow nematode: A preliminary report. *Plant Disease Reporter, 28*, 395–397.

Jenkins, W. A. (1948). A root disease complex of small grains in Virginia. *Phytopathology, 38*, 519–527.

Johnson, C. S. (1998). Tobacco. In K. R. Barker, G. A. Pederson, & G. L. Windham (Eds.), *Plant and nematode interactions, agronomy monograph 36*. Madison: American Society of Agronomy, Crop Science Society of America, and Soil Science Society of America.

Johnson, C. S. (2017). Diseases and nematode management in field crops: Tobacco. In *Pest management guide field crops*. Blacksburg: Virginia Cooperative Extension; Communications and Marketing, Virginia Tech.

Johnson, C. S., Komm, D. A., & Jones, J. L. (1989). Control of *Globodera tabacum solanacearum* by alternating host resistance and nematicide. *Journal of Nematology, 21*, 16–23.

Johnson, C. S., Leslie, R. G., & Watson, J. W. (1994). Fumigation vs. contact nematicides for control of tobacco cyst nematode in Virginia. *Fungicide and Nematicide Tests, 49*, 185.

Johnson, C. S., Way, J., & Barker, K. R. (2005). Nematode parasites of tobacco. In M. Luc, R. A. Sikora, & J. Bridge (Eds.), *Plant parasitic nematodes in tropical agriculture* (pp. 675–708). Wallingford: CAB International.

King, T. N., & Kotcon, J. (2010). Reproduction of *Pratylenchus penetrans* in soils with and without *Pasteuria* spp. *Journal of Nematology, 42*, 250.

Kish, A. J. (1973). *A biotype of Meloidogyne incognita that attacks root-knot resistant tomatoes*. M.S. thesis, West Virginia University.

Koenning, S. (2000). *Management of soybean cyst nematode, Plant Pathology Information Note Soybean No. 6*. Raleigh: North Carolina State University https://www.ces.ncsu.edu/depts/pp/notes/Soybean/soy001/soy001.htm.

Komm, D. A., Reilly, J. J., & Elliott, A. P. (1983). Epidemiology of a tobacco cyst nematode (*Globodera solanacearum*) in Virginia. *Plant Disease, 67*, 1249–1251.

Kotcon, J. B. (1984). Dynamics of root growth in potato fields affected by the early dying syndrome. *Phytopathology, 74*, 462–467.

Kotcon, J. B. (1985). Interactions of *Verticillium dahliae, Colletotrichum coccodes, Rhizoctonia solani*, and *Pratylenchus penetrans* in the early dying syndrome of Russet Burbank potatoes. *Phytopathology, 75*, 68–74.

Kotcon, J. B. (1987). Fall fumigation of potato with 1,3-dichloropropene: Efficacy against *Pratylenchus crenatus*, yield response, and groundwater contamination potential. *Plant Disease, 71*, 1122–1124.

Kotcon, J. B. (1990). Distribution, frequency, and population density of nematodes in West Virginia peach orchards. *Journal of Nematology, 22*, 712–717.

Kotcon, J. B. (2018). *Intercropping with resistant varieties for management of plant diseases in organic tomato production*. Organic farming research foundation project report. http://ofrf.org/research/grants/outcome-intercropping-resistant-varieties-management-plant-diseases-organic-tomato

Kotcon, J. B., & Loria, R. (1987). Fall fumigation of potato with 1,3-dichloropropene: Efficacy against *Pratylenchus crenatus*, yield response, and groundwater contamination potential. *Plant Disease, 71*, 1122–1124.

Kotcon, J. B., & Rouse, D. I. (1984). Root deterioration in the potato early dying syndrome: Causes and effects of root biomass reductions associated with colonization by *Verticillium dahliae*. *American Potato Journal, 61*, 557–568.

Kotcon, J. B., Bird, G. W., Rose, L. M., & Dimoff, K. (1985). Influence of *Glomus fasciculatum* and *Meloidogyne hapla* on *Allium cepa* in organic soils. *Journal of Nematology, 17*, 55–60.

Kuepper, G., & Thomas, R. (2008). *Organic tobacco production*. A publication of ATTRA, National Sustainable Agriculture Information Service. https://attra.ncat.org/attra-pub/viewhtml.php?id=94

Lapp, N. A. (1967). Some aspects of the biology, morphology, and population dynamics of *Dolichodorus silvestris*. M.S. thesis, West Virginia University.

Leukel, R. W. (1931). Observations on the nematode disease (*Tylenchus tritici*) of wheat in the vicinity of Luray, Virginia. *Plant Disease Reporter, 15*, 129–130.

Libman, G. J., Leach, G., & Adams, R. E. (1964). Role of certain plant parasitic nematodes in infection of tomatoes by *Pseudomonas solanacearum*. *Phytopathology, 54*, 151–153.

Mehl, H. L., P. M. Phipps, and J. D. Eisenback. (2013). Susceptibity of cotton varieties to root-knot nematode, 2013. Plant Disease Management Reports 7:N003.

Mehl, H. (2017). Disease and nematode management in field crops. In *Pest management guide field crops 2017* (pp. 3–1 – 3–58). Blacksburg: Virginia Cooperative Extension; Communications and Marketing, Virginia Tech.

Miller, L. I. (1952). Control of the sting nematode on peanuts in Virginia. *Phytopathology, 42*, 470.

Miller, L. I. (1983). Diversity of selected taxa of *Globodera* and *Heterodera* and their interspecific and intergeneric hybrids. In A. R. Stone, H. M. Platt, L. F. Hhalil, (Eds), *Concepts in nematode systematics*. Proceeding of International Symposium, (pp. 207–220), Cambridge University, 2–4 Sept. 1981. Systematics Association Special Volume, 22.

Miller, L. I., & Gray, B. J. (1968). Horsenettle cyst nematode, *Heterodera virginiae* n. sp., a parasite of solanaceous plants. *Nematologica, 14*, 535–543.

Miller, L. I., & Gray, B. J. (1972). *Heterodera solanaciarum* n. sp., a parasite of solanaceaus plants. *Nematologica, 18*, 404–413.

Nelson, K. A., & Kotcon, J. (2005). In vitro culturing of the predatory soil nematode *Clarkus papillatus*. *Nematology, 7*, 5–9.

Nesiius, E. J. (1988). *The first 100 years a history of the West Virginia agricultural and forestry experiment station*. Parsons: McClain Printing.

Osborne, W. W. (1961). Tobacco attacked by a cyst-forming nematode in Virginia. *Plant Disease Reporter, 45*, 812–813.

Panaccione, D. G., Kotcon, J., Schardl, C. L., Mortonn, J. B., Edenborn, S., & Sexstone, A. J. (2006). Ergot alkaloids are not essential for endophytic fungus-associated population suppression of the lesion nematode, *Pratylenchus scribneri*, on perennial ryegrass. *Nematology, 8*, 583–590.

Parkunan, V., Johnson, C. S., & Eisenback, J. D. (2010). Effects of Ph_p gene-associated versus induced resistance to tobacco cyst nematode in flue-cured tobacco. *Journal of Nematology, 42*, 261–266.

Parris, G. K. (1943). Reduction in the yield of celery caused by the root-knot nematode. *Plant Disease Reporter, 27*, 234.

Parris, G. K., & Jehle, R. A. (1943). Root-knot nematode on lima beans in Maryland. *Plant Disease Reporter, 27*, 235.

Phipps, P. (2013). *Virginia – Summary of nematode survey activity in Virginia*. Cordova: National cotton council of America http://www.cotton.org/tech/pest/nematode/survey/virginia.cfm.

Phipps, P. M., & Eisenback, J. D. (2012a). The comparison of cotton seed and in-furrow treatments for control of nematodes. *Plant Disease Management Reports, 6*, N004.

Phipps, P. M., & Eisenback, J. D. (2012b). Response of peanut cultivars to proline, propulse, and sectagon-42 for disease control in peanut. *Plant Disease Management Reports, 6*, N003.

Pollok, J. R., Johnson, C. S., Eisenback, J. D., & David Reed, T. (2016). Reproduction of *M. incognita* race 3 on flue-cured tobacco homozygous for *Rk1* and/or *Rk2* resistance genes. *Journal of Nematology., 48*, 79–86.

Potenza, C. L., Thomas, S. H., Higgins, E. A., & Sengupta-Gopalan, C. (1996). Early root response to *Meloidogyne incognita* in resistant and susceptible alfalfa cultivars. *Journal of Nematology, 28*, 475–484.

Rephann, T. (2017). *The economic impact Virginia's agriculture and forest industries*. Charlottesville: Weldon Cooper Center for Public Service, University of Virginia.

Rideout, S. L., Johnson, C. S., Eisenback, J. D., & Wilkinson, C. A. (2000). Development of selected tobacco cyst nematode (*Globodera tabacum solanacearum*) populations on resistant and susceptible cultivars of flue-cured tobacco. *Journal of Nematology, 32*, 62–69.

Rodriques, E. R. (1983). *Paeciliomyces lilacinus* as a biological control agent of *Meloidogyne* sp., the root knot nematode. M.S. thesis, West Virginia University.

Sahebani, N., & Hadavi, N. (2008). Biological control of the root-knot nematode *Meloidogyne javanica* by *Trichoderma harzianum*. *Soil Biology and Biochemistry, 40*, 2016–2020.

Salinas, K. A., & Kotcon, J. B. (2007). Effect of Kodiak© (*Bacillus subtilis* Strain GB03) on soil-inhabiting nematodes near the rhizosphere of treated versus untreated snap bean seeds in situ. *Journal of Sustainable Agriculture, 29*, 5–12.

Salinas, K. A., Edenborn, S. L., Saxstone, A. J., & Kotcon, J. B. (2007). Bacterial preferences of the bacteiovorus soil nematode *Cephalobus brevicauda* (Cephalobidae): Effect of bacterial type and size. *Pedobiologia – Journal of Soil Ecology, 51*, 55–64.

Samac, D. A., Rhodes, L. H., & Lamp, W. O. (2015). *Compendium of alfalfa diseases and pests*. St. Paul: APS Press.

Schooley T. N. (2004). *Crop profile for alfalfa in Virginia*. https://ipmdata.ipmcenters.org/documents/cropprofiles/VAalfalfa.pdf

Schooley T. N. (2013). *Crop profile for pumpkins in Virginia*. https://ipmdata.ipmcenters.org/documents/cropprofiles/VApumpkins2013.pdf

Schooley, T. N., Tuckey, D., Alexander, S., Kuhar, T., Weaver, M., & Wilson, H. (2003). *Crop profile for white potatoes in Virginia*. Blacksburg: Virginia Polytechnic Institute and State University, Department of Entomology, Virginia Tech Pesticide Programs 24061. (https://ipmdata.ipmcenters.org/documents/cropprofiles/VApotato.pdf).

Smart, G. C. (1961). Culture of the soybean cyst nematode. *Virginia Journal of Science, 12*, 15.

Smart, G. C. (1964). The effect on yield of soybeans infested with the soybean cyst nematode, *Heterodera glycines*, from Virginia. *Virginia Journal of Science, 15*, 265.

Spasoff, L., Fox, J. A., & Miller, L. I. (1971). Multigenic inheritance of resistance to Osborne's cyst nematode. *Journal of Nematology, 3*, 329–330.

Starr, J. L., Koenning, S. R., Kirkpatrick, T. L., Robinson, A. F., Roberts, P. A., & Nichols, R. L. (2007). The future of nematode management in cotton. *Journal of Nematology, 39*, 283–294.

Steiner, G. (1931). The finding of *Heterodera schachtii*, the sugar beet nema, on *Polygonum* in Virginia. *Plant Disease Reporter, 15*, 145.

Sutherland, J. R. (1967). Parasitism of *Tylenchus emarginatus* on conifer seedling roots and some observations on the biology of the nematode. *Nematologica, 13*, 191–196.

Sutherland, J. R., & Adams, R. E. (1964). Host range of *Tylenchorhynchus claytoni* on some forest nursery seedlings. *Phytopathology, 54*, 749.

Sutherland, J. R., & Adams, R. E. (1965). Stand, growth, and nitrogen content of red pine seedlings following chemical treatment of the soil to control disease. *Tree Planters' Notes, 16*(4), 7–10.

Sutherland, J. R., & Adams, R. E. (1966). Population fluctuations of nematodes associated with red pine seedlings following chemical treatment of the soil. *Nematologica, 12*, 122–128.

Tamburo, S. E., & Adams, R. E. (1962). Nematode populations and growth of apple trees. *Plant Disease Reporter, 46*, 281–284.

Taylor, C. F. (1944). Distribution of the meadow nematode in Virginia. I. On boxwood. *Plant Disease Reporter, 28*, 339–340.

Thomas, S., Schreoder, J., & Murray, L. (2005). The role of weeds in nematode management. *Weed Science, 53*, 923–928.

USDA Census of Agriculture. (2016). https://www.nass.usda.gov/Quick_Stats/Ag_Overview/stateOverview.php?state=VIRGINIA

VDAI (1940). Virginia Department of Agriculture and Immigration Bulletin 1940. 379:10.

VDAI (1956). Virginia Department of Agriculture and Immigration Bulletin 1956. 560:10–11.

Verdegaal, P. (2015). *Nematodes in grapes*. http://www.lodigrowers.com/wp-content/uploads/2015/02/Nematodes-inGrapes2015LWC1.pdf

Walters, S. A., & Barker, K. R. (1994). Current distribution of five major *Meloidogyne* species in the United States. *Plant Disease, 78*, 772–774.

Wang, J., Johnson, C. S., & Eisenback, J. D. (1997). Enhanced hatching of *Globodera tabacum solanacearum* juveniles by root exudates of flue-cured tobacco. *Journal of Nematology, 29*, 484–490.

Yang, B., & Eisenback, J. D. (1983). *Meloidogyne enterolobii* n. sp. (Meloidogynidae), a root-knot nematode parasitizing pacara earpod tree in China. *Journal of Nematology, 15*, 381–391.

Ye, W. M., Koenning, S. R., Zhuo, K., & Liao, J. L. (2013). First report of *Meloidogyne enterolobii* on cotton and soybean in North Carolina, United States. *Plant Disease, 97*, 1262.

Zimmerley, H. H. (1919). *Greenhouse tomato growing in Virginia*. Virginia Truck Experiment Station Bulletin 26.

Chapter 12
Plant Parasitic Nematodes of Tennessee and Kentucky

Ernest C. Bernard

12.1 Introduction

Tennessee and Kentucky are often paired by the general public due to their long common border, location within the Central Appalachian-Upper South States, heavily traversed major north-south highways and similar agricultural histories. Both states are well-watered, with the Ohio, Cumberland and Tennessee River systems from north to south and the Mississippi River forming their western boundaries. The two states also share several geographical provinces that influence agricultural practices in similar ways.

The geology and topography of these two states are complex and can only be summarized here. The regions of each state often have alternative terms for the same geographical regions. For complete information on Tennessee and Kentucky see USGS (2002) and USEPA (2003). Instructive maps on agricultural intensity in Kentucky and Tennessee are available from AFT (2012).

The Gulf Coastal Plain extends over the western fourth of Tennessee and corresponds to the traditional "West Tennessee" concept. In Kentucky this province (the Jackson Purchase) covers a much smaller area south of Illinois and the Ohio River. The Gulf Coastal Plain is the most intensively farmed land in the two states, consisting of row crops such as soybean, cotton, corn and wheat. Central ("Middle") Tennessee and most of Kentucky comprise the Interior Low Plateau (USGS 2001), an expanse productive in soybean, corn and wheat. The Tennessee portion is called the Highland Rim; at its center is the Nashville Basin. Kentucky portions are the Pennyroyal, Western Coalfields and Bluegrass Regions. The Bluegrass has long been the top horse-production region in the country, and therefore, forages and hay

E. C. Bernard (✉)
Department of Entomology and Plant Pathology, University of Tennessee,
Knoxville, TN, USA
e-mail: ebernard@utk.edu

S. A. Subbotin, J. J. Chitambar (eds.), *Plant Parasitic Nematodes in Sustainable Agriculture of North America*, Sustainability in Plant and Crop Protection,
https://doi.org/10.1007/978-3-319-99588-5_12

crops are also important. The Appalachian Plateau occupies a band east of the Highland Rim and most of the eastern third of Kentucky (Eastern Coalfields). Farming is less intensive in this region, but fertile pockets are particularly suitable for vegetable production as well as corn and forage. The Blue Ridge Mountains, part of the Appalachian chain, run along the easternmost edge of Tennessee, and between these mountains and the Highland Rim is the Ridge and Valley Province, which also touches Kentucky on its extreme eastern side. Fresh vegetables are a major crop in alluvial deposits in several Eastern Tennessee counties. The most eastern part of Kentucky, the Eastern Coalfields, has relatively little commercial agriculture due to the prevalence of coal surface-mining activities.

12.2 Historical Background

Nematology in the U.S. nearly became a science in 1889, with the publication of three significant papers from researchers in Alabama (Atkinson 1889), Florida (Neal 1889) and Tennessee (Scribner 1889). All three papers dealt in whole or in part with root knot nematodes. That of Scribner included accurate descriptions and simple but diagnostic sketches of both *Meloidogyne* and *Pratylenchus* juveniles and symptoms on potato tubers. The lesion nematode was described more than 50 years later by Steiner (1943) as *Pratylenchus scribneri*, a plant-pathogenic species widespread in North America (Society of Nematologists 1984). Specialized nematology research did not spring from this promising beginning, but the noted Tennessee plant pathologist C. D. Sherbakoff (1939), in a study of differential reproduction of *Meloidogyne incognita* on tomato and cotton, hypothesized that host specialization was due to selection of genetic races. This hypothesis eventually helped lead to current concepts of nematode races identified by their reaction to plants with resistant genes. The first trained nematologist hired, in 1965, at the University of Tennessee was Carroll J. Southards. Soon after Dr. Southards was appointed head of what is now the Entomology and Plant Pathology Department, Ernest C. Bernard joined the faculty as a nematologist in 1977. Directed nematology programs in Tennessee, however, began in 1956, with the discovery of *Heterodera glycines* in Lake County (Epps 1957). This isolate produced milder symptoms than the North Carolina isolate that had been reported in 1954; these differences led to intensive analysis of *H. glycines*-soybean interactions and the use of resistance genes that continues unabated. Soon after this discovery, USDA established a research station at Jackson, Tennessee, with J. M. Epps on the staff until 1979. Working with E. E. Hartwig and others, the scientists at Jackson established a nationally significant program on the use of rotations and resistant cultivars for management of soybean cyst nematode (SCN). During and following retirement of Mr. Epps in 1979, several distinguished nematologists continued to expand USDA-SCN research at Jackson: B. Y. Endo

(1958–1963), G. R. Noel (1979–1981), L. D. Young (1982–2002), P. A. Donald (2003–2013) and P. R. Arelli (2013–present).

Nematology in Kentucky began much later, but concerns in the State about protection of the tobacco crop, then the State's most valuable agricultural commodity, led to its participation in the Tobacco Disease Council of the 1930s and 1940s, which is credited with raising the interest in nematodes throughout the Southern U.S. (Wilson 1982). The University of Kentucky plant scientist W. D. Valleau chaired the 1947 meeting of the Tobacco Disease Council, from which emerged a project proposal that eventually, after much modification, was approved as Regional Research Project S-19 (Wilson 1982). In 1950 Richard A. Chapman was hired as the University of Kentucky's first trained nematologist (Smith 1995). Dr. Chapman was the mentor for several outstanding nematologists, most notably Diana H. Wall and Edward C. McGawley.

12.3 Crop Production in Kentucky and Tennessee

With similar topographies and soils, Kentucky and Tennessee grow many of the same row and field crops (Tables 12.1 and 12.2). However, Tennessee is a significant cotton production state, whereas Kentucky's production since the 1970s has been negligible. Hemp production is seen by some as a replacement crop for cotton. For its size, Kentucky has an enormous production of hay and alfalfa, largely in support of its horse industry, which is the largest in the U.S. The other major difference is that Tennessee is a much larger producer of commercial vegetables, with major fresh tomato and snap bean production areas. Many more details of land used in these states can be found by reference to on-line documents posted by the U.S. National Agricultural Statistics Service and the Departments of Agriculture of the two states.

Table 12.1 Most important crops grown in Kentucky

	2016		2012	
Crop	Hectares	Value (dollars)	Hectares	National Rank
Corn	607,028	824,000,000	619,169	14
Hay + alfalfa	1,821,085	674,000,000	826,368	10
Soybean	724,387	881,000,000	594,078	16
Tobacco	30,472	283,000,000	N/A	2
Wheat	206,390	144,000,000	189,393	19

From: https://www.nass.usda.gov/Quick_Stats/Ag_Overview/stateOverview.php?state=KENTUCKY

Table 12.2 Most important crops grown in Tennessee

	2016		2012	
Crop	Hectares	Value (dollars)	Hectares	National rank
Corn	356,123	457,000,000	420,873	17
Cotton	103,195	457,000,000	153,781	6
Fresh tomatoes	1578	47,000,000	1497	5
Hay + alfalfa	734,504	438,000,000	714,270	12
Snap beans	4290	25,000,000	3439	7
Soybeans	659,637	715,000,000	497,763	17
Tobacco	9813	121,000,000	9672	3
Wheat	161,874	114,000,000	169,968	17

12.4 Nematodes of Importance in Kentucky and Tennessee

The two states have many species of plant parasitic nematodes in common, including several of the most damaging species: *Heterodera glycines* (SCN, soybean cyst nematode), *Meloidogyne incognita* (MI, southern root knot nematode) and *M. hapla* (MH, northern root knot nematode). *Rotylenchulus reniformis* (reniform nematode) occurs in several West and Middle Tennessee counties but has not been reported in Kentucky. General surveys of plant parasitic nematodes have been conducted in both states (Bernard 1980; Chapman 1957), but those surveys are quite old and may not accurately reflect current nematode distributions.

In the discussions below, emphasis is placed on nematological research that has been conducted in Kentucky and Tennessee. The occurrence (or lack) of nematode problems in these two states does not imply the same elsewhere. For instance, tobacco is damaged by several different species in other states, but in the present two they are not a problem (Seebold et al. 2013). Unless stated, no explicit claims of first findings should be assumed. For instance, Wartman and Bernard (1985) reported on *Pratylenchus alleni* relationships with various crop species including soybean, but V. R. Ferris and R. L. Bernard (no relation) (1962) provided the first information on the *P. alleni*-soybean host-parasite relationship. Reviews and important papers from other states, regions or countries may be cited for the sake of perspective, to emphasize a point or to provide entrée to more comprehensive study, but are not intended to provide a complete analysis of the general literature for a species.

Accelerated climate change is exhaustively documented (USGCRP 2009; IPCC 2013). Changes in temperature and precipitation likely will have major impacts on nematode pests. Temperature, precipitation and evaporation in this area of North America are projected to increase, while soil moisture will decrease. Changing conditions may suppress some nematode pests (see Jones et al. 2017) but encourage others that are adapted to higher temperatures and have resistant life stages such as *R. reniformis*.

12.4.1 Soybean Cyst Nematode, **Heterodera glycines**

The soybean cyst nematode is the most important nematode on soybean in North America. Average annual yield losses in the U.S. for 2003–2005 were approximately 2.8 million tons (Wrather and Koenning 2006). In 2 of the 3 years *H. glycines* was the most damaging pathogen on soybean in Kentucky, whereas for all 3 years in Tennessee the nematode was second only to frogeye leaf spot caused by the fungal pathogen, *Cercospora sojina*. Tennessee losses in 2015 due to *H. glycines* were 2.5%, just below 2.6% for frogeye leaf spot (Kelly 2016). Nematode effects on yield can vary widely from year to year (Young 1996a). Management of *H. glycines* is a complex problem reflected in the huge body of literature devoted to this nematode. As of July 2017, the Web of Science database returned 4597 hits in response to a search for "*Heterodera glycines*". Schmitt et al. (2004) provide a full account of *H. glycines* and the multiple options for managing it. This volume should be consulted for an overview of the *H. glycines* problem in the U.S. and elsewhere in the world. The fifth edition of the SCN Management Guide (Niblack and Tylka 2008) is a recent on-line publication with much detail about managing *H. glycines*. The account given below concerns research conducted on the soybean cyst nematode in Tennessee and Kentucky.

As noted above, *H. glycines* was found in Lake County, Tennessee, in 1956 by Epps (1957). This report was the second for the U.S., after the initial collection in North Carolina in 1954 (Winstead et al. 1955). By 1957, the nematode had been collected from three Northwestern Tennessee counties (Dyer, Lake, Lauderdale), and by the early 1970s, had been found in all of the coastal plains counties of West Tennessee as well as Lincoln County in the south-central part of the state. By this time *H. glycines* was also established in seven counties in Kentucky: the four most western in the Jackson Purchase coastal plain (Ballard, Carlisle, Hickman, Fulton) and three along the Western Ohio River (Daviess, Henderson, Union). This nematode continued to spread inexorably eastward across the soybean-growing regions of both states (for maps see Tylka and Marett 2014). The damage potential of this nematode was promptly realized and major federal and state resources throughout the affected states were, and continue to be, devoted to management of the nematode and protection of the U.S. soybean crop. Strong collaboration among investigators has been a key to development of effective, sustainable approaches to management of *H. glycines*.

Immediately after the discovery of *H. glycines* in Tennessee, a USDA research station was established at Jackson, TN, in cooperation with the University of Tennessee. Research at Jackson was quickly focused on distribution surveys, the means by which the nematode becomes disseminated, and the evaluation of resistant soybean lines. By the early 1970s, it was clear that most soybean fields in the Western Tennessee counties were already infested (Klobe 1976). The role of birds in disseminating cysts was studied (Epps 1971). Caged starlings (Sturnidae), cowbirds and grackles (Icteridae) were force-fed 300–400 cysts each, and droppings were collected in pans of water. Cysts were recovered, crushed and added to soil around soybean seedlings or examined directly for eggs and juveniles. Cysts were obtained from all birds and new cysts developed on inoculated soybeans. More realistically, Epps (1971) dissected 54 starlings caught in an *H. glycines*-infested field. The digestive tracts of 7 of the 54 birds contained cysts, some of which had viable eggs that hatched and matured on soybean plants. It is reasonable to conclude that migratory birds that feed on growing plants, especially increasingly troublesome Canada geese, could have a role to play in long-distance aerial dispersal. Nevertheless, since the soybean-growing regions of North America are now fully infested nearly to their limits, this means of dispersal, if ever significant, no longer is.

Cysts incorporated into soil peds, collected at seed-cleaning stations (Epps 1968) or included as debris in seed bags (Epps 1969), contained viable eggs 6–8 months after harvest, long enough to be planted in fields the next season. A previous set of experiments (Epps 1958) under less-than-controlled conditions had indicated that eggs and juveniles in cysts were nonviable after a 2-month storage period. These results were contradicted by Endo (1962a, b) in a series of controlled experiments, in which juveniles remained viable in cysts for up to 5 months at a relative humidity of 3.2%. These results proved that sanitation and proper seed storage have roles to play in preventing the spread of *H. glycines*. Related to this work is the possibility of dispersal on plant nursery stock grown on old soybean land. Old cysts could be transported in soil clinging to roots, especially in ball-and-burlaped operations; but more likely, *H. glycines* could develop on various crops and weeds in in-ground nurseries (Epps and Chambers 1958, 1959). Ward et al. (2011) provided a list of *H. glycines* hosts and recommendations for avoiding cyst contamination of materials. The problem should be minimal in containerized operations, especially if the plants are on plastic.

12.4.1.1 Management

Combinations of crop rotations, use of non-hosts and nematode-resistant cultivars for managing *H. glycines*, often incorporated into reduced or no-till cropping for soil conservation, are the most important approaches for managing this nematode. Soybean cyst nematode populations have a particular virulence phenotype based on soybean indicator lines that include the known sources of resistance (Niblack and Riggs 2004). Recognition of the particular HG type predominant in a field (Niblack et al. 2001; Tylka 2016) guides the choice of cultivars to plant, and thus is an essential step to implementation of a successful management plan. The HG Type Test has supplanted the older *H. glycines* race determination test of Riggs and Schmitt (1988). Even so, the older scheme and its progenitors provided a wealth of information about the genetic variability within *H. glycines* in the mid-south region as well as its application to suppression of the nematode (Epps and Duclos 1970; Young 1990, 1992, 1996b; Young and Kilen 1994).

Management guidelines for *H. glycines* in Tennessee (University of Tennessee 2017) include selection and rotation of cultivars with nematode resistance, based on yearly disease ratings published every year in a searchable database; rotation with non-hosts (alfalfa, barley, corn, oat, potato, sorghum, sugar beet, sunflower and wheat); nematicidal seed treatments; and cultural practices to reduce plant stress such as sufficient fertility and weed management. These recommendations mirror closely those given in the SCN Management Guide referenced above. The Kentucky Extension Service (2017) also recommends the SCN Management Guide on its web site. Management recommendation publications by Hershman (2015), Hershman et al. (2009) and Johnson et al. (2015) are particularly useful.

Preplant nematicides are not a viable management option, as they rarely perform as well as a well-chosen resistant cultivar and are far more expensive than the cost of *H. glycines*-resistant seed. In a series of nematicide evaluations in Kentucky Hershman et al. (1986, 1987), Stuckey and Chapman (1979) and Stuckey et al. (1985) found that resistant cultivars outperformed nematicide-treated susceptible cultivars.

Research on crop rotations and resistant cultivars to manage *H. glycines* went hand-in-hand, since it was recognized early on that the nematode persisted for years in soil and that resistant cultivars would select for nematodes able to mature on them. Building on the work of Triantaphyllou (1976) and others, Lawrence Young, USDA-Jackson, studied intensively the adaptation of *H. glycines* to resistant soybean cultivars and how the effect could be mediated with introduction of new soybean germplasm, susceptible-resistant soybean rotations, inclusion of non-hosts in rotations, and evaluation of resistant and susceptible soybean plantings. Many new soybean cultivars have been released that have resistance to *H. glycines*, but resistance to *H. glycines* is dependent on a limited number of genes to which nematode populations can adapt (Shannon et al. 2004). In a greenhouse experiment, Young (1982a) found that *H. glycines* developed in high numbers (74–100% of susceptible control) on cv. "Bedford" and PI88788 after 11 generations of selection, but not on three other PIs. These results were mirrored in a separate greenhouse experiment (Young 1982b). In a field study, where nematode reproduction was followed in fields of continuous Bedford soybean, development on Bedford was 38–70% of that on a susceptible cultivar (Young 1984a). Selection of more virulent nematode strains continues to be a central consideration in management. Hershman (2008) reported that *H. glycines* was adapting to the resistant soybean cultivars grown in Kentucky and recommended a switch to lines with resistance from other genetic sources. Gene pyramiding (Arelli et al. 2015) promises to provide more durable resistance to multiple HG types.

Because resistant cultivars generally have a limited useful life, nematologists reasoned that *H. glycines* could be kept off-balance by rotation with susceptible cultivars and non-hosts (Epps and Chambers 1965). A resistant cultivar would strongly reduce nematode field densities, making possible a successful crop with the susceptible cultivar the next year (Young 1984b, 1994). Nematode rebound would be countered the 3rd year with the resistant cultivar. Adding a non-host into the rotation could provide even better suppression. Young and Hartwig (1992) demonstrated that rotations of resistant and susceptible cultivars gave greater yields than continuous susceptible or resistant cultivars, but corn-resistant soybean rotations were even better. A blend of resistant/susceptible seed (70/30) gave results similar to the rotations with corn. In a later study, a 3-year rotation of corn-susceptible-resistant gave better soybean yields than any other continuous or combination treatment (Young 1998). The combined resistant cultivar-crop rotation approach has been summarized by Young (1992), with important caveats regarding other nematodes such as root knot species (*Meloidogyne* spp.) that could increase on *H. glycines*-resistant cultivars.

Concurrent adoption of reduced or no-till agriculture in much of the soybean belt, especially coupled with double-cropping with a winter annual crop such as wheat, is often thought to contribute to more effective management of *H. glycines.* Flinchum (2001) listed the benefits of no-till soybean production: (1) lower production cost due to reduced machinery, labor and energy requirements; (2) reduced soil erosion; (3) yields equal to or greater than conventionally planted soybeans; (4) more intensive use of resources and opportunity for expanded farming operations with the surplus labor and equipment. Effects on *H. glycines,* however, are variable and difficult to separate from other factors. Tyler et al. (1987) found that cyst numbers generally were lower in long-term no-till vs. conventionally tilled soybean following a wheat crop, although yields were not necessarily higher in the no-till plots. Short-term no-till + wheat of 1 or 2 years did not influence cyst numbers. However, in a later study, cyst densities were much lower in no-till field plots than in more conventional treatments (Tyler et al. 1983). The authors hypothesized that reduced soil manipulation may have lessened the introduction of nematodes into plots; the no-till mulch may have reduced temperatures and thereby slowed nematode development; or the increase in organic matter may have stimulated soil organisms that feed on nematodes. In another no-till soybean and soybean-wheat field experiment, Baird and Bernard (1984) found that *H. glycines* infective juvenile soil densities in conventional systems were significantly higher in July than in May or October, but that cyst numbers did not differ. These results support the idea that cooler soil temperatures may delay egg hatch, but do not deter invasion and maturation. Results also suggested that the major effect on plant parasitic nematode communities was the effect of wheat rather than no-till. This conclusion was strengthened by Jennings and Bernard (1986). Infection rates in soybean seedlings were much lower in pots previously grown with wheat than in soybean when wheat was co-planted with soybean. In field experiments, Hershman and Bachi (1995) determined that wheat residue was correlated with a decrease in cyst and egg numbers at the end of the next growing season. Taken together, these reports reinforce the efficacy of wheat in double-cropping. Hershman (2009) provided a succinct summary of the value of wheat in a cyst nematode management program. Finally, with regard to increasing organisms antagonistic to *H. glycines*, Bernard et al. (1996) studied incidence and diversity of soil fungi parasitizing eggs, females and cysts in several long-term (up to 7 years) production systems including wheat-soybean no-till treatments. None of the treatments resulted in significant differences, indicating that double-cropped no-till systems are not likely to enhance biological control of *H. glycines*.

12.4.2 Root Knot Nematodes, Meloidogyne hapla *and* M. incognita

Root knot nematodes occur worldwide on thousands of different plant species. At least two species, *Meloidogyne hapla* (Fig. 12.1c) and *M. incognita,* are widespread and probably ubiquitous in both Kentucky and Tennessee, although perhaps *M. hapla* is more common in Kentucky and *M. incognita* in Tennessee. These two species are quite different and results from work on one species cannot be applied to the other. As one of many possible examples, Chapman (1963c) demonstrated that *M. hapla* was much more virulent than *M. incognita* on alfalfa. Still other

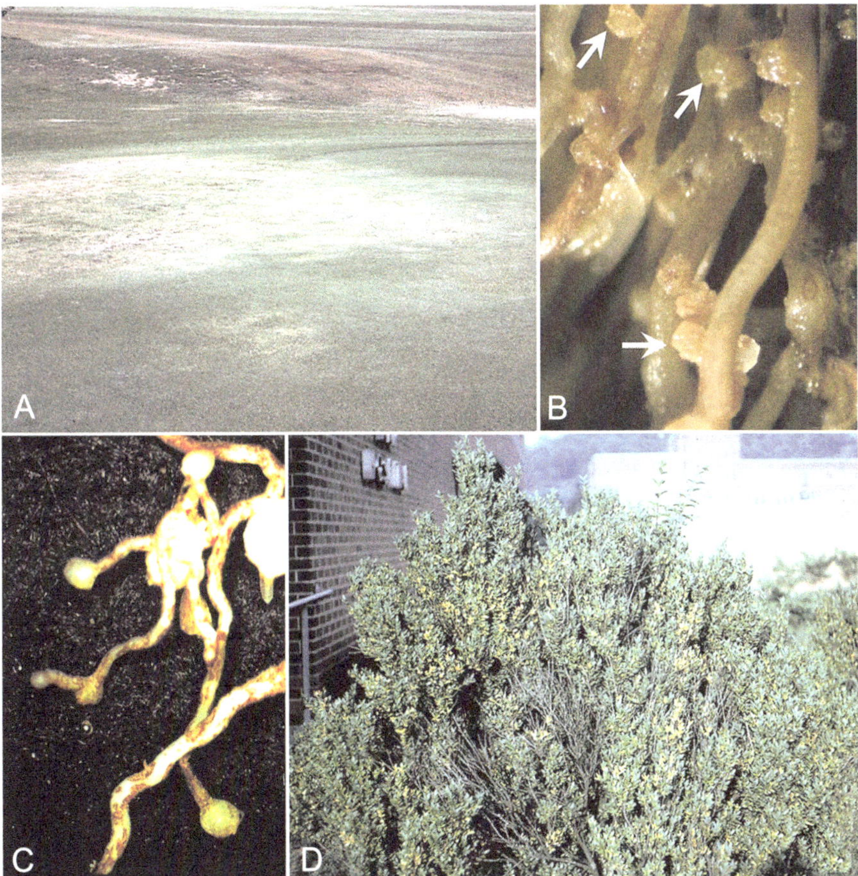

Fig. 12.1 (**a**) *Belonolaimus longicaudatus* damage to golf green; (**b**) *Rotylenchulus reniformis* reproduction on cotton roots. Arrows indicate egg masses; (**c**) Galling by *Meloidogyne hapla* on dogwood roots; (**d**) Damage to boxwood caused by *Pratylenchus vulnus* and *Rotylenchus buxophilus*. (Photo by C.H. Hadden)

species have been found in these two states and are briefly mentioned in the "Other Endoparasites" section below.

Despite the prominence of these nematodes in world agriculture, they have generally not been considered significant pests in Kentucky and Tennessee except in some specialty crop situations. In some parts of the U.S. *Meloidogyne* spp. are constraints on tobacco production, but in these two big tobacco production states nematodes are not considered a serious problem. However, *M. incognita* was able to increase seven-fold over a 2-year period on tobacco; fall tillage reduced nematode numbers compared to non-tillage (Southards 1971). Newman (1998) reported that root knot nematodes were of negligible importance on cotton in Tennessee. On soybean, considerable research elsewhere has demonstrated that MI is outcompeted by *H. glycines* on soybean and therefore, accounts for relatively little damage. In the 2003–2005 period, plant parasitic nematodes (including *Meloidogyne* spp. but excluding *H. glycines*) caused no loss to soybean in Kentucky and only negligible losses in Tennessee (Wrather and Koenning 2006).

The major threat in these states is to commercial vegetable production, home gardens and commercial nurseries. Tennessee is a major producer of snap beans, fresh tomatoes and tomato seedlings for interstate shipment, all of which can be seriously damaged by root knot nematodes. Historically, Tennessee produced tomato seedlings in field soil for interstate shipment northward, risking root knot nematode infection that would result in rejection of the lot by plant inspectors (Chambers and Reed 1961). Tomato seedlings produced for sale now are seeded into growing containers or trays in an artificial soil mix (Rutledge et al. 1999), thereby avoiding the chance for infection. In the fresh tomato production area of Eastern Tennessee counties, *M. incognita* was identified from 94% of bioassayed samples, while *M. hapla* was found in the other 6% (Bernard 1981). Juvenile densities in a fourth of these samples were greater than 1000/100 cm^3 of soil and in a separate study, juvenile numbers exceeded 5000/100 cm^3 soil in several fields in summer (Stockdale 1985).

12.4.2.1 Management

Management of root knot nematodes in commercial tomato fields has been an important component of nematology programs in Tennessee and Kentucky for at least 50 years. Johnson et al. (1967) applied finely chopped hay residues (alfalfa, oat, lespedeza, flax) to field plots infested with *M. incognita*. All residues suppressed infection, but a 10-ton rate applied 8 months prior to assay was more effective than a 5-ton rate or a shorter assay time. Nevertheless, nematicides were more effective than any residue application. Suppression of *M. incognita* with nematicides in the most heavily infested fields increased yields by a factor of 2.5 over the average yield per hectare (Bernard and Hadden 1981). Although several nematicides are still available to commercial operators, resistant tomato cultivars and/or rotations with non-hosts were recognized early on as a viable alternative to chemical management (Southards 1973). Non-nematidical approaches are now preferred and increasingly widely practiced in both states (Bost 2015; Seebold 2010). Recommendations for home vegetable gardens are similar (Bost 2013; Seebold 2010) and in Kentucky newer approaches such as biological control, solarization and suppressive crops such as marigold and rapeseed (canola), have been proposed. Rapeseed as a growing plant is a good host for *M. hapla* and *M. incognita* (Bernard and Montgomery-Dee 1993), but incorporation of plants as green manure releases nematicidal chemicals that kill nematodes (Halbrendt 1996; Mojtahedi et al. 1991).

Root knot nematodes should also be considered a potentially major constraint on production of woody ornamentals. In an extensive survey of Tennessee dogwood, maple and peach nurseries, *M. hapla* was found in 23% of surveyed nursery blocks (21 of 92), while *M. incognita* was found in only 1 site (1%) (Niblack and Bernard 1985). Interestingly, blocks separated geographically but operated by the same grower tended to have similar nematode communities. In a greenhouse host range study, *M. hapla* heavily galled and reproduced well on 7 of the 35 species and cultivars tested including flowering dogwood, hydrangea and spirea (Bernard and Witte 1987). In a similar host range experiment, 17 holly species and cultivars were rated for galling by Tennessee and North Carolina isolates of *M. hapla* and one Tennessee isolate of *M. incognita* (Bernard et al. 1994). Four cultivars were heavily galled by all three isolates; the *M. incognita* isolate produced numerous galls on 9 of the 17 hollies, while the two *M. hapla* isolates were variable. Yaupon holly (*Ilex vomitoria*) appeared to be immune to *M. hapla* but was moderately galled by *M. incognita*. The ability of *M. hapla*, in particular, to parasitize many woody plants is worthy of further investigation to prevent spread of the nematode on infected stock, both bare-rooted and ball-and-burlapped.

12.4.3 Root Lesion Nematodes, **Pratylenchus** *spp.*

Lesion nematodes are common, destructive migratory endoparasites on many crops (Castillo and Vovlas 2007; Thames 1982). Some *Pratylenchus* spp., especially *P. penetrans*, are members of disease complexes (Mai et al. 1977; Thames 1982). Together, Kentucky and Tennessee have at least ten species of *Pratylenchus*: *P. alleni, P. brachyurus, P. coffeae, P. crenatus, P. hexincisus, P. neglectus, P. penetrans, P. scribneri, P. vulnus* and *P. zeae* (Bernard 1980; Chapman 1957; Freckman and Chapman 1972; Society of Nematologists 1984; Wartman and Bernard 1985). Chapman (1956) found four species associated with strawberry in Kentucky, Niblack and Bernard (1985) identified six species from Tennessee woody ornamentals nurseries and Inserra et al. (2007) collected four species at the type locality of *P. scribneri* (but not *P. scribneri* itself). Despite their ubiquity and importance in phytopathology, they have not been studied extensively in either Kentucky or Tennessee. In Kentucky, however, Valleau and Johnson (1947) associated a *Pratylenchus* sp. with brown rot of tobacco. Later, R.A. Chapman and his students devoted significant effort to the population dynamics of *P. penetrans* on red clover and alfalfa, important forage crops for the Kentucky horse industry. Meadow (lesion) nematodes caused extensive damage to alfalfa roots planted in the spring, especially following winter wheat, with means of 12,000–27,000 nematodes/g of fresh alfalfa root. Conversely, means for fall-planted alfalfa were 25–440 nematodes/g of root (Chapman 1954). Similar results were obtained in later greenhouse experiments (Chapman 1958, 1959). Red clover was an excellent host for *P. penetrans* in these studies, with means of 3000–46,000 nematodes per gram of root. Symptoms of poor growth were worse in summer due to high soil temperatures and root damage. Chapman (1958) concluded that the repeated harvest of infected alfalfa and red clover would lead to loss of stands. In a study of *P. penetrans* and *Heterodera trifolii* on red clover, Freckman and Chapman (1972) found little interaction of the two species with regard to root invasion, with the lesion nematode penetrating roots at the same rate regardless of the presence of *H. trifolii*. However, Turner and Chapman (1972) examined the reciprocal influence of *P. penetrans* and *M. incognita* on alfalfa and red clover. In both plants penetration was not inhibited when initial nematode numbers were equal; but when *P. penetrans* numbers were four times higher *M. incognita* penetration was strongly reduced. When the effect of inoculation of *M. incognita* before *P. penetrans* was studied, *P. penetrans* tended to produce fewer eggs than in simultaneous inoculations (Chapman and Turner 1975). These interesting studies are indications of the dynamic relationships nematode species have with one another in much more complicated natural environments.

Research on *Pratylenchus* in Tennessee has been limited. *Pratylenchus vulnus* is commonly found in holly (*Ilex* spp.) and boxwood (*Buxus sempervirens*) rhizospheres in Tennessee, often concomitant with the ectoparasitic boxwood nematode, *Rotylenchus buxophilus* (Fig. 12.1d). Osborne and Jenkins (1962) provided proof of the pathogenicity of *P. vulnus* to boxwood, which consisted of chlorosis and branch dieback. *Pratylenchus alleni* may be an overlooked pathogen on a number of crops. In a greenhouse experiment, nematode reproduction varied widely on a range of soybean cultivars but all were suitable hosts (Wartman and Bernard 1985). Snap bean, tomato, wheat, cowpea and corn were also suitable hosts, but Lima bean, alfalfa, cabbage and cotton were highly resistant, suggesting that crop rotations of certain vegetables or rotations, including cotton or alfalfa rotations, would be successful in managing this nematode. In a greenhouse experiment, *Pratylenchus scribneri* reproduced very well on all tested sunflower cultivars in comparison to soybean (Bernard and Keyserling 1985) and thus would not be a suitable rotation crop on *P. scribneri*-infested land. Endophyte-free tall fescue was not a suitable host for *P. scribneri* although some reproduction occurred, but less than ten individuals were extracted from all endophyte-infected root systems (Kimmons et al. 1990). Therefore, tall fescue may be a suitable pasture and range forage in the presence of *P. scribneri* regardless of its endophyte status.

12.4.4 *Reniform Nematode,* Rotylenchulus reniformis

Reniform nematode is a major pathogen of cotton and soybean in the Southern U.S. and parasitizes hundreds of other plant species (Robinson et al. 1997). This nematode occurs in cotton fields (Fig. 12.1b) in 16 Tennessee counties, mostly in the western part of the state but also in two southcentral counties (Stebbins et al. 1998; Koenning et al, 2004; Kelly 2015). Surveys conducted by Newman (1998) in West Tennessee detected *R. reniformis* in 70 of 1580 soil samples, with densities up to 268 per 200 cm^3 of soil. Newman (1998) considered economic thresholds for *R. reniformis* in Tennessee to be 210 juveniles/100 cm^3 at planting or 1050 adults + juveniles/100 cm^3 at harvest. Therefore, the nematode was considered to not be of immediate concern on cotton, although average losses per year for *R. reniformis* plus minor ectoparasites for 1992–1998 were estimated to be 0.23%, or 1778 bales valued at $675,724. In recent years the known range of reniform nematode has expanded to several surrounding counties as well as two southcentral counties. In addition, densities in some fields have increased substantially. In a survey of fields in 2014–2015, *R. reniformis* numbers in several samples were as high as 5000/100 cm^3 soil, indicating that this nematode is potentially a significant pathogen in the state. In the same survey, *R. reniformis* numbers under corn were less than 150/100 cm^3 soil. Management of reniform nematode in cotton has not been exhaustively studied in Tennessee, but rotations of corn or grain sorghum were recommended by Newman (1998).

Reniform nematode also occurs on soybean in Tennessee and will be found in the same coastal plain soils where it now attacks cotton. However, it is yet to be a significant problem; in 2015 *R. reniformis* was estimated to have reduced total soybean yields by only 0.01% (Kelly 2016). In Mississippi the damage threshold for reniform nematode on cotton is 21/100 cm^3 soil (Stetina et al. 2014), suggesting that susceptible soybean rotated with cotton on infested land will be heavily damaged. Although relatively little attention has been paid to the possible effects of this nematode on soybean in Tennessee, resistance has been incorporated into many cultivars and lines (e.g., Robbins et al. 2013; Stetina et al. 2014). Some of these newer lines have resistance to multiple pests and pathogens, especially *Heterodera glycines* (Arelli et al. 2015, 2017).

12.4.5 *Cyst Nematode,* Vittatidera zeaphila, *on Corn*

The recently described corn cyst nematode, *Vittatidera zeaphila* (Bernard et al. 2010) occurs in the Gulf Coastal Plain in Obion and Lauderdale Counties in northwestern Tennessee and Hickman County in southwestern Kentucky (Donald et al. 2012). These counties all border the Mississippi River but are not completely contiguous; therefore, the range of this nematode probably is larger than presently known. The only good hosts of this nematode are goosegrass (*Eleusine indica*) and corn hybrids including popcorn cultivars, although the nematode can just maintain itself on barley and oat (Donald et al. 2012). Soybean, wheat, native grasses and teosinte, among others, are non-hosts. However, several inbred corn lines exhibited moderate to complete resistance as measured by the ability of nematodes to reproduce on them. Crosses of several inbred lines produced results suggesting that the corn cytoplasm is potentially involved in conferring resistance (Donald et al. 2012).

Although *V. zeaphila* is not considered a significant threat to crop production at this time, it has the ability to reproduce well on many corn cultivars and selections. *Vittatidera zeaphila* appears to benefit from warmer soil temperatures; Donald et al. (2012) increased inoculum in water baths set at 27.5 °C. Therefore, with continued global warming this nematode may well become an economic constraint on corn production, necessitating development of resistant selections.

12.4.6 Other Endoparasitic Nematodes

Several additional root knot nematodes have been studied in Tennessee and Kentucky, but do not appear to be of concern in agriculture at this time. Chapman (1963b), describing the population development of *M. arenaria* on red clover in the greenhouse, reported that high initial numbers of juveniles reduced root and shoot weights. Chapman's nematode identification may have been in error since *M. arenaria* is a more southern species and unlikely to occur in Kentucky. However, he did not give the provenance of his isolate. *Meloidogyne graminis* was found on seven golf courses in West Tennessee, in both the greens and fairways. Bermuda grass was particularly affected (Southards 1967). This nematode may well have been *M. marylandi*, a similar grass-parasitic species undescribed at the time. Numbers of *M. marylandi* were higher in an endophyte-free perennial ryegrass selection than in those of endophyte-infected ryegrass (Ball et al. 1997). Presence or absence of peramine and lolitrem B did not seem to influence the host-parasite relationship. *Meloidogyne trifoliophila*, described from white clover in West Tennessee (Bernard and Eisenback 1997), was found to have a wide experimental host range including all 38 tested clover species and cultivars, several Fabaceae including some soybean cultivars and most Apiaceae and Brassicaceae (Bernard and Jennings 1997). However, cotton, tomato and most monocotyledons including corn, were non-hosts. This nematode should occur widely in legume-based fields and home gardens, but has been infrequently reported in the U.S.

Other than *H. glycines*, the only other cyst nematode of note in the region is the clover cyst nematode, *H. trifolii*. This common and widespread species has been studied extensively in Kentucky to determine its pathogenicity on forages vital to the livestock industry, especially horses. Chapman (1964) investigated the suitability of common clovers as hosts for *H. trifolii*. In this greenhouse test, top weights of all three tested clovers (white, ladino, red) were reduced in nematode treatments, with red clover being severely damaged. Conversely, nematode reproduction on white and Ladino clovers was much higher than on red clover. Chapman attributed this response to the severe response of red clover to invasion, which reduced the capacity of the roots to support more nematodes. Freckman and Chapman (1972) demonstrated that 25–30% of inoculated juveniles penetrated red clover roots, but that they did so promptly, unlike results in earlier reports from elsewhere. These authors concluded that some populations of *H. trifolii* were better adapted than others for parasitizing red clover.

12.4.7 Ectoparasitic Nematodes

The pathogenicity of ectoparasitic nematodes in Kentucky and Tennessee has received little attention despite their taxonomic diversity. Most reports of these nematodes are derived from surveys on agricultural and horticultural crops (e.g., Chapman 1956; Niblack and Bernard 1985), not controlled experiments. Some species such as *Hoplolaimus magnistylus*, are of possible importance due to their similarity to congeneric species (Donald et al. 2013), but much additional experimentation is needed to establish the pathogenicity of these nematodes.

The attention paid to ectoparasites in the two states is roughly related to their importance in agriculture compared to root knot and cyst nematodes. Even in high numbers, they usually do not cause noticeable problems. For example, a spiral nematode, *Helicotylenchus pseudorobustus*, can increase to enormous numbers (>2000/100 cm^3 soil) on sunflower and rapeseed (Bernard and Keyserling 1985; Bernard and Montgomery-Dee 1993) without causing noticeable plant symptoms. McGawley and Chapman (1983) conducted a complex series of experiments comparing reproduction on soybean of three ectoparasitic species: *Mesocriconema simile* (= *Criconemoides simile*), *H. pseudorobustus* and *Paratylenchus projectus*, alone or in combination, on soybean cultivars. Nematode reproduction was not proportional to root weights, which did not differ among the treatments. Numbers of *P. projectus* were suppressed by half in the presence of the other two nematodes; this result suggests the competition among species that regulates nematode densities in soil. A stunt nematode, *Tylenchorhynchus martini*, developed well on alfalfa and red clover but did not induce symptoms (Chapman 1959). Its numbers, however, were suppressed in the presence of *P. penetrans*.

Nevertheless, some ectoparasites have the potential to cause economic damage either as native species or by introduction. McGawley and Chapman (1982) were able to reduce root weights of soybean by adding large numbers of *M. simile* to transplanted seedlings but recognized that under natural conditions nematode numbers might not approach their experimental densities. The potential for introduction of pathogenic species should be kept in mind when examining unfamiliar symptoms. For instance, *Belonolaimus longicaudatus*, the sting nematode (Fig 12.1a), is not found naturally in these states; yet, in the early 1980s it caused severe damage to greens on a golf course in West Tennessee, requiring complete renovation (unpubl.). This infestation was caused by renovation of the greens with infested turf from out-of-state. In another interesting occurrence, Chapman (1963a) obtained *Scutellonema brachyurum* from *Clivia miniata* (Kaffir lily) growing in a University of Kentucky greenhouse. Although it was capable of increasing to high numbers on red clover in the greenhouse, it did not induce noticeable plant symptoms.

References

AFT (American Farmland Trust). (2012). *Farming on the edge: State maps*. https://www.farmland.org/farming-on-the-edge-state-maps. Accessed July 2017.

Arelli, P. R., Pantalone, V. R., Allen, F. L., Mengistu, A., & Fritz, L. A. (2015). Registration of JTN-5203 soybean germplasm with resistance to multiple cyst nematode populations. *Journal of Plant Registrations, 9*, 108–114.

Arelli, P. R., Shannon, J. G., Mengistu, A., Gillen, A. M., & Fritz, L. A. (2017). Registration of conventional soybean germplasm JTN-4307 with resistance to nematodes and fungal diseases. *Journal of Plant Registrations, 11*, 192–199.

Atkinson, G. F. (1889). *A preliminary report upon the life history and metamorphoses of a root-gall nematode, Heterodera radicicola (Greeff) Müll., and the injuries caused by it upon the roots of various plants*. Science Contributions from the Agricultural Experiment Station, Alabama Polytechnic Institute (Vol. 1, pp. 177–222).

Baird, S. M., & Bernard, E. C. (1984). Nematode population and community dynamics in soybean-wheat cropping and tillage regimes. *Journal of Nematology, 16*, 379–386.

Ball, O. J.-P., Bernard, E. C., & Gwinn, K. D. (1997). *Effect of selected Neotyphodium lolii isolates on root knot nematode (Meloidogyne marylandi) numbers in perennial ryegrass*. In Proceedings of the 50th New Zealand plant protection conference, pp. 65–68.

Bernard, E. C. (1980). *Identification, distribution, and plant associations of plant parasitic nematodes in Tennessee* (18 pp). University of Tennessee Agricultural Experiment Station Bulletin 594.

Bernard, E. C. (1981). Distribution of root knot nematodes (*Meloidogyne* spp.) in commercial tomato fields of Cocke County, Tennessee. *Tennessee Farm and Home Science, 118*, 2–3.

Bernard, E. C., & Eisenback, J. D. (1997). *Meloidogyne trifoliophila* n. sp. (Nemata: Meloidogynidae), a parasite of clover from Tennessee. *Journal of Nematology, 29*, 43–54.

Bernard, E. C., & Hadden, C. H. (1981). Influence of nematicides on southern root knot nematode and tomato yields. 1981. *Fungicide and Nematicide Tests, 36*, 182–183.

Bernard, E. C., & Jennings, P. L. (1997). Host range and distribution of the clover root knot nematode, *Meloidogyne trifoliophila*. *Journal of Nematology, 29*(4S), 662–672.

Bernard, E. C., & Keyserling, M. L. (1985). Reproduction of root knot, lesion, spiral, and soybean cyst nematodes on sunflower. *Plant Disease, 69*, 103–105.

Bernard, E. C., & Montgomery-Dee, M. E. (1993). Reproduction of plant parasitic nematodes on winter rapeseed (*Brassica napus* ssp. *oleifera*). *Journal of Nematology, 25*(4S), 863–868.

Bernard, E. C., & Witte, W. T. (1987). Parasitism of woody ornamentals by *Meloidogyne hapla*. *Annals of Applied Nematology, 1*, 41–45.

Bernard, E. C., Witte, W. T., Dee, M. M., & Jennings, P. L. (1994). Parasitism of holly cultivars by three root knot nematode isolates. *Holly Society Journal, 12*, 12–17.

Bernard, E. C., Self, L. H., & Tyler, D. D. (1996). Fungal parasitism of soybean cyst nematode, *Heterodera glycines* (Nemata: Heteroderidae), in differing cropping-tillage regimes. *Applied Soil Ecology, 5*, 57–70.

Bernard, E. C., Handoo, Z. A., Powers, T. O., Donald, P. A., & Heinz, R. D. (2010). *Vittatidera zeaphila* (Nematoda: Heteroderidae), a new genus and species of cyst nematode parasitic on corn (*Zea mays*). *Journal of Nematology, 42*, 139–150.

Bost, S. (2013). *Nematode control in the home garden*. University of Tennessee Extension publication SP341-L. http://trace.tennessee.edu/utk_agexgard/112. Accessed July 2017.

Bost, S. (2015). *Commercial vegetable disease guide* (50 pp). University of Tennessee Extension Publication W141. http://extension.tennessee.edu/publications/Documents/W141.pdf. Accessed July 2017.

Castillo, P., & Vovlas, N. (2007). *Pratylenchus (Nematoda: Pratylenchidae): Diagnosis, biology, pathogenicity and management, Nematology Monographs and Perspectives 6* (p. 529). Leiden: Brill Academic Publishers.

Chambers, A. Y., & Reed, H. E. (1961). The aim: Better nematode control on tomato plants. *Tennessee Farm and Home Science, Progress Report, 40*, 2–3.

Chapman, R. A. (1954). Meadow nematodes associated with failure of spring-sown alfalfa. *Phytopathology, 44*, 542–545.

Chapman, R. A. (1956). Plant parasitic nematodes associated with strawberries in Kentucky. *Plant Disease Reporter, 40*, 179–181.

Chapman, R. A. (1957). Species of plant parasitic nematodes in Kentucky. *Transactions of the Kentucky Academy of Science, 18*, 70–74.

Chapman, R. A. (1958). The effect of root lesion nematodes on the growth of red clover and alfalfa under greenhouse conditions. *Phytopathology, 48*, 525–530.

Chapman, R. A. (1959). Development of *Pratylenchus penetrans* and *Tylenchorhynchus martini* on red clover and alfalfa. *Phytopathology, 49*, 357–359.

Chapman, R. A. (1963a). Population development of the plant parasitic nematode *Scutellonema brachyurum* on red clover. *Proceedings of the Helminthological Society of Washington, 30*, 169–173.

Chapman, R. A. (1963b). Population development of *Meloidogyne arenaria* in red clover. *Proceedings of the Helminthological Society of Washington, 30*, 233–236.

Chapman, R. A. (1963c). Development of *Meloidogyne hapla* and *M. incognita* in alfalfa. *Phytopathology, 53*, 1003–1005.

Chapman, R. A. (1964). Effect of clover cyst nematode on growth of red and white clovers. *Phytopathology, 54*, 417–418.

Chapman, R. A., & Turner, D. R. (1975). Effect of *Meloidogyne incognita* on reproduction of *Pratylenchus penetrans* in red clover and alfalfa. *Journal of Nematology, 7*, 6–10.

Donald, P. A., Heinz, R., Bernard, E., Hershman, D., Hensley, D., Flint-Garcia, S., & Joost, R. (2012). Distribution, host status and potential sources of resistance to *Vittatidera zeaphila*. *Nematropica, 42*, 91–95.

Donald, P. A., Holguin, C. M., & Agudelo, P. A. (2013). First report of lance nematode (*Hoplolaimus magnistylus*) on corn, soybean, and cotton in Tennessee. *Plant Disease, 97*, 1389.

Endo, B. Y. (1962a). Survival of *Heterodera glycines* at controlled relative humidities. *Phytopathology, 52*, 80–88.

Endo, B. Y. (1962b). Studies on the dessication of the soybean cyst nematode under controlled relative humidity conditions. *Tennessee Farm and Home Science Progress Report, 42*, 12–13.

Epps, J. M. (1957). Soybean cyst nematode found in Tennessee. *Plant Disease Reporter, 41*, 33.

Epps, J. M. (1958). Viability of air-dried *Heterodera glycines* cysts. *Plant Disease Reporter, 42*, 594–595.

Epps, J. M. (1968). Survival of soybean cyst nematodes in seed bags. *Plant Disease Reporter, 52*, 45.

Epps, J. M. (1969). Survival of the soybean cyst nematode in seed stocks. *Plant Disease Reporter, 53*, 403–405.

Epps, J. M. (1971). Recovery of soybean cyst nematodes (*Heterodera glycines*) from the digestive tract of blackbirds. *Journal of Nematology, 3*, 417–419.

Epps, J. M., & Chambers, A. Y. (1958). New host records for *Heterodera glycines*; including one host in the Labiatae. *Plant Disease Reporter, 42*, 194.

Epps, J. M., & Chambers, A. Y. (1959). Mung bean (*Phaseolus aureus*), a host of the soybean cyst nematode (*Heterodera glycines*). *Plant Disease Reporter, 43*, 981–982.

Epps, J. M., & Chambers, A. Y. (1965). *Behavior of soybean-cyst-nematode populations under different cropping systems*. Tennessee Farm and Home Science Progress Report, 53.

Epps, J. M., & Duclos, L. A. (1970). Races of the soybean cyst nematode in Missouri and Tennessee. *Plant Disease Reporter, 54*, 319–320.

Ferris, V. R., & Bernard, R. L. (1962). Injury to soybeans caused by *Pratylenchus alleni*. *Plant Disease Reporter, 46*, 181–184.

Flinchum, W. T. (2001). *Soybean production in Tennessee* (32 pp). University of Tennessee Agricultural Extension Service PB 1608. https://extension.tennessee.edu/publications/Documents/PB1608.pdf. Accessed July 2017.

Freckman, D. W., & Chapman, R. A. (1972). Infection of red clover seedlings by *Heterodera trifolii* Goffart and *Pratylenchus penetrans* (Cobb). *Journal of Nematology, 4*, 23–28.

Halbrendt, J. M. (1996). Allelopathy in the management of plant parasitic nematodes. *Journal of Nematology, 18*, 1–8.

Hershman, D. E. (2008). Soybean cyst nematode, *Heterodera glycines*, populations adapting to resistant soybean cultivars in Kentucky. *Plant Disease, 92*, 1475.

Hershman, D. E. (2009). *The value of wheat residue in soybean cyst nematode management programs* (3 p). University of Tennessee Cooperative Extension Service. PPFS-AG-S-08. http://plantpathology.ca.uky.edu/files/ppfs-ag-s-08.pdf. Accessed July 2017.

Hershman, D. E. (2015). *2015 soybean cyst nematode (SCN) recommendations for Kentucky*. University of Kentucky Plant Pathology Fact Sheet PPFS-AG-S-24. http://plantpathology.ca.uky.edu/extension/soybean-cyst-nematode . Accessed July 2017.

Hershman, D. E., & Bachi, P. R. (1995). Effect of wheat residue and tillage on *Heterodera glycines* and yield of doublecrop soybean in Kentucky. *Plant Disease, 79*, 631–633.

Hershman, D. E., Bachi, P. R., Stuckey, R. E., Clinton, W., & Henson, G. (1986). Efficacy of nematicides and method of application on soybean yield and soybean cyst nematode populations, 1985. *Fungicide and Nematicide Tests, 41*, 79.

Hershman, D. E., Bachi, P. R., Stuckey, R. E., Clinton, W., & Henson, G. (1987). Effect of reduced rates of Temik 15G and soybean cultivar on soybean yield and soybean cyst nematode populations, 1986. *Fungicide and Nematicide Tests, 42*, 169.

Hershman, D. E., Kennedy, B., Yielding, T., Bachi, P., Laurent, C., & Kelley, G. (2009). *Host resistance and seed treatments for management of soybean cyst nematode in Hopkins County, Kentucky, 2009*. Plant Disease Management Reports 4:N029. https://doi.org/10.1094/PDMR03. Online publication. http://www.plantmanagementnetwork.org/pub/trial/pdmr/reports/2010/N029.pdf. Accessed July 2017.

Inserra, R. N., Troccoli, A., Gozel, U., Bernard, E. C., Dunn, D., & Duncan, L. W. (2007). *Pratylenchus hippeastri* n. sp. (Nematoda: Pratylenchidae) from amaryllis in Florida with notes on *P. scribneri* and *P. hexincisus*. *Nematology, 9*, 29–42.

IPCC (Intergovernmental Panel on Climate Change). (2013). Climate change 2013: The physical science basis. In T. F. Stocker, D. Qin, G.-K. Plattner, M. Tignor, S. K. Allen, J. Boschung, A. Nauels, Y. Xia, V. Bex, & P. M. Midgley(Eds.), *Contribution of working group I to the fifth assessment report of the intergovernmental panel on climate change* (1535 pp). Cambridge: Cambridge University Press.

Jennings, P. L., & Bernard, E. C. (1986). Cover crop effects on *Heterodera glycines*. *Journal of Nematology, 18*, 616.

Johnson, L. F., Chambers, A. Y., & Reed, H. E. (1967). Reduction of root knot of tomatoes with crop residue amendments in field experiments. *Plant Disease Reporter, 51*, 219–222.

Johnson, D. W., Townsend, L. H., Green, J. D., Martin, J. R., Witt, W. W., Hershman, D. E., Murdock, L., & Herbek, J. (2015). *Kentucky integrated crop management manual for soybeans, IPM-3*. https://ipm.ca.uky.edu/files/ipm3soy2.pdf. Accessed July 2017.

Jones, L. M., Koehler, A.-K., Trnka, M., Balek, J., Challinor, A. J., Atkinson, H. J., & Urwin, P. E. (2017). Climate change is predicted to alter the current pest status of *Globodera pallida* and *G.rostochiensis* in the United Kingdom. *Global Change Biology, 2017*, 1–11. https://doi.org/10.1111/gcb.13676.

Kelly, H. M. (2015). *Cotton disease and nematode control 2016*. University of Tennessee Extension Publication, 15 pp. https://ag.tennessee.edu/. Accessed July 2017.

Kelly, H. M. Y. (2016). *Soybean disease and nematode ratings and yields. 2015 variety and fungicide trial summaries*. University of Tennessee Extension Bulletin W369, 12 pp.

Kentucky Extension Service. (2017). *Extension publications*. http://plantpathology.ca.uky.edu/extension/soybean-cyst-nematode. Accessed 9 Mar 2018.

Kimmons, C. A., Gwinn, K. D., & Bernard, E. C. (1990). Nematode reproduction on endophyte-infected and endophyte-free tall fescue. *Plant Disease, 74*, 757–761.

Klobe, W. D. (1976). Occurrence of plant parasitic nematodes in seven West Tennessee counties. *Tennessee Farm and Home Science Progress Report, 99*, 14–15.

Koenning, S. R., Kirkpatrick, T. L., Starr, J. L., Wrather, J. A., Walker, N. A., & Mueller, J. D. (2004). Plant parasitic nematodes attacking cotton in the United States: Old and emerging challenges. *Plant Disease, 88*, 100–113.

Mai, W. F., Bloom, J. R., & Chen, T. A. (Eds.). (1977). Biology and ecology of the plant parasitic nematode *Pratylenchus penetrans*. *Northeastern Regional Research Publication Bulletin, 815*, 64.

McGawley, E. C., & Chapman, R. A. (1982). Population dynamics of *Criconemoides simile* on soybean. *Journal of Nematology, 14*, 572–575.

McGawley, E. C., & Chapman, R. A. (1983). Reproduction of *Criconemoides simile*, *Helicotylenchus pseudorobustus*, and *Paratylenchus projectus* on soybean. *Journal of Nematology, 15*, 87–91.

Mojtahedi, H., Santo, G. S., Hang, A. N., & Wilson, J. H. (1991). Suppression of root knot nematode populations with selected rapeseed cultivars as green manure. *Journal of Nematology, 23*, 170–174.

Neal, J. C. (1889). The root knot disease of peach, orange, and other plants in Florida, due to the work of *Anguillula*. *USDA Division of Entomology Bulletin, 20*, 1–31.

Newman, M. A. (1998). *Summary of nematode survey activity in Tennessee*. National Cotton Council of America. http://www.cotton.org/tech/pest/nematode/survey/tennessee.cfm. Accessed July 2017.

Niblack, T. L., & Bernard, E. C. (1985). Plant parasitic nematode communities in dogwood, maple, and peach nurseries in Tennessee. *Journal of Nematology, 17*, 132–139.

Niblack, T. L., & Riggs, R. D. (2004). Variation in virulence phenotypes. In D. P. Schmitt, J. A. Wrather, & R. D. Riggs (Eds.), *Biology and management of soybean cyst nematode* (2nd ed., pp. 57–71). Marceline: Schmitt and Associates of Marceline.

Niblack, T. L., & Tylka, G. L. (2008). *SCN soybean cyst nematode management guide*, 5th edition. www.ncsrp.com/pdf_doc/SCN_Management.pdf. Accessed July 2017.

Niblack, T. L., Arelli, P. R., Noel, G. R., Opperman, C. H., Orf, J. H., Schmitt, D. P., Shannon, J. G., & Tylka, G. L. (2001). A revised classification scheme for genetically diverse populations of *Heterodera glycines*. *Journal of Nematology, 34*, 279–288.

Osborne, W. W., & Jenkins, W. R. (1962). Pathogenicity of *Pratylenchus vulnus* on boxwood, *Buxus sempervirens* var. *arborescens*. *Plant Disease Reporter, 46*, 712–714.

Riggs, R. D., & Schmitt, D. P. (1988). Complete characterization of the race scheme for *Heterodera glycines*. *Journal of Nematology, 20*, 392–395.

Robbins, R. T., Shipe, E., Shannon, G., Chen, P., Kantartzi, S. K., Jackson, L. E., Gbur, E. E., Dombek, D. G., & Velie, J. T. (2013, January). *Reniform nematode reproduction on soybean cultivars and breeding lines in 2012*. In Proceedings of the Beltwide cotton conference, San Antonio, TX.

Robinson, A. F., Inserra, R. N., Caswell-Chen, E. P., Vovlas, N., & Troccoli, A. (1997). *Rotylenchulus* species: Identification, distribution, host ranges, and crop plant resistance. *Nematropica, 27*, 127–180.

Rutledge, A. D., Wills, J. B., & Bost, S. (1999). *Commercial tomato production*. University of Tennessee Agricultural Extension Service Publication PB 737, 32 pp. https://extension.tennessee.edu/publications/Documents/pb737.pdf. Accessed July 2017.

Schmitt, D. P., Wrather, J. A., & Riggs, R. D. (Eds.). (2004). *Biology and management of soybean cyst nematode* (2nd ed.p. 262). Marceline: Schmitt and Associates of Marceline.

Scribner, F. L. (1889). Diseases of the Irish potato. *Bulletin of the Tennessee Agricultural Experiment Station, 2*, 27–45.

Seebold, K. W. (2010). *Root-knot nematode in commercial and residential crops*. University of Kentucky Plant Pathology Fact Sheet PPFS-GEN-10. https://plantpathology.ca.uky.edu/files/ppfs-gen-10.pdf. Accessed July 2017.

Seebold, K. W., Pearce, R. C., Bailey, W. A., Bush, L. P., Green, J. D., Miller, R. D., Powers, L. A., Snell, W. M., Townsend, L. H., Purschwitz, M., Wells, L. G., Wilhoit, J. H., Velandia, M., Bost, S., Burgess, G., Hale, F., Hensley, D., Denton, P., & Rhodes, N. (2013). *2013–2014 Kentucky and Tennessee tobacco production guide*. University of Kentucky Agriculture and Natural Resources Publication 74, 71 pp. http://uknowledge.uky.edu/anr_reports/74. Accessed July 2017.

Shannon, J. G., Arelli, P. R., & Young, L. D. (2004). Breeding for resistance and tolerance. In D. P. Schmitt, J. A. Wrather, & R. D. Riggs (Eds.)., 2004 *Biology and management of soybean cyst nematode* (2nd ed., pp. 155–180). Marceline: Schmitt and Associates of Marceline.

Sherbakoff, C. D. (1939). Root knot nematodes on cotton and tomatoes in Tennessee. *Phytopathology, 29*, 751–752.

Smith, D. A. (1995). Richard A. Chapman, 1918 to 1994. *Phytopathology, 85*, 1052.

Society of Nematologists. (1984). *Distribution of plant parasitic nematode species in North America* (p. 205). Gainesville: Society of Nematologists.

Southards, C. J. (1967). The pseudo-root knot nematode of Bermuda grass in Tennessee. *Plant Disease Reporter, 51*, 455.

Southards, C. J. (1971). Effect of fall tillage and selected hosts on the population density of *Meloidogyne incognita* and *Pratylenchus zeae*. *Plant Disease Reporter, 55*, 41–44.

Southards, C. J. (1973). A field evaluation of nematode-resistant tomato varieties for vine-ripe tomato production. *Tennessee Farm and Home Science, 85*, 18–20.

Stebbins, T. C., Newman, M. A., & Cook, D. L. (1998). Incidence of the reniform nematode in West Tennessee cotton fields. *Phytopathology, 88*(9 Suppl), S85.

Steiner, G. (1943). In C. D. Sherbakoff, & W. W. Stanley (Eds.), *The more important diseases and insect pests of crops in Tennessee*. University of Tennessee Agricultural Experiment Station Bulletin 186, 142 pp.

Stetina, S. R., Smith, J. R., & Ray, J. D. (2014). Identification of *Rotylenchulus reniformis*-resistant *Glycine* lines. *Journal of Nematology, 46*, 1–7.

Stockdale, W. R. (1985). *Relationship of sampling date and soil type to southern root knot nematode (Meloidogyne incognita) populations in selected fields in Cocke County, Tennessee*. M.S. thesis, University of Tennessee, 29 pp.

Stuckey, R. E., & Chapman, R. A. (1979). *The influence of nematicides and varieties on soybean cyst nematode populations and soybean yield in Kentucky*. Ninth International Congress of Plant Protection, IX International Congress of Plant Protection, Washington, D.C., Abstract 867.

Stuckey, R. E., Clinton, W., Hershman, D. E., Bachi, P. R., & Henson, G. (1985). Effectiveness of nematicides and application method on the soybean cyst nematode and soybean yield, 1984. *Fungicide and Nematicide Tests, 40*, 110.

Thames, W.H. (1982). The genus *Pratylenchus*. In R. D. Riggs (Ed.), *Nematology in the Southern Region of the United States* (pp. 108–126). Southern Cooperative Series Bulletin 276, 206 pp.

Triantaphyllou, A. C. (1976). Genetic structure of races of *Heterodera glycines* and inheritance of ability to reproduce on resistant soybeans. *Journal of Nematology, 7*, 356–364.

Turner, D. R., & Chapman, R. A. (1972). Infection of seedlings of alfalfa and red clover by concomitant populations of *Meloidogyne incognita* and *Pratylenchus penetrans*. *Journal of Nematology, 4*, 280–286.

Tyler, D. D., Overton, J. R., & Chambers, A. Y. (1983). Tillage effects on soil properties, diseases, cyst nematodes, and yields. *Journal of Soil and Water Conservation, 38*, 374–376.

Tyler, D. D., Chambers, A. Y., & Young, L. D. (1987). No-tillage effects on population dynamics of soybean cyst nematode. *Agronomy Journal, 79*, 799–802.

Tylka, G. L. (2016). Understanding soybean cyst nematode HG types and races. *Plant Health Progress, 17*, 149–151.

Tylka, G. L., & Marett, C. C. (2014). Distribution of the soybean cyst nematode, *Heterodera glycines*, in the United States and Canada: 1954 to 2014. *Plant Health Progress, 15*, 13. https://doi.org/10.1094/PHP-BR-14-0006.

University of Tennessee. (2017). *Soybean diseases and nematodes*. http://utcrops.com/soybean/diseases.htm. Accessed 9 Mar 2018.

USEPA (U.S. Environmental Protection Agency). (2003). *Ecoregions of Kentucky*. ftp://newftp.epa.gov/EPADataCommons/ORD/Ecoregions/ky/ky_eco_pg.pdf. Accessed July 2017.

USGCRP (U.S. Global Change Research Program). (2009). *Global climate change impacts in the United States*. https://nca2009.globalchange.gov/index.html. Accessed July 2017.

USGS (U.S. Geological Survey). (2001). *Geologic provinces of the United States: Interior Plain Province*. https://geomaps.wr.usgs.gov/parks/province/intplain.html. Accessed July 2017.

USGS (U.S. Geological Survey). (2002). *Ecoregions of Tennessee*. https://store.usgs.gov/assets/MOD/StoreFiles/Ecoregion/21632_tn_front.pdf. Accessed July 2017.

Valleau, W. D., & Johnson, E. M. (1947). The relation of meadow nematodes to brown root rot of tobacco. *Phytopathology, 37*, 838–841.

Ward, N., Hershman, D., & Dunwell, W. (2011). Soybean cyst nematode: A potential problem for nurseries. *Kentucky Cooperative Extension Service, ID-110*, 4.

Wartman, F. S., & Bernard, E. C. (1985). Reproduction of *Pratylenchus alleni* (Nematoda) on soybean and other field and vegetable crops. *Tennessee Farm and Home Science, 135*, 3–5.

Wilson, C. T. (1982). Development of an expanded program of teaching and research in nematology in the Southern Region. In R. D. Riggs (Ed.). *Nematology in the Southern Region of the United States* (pp. 1–7). Southern Cooperative Series Bulletin 276, 206 pp.

Winstead, N. N., Skotland, S. B., & Sasser, J. N. (1955). Soybean-cyst nematode in North Carolina. *Plant Disease Reporter, 39*, 9–11.

Wrather, J. A., & Koenning, S. R. (2006). Estimates of disease effects on soybean yields in the United States 2003 to 2005. *Journal of Nematology, 38*, 173–180.

Young, L. D. (1982a). Reproduction of differentially selected soybean cyst nematode populations on soybean. *Crop Science, 22*, 385–388.

Young, L. D. (1982b). Reproduction of Tennessee soybean cyst nematode population on cultivars resistant to Race 4. *Plant Disease, 66*, 251–252.

Young, L. D. (1984a). Effects of continuous culture of resistant soybean cultivars on soybean cyst nematode reproduction. *Plant Disease, 68*, 237–239.

Young, L. D. (1984b). Changes in the reproduction of *Heterodera glycines* on different lines of *Glycine max*. *Journal of Nematology, 16*, 304–309.

Young, L. D. (1990). Survey of soybean cyst nematode races in Tennessee. *Journal of Nematology, 22*(4S), 672–675.

Young, L. D. (1992). Problems and strategies associated with long-term use of nematode resistant cultivars. *Journal of Nematology, 24*, 228–233.

Young, L. D. (1994). Changes in the *Heterodera glycines* female index as affected by ten-year cropping sequences. *Journal of Nematology, 26*, 505–510.

Young, L. D. (1996a). Yield loss in soybean caused by *Heterodera glycines*. *Journal of Nematology, 28*(4S), 604–607.

Young, L. D. (1996b). Advising growers in selection of *Heterodera glycines*-resistant soybean cultivars. *Journal of Nematology, 28*(4S), 608–611.

Young, L. D. (1998). Influence of soybean cropping sequences on seed yield and female index of the soybean cyst nematode. *Plant Disease, 82*, 615–619.

Young, L. D., & Hartwig, E. E. (1992). Cropping sequence effects on soybean and *Heterodera glycines*. *Plant Disease, 76*, 78–81.

Young, L. D., & Kilen, T. C. (1994). Genetic relationships among plant introductions for resistance to soybean cyst nematode Race 5. *Crop Science, 34*, 936–939.

Chapter 13
Nematodes in Maryland and Delaware Crops

Ramesh R. Pokharel

13.1 Introduction

In Maryland, the green industry currently ranks second among agricultural commodities with a total of approximately $2 billion in gross receipts, occupying 8458 ha, including 1,765,158 m^2 of greenhouse space and employing more than 18,500 people with wages totaling $451 million. These businesses deal with several problems each year, including plant parasitic nematodes (PPN) that cause tremendous challenges because of difficulties of diagnosis and management.

The State of Maryland is divided into 24 counties and separated by Chesapeake Bay into two major production regions with variations in soil and microclimate. The Eastern Shore region, producing more agricultural crops, is shared with the Delaware Peninsula and has predominantly sandy soil, while the western region of the state has clay soil. Because of microclimatic differences throughout the state, a good variation of crops is grown including cereals, vegetables and fruit. In 2015, 15,378 ha of land were planted to corn producing 1,574,883 tons. Winter wheat yielded 4.3 tons per ha, with 431,823 ha harvested. Barley production has grown to 50,803 tons, averaging 4.64 tons per ha (MDA 2015). The State of Maryland also has significant acreage in fresh market vegetables including watermelon, snap bean and cucumber, with a value of $35.6 million in 2015, while the potato crop was worth $8.31 million. Sweet corn for fresh corn-on-the-cob is grown on over 1416 ha. A wide variety of fruits and vegetables are grown for direct sales to the public at farm stands, roadside markets, pick-your-own operations and farmers' markets throughout the state. Watermelon, cantaloupe, sweet corn, cabbage, green bean, potato, pepper, tomato, pumpkin, peach, apple and strawberry are the major crops

R. R. Pokharel (✉)
Maryland Department of Agriculture, Harry S Truman PKWY, Annapolis, MD, USA
e-mail: ramesh.pokharel@maryland.gov

S. A. Subbotin, J. J. Chitambar (eds.), *Plant Parasitic Nematodes in Sustainable Agriculture of North America*, Sustainability in Plant and Crop Protection,
https://doi.org/10.1007/978-3-319-99588-5_13

grown for the fresh market. This state is a major watermelon producing region with 1214 ha under cultivation.

Orchards cover 1610 ha in Maryland, with apples and peaches as the most productive crops. In 2015, about 728 ha of apple orchards and 724 ha of peach orchards were valued at nearly $12.2 million (USDA 2017). In addition, grape cultivation has been increasing with about 347 ha of vineyard and wineries in the state (MGGA 2014). A small number of cover crops grown in the state include rye, barley and other cereal grains. These are planted in the fall after summer crops are harvested.

Delaware is a small state of three counties and lies in the east of Maryland, in the Eastern Shore region near Delaware Bay and Atlantic Ocean. Agriculture is Delaware's largest single land use, with 41% of the state's land hectares in farming. The annual aggregate (direct, indirect, and induced) economic contribution of agriculture to Delaware's economy is estimated at nearly $8 billion (Cadwallader 2010). The annual value of agricultural production is over $1 billion. The value of agricultural products sold annually directly to consumers is over $3.5 million. Kent and Sussex are among top 2% of U.S. counties in value of vegetables sold. Delaware ranks number one nationally in value of agricultural products sold per farm at $425,387 and value of agricultural production produced per acre of land in farms at $2,123 (Cadwallader 2010). The state has about 2450 farms and more than 46,539 ha of farmland permanently preserved for agriculture. Soybean is the state's most important crop, followed by corn. Farmers also grow barley and wheat among grain crops, whereas potato and pea are the state's largest vegetable crop. Apple is Delaware's greatest fruit crop. Greenhouse and nursery products (flowers, ornamental shrubs, young plants) provide some income. About 2,500 farms spread across 206,491 ha of farmland benefit from some natural advantages such as presence of the state's natural soil, 'Greenwich'. This soil is classified as "Prime Farmland Soil," meaning it is one of the most productive soils for the state's agriculture and forestry (Do 2014).

13.2 Plant Parasitic Nematodes of Importance

13.2.1 Maryland

Several species of plant parasitic nematodes occur in Maryland soils, generally as mixed populations that are unevenly distributed throughout a field. Nematodes were estimated to cause 20% reduction in crop yields 60 years ago (Jenkins et al. 1957). It is hard to predict this figure at present because of a lack of recent data. Limited numbers of PPN-related publications exist for Maryland. Most regional publications have been based on surveys and nematode diagnostics. Research and diagnostic studies have been based on the traditional morphological taxonomic approach rather than molecular techniques. Available information indicates that most targeted, major areawide research on PPN was conducted in the 1950s. Jenkins et al. (1957) did extensive surveys of 1,210 farms and gardens with different crop plants. This was probably the only thorough published survey covering all counties in Maryland. These researchers collected variable numbers of samples in each county, with a minimum of 20 each in Kent and Hartford County and a maximum of 121 in Wicomico County. They found PPN in all samples examined and recorded 34 verified and possible PPN genera.

Plant parasitic nematodes of the genera *Xiphinema* (dagger nematodes) followed by *Pratylenchus* (root lesion nematodes), *Tylenchorhynchus* (stunt nematode) and *Ditylenchus* (stem and bulb nematode) were the most commonly recorded nematodes in Maryland (Jenkins et al. 1957). However, at present, soybean cyst (*Heterodera glycines*), root knot (*Meloidogyne* spp.), root lesion, stem and bulb and dagger nematodes are economically important in the state. Research projects have focused on crops and/or common and important associated nematode genera.

13.2.2 Delaware

Limited numbers of PPN-related publications for Delaware exist, similar to Maryland. Unlike Maryland, extensive nematode related work in the past creates a gap for the importance of plant parasitic nematodes in the state. Most regional publications only appeared recently and have been based on diagnostic and bioassay works. As in Maryland, research and diagnostic works have been based on the traditional taxonomic approach rather than modern molecular techniques. However, the important nematode genera within the state vary with crop and location and need further investigation. In Delaware, there is evidence of a shift in nematode populations, but more detailed investigations are needed. Similar to Maryland, at present, soybean cyst, root knot, and root lesion, stem and bulb and dagger nematodes appear to be economically important.

In this chapter, only known plant parasitic nematode genera are discussed for both states. Emphasis has been given to those genera/species which are more important or common or may be important for plant production. A nematode genus and/or species encountered more than once (sample, location or published paper) is considered common; populations causing significant damage or impact in plant production are considered important. Based on available records, nematode genera or species found in Maryland and Delaware are listed in Table 13.1. However, nematode genera and, or species, described below are based on published records in journals, extension publications, websites and personal communications for Maryland and Delaware. This paper does not include write-ups for all species or genera listed in 'Widely Prevalent Nematodes in the USA' (USDA 2014). Some references identified species that are listed as occurring in Maryland from specimens present in the USDA Nematological Laboratory collections.

Table 13.1 Some plant parasitic nematode species encountered in Maryland and Delaware

Species	Crop or plant	References
Anguina tritici	Wheat	Jenkins et al. (1957)
Aphelenchoides fragariae	Strawberry, butterfly bush	Jenkins et al. (1957), Esser (1966), and USDA (2014)
A. ritzemabosi	Chrysanthemum, strawberry	Jenkins et al. (1957) and USDA (2014)
A. besseyi	Chrysanthemum, strawberries	Jenkins et al. (1956), Esser (1966), and Cavigelli et al. (2005)
Belonolaimus longicaudatus	Soybean, turfgrass	Handoo et al. (2010)
Criconema mutuabile	Unknown	USDA (2014)
Ditylenchus dipsaci	Phlox, tulip, narcissus, hyacinth, onion, garlic	USDA (2014)
D. myceliophagus	Corn	Cavigelli et al. (2005)
Dolichodorus similis	Celery, sweet corn	Feldmesser and Golden (1972)
D. marylandicus	Perennial bluegrass	Lewis and Golden (1980)
H. dihystera	Alfalfa, asparagus, barley, clover, corn, oat, soybean, timothy, tobacco, tomato, vetch, wheat	Jenkins et al. (1957), Golden and Rebois (1978), and USDA (2014)
H. digonicus	Corn, turfgrass	Feldmesser and Golden (1972), Cavigelli et al. (2005), and USDA (2014)
H. erythrinae	Chrysanthemum, cucumber, lespedeza, muskmelon, pea, strawberry, watermelon	Jenkins et al. (1957) and USDA (2014)
H. microlobus	Unknown	USDA (2014)
H. multicinctus	Some plants	Bernard and Keyserling (1985)
H. pseudorobustus	Soybean, corn, turfgrass, tobacco	Feldmesser and Golden (1972), Golden and Rebois (1978), Cavigelli et al. (2005), Kaplan et al. (2008), and USDA (2014)
H. platyurus	Unknown	USDA (2014)
Hemicycliophora spp.	Barley, corn, grasses, raspberry	Jenkins et al. (1956, 1957)
Heterodera glycines	Soybean	Sindermann et al. (1993) and USDA (2014)
H. schachtii	Clover	Jenkins et al. (1957) and Golden and Rebois (1978)

(continued)

Table 13.1 (continued)

Species	Crop or plant	References
H. zeae	Corn, barley, oat, rice, sorghum, sugar cane, wheat, fall panicum, meadow foxtail, green sprangletop, witchgrass, broomcorn, fountain grass, reed canary grass, common reed, eastern gamagrass, teosinte	McGrown (1981), Ringer et al. (1987), and Sindermann et al. (1993)
Hoplolaimus galeatus	Carnation, clover, corn, grape, grass, lespedeza, oat, pea, pepper, ryegrass, soybean, sweet potato, timothy, turfgrassn tomato, tobacco, wheat	Jenkins et al. (1957), Feldmesser and Golden (1972), Golden and Rebois (1978), and USDA (2014)
H. columbus	Cotton, soybean, corn	Jenkins et al. (1957)
Longidorus spp.	Corn, soybean, tobacco	Jenkins et al. (1956, 1957)
Meloidogyne graminis	Turfgrass	Jenkins et al. (1956) and Feldmesser and Golden (1972)
M. hapla	Alfalfa, clover, tobacco, soybean, vetch	Sasser (1954), Jenkins et al. (1957), Golden and Rebois (1978), and USDA (2014)
M. javanica	Snapdragon plants	Jenkins et al. (1957) and Golden and Rebois (1978)
M. incognita	Corn, muskmelon, soybean, sweet potato, tobacco, tomato, vetch, wheat	Jenkins et al. (1957) and USDA (2014)
M. sasseri	American beach grass	Handoo et al. (1993)
Meloidogyne spp.	Snap bean, watermelon, cucumber, tomato, pepper, sorghum, sudangrass, lima bean, pea, cantaloupe, muskmelon, pumpkin, squash, potato, corn, as well as weeds such as, dandelion, mallow, purslane, pigweed, prickly sida, morning glory	Sasser (1954) and Golden and Rebois (1978)
Merlinius brevidens	Alfalfa, barley, corn, clover, grasses, oat, pea, timothy, wheat	Jenkins et al. (1957), Cavigelli et al. (2005), and USDA (2014)
Mesocriconema curvatum	Unknown	USDA (2014)
M. ornatum	Turfgrass	USDA (2014)
M. rusticum	Unknown	USDA (2014)
M. xenoplax	Unknown	USDA (2014)
M. simile	Peach	Jenkins et al. (1957), 4 and USDA (2014)
Nanidorus minor	Soybean	Golden and Rebois (1978)
Paratrichodorus pachydermus	Turfgrass	Jenkins et al. (1957)
Paratylenchus dianthus	Carnation, clover, corn, timothy, vetch, wheat	Jenkins et al. (1957)
P. hamatus	Clover	Jenkins et al. (1957)
P. nanus	Wheat	Jenkins et al. (1957)

(continued)

Table 13.1 (continued)

Species	Crop or plant	References
P. projectus	Alfalfa, bean, clover, corn, grass, lespedeza, soybean	Jenkins et al. (1957), Golden and Rebois (1978), Cavigelli et al. (2005), and USDA (2014)
Pratylenchus agilis	Soybean	Golden and Rebois (1978) and USDA (2014)
P. brachyurus	Alfalfa, asparagus, barley, clover, corn, grass, oat, rye, sorghum, soybean, timothy, tobacco, tomato, wheat	Jenkins et al. (1957) and Golden and Rebois (1978)
P. coffeae	Soybean	Golden and Rebois (1978) and Cavigelli et al. (2005)
P. crenatus	Soybean	Golden and Rebois (1978)
P. hexincisus	Soybean	Jenkins et al. (1957) and Golden and Rebois (1978)
P. neglectus	Corn	Cavigelli et al. (2005)
P. penetrans	Corn, soybean, bean, clover, tobacco, tomato	Jenkins et al. (1957), Golden and Rebois (1978), Cavigelli et al. (2005), and USDA (2014)
P. pinguicaudatus	Corn	Cavigelli et al. (2005)
P. pratensis	Alfalfa, barley, bean, clover, corn, grass, oat, pea, rye, sorghum, soybean, timothy, tobacco, tomato, vetch, wheat	Jenkins et al. (1957)
P. scribneri	Soybean	Golden and Rebois (1978) and USDA (2014)
P. subpenetrans	Boxwoods, broccoli, carnation, lespedeza, muskmelon, peach, potato, snapdragon, sweet potato, watermelon	Jenkins et al. (1957)
P. thornei	Corn	Cavigelli et al. (2005) and USDA (2014)
P. vulnus	Boxwood, ornamentals, fruits, nuts, vegetables	Jenkins et al. (1957), McGrown (1981), and USDA (2014)
P. zeae	Soybean	Jenkins et al. (1957), Golden and Rebois (1978), and USDA (2014)
Quinisulcius acutus	Soybean	Golden and Rebois (1978)
Rotylenchus buxophilus	Alfalfa, corn, grass, tobacco, wheat	Jenkins et al. (1957)
R. robustus	Alfalfa	Jenkins et al. (1957)
Scutellonema brachyurus	Unknown	USDA (2014)
Trichodorus primitivus	Mimosa	Jenkins et al. (1957) and USDA (2014)

(continued)

Table 13.1 (continued)

Species	Crop or plant	References
Tylenchorhynchus capitatus	Unknown	Jenkins et al. (1957)
T. clarus	Unknown	USDA (2014)
T. clatytoni	Alfalfa, barley, bean, clover, corn, grasses, oat, pepper, rye, soybean, strawberry, sweet potato, Timothy grass, tomato, tobacco, soybean, grasses, vetch, and wheat	Jenkins et al. (1956, 1957), Feldmesser and Golden (1972), Golden and Rebois (1978), and USDA (2014)
T. dubius	Barley and grasses, clover, grasses, oat, vetch, and wheat	Jenkins et al. (1957), Feldmesser and Golden (1972) and USDA (2014)
Geocenamus ornatus	Barley	Jenkins et al. (1957)
Xenocriconemella macrodora	Unknown	USDA (2014)
Xiphinema americanum	Tomato, turfgrass, soybean	Jenkins et al. (1957), Sindermann et al. (1993), Golden and Rebois (1978), Jenkins et al. (1956), Cavigelli et al. (2005), USDA (2014), and Evans et al. (2007)
X. chambersi	Unknown	USDA (2014)
X. rivesi	Apple, corn, potato, tobacco	Cavigelli et al. (2005) and USDA (2014)

13.3 Major Plant Parasitic Nematodes of Maryland and Delaware

13.3.1 Cyst Nematodes, Heterodera *spp.*

This important group of plant parasitic nematodes is becoming common in many areas within Maryland and Delaware. In Maryland, during the 1957 survey, Jenkins et al. (1957) identified the genus *Heterodera* in 7% of the total number of samples examined, mostly from northern and western counties. Five species of cyst nematodes were recorded from Maryland namely, *H. schachtii*, *H. trifolii*, *H. glycines* and *Cactodera* sp. At that time, *H. schachtii* group was the most commonly found species in Maryland. *Heterodera trifolii* was identified from clover in the state. This crop had frequently shown poor growth that initially was suspected to be caused by nematodes. *Heterodera trifolii* was also observed from fields with alfalfa, barley, bean, clover, corn, grasses, oat, pea, rye, soybean, Timothy grass, tomato, vetch and wheat in all Maryland counties, except the five southern counties (Anne Arundel, Prince George's, Charles, Calvert, and St. Mary's) and the Lower Eastern Shore counties (Wicomico, Somerset, and Worcester). Later, Golden and Rebois (1978) reported *H. schachtii* and *H. trifolii* in 15% of samples from soybean fields. In Maryland, the cyst nematodes have been regularly observed each year during Maryland Department of Agriculture soybean cyst nematode surveys (Maryland Department of Agriculture survey records) and are known to cause severe crop loss and poor soybean plant growth in Maryland. '*Heterodera cacti*' identified by Jenkins et al. (1957) from soybean field samples, should be considered as *Cactodera weissi* described later by Mulvey and Golden (1983). In Delaware, *H. glycines* is the only cyst nematode species reported.

13.3.2 Soybean Cyst Nematode (SCN), Heterodera glycines

Heterodera glycines damages soybean roots, reduces yield, and may cause reduction in plant height. Soybean cyst nematode is the most economically important nematode pest of soybean in the U.S., with yield losses estimated at $1.5 billion annually. From 2010 to 2014, surveys of soybean yield losses in 28 soybean-producing US states and Canada found that SCN was estimated to have caused more than twice as much yield loss to any other disease (Allen et al. 2017).

The SCN was first found in North America in North Carolina in 1954, and since then has spread to at least 31 soybean-producing states, and Canada. In Maryland, this nematode was first detected in Worcester County in 1980 (Sardanelli et al. 1982). The county was placed under quarantine by the Maryland Department of Agriculture. In 1990, more extensive surveys were conducted in various counties and eight were found positive for the soybean cyst nematode. Percentages of positive samples for each county are given in parenthesis: Caroline (31%), Dorchester (21%), Kent (3%), Queen Anne's (7%), Somerset (35%), Talbot (20%), Wicomico (51%) and Worcester (36%) (Sindermann et al. 1993). Further spread of this nematode to uninfected fields was suspected and races 1 and 3 of SCN were identified in a small number of fields in Maryland (Sindermann et al. 1993). During a CAPS (the Cooperative Agricultural Pest Survey program of the USDA APHIS PPQ), the Maryland Department of Agriculture detected SCN in two more counties: Charles and St Mary's Counties in 1993, and 1996. Surveys conducted in Maryland in 1993 and 1999 found that other than the ten counties mentioned above, other parts of the state were found free from SCN. However, in 2012, the Maryland Department of Agriculture detected SCN in two additional counties (Cecil and Harford) during a regular survey for the nematode.

In Delaware, the nematode was first discovered in the fall of 1979 when soybean cyst nematode was widespread in Sussex County (Mulrooney 2011). Although found in Kent County just a few years later, SCN was not discovered until 1991 in the adjacent Newcastle County, near Clayton. Soybean cyst nematode, which is not restricted by soil type, can be found anywhere soybeans have been grown for a long time. The symptoms of infection are not always obvious so it is difficult to determine incidence of the disease. It can go undetected for years until severe stunting or yield losses are experienced during harvest. Often growers can be unaware of the presence of SCN in fields and therefore need to understand how to detect and manage the nematode.

In Delaware, during 1993 and 1994, a major effort was made to survey soybean acreage for SCN and determines prevalent race composition of field populations present. During that time, about 60% and 30% samples belonged to race 3 and race 1 respectively. No similar surveys for SCN were conducted in Delaware from 1996 to 2009. In 2009, a second survey conducted in Delaware to determine the SCN race in fewer fields found a shift in composition. This time seven populations belonged to race 1, representing 47% of samples, five populations of race 5, representing 33% of the samples and three populations of race 2 accounted for 20% of the samples. Race 3 populations were not observed. This study also found evidence that soybean cv. 'PI88788' (a Round-up® ready cultivar) was no different from a susceptible one. In 2010, testing of a small set of samples indicated that a majority of the tested populations were race 1. This may have been due to the use of PI88788 which allowed reproduction of race 1 populations (Mulrooney and Gregory 2010). These two studies further confirmed that there was a shift in race composition of SCN in Delaware. In addition, Mulrooney and Gregory (2010) indicated that SCN was the most limiting biotic factor of soybean production in Delaware during that period. Mulrooney (2011) further suggested that growers may need to plant soybean cultivars derived from non-PI88788 resistance sources in order to successfully manage soybean cyst nematode in the future. Furthermore, in Delaware, soybean cyst nematode was observed only in 32% of 38 samples obtained in a Nematode Assay Service conducted by the University of Delaware Extension Service in 2015 (Kness and Kleczewski 2015). There is a current need for more extensive surveys to determine the shift in race composition and virulence of the nematode populations, since periodic checks of race and virulence of pathogen is important for the efficient management of the pathogen. No such information is presently available for Maryland.

13.3.2.1 Management

In Wicomico County, Maryland, soil amendment with poultry litter at V2 and V5 stages of soybean development in Manokin (resistant) and Essex (susceptible) soybean cultivars, at rates of 0, 5 or 10 t/ha found that 5 t/ha poultry litter was an effective means of improving productivity when soybean is grown in SCN-infested soil and the poultry litter had a greater impact on reducing cyst infestation in Essex than Manokin (Mervalin et al. 1997).

In Delaware, cropping without soybean is recommended in fields with a history of SCN. Due to limited sources of SCN resistance available for the area, avoiding continuous planting of soybeans and rotating with a crop such as corn, for at least one season between soybean plantings is recommended. Fields with higher numbers of nematodes should crop more years without soybean (Ernest and Johnson 2014). In the past, crop rotation and use of resistant varieties of soybean such as PI88788, were recommended for the control of this nematode. PI88788, the major source of resistance to SCN for the last 25 years, was very effective against the common races of SCN, and its resistance was easily incorporated into new varieties. In fact, there are few modern soybean varieties without SCN resistance (Mulrooney 2011).

13.3.3 Corn Cyst Nematode, Heterodera zeae

Heterodera zeae feeds on corn (*Zea mays*) causing stunted plant growth and reduced yield (Hashmi et al. 1993; Krusberg et al. 1997). In addition, this species is of phytosanitary concern.

In early 1981, corn cyst nematode (CSN) was discovered for the first time in the Western Hemisphere, in corn fields in Kent County, Maryland (Sardanelli et al. 1981). This species was found only in four counties, Cecil, Harford, Kent, and Queen Anne's Counties. Fields known to be infested with the corn cyst nematode were quarantined by the Maryland Department of Agriculture in 1986, and the quarantine was lifted in 1996. The nematode was later identified from Cumberland County, Virginia, over 274 km from the nearest known infested field in Maryland.

In 1987, the host range of CSN was investigated. This nematode infected all 22 corn cultivars tested, along with certain barley (*Hordeum vulgare*) cultivars, oat (*Avena sativa*), rice (*Oryza sativa*), sorghum (*Sorghum bicolor*), sugarcane (*Saccharum* interspecific hybrid) and wheat (*Triticum aestivum*). Fall panicum (*Panicum dichotomiflorum*), a weed species common to cultivated fields in Maryland, was also a host for *Heterodera zeae*. Other hosts included meadow foxtail (*Alopecurus pratensis*), *Calamagrostis eipgeios*, Job's tears (*Coix lachryma-jobi*), green sprangletop (*Leptochloa dubia*), witchgrass (*Panicum capillare*), broomcorn (*Panicum miliaceum*), fountain grass (*Pennisetum rueppeli*), reed canary grass (*Phalaris arundinacea*), common reed (*Phragmites australis*), eastern gamagrass (*Tripsacum dactyloides*), corn (*Zea mays*) and teosinte (*Zea mexicana*) (MacGrown 1981; Ringer et al. 1987).

The optimum temperature for development of a Maryland population of *H. zeae* was reported to be 36° C (Hutzell and Krusberg 1990). Reproduction of *H. zeae* increased with temperature increasing from 24 to 36 °C. (Hutzell and Krusberg 1990). They also showed that temperature affected nematode-induced suppression of plant growth. Females were produced in bioassays of cysts recovered from soil which had been stored for 38 months at 24 °C and for 32 months at 2 °C. No second stage juveniles were recovered from soil after 1 month in storage at −18 °C, but even after 7-month storage, second stage juveniles emerged from cysts and developed into females (Krusberg and Sardanelli 1989).

Dry weight and yield responses of corn plants to *H. zeae* were greater in coarse-textured soil than in fine-textured soil. Fertilizer amendments did not alleviate suppression of plant growth by *H. zeae*. The nematodes suppressed corn yields to a greater degree and more consistently in sandy soil than in silty soil and caused more plant damage in hot and dry than in cool and wet seasons (Krusberg et al. 1997). Because of a higher temperature requirement to complete a life cycle, *H. zeae* was considered economically unimportant in Maryland. The present situation of corn cyst nematode in Maryland including original positive counties, is unknown. No report on CSN exists for Delaware, even though some counties of Maryland adjoin corn cyst nematode-positive counties.

13.3.3.1 Management

In Maryland, from 1982 to 1984, several granular nematicides and one fumigant were applied in experimental field plots with populations of 50–300 cysts/250 cm^3 in Kent and Harford Counties. Fumigation greatly lowered nematode population densities in soil without any increase in corn yield, compared to unfumigated soil (Krusberg et al. 1997).

13.3.4 Root Knot Nematodes, **Meloidogyne** *spp.*

Root knot nematodes are endoparasitic nematodes that infect several plant species worldwide and cause approximately 5% of global crop loss (Sasser and Carter 1985). Root knot nematode juveniles infect plant roots causing the development of

Fig. 13.1 (**a**) Root knot nematode infection of carrot. (**b**) Dagger nematode

galls (Fig 13.1a) that interfere with proper intake and utilization of the plant's photosynthate and nutrients. Infection of young plants may be lethal, while infection of mature plants causes decreased yield and quality.

In Maryland, *Meloidogyne* spp. were reported from 12 counties (Sasser 1954). Very low percentages of 1%, 3%, 10% of 143 corn, 111 tobacco and 74 soybean samples, respectively, contained *Meloidogyne* spp. (Jenkins et al. 1956). However, a year later, Jenkins et al. (1957) described three species from their survey. They suggested that these nematodes might be important only in lighter soils of the Eastern Shore and Southern Maryland. *Meloidogyne* spp. were observed in 19% and 22% of soil and root samples of soybean (Golden and Rebois 1978) and in 2% of samples examined from soybean fields in eight counties of Maryland (Sindermann et al. 1993).

Over the past 20 years, significant numbers of root knot nematode populations have been found regularly in commercial plant and soil samples submitted to the University of Maryland Nematology Laboratory (Everts et al. 2006). In Delaware, root knot nematode was the most common plant parasitic nematode, occurring in 24% of 38 samples in a University of Delaware Extension Nematode Assay Service in 2015 (Kness and Kleczewski 2015). In Baltimore, Maryland, *Meloidogyne*, spp. were found in Swiss chard exhibiting heavy root galling, and were detected in less than 1% of experimental samples containing 500 plants examined at the Central Maryland Research and Education Center, University of Maryland (Kaplan et al. 2008).

The genus *Meloidogyne* has about 100 valid species (Elling 2013) worldwide, but only a few are reported from Maryland and Delaware. Two species, *M. incognita* (southern root knot nematode, SRKN) and *M. hapla* (northern root knot nematode, NRKN), are common in Maryland and Delaware (Sasser 1954). Jenkins et al. (1957) found a single incidence of an additional species *M. javanica*, from snapdragon plants (*Antirrhinum majus*) in a greenhouse in Baltimore County, Maryland. *Meloidogyne incognita* was found in three more counties (Jenkins et al. 1957). Feldmesser and Golden (1972) found *M. graminis* occasionally in eight *Zoysia japonica* and one *Poa pratensis* turf lawns in several locations of Maryland.

Meloidogyne incognita is the most common root knot nematode species in Maryland and Delaware (Everts et al. 2006). *Meloidogyne incognita* produces larger galls and causes more severe stunting, yellowing and wilting symptoms than *M. hapla* (Traunfeld 1998). This species causes severe damage on the Eastern Shore and in Southern Maryland, and is capable of overwintering in Maryland and Delaware soils (Traunfeld 1998). In Maryland, the species is commonly found associated with vegetable crops, corn, muskmelon (*Cucumis melo*), soybean (*Glycine max*), sweet potato (*Ipomoea batatas*), tobacco (*Nicotiana tabacum*), tomato (*Lycopersion esculentum*), vetch (*Vicia villosa*) and wheat (*Triticum aestivum*) (Jenkins et al. 1957). Similarly, *M. incognita acrita* was recognized in 5% of 42 soybean samples (Golden and Rebois 1978). These researchers also found an undescribed *Meloidogyne* sp. in 2% of 42 field samples. Potato (*Solanum tuberosum*), processing cucumber (*C. sativus*), sweet corn, green bean (*Phaseolus vulgaris*) and other vegetables are grown throughout the region and also have experienced significant losses due to this unknown species (Everts et al. 2006).

Meloidogyne hapla produces tiny galls on a wide variety of plants, compared to *M. incognita*. In 1957, it was associated with alfalfa, clover and vetch and categorized as a less common, but still important, species (Jenkins et al. 1957). High populations were associated with stunting and yield loss in tobacco in Charles and Calvert Counties in Maryland. Golden and Rebois (1978) recorded it in 14% of soil and 19% of 42 soybean root samples.

Meloidogyne sasseri was described and illustrated from American beachgrass (*Ammophila breviliffulata*) originally collected from Henlopen State Park and Fenwick Island near the Maryland state line in Delaware (Handoo et al. 1993).

The root knot nematodes cause yield and quality losses for most vegetable and field crops in Maryland when they exceed certain threshold levels and control measures are not applied. Growers need to determine nematode field population levels for efficient soil management practices. Summer or early fall sampling is more effective than mid-spring sampling. However, a degree days-based sampling has been suggested for Maryland (Kratochvil et al. 2004). The SRKN and root lesion nematodes prevalent in Maryland and Delaware cause severe damage in areas with sandy soils and in crops including tobacco, and vegetables such as sweet potatoes, tomatoes, potatoes, cucumber and green beans. Fields planted repeatedly with these crops have experienced significant losses due to RKN. Corn and wheat, common crops in the region, are reproductive hosts for RKN.

In Delaware, root knot nematode is found associated with soybean, snap bean (*Phaseolus vulgaris*), watermelon (*Citrullus lunatus*), cucumber (*Cucumis sativus*), tomato, pepper (*Capsicum* spp.), sorghum and sudangrass (*Sorghum* spp.), lima bean (*Phaseolus lunatus*), pea (*Pisum sativum*), cantaloupe and muskmelon (*Cucumis melo*), pumpkin and squash (*Cucurbita* spp.), potato, corn as well as weeds, including dandelion (*Taraxacum* spp.), mallow (*Malva neglecta*), purslane (*Portulaca oleracea*), pigweed (*Amaranthus* spp.), prickly sida (*Sida spinosa*) and morning glory (*Ipomoea* spp.).

In Delaware, root knot nematode economic threshold levels are considered to have a two RGS rating (1–4 galls in each whole root) for carrots and three RGS rating (5–12 galls) for other vegetables (Ernest and Johnson 2014). Economic threshold levels of root knot nematode, when combined with other nematodes such as root lesion nematode in fall (1 + 2.2 nematodes per cm^3 soil) and in spring (4 + 1.6 nematodes per cm^3 soil) is lower than those for either root knot nematode or lesion nematode alone (Mulrooney 2012a, b).

13.3.4.1 Management

Currently, in both states, limited control measures for root knot nematode exist. Limited availability of effective broad spectrum nematicides such as methyl bromide, is a growing concern for crop producers, primarily vegetable producers. Also, due to a present limited availability of experts with an understanding of field biology of nematodes, a shift in nematode population dynamics and constant change in crop cultivars and crop rotation, the economic importance and damages caused by nematodes may increase in Maryland and Delaware.

Increased use of nematicides in many different crops by the growers has been observed in Delaware. Application of nematicides such as NIMITZ™ (fluensulfone), is highly effective in controlling plant-parasitic nematodes, especially root knot nematodes in tomato, other fruiting vegetables and cucurbits. Registration for crops such as carrots, strawberries and other crops is expected to follow. Some growers in the region have started using nematicide seed treatments (Avicta, Votivo) on field corn (Ernest and Johnson 2014). However, there are conflicting reports of increased yield due to seed treatments as no differences in nematode population between treated and nontreated controls have been observed later in the season. Thus, use of seed treatment in corn to reduce root knot nematode populations in succeeding vegetable crops, is still under debate (Ernest and Johnson 2014). However, such chemicals should be used when the nematode populations cross the economic threshold level.

Rotation is often a limited control strategy for root knot nematode because of its wide host range. Crops such as alfalfa or oats may be the safer crops to use in rotation in order to reduce root knot nematode populations. Increasing organic matter in fields with low organic matter and high root knot populations or other plant parasitic nematodes, can have a suppressing effect on root knot populations. Planting rape and other mustards may help suppress root knot populations by releasing isocythicynite, a toxic gas, by plowing green plant materials during flowering, before they go to seed in the spring prior to planting the next crop. Pokharel and Reighard (2015) reported increased efficacy by covering the plot after incorporating mustard green plant materials into the soil. The nematode may infect rape if the populations are high and soil temperatures are above 18.3 °C, at planting or in fall (Christy and Mulrooney 2011*)*. Use of cover crops and poultry litter compost are effective methods to reduce nematode populations only if successively incorporated into rotational cropping sequences (Everts et al. 2006).

Soybean does not produce visible symptoms with low root knot nematode populations but yield loss can occur depending on growing conditions, especially low rainfall. High populations and adverse growing conditions causing plant stress, can cause stunting as severe as that produced by soybean cyst nematode. Root knot nematode resistance has been available for a long time but such varieties (group 4 soybeans) still are limited in use. Root knot nematode resistant soybeans would be an excellent rotation crop for vegetable growers who plant susceptible, free market or processing vegetables (Ernest and Johnson 2014).

Though several alternative techniques such as sanitation, soil management, organic amendments, fertilization, biological control and heat-based methods are recommended, their use alone has limited implications for yield loss management in comparison to nematicides. Combining control methods in a systemic analysis presents a challenge; sustainable management of root knot nematode is only possible with integration of several approaches, including maintenance of constant and rigorous research for each local nematode population and agro-climatic region.

13.3.5 Root Lesion Nematodes, Pratylenchus *spp.*

One of the most important nematode genera, *Pratylenchus*, has been found in almost all studies reported from Maryland and Delaware. *Pratylenchus* spp. were found in 40% of total number of samples collected from Maryland and in 60% of total numbers of samples collected from the four western counties of Frederick, Washington, Allegany, and Garrett. Seven species of *Pratylenchus* were reported from different crops (Jenkins et al. 1957). Lesion nematode was found in 78% of 362 samples of soybean fields in eight counties of Maryland (Sindermann et al. 1993) and in 29% tobacco samples analyzed at the Central Maryland Research and Education Center, University of Maryland (Kaplan et al. 2008). Jenkins et al. (1957) reported that *Pratylenchus* spp. and *Tylenchorhynchus* spp. are probably the most serious plant crop pests in Maryland based on their study of plant parasitic nematodes in 1,210 farms and gardens throughout the state.

Several species of *Pratylenchus* are described in Maryland, but not all are important. Species found in Maryland and Delaware are discussed below. Their importance varies with the targeted crop and location.

Pratylenchus pratensis was the most common species found in 9% of 1,210 survey samples by Jenkins et al. (1957). It was commonly found in the northern part of Maryland possibly because the nematode is better adapted to heavier clay-loam soil and somewhat cooler temperatures. The nematode was found during the latter part of July and August and was associated with alfalfa, barley, bean, clover, corn, grass, oat, pea, rye, sorghum, soybean, timothy, tobacco, tomato, vetch and wheat. Also, it was frequently detected in clover and grass roots in pasture fields (Jenkins et al. 1957).

Pratylenchus brachyurus was discovered in 4% of 1,210 soil samples collected from Northern and Western Maryland (Jenkins et al. 1957). This was the second most frequently detected species and was associated with alfalfa, asparagus, barley, clover, corn, grass, oat, rye, sorghum, soybean, timothy, tobacco, tomato and wheat. It was most often found in clover and grass roots in pastures.

Pratylenchus zeae was identified in about 4% of the samples from southern counties where light sandy soil is common. The largest population of this species was detected somewhat earlier than *P. pratensis* and *P. brachyurus*, which might be due to an earlier planting time. This species was associated with alfalfa, bean, corn, cucumber, grass, tobacco and tomato (Jenkins et al. 1957). Higher numbers of corn, followed by soybean samples, contained this species (Jenkins et al. 1956). Golden and Rebois (1978) found this species in 2% of 42 soybean fields in Maryland.

Pratylenchus penetrans was the fourth most commonly found species during the survey (Jenkins et al. 1957). It was found in about 2% of 1,210 samples distributed throughout the state with no soil type preference. This species was found in alfalfa, bean, clover, grasses, raspberry, soybean, strawberry, tobacco, tomato, vetch and wheat. It was most often observed in clover and grass roots in pasture (Jenkins et al. 1957). In Maryland, *P. penetrans* was found in 33% of 42 soybean fields (Golden and Rebois 1978). Sindermann et al. (1993) reported *P. penetrans*, a common species in Maryland and Delaware soybean fields capable of causing severe damage to potato and cucumber. Symptoms of root lesion nematode infections in potato range from poor crop growth and chlorotic foliage to root cell death, resulting from nematode feeding, and tubers with scabby or shrunken areas. Yield losses due to *P. penetrans* are highly variable and influenced by environmental conditions and the presence of the fungus *Verticillium dahlia* (Everts et al. 2006).

Pratylenchus hexincisus was found in less than 1% of 1,210 samples (Jenkins et al. 1957). No particular distribution pattern of this nematode was found. It was found associated with clover, corn, millet, rye, soybean and tomato. The largest population was observed during August, without any soil type preference. This species was found in 12% of 42 soybean fields in Maryland (Golden and Rebois 1978).

Pratylenchus vulnus was observed in boxwood samples collected from Montgomery County, Maryland (Jenkins et al. 1957).

Pratylenchus subpenetrans was found in pasture grasses in Prince George's County (Jenkins et al. 1957). They also reported the species to be associated with broccoli, carnation, lespedeza, muskmelon, peach, potato, snapdragon, sweet potato and watermelon. This nematode caused significant stunting of boxwood (Golden 1956).

Based on a statewide survey, Jenkins et al. (1957) described seven *Pratylenchus* species of which *P. pratensis* was the most common species detected. Golden and Rebois (1978) reported 7 species which, at that time included the new species, *P. agilis*, *P. coffeae*, *P. crenatus* and *P. scribneri,* from 42 soybean fields in Maryland with *P. agilis* being the most common. Similarly, seven species of *Pratylenchus* (*P. pinguicaudatus*, *P. neglectus*, *P. projectus*, *P. thornei*, *P. penetrans* and *P. zeae*.) were recorded from a no-till corn field in Beltsville, Maryland, where the last two species were previously reported and *P. thornei* was the major species detected (Cavigelli et al. 2005).

Pratylenchus was one of two important genera found in potatoes in Maryland, however, presently, root lesion nematode damage to the Maryland potato crop is low. This may be because processing potatoes are a relatively new crop in Maryland, and nematode populations have not yet reached economic levels (Edward et al. 2002). The lesion nematode in corn prefers high clay content. Yield loss trials conducted in the U.S. Corn Belt using NemaStrike™ Technology for 3-Year Average (2014–2016) in 264 trials in several states, including Maryland, demonstrated an average yield protection advantage of 0.18 ton, 0.08 ton in soybean, and varying results in wheat based on nematode pressure in each field (AgWeb 2017).

In Delaware, root lesion nematode was the most common plant parasitic nematode, occurring in 68% of 38 samples in a University of Delaware Extension Nematode Assay Service in 2015 (Kness and Kleczewski 2015). Two species of root lesion nematode, *P. thornei* and *P. neglectus*, may be present at damaging levels in wheat. In a low-precipitation environment, winter wheat losses can be up to 36% at a level of 10,000 *P. neglectus*/kg soil. Spring wheat losses are lower in high precipitation environments, but are still 14% at a level of 4,000 *P. neglectus*/kg soil (Smiley et al. 2004). Wheat and corn are important crops but no extensive survey of nematodes associated with these crops has been conducted even though both lesion nematode species are present in Maryland (and probably in Delaware as well). Losses caused by these two species, especially on wheat, are not known. In Delaware, the economic threshold levels for root lesion nematode in corn are 3.2 (juveniles, females or eggs) per cm^3 soil in fall and 2 (juveniles, females or eggs) per cm^3 soil in spring (Mulrooney 2012a, b), which may be comparable for Maryland growers.

13.3.5.1 Management

Growers use costly nematicides to control root knot and root lesion nematode on potatoes and cucumbers. In a typical year, as in 1998, only 25% of pickling cucumber acreage in Maryland and Delaware was fumigated for the nematode. Root lesion nematode has also caused severe damage on potato and cucumber, although yield losses due to the nematode are highly variable and influenced by environmental conditions (SARE 2005).

The use of sorghum-sudangrass in a regular rotation with susceptible vegetable and agronomic crops can be a suitable bio-control management practice for *M. incognita* and *Pratylenchus* spp. Late summer and early fall production of this high-biomass yielding crop should effectively manage residual nitrogen. Sorghum-sudangrass has the potential to be both a good cover crop for nutrient management and a good bio-control option for managing parasitic root nematodes (Kratochvil et al. 2004).

Soil amendment of cover crops, poultry litter (PL) and PL compost was tested in microplots to suppress root knot (*M. incognita*) and root lesion nematodes (*P. penetran)* in 1, 2, 3-year rotational sequences comprising potato (year 1), cucumber (year 2), followed by a moderately RKN-resistant or susceptible soybean cultivar, castor bean, grain sorghum, or sorghum-sudangrass; PL or PL compost were amended into some of the RKN-susceptible soybean and sorghum-sudangrass plots. In the 3rd year of the rotation, potato followed by soybean was planted in all 12 treatments. The RKN-resistant soybean, castor bean, sorghum-sudangrass, and fallow or tillage decreased populations of *M. incognita* compared with microplots where RKN-susceptible soybean had been grown. However, RKN populations quickly recovered. Root lesion nematode was reduced in the spring of year 1 following application of high rates of PL and PL compost the previous year. In the fall of year 1, sorghum-sudangrass alone or in combination with PL or PL compost, grain sorghum, or fallow or tillage reduced root lesion nematodes compared with either soybean cultivar. No treatment affected root lesion nematode the following year. The use of cover crops and PL compost is an effective method to reduce nematode populations only if successively incorporated into rotational cropping sequences (Everts et al. 2006).

13.3.6 Stem and Bulb Nematode, Ditylenchus dipsaci

Ditylenchus dipsaci, also commonly known as garlic bloat nematode, was observed in 57% of samples collected throughout Maryland, however, no large populations of this nematode were associated with plant injury (Jenkins et al. 1957). The genus was associated with alfalfa, asparagus, barley, bean, carnation, clover, corn, cucumber, grass, lespedeza, oat, pea, pepper, potato, rye, soybean, strawberry, sweet potato, timothy, tobacco, tomato, vetch and wheat. Golden and Rebois, (1978) found the genus in 4% of 42 soybean samples. Brust and Rane (2012) found garlic bloat nematode in several garlic samples submitted from growers with bloated, twisted, swollen leaves, distorted and cracked bulbs with dark rings. *Ditylenchus dipsaci* and *D. myceliophagus*, a mushroom parasitic species, are reported from Maryland.

Ditylenchus dipsaci is probably one of the important parasitic nematodes of several agriculture crops in the world. It is one of the destructive pests of several economically important crops such as alfalfa, garlic, onion, and tulips. However, damage found in many crops is often misidentified or not reported. This species causes twisted, stunted and abnormal plant growth, with symptoms in garlic and onion often starting early in the growing season. The species is often found in ornamentals including phlox, tulip, narcissus and hyacinth, but many other crops and weeds can also support it. Infected bulbs have distorted leaves and browned bulb scales. Infected phlox shows distorted growth, stunting and plant death. Large numbers of nematodes may be present in symptomatic plant tissues and can be seen through a microscope as a writhing white mass called "nematode cotton". In the past few years, this nematode has been found in garlic samples (Brust and Rane 2012) and many past incidences of decline observed in Maryland could be due to *D. dipsaci.*

13.3.7 Dagger Nematodes, Xiphinema *spp.*

Dagger nematodes belong to the economically important ectoparasitic nematode genus *Xiphinema* (Fig 13.1b) comprising certain species that transmit plant viruses. Jenkins et al. (1956) found *Xiphinema* sp. in 39%, 44% and 39% of 143 corn, 111 tobacco and 74 soybean samples, respectively, and reported the genus *Xiphinema* as one of the most commonly found groups in 41% of all soil samples collected in their extensive survey of 1,210 farms and gardens in Maryland. It has been reported that populations tend to be highest with outbreaks more likely in light, sandier soils, as opposed to heavier clay soils. Dorchester County experiences high levels of nematodes that may be due to favorable soil conditions. Population levels are prone to drastic fluctuations. Dagger nematode was found in 54% of 12 soybean fields in 8 Maryland counties (Sindermann et al. 1993).

Xiphinema americanum was the only species found in the extensive survey of 1,210 farms and gardens throughout Maryland, associated with stunting and poor growth of tomato plants in a field in Worcester County. This species was also found associated with turf lawns (*Zoysia japonica* and *Poa pratensis*), in several locations of Maryland (Feldmesser and Golden 1972). Later, Golden and Rebois (1978) found 26% of 42 soybean fields surveyed infected with *X. americanum*. This group is known to transmit different NEPO viruses and may be important for many crop growers in the region.

Xiphinema rivesi was found associated with apple (Lamberti and Bleve-Zacheo 1979; Wojtowlcz et al. 1982) and no-till corn in Beltsville, MD (Cavigelli et al. 2005).

Xiphinema spp. are a potentially serious pest of potatoes in Maryland, although at present no yield loss has been directly attributed to these nematodes. Potatoes are good hosts, causing increased populations. While yields don't seem to be affected, subsequent crops can be damaged. Dagger nematode was associated with tobacco in 44.17% of samples from Central Maryland Research and Education Center (Kaplan et al. 2008). In Delaware, dagger root nematode was observed only in 8% of samples (out of 38 samples) in a Nematode Assay Service in 2015 (Kness and Kleczewski 2015).

In Delaware and Maryland, dagger nematodes (*Xiphinema* spp.) were found at damaging levels in several samples that were to be planted to wine grapes (Mulrooney and Gregory, 1210). Minor species reported from Maryland include *X. chambersi* (Chamber's dagger nematode). While studies show that *Xiphinema* is commonly present in different crops in Maryland and Delaware, occurrence and densities vary with location and time. Thus, their role in crop production, including transmission of plant viruses in this area, is not known.

13.3.7.1 Management

In Sussex County, Delaware, more than 60 fields with severely stunted soybean plants with thickened, dark-green leaves have been observed to produce little to no yield over several years. These soybean plantings are infested with *Soybean severe stunt virus (SSSV)* which is transmitted by *X. americanum,* and symptomatic plants in the field are consistently associated with the dagger nematode. Greenhouse studies indicated that corn, wheat, marigold, castor and fallow treatments reduced dagger nematodes the most after 14 weeks compared growing with 'Essex' and 'HT 5203' soybean cultivars (Evans et al. 2007).

13.4 Other Plant Parasitic Nematodes of Lesser or Undetermined Importance in Maryland and Delaware

13.4.1 Foliar Nematodes, Aphelenchoides *spp.*

These above-ground plant-feeding nematodes cause damage to many landscape plants, affecting their esthetic value. As foliar feeders, these nematodes are known to infect ferns, peonies, begonias, anemones, *Baptisia*, *Hepatica*, *Heuchera*, hostas, *Hypericum*, *Ipomoea aculeatum*, iris, lilies, *Ligularia*, orchids, *Papaver*, *Orientale*, *Phlox*, *Polygonatum*, *Rogersia*, salvias and *Tricyrtus* (Kohl 2011). Privet and azalea are among the woody plants commonly infected with foliar nematodes. Foliar nematodes can easily spread from these woody hosts into herbaceous ones. This genus was recorded in 29% of 42 soybean fields surveyed (Golden and Rebois 1978) as well as in no-till corn samples in Beltsville, Maryland (Cavigelli et al. 2005). Three species have been reported in Maryland: *A. besseyi*, *A. ritzemabosi* and *A. fragariae*.

Aphelenchoides besseyi is mostly a pest of rice in developing countries causing white tip disease, however, in Maryland, it was found associated with chrysanthemum (Jenkins et al. 1957). Also known as 'summer dwarf nematode of strawberry', *A. besseyi* was found in wild strawberry in Maryland and considered widespread in commercial strawberries cultivated in Maryland and Delaware (Esser 1966).

Aphelenchoides ritzemabosi was observed in chrysanthemum from Montgomery County, Maryland (Jenkins et al. 1957). This species produces 'summer crimp' which appears in hot summer weather, intensifies with rain, and slows down by cool weather (Esser 1966).

Aphelenchoides fragariae was extracted from strawberry leaves in three locations on the Eastern Shore (Jenkins et al. 1957). This species, along with *A. besseyi*, was found in Maryland and Delaware in strawberry plants. It causes 'spring crimp' in strawberry (Esser 1966). The nematode can survive mild winters and hot summers in Maryland and may be a serious threat to strawberry production. In Delaware, the nematode was found associated with butterfly bush (*Buddleja* spp.) (Kunkel 2010). Information is limited on incidence, host-parasite relationships and importance of this nematode in strawberry as well as ornamental plants in Maryland and Delaware.

13.4.2 Spiral Nematodes

Spiral nematodes of the genus *Helicotylenchus* are frequently found in Maryland and Delaware. This nematode was observed in 57%, 46% and 47% of 143 corn, 111 tobacco and 74 soybean samples, respectively (Jenkins et al. 1956). The exact role of this external feeding nematode group in various crop plants is not known unless very high population levels exist. This nematode, recorded in 20% of the samples examined during a survey of 1210 farms in Maryland, was considered less common but still important (Jenkins et al. 1957). This genus is best suited to clay loam soils that are more common in the northern and western regions of the state. *Helicotylenchus* spp. were observed in 36% and 7% of 42 soybean soil and root samples (Golden and Rebois 1978). The following species of this genus are reported from Maryland in different crops.

Helicotylenchus erythrine was found associated with chrysanthemum, cucumber, lespedeza, muskmelon, pea strawberry and watermelon. Large populations of this species are associated with root injury and stunting of plants. This species is also regarded as one of the important parasites or pathogens of corn.

Helicotylenchus digonicus was occasionally found, in low numbers, to be associated with turf lawns (eight *Zoysia japonica* and one *Poa pratensis*) in several locations of Maryland (Feldmesser and Golden 1972) and in no-till corn in Beltsville, Maryland (Cavigelli et al. 2005).

Helicotylenchus dihystera was the most common species observed in Maryland. It was found associated with alfalfa, asparagus, barley, clover, corn, oat, soybean, timothy, tobacco, tomato, vetch, and wheat (Jenkins et al. 1957). *Helicotylenchus dihystera* was observed in 12% of 42 soil samples during the survey (Golden and Rebois 1978).

Helicotylenchus pseudorobustus was observed in 29% and 7% of soybean soil and root samples, respectively (Golden and Rebois 1978) and in low numbers in no-till corn in Beltsville, Maryland (Cavigelli et al. 2005). In soybean, this genus was found in 58% of 269 samples collected and tested from eight counties of Maryland (Sindermann et al. 1993). It was also observed associated with turf lawns (eight *Zoysia japonica* and one *Poa pratensis*) in several locations of Maryland (Feldmesser and Golden 1972). Similarly, it was also found in 91% samples of tobacco at the Central Maryland Research and Education Center (University of Maryland) (Kaplan et al. 2008).

Helicotylenchus multicinctus can infect a wide range of cultivated plants and is reported from Maryland (Bernard and Keyserling 1985).

In Delaware, spiral nematodes of the genus *Helicotylenchus* were the second most common plant parasitic genus occurring in 55% of 38 samples in a Nematode Assay Service in 2015 (Kness and Kleczewski 2015).

Other spiral nematode, genus *Rotylenchus* was found in 3% of samples from all over Maryland, in both low frequency and numbers. Preference for soil type was not determined (Jenkins et al. 1957). Only two species, *R. robustus* and *R. buxophilus*, are found in Maryland.

Rotylenchus robustus was detected in small numbers in soil around alfalfa roots in Frederick County (Jenkins et al. 1957).

Rotylenchus buxophilus was associated with alfalfa, corn, grass, oat, tobacco and wheat. Later, this genus was also recorded in 4.2% of 143 corn and 2.7% of 74 tobacco, but not in soybean samples in Maryland (Golden and Rebois 1978). A large population was found in boxwood samples affecting the plants in wider areas in Maryland (Sasser 1954). Sasser (1954) also found it to cause significant reduction of root growth in inoculated plants.

Spiral nematodes of the genus *Scutellonema* spp. infect several plant species, grass being more important than others. This nematode is recorded from Maryland in grass and apple samples (Horst 2008). No species is reported from these states.

13.4.3 Stunt Nematodes

This genus *Tylenchorhynchus* is the second most important, after *Pratylenchus* in Maryland. *Tylenchorhynchus* spp. were estimated to be the most serious pests of agricultural crops in Maryland in the 1950s and was recorded in 51%, 68% and 39% of corn, tobacco and soybean, respectively (Jenkins et al. 1956). Later, the genus was recorded in 60% soil and 7% of root samples in soybean, with two species identified in 42 soybean samples (Golden and Rebois 1978). The stunt nematode was observed in 83% of samples examined from the Central Maryland Research and Education Center (University of Maryland) (Kaplan et al. 2008). Five species are described from Maryland.

Tylenchorhynchus claytoni was the most common species found in 15% of samples from all counties, except, Kent, Queen Anne's, Talbot, and Baltimore. It was commonly found in Anne Arundel, Charles, Calvert, St. Mary's and Prince George's Counties. This species was associated with alfalfa, barley, bean, clover, corn, grasses, oat, pepper, rye, soybean, strawberry, sweet potato, timothy grass, tomato, vetch and wheat. It was also frequently associated with tobacco (Jenkins et al. 1957), soybean in 55% of 42 samples (Golden and Rebois 1978) and grasses including 8 *Zoysia japonica* and one *Poa pratensis*, in several Maryland locations (Feldmesser and Golden 1972).

Tylenchorhynchus dubius, the second most frequently encountered species, was observed in about 1% of 1210 soil samples, limited to northern and western counties (Howard, Montgomery, Frederick, Washington, Allegany, and Garrett) where soil is clay-loam. Dense nematode populations were associated with stunting and chlorosis of barley and the nematodes were also recovered from clover, grasses, oat, vetch and wheat (Jenkins et al. 1957). It was also occasionally associated with eight *Zoysia japonica* and one *Poa pratensis* turf lawn in several locations of Maryland (Feldmesser and Golden 1972).

Merlinius brevidens was identified in less than 1% of 1,210 samples. It was found in alfalfa, barley, clover, grasses, oat, pea, timothy, and wheat in clay loam soils of Carroll, Frederick, and Howard Counties (Jenkins et al. 1957). Other minor species, *Geocenamus ornatus* in a single sample from barley in Howard County (Jenkins et al. 1957), and *Quinisulcius acutus* in 2% of 42 soybean samples (Golden and Rebois 1978) have also been identified from Maryland.

13.4.4 Lance Nematodes, **Hoplolaimus** *spp.*

The lance nematode feeds on plant roots, externally (body remaining outside), internally (body completely within roots) or partially outside or inside (head region only within roots). In Maryland, this genus was observed in 18%, 15% and 23% of 143 corn, 111 tobacco and 74 soybean samples respectively (Jenkins et al. 1956) and in 43% of 199 soybean fields in 8 counties of Maryland (Sindermann et al. 1993). It was observed in 23.33% samples of tobacco at the Central Maryland Research and Education Center (University of Maryland) (Kaplan et al. 2008).

Jenkins et al. (1957) found lance nematodes in 16% of 1,210 samples collected in all counties, except Charles County. Despite occasional high populations in sandy soils, clay or clay-loam soil was the most frequent habitat. This nematode was associated with grass in the majority (about 3%) of total samples, and in tobacco and clover about seven times less. The nematode was also observed in grape roots in Montgomery County. In addition, it was associated with alfalfa, barley, carnation, clover, corn, grass, lespedeza, oat, pea, pepper, rye, soybean, sweet potato, timothy, tobacco, tomato and wheat; no difference in the preference of soil type was found. *Hoplolaimus* spp. were considered less common than other plant parasitic nematodes of the region. Three *Hoplolaimus* species important to agricultural pathogens are found in Maryland:

Hoplolaimus galeatus (= *H. coronatus*) was found in an extensive survey of 1,210 fields and garden (Jenkins et al. 1957), where the single species was present in 11% of samples examined. It was also found associated with alfalfa, barley, carnation, clover, corn, grape, grass, lespedeza, oat, pea, pepper, rye, soybean, sweet potato, timothy, tobacco, tomato and wheat. This species should be considered as a potentially important nematode pest in Maryland crops (Jenkins et al. 1957).

Hoplolaimus galeatus is important primarily in turf grasses and may be the major nematode pest of turf grass, after sting nematodes, in many warmer areas. It was reported associated with eight *Zoysia japonica* and one *Poa pratensis* in several locations of Maryland (Feldmesser and Golden 1972). Golden and Rebois (1978) found this species in 38% and 41% of 42 soybean soil and root samples. This species was also found in no-till corn with low in numbers in Beltsville, Maryland (Cavigelli et al. 2005). It also can be found in many crops, along with pine trees and grasses. The nematode, commonly found in lawns, is the major nematode pest of St. Augustine grass. Low numbers of this nematode can cause damage to turf grasses. It is common on the East Coast, from New England to Florida (Crow and Brammer 2015) including Maryland and Delaware.

In Delaware, lance nematode was the most common plant parasitic nematode, occurring in 29% of 38 samples in a Nematode Assay Service in 2015 (Kness and Kleczewski 2015). Because of limited study, it was hard to ascertain the current situation and potential threat of *Hoplolaimus* in different plants, especially landscape plants, in Maryland and Delaware.

13.4.5 Stubby Root Nematodes

Stubby root nematodes feed on plant roots and cause economic loss by direct feeding and indirectly by transmitting plant viruses. Populations of this nematode were identified in about 3% of 1210 samples, primarily from the lower Eastern Shore, where three species were reported (Jenkins et al. 1957). They found the nematode associated with bean, corn, grass, pepper, rye, soybean, strawberry, sweet potato, tobacco, tomato and wheat. Four species have been identified from Maryland. *Trichodorus* sp. was occasionally noted from eight *Zoysia japonica* and one *Poa pratensis* lawn, in several locations of Maryland (Feldmesser and Golden 1972).

Nanidorus minor (= *Paratrichodorus christiei*) was identified in 14% of 74 soybean soil samples (Golden and Rebois 1978).

Paratrichodorus obtusus was identified by Crow (2005) in large numbers. Later, Kaplan et al. (2008) also observed this nematode associated with tobacco during a field experiment at the Central Maryland Research and Education Center).

Paratrichodorus pachydermus was found in a single turf sample from Prince George's County, Maryland (Jenkins et al. 1957).

Trichodorus primitivus was found in soils from mimosa in Talbot County, Maryland (Jenkins et al. 1957). This nematode was found in 8%, 5% and 14% of 143 corn, 111 tobacco and 74 soybean samples, respectively (Jenkins et al. 1956). This genus was observed in 9% of 500 tobacco plant samples at the Central Maryland Research and Education Center (Upper Marlboro, MD (Kaplan et al. 2008) during a field research.

In Delaware, stubby root nematode was observed only in 16% of 38 samples in a Nematode Assay Service in 2015 (Kness and Kleczewski 2015).

13.4.6 Ring Nematodes, **Mesocriconema** *spp.*

In Maryland, these nematodes were reported from 4%, 2% and 3% of 143 of corn, 111 tobacco and 74 soybean soil samples, respectively (Jenkins et al. 1956). The genus, observed in about 3% of 1,210 soil samples, was found throughout the state associated with barley, bean, clover, corn, cucumber, grasses, pepper, rye, tobacco and soybean (Jenkins et al. 1957). Also ring nematodes were detected in 7% of 42 soil samples collected from soybean fields (Golden and Rebois 1978). Because of a lack of serious damage problems caused by ring nematodes in any crop plant, these nematodes were not considered an agricultural importance. However, it has been found that this group can play an important role in peach tree short life (PTSL) which is a severe problem in Georgia and the Carolinas (Nyczepir et al. 1985) but not in Maryland and Delaware, even though it has been reported from these states.

Mesocriconema ornatum was found in Maryland and was reported to be associated with turf lawns (eight *Zoysia japonica* and one *Poa pratensis*) several locations (Feldmesser and Golden 1972).

Mesocriconema simile was found associated with peach decline in some states including, Maryland (Horst 2008).

13.4.7 Sting Nematode, **Belonolaimus longicaudatus**

The sting nematode was found in irregular areas of severely chlorotic, stunted and dead soybean plants in Delaware, with a population density of 216 nematodes per 250 cm^3 of soil (Handoo et al. 2010). No other published information is available for this nematode in Maryland. Since *Belonolaimus longicaudatus* can be a serious problem in many lawns and golf grass, it may also be a problem in Maryland and Delaware turf lawns.

13.4.8 Sheath Nematodes, **Hemicyclophora** *spp.*

This genus is not common in Maryland and Delaware. This nematode was detected in only 5 of 1,210 corn soil samples collected from Maryland (Jenkins et al. 1957). The genus was also associated with barley, corn and grass, and a large population in raspberry was associated with chlorosis, reduced yield and general decline, thereby indicating a potential problem in raspberry production (Jenkins et al. 1957).

13.4.9 Needle Nematodes, **Longidorus** *spp.*

Longidorus spp. is an important group of plant parasitic nematodes, largely because they are known to transmit many plant viruses., However, the present incidence of this nematode in Maryland and Delaware is unknown. In earlier studies, it was present in low numbers in 6% and 7% of corn and soybean samples, respectively, but not in tobacco in Maryland (Jenkins et al. 1956). Some individual nematodes collected from stunted corn in three counties of Southeastern Iowa in 1971 and 1972, were similar to those collected from Delaware, but morphologically dissimilar to any recorded account, and the nematode was described as a new species, *Longidorus breviannulatus* (Norton and Hoffmann 1975).

13.4.10 Ear Cockle Nematode, **Anguina tritici**

Anguina tritici causes a disease in wheat and rye called, "ear-cockle" or "seed gall". There is a brief account of the symptoms of "head nematode disease" (*Anguina tritici*) in wheat, which is important in underdeveloped countries, causing up to 70% losses (Leukel 1957), and as a phytosanitary pathogen in developed counties. This nematode was first found in the United States in 1909 in California, and subsequently in several states including Maryland, Virginia, and Georgia, primarily in wheat, but also in rye to a lesser extent (Leukel 1957). In Maryland, it was found in one sample with 20% infection from St. Mary's County (Jenkins et al. 1957). However, modern mechanized agricultural seed grading practices that result in separation of clean seed from galls, along with crop rotation, have practically eliminated *A. tritici* from countries which have adopted these practices, and the nematode has not been found in the United States since 1975 (Randhawa 2017).

13.4.11 Pin Nematodes, **Paratylenchus** *spp.*

This nematode was found in 8%, 12% and 3% of corn, tobacco and soybean samples, respectively in Maryland, and was observed in 14% of 1,210 samples in various other crops without any preference of soil type (Jenkins et al. 1957). *Paratylenchus* sp. was found associated with strawberry in the wild (wooded areas) and this genus is also frequently found in commercial strawberry plantings (Crow and MacDonald 1976).

Paratylenchus dianthus was identified from 6 samples of 1,210 farms and gardens in Maryland from Carnation, clover, corn, timothy, vetch and wheat from Allegany, Washington, Prince George's and Baltimore Counties (Jenkins et al. 1957) and currently is believed to be distributed throughout the state (Horst 2008).

Paratylenchus hamatus was found in Allegany and Carroll Counties, associated with clover showing stunted growth and low yield record (Jenkins et al. 1957).

Paratylenchus nanus was recorded in Garrett County from asymptomatic wheat (Jenkins et al. 1957).

Paratylenchus projectus was the most common species distributed throughout the state, associated with alfalfa, bean, clover, corn, grass, lespedeza and soybean (Jenkins et al. 1957). Furthermore, Jenkins et al. (1957) also recorded an unidentified species of this genus associated with alfalfa, barley, bean, carnation, clover, corn, grass, oat, pea, rye, strawberry, soybean, sweet potato, Timothy-grass, tobacco, tomato, vetch and wheat.

References

AgWeb. (2017). The lesion nematode, a major nematode species in corn in the U.S., prefers a higher clay content (Nematodes are everywhere, including in corn. https://www.agweb.com/article/nematodes-are-everywhere-including-in-corn-naa-sponsored-content/.

Allen, T.W., Bradley, C.A., Sisson, A. J., Byamukama, E., Chilvers, M. I., Coker, C. M., Collins, A. A., Damicone, J. P., Dorrance, A. E., Dufault, N. S., Esker, P. D., Faske, T. R., Giesler, L. J., Grybauskas, A. P., Hershman, D. E., Hollier, C. A., Isakeit, T., Jardine, D. J., Kelly, H. M., Kemerait, R. C., Kleczewski, N. M., Koenning, S. R., Kurle, J. E., Malvick, D. K., Markell, S. G., Mehl, H. L., Mueller, D. S., Mueller, J. D., Mulrooney, R. P., Nelson, B. D., Newman, M. A., Osborne, L., Overstreet, C., Padgett, G. B., Phipps, P. M., Price, P. P., Sikora, E. J., Smith, D. L., Spurlock, T. N., Tande, C. A., Tenuta, A. U., Wise, K. A., Wrather, J. A. (2017). Soybean yield loss estimates due to diseases in the United States and Ontario, Canada, from 2010 to 2014. *Plant Health Progress*. April 2017. 13 April 2018. St. Paul MN USA, ISSN 1535-1025.

Bernard, E. C., & Keyserling, M. L. (1985). Reproduction of root knot, lesion, spiral and soybean cyst on sunflower. *Plant Disease, 69*, 103–105.

Brust, G., & Rane, K. (2012). Garlic bloat nematode found in several garlic samples. University of Maryland Extension Bulletin.

Cadwallader, C. (2010). Delaware agriculture. https://www.nass.usda.gov/Statistics_by_State/Delaware/Publications/DE%20Ag%20Brochure_web.pdf.

Cavigelli, M. A., Lengick, L. L., Buyer, J. S., Fravel, D., Handoo, Z., McCarty, G., Millner, P., Sikora, L., Wright, S., Vinyad, B., Rabenhorst, R., et al. (2005). Landscape level variation in soil resources and microbial properties in a no-till corn field. *Applied Soil Ecology, 29*, 99–123.

Christy, M., & Mulrooney, R. P. (2011). *Managing root knot nematodes*. University of Delaware Cooperative Extension Online https://extension.udel.edu/weeklycropupdate/?p=2860.

Crow, W. T. (2005). Diagnosis of *Trichodorus obtusus* and *Paratrichodorus minor* on turf grasses in the Southeastern United States. Plant Management Network.

Crow, W. T., & Brammer, A. S. (2015). *Common name: Lance nematode scientific name: Hoplolaimus galeatus (Cobb, 1913) Thorne, 1935 (Nematoda: Tylenchida: Tylenchoidea: Hoplolaimidae)*. UF/IFAS.

Crow, R. V., & MacDonald, D. H. (1976). Phytoparasitic nematodes adjacent to established strawberry plantations 1, Paper No. 9661, Scientific Journal Series, Minnesota Agricultural Experiment Station. St. Paul: University of Minnesota.

Do. (2014). Delaware agriculture: A to Z. https://www.delawareonline.com/story/life/did-you-know/2014/11/12/delaware-agriculture-z/18899007/.

Edward, B., Dively, G., Dutky, E., Mulrooney, R., Sardanelli, S., Whalen, J. (2002). *Crop profile for potatoes in Maryland*. http://www.ipmcenters.org/cropprofiles/docs/mdpotato.pdf.

Elling, A. A. (2013). Major emerging problems with minor *Meloidogyne* species. *Phytopathology, 103*, 1092–1102.

Ernest, E., & Johnson, G. (2014). Managing root knot nematode. University of Delaware, Extension. Bulletin.

Esser, R. P. (1966). Nematode attacking plants above the soil surface. Nematology Circulation no 5. Florida Department of Agriculture.

Evans, T. A., Miller, L. C., Vasilas, B. L., Taylor, R. W., & Mulrooney, R. P. (2007). Management of *Xiphinema americanum* and soybean severe stunt in soybean using crop rotation. *Plant Disease, 91*, 216–219.

Everts, K. L., Sardanelli, S., Kratochvil, R. J., Armentrout, D. K., & Gallagher, L. E. (2006). Root knot and root lesion nematode suppression by cover crops, poultry litter, and poultry litter compost. *Plant Disease, 90*, 487–492.

Feldmesser, J., & Golden, A. M. (1972). Control of nematodes damaging home lawn grass in two counties of Maryland. *Plant Disease Report, 56*, 476–480.

Golden, A. M. (1956). Taxonomy of the spiral nematodes (*Rotylenchus* and *Helicotylenchus*) and the developmental stages and host parasite relationships of *R. buxophilus* n. sp, attacking boxwood. Maryland Agri. Experiment Station Bulletin A. 85. 28 P.

Golden, A. M., & Rebois, R. V. (1978). Nematodes on soybean in Maryland. *Plant Disease Report, 62*, 430–432.

Handoo, Z. A., Huettel, R. N., & Golden, A. M. (1993). Description and SEM observations of *Meloidogyne sasseri* n. sp. (Nematoda: Meloidogynidae), parasitizing beachgrasses. *Journal of Nematology, 25*(4), 628–641.

Handoo, Z. A., Skantar, A. M., & Mulrooney, R. P. (2010). First report of the sting nematode *Belonolaimus longicaudatus* on soybean in Delaware. *Plant disease: an international journal of applied plant pathology, 94*(1). https://doi.org/10.1094/PDIS-94-1-0133B.

Hashmi, S., Krusberg, L. R., Sardanelli, S., et al. (1993). Reproduction of *Heterodera zeae* and its suppression of corn plant growth as affected by temperature. *Journal of Nematology, 25*, 55–58.

Horst, R. K. (2008). Plant diseases and their pathogens. *Westcott's Plant Disease Hand Book 6th edition*.

Hutzell, P. A., & Krusberg, L. R. (1990). Temperature and the life cycle of *Heterodera zeae*. *Journal of Nematology, 29*, 414–417.

Jenkins, W. R., Taylor, D. P., & Rohde, R. A. (1956). A preliminary report of nematodes found on corn, tobacco, and soybean in Maryland. *Plant Disease Report, 40*, 37–38.

Jenkins, W. R., Taylor, D. P., Rohde, R. A., Coursen, B. W, et al. (1957). Nematodes associated with crop plants in Maryland. Bulletin/University of Maryland, Agricultural Experiment Station. A-89(1957). P25.

Kaplan, I., Sardanelli, S., & Denno, R. F. (2008). Field evidence for indirect interactions between foliar-feeding insect and root-feeding nematode communities on *Nicotiana tabacum* journal compilation. *Ecological Entomology, 34*, 262–270.

Kness, A., & Kleczewski, N. M. (2015). UD nematode assay service summary. In: N. Kleczewski & K. Everts (Eds.), *Applied research results on field crop and vegetable disease control P49*. University of Extension. https://desoybeans.org/hmi/wp-content/uploads/2017/02/KleczewskiAppliedResearchBooklet2015v2.pdf.

Kohl, L. M. (2011). Foliar nematodes: A summary of biology and control with a compilation of host range. *Plant Health Progress*. https://doi.org/10.1094/PHP-2011-1129-01-RV.

Kratochvil, R. J., Sardanelli, S., Everts, K., & Gallagher, E. (2004). Evaluation of crop rotation and other cultural practices for management of root knot and lesion nematodes. *Agronomy Journal, 96*, 1419–1428.

Krusberg, L., & Sardnelli, S. (1989). Survival of *Heterodera zeae* in soil in the field and in the laboratory 1. *Journal of Nematology, 21*(3), 347–355.

Krusberg, L. R., Sardanelli, S., Grybauskas, A. P., et al. (1997). Damage potential of *Heterodera zeae* to *Zea mays* as affected by edaphic factors. *Fundamentals of Applied Nematology, 20*, 593–599.

Kunkel, B. (2010). Foliar nematodes (*Aphelenchoides* sp.). https://www.forestryimages.org/browse/detail.cfm?imgnum=5429861.

Lamberti, F., & Bleve-Zacheo, T. (1979). Studies on *Xiphinema americanum sensu lato* with descriptions of 15 new species (Nematoda, Longidoridae). *Nematologia Mediterranea, 7*, 51–106.

Leukel, R. W. (1957). Nematode disease of wheat and rye. USDA Farmers Bulletin, 1607.

Lewis, S. A., & Golden, A. M. (1980). Description and SEM observations of *Dolichodorus marylandicus* sp. n. with a key to species of *Dolichodorus*. Contribution No. 1800 of the South Carolina Agricultural Experiment Station.

MacGrown. (1981). *Heterodera zeae*, a cyst nematode of corn. Nematology Circulation no 77. University of Florida.

MDA. (2015). Annual report 2015. Maryland Department of Agriculture. http://mda.maryland.gov/Documents/15mda_ar.pdf.

Mervalin, A., Morant, J. L., Casola, C. B., Brooks, E., Philip, V., & Mitchell, G. (1997). Poultry litter enhances soybean productivity in field infested with soybean cyst nematode. *Journal of Sustainable Agriculture, 11*, 39–51.

MGGA. (2014). Maryland grape growers association. Annual report. http://www.marylandgrapes.org/ https://www.agcensus.usda.gov/Publications/2012/Full_Report/Volume_1,_Chapter_1_State_Level/Maryland/.

Mulrooney, R. P. (2011). *Soybean cyst nematode*. University of Delaware, Extension bulletin PP 02. http://extension.udel.edu/factsheets/soybean-cyst-nematode/.

Mulrooney, R. P. (2012a). Nematode control recommendations for soybean 112012, nematode control recommendations for corn. University of Delaware Extension Bulletin.

Mulrooney, R. P. (2012b). Nematode control recommendation for corn. http://webcache.googleusercontent.com/search?q=cache:42Y5cXpPP3oJ:s3.amazonaws.com/udextension/ag/files/2012/05/Nematode-Control-Corn-11-2012.doc+andcd=1andhl=enandct=clnkandgl=us.

Mulrooney, R., & Gregory, N. (2010). Nematode Assay Service 2010 Report Cooperative Extension Service Department of Plant and Soil Science University of Delaware. In R. P. Mulrooney (Ed.), Plant pathology field trial results 2010 plant diagnostic clinic report nematode assay service report. Extension Plant Pathologist Nancy F. Gregory, Plant Diagnostician.

Mulvey, R. H., & Golden, A. M. (1983). An illustrated key to the cyst-forming genera and species of Heteroderidae in the Western Hemisphere with species morphometrics and distribution. *Journal of Nematology, 15*, 1–59.

Norton, D. C., & Hoffmann, J. K. (1975). *Longidorus breviannulatus* n. sp. (Nematoda: Longidoridae) associated with stunted corn in Iowa. *Journal of Nematology, 7*, 168–171.

Nyczepir, A. P., Bertrand, P. F., Miller, R. W., Motsinger, R. E., et al. (1985). Incidence of *Criconemella* spp . and peach orchards histories in short-life and non-short life sites in Georgia and South Carolina. *Plant Disease, 69*, 847–877.

Pokharel, R. R., & Reighard, G. L. (2015). Evaluation of biofumigation, soil solarization and rootstock on peach replant disease. https://doi.org/10.17660/ActaHortic.2015.1084.78.

Randhawa, R. (2017). *Anguina tritici* (Steinbach, 1799), Chitwood, 1935. Wheat seed gall nematode. California Pest Rating. California Department of Agriculture. http://blogs.cdfa.ca.gov/S/?p=3446.

Ringer, C. E., Sardanelli, S., & Krusberg, L. R. (1987). Investigations of the host range of the corn cyst nematode, *Heterodera zeae*, from Maryland. *Journal of Nematology, 19*, 97–106.

Sardanelli, S., Krusberg, L. R., & Golden, A. M. (1981). Corn cyst nematode, *Heterodera zeae*, in the United States. *Plant Disease, 65*, 622.

Sardanelli, S. L., Krusberg, L. R., Kantzes, J. G., & Hutzell, P. A. (1982). Soybean cyst nematode, fact sheet 340. College Park: Maryland Cooperative Extension Service.

SARE. (2005). Cultural practices for root knot and root lesion nematode suppression in vegetable crop rotations. Sustainable Agriculture and Research and Extension. http://www.sare.org/Learning-Center/Fact-Sheets/Cultural-Practices-for-Root knot-and-Root lesion-Nematode-Suppression-in-Vegetable-Crop-Rotations.

Sasser, J. N. (1954). Identification and host parasite relationship of certain root knot nematodes (*Meloidogyne*). *University of Maryland Agricultural Experiment Station Bulletin A, 77*:1–31.

Sasser, J. N., & Carter, C. C. (1985). Overview of the international *Meloidogyne* project 1975–1984. In J. N. Sasser & C. C. Carter (Eds.), *An advanced treatise on Meloidogyne* (pp. 19–24). Raleigh: North Carolina State University Graphics.

Sindermann, A., Williams, G., Sardanelli, S., & Krusberg, L. R. (1993). Survey for *Heterodera glycines* in Maryland. *Journal of Nematology, 25*, 887–889.

Smiley, R. W., Merrifield, K., Patterson, L. M., Whittaker, R. G., Gourlie, J. A., Easley, S. A., et al. (2004). Nematodes in dryland field crops in the Semiarid Pacific Northwest United States. *Journal of Nematology, 36*, 54–68.

Traunfeld, J. (1998). Root knot nematodes and vegetable crops. University of Maryland. Extension Bulletion HG 77 1999.

USDA. (2014). Widely prevalent nematodes in the United States. Data provided by Society of Nematology Widely Prevalent Plant Pathogenic Nematode Committee. https://www.prevalentnematodes.org/about.cfm.

USDA. (2017). U.S. Department of Agriculture's 2012 census of agriculture. https://www.agcensus.usda.gov/.

Wojtowlcz, M. R., Golden, A. M., Forer, L. B., & Stouffer, R. F. (1982). Morphological comparisons between *Xiphinema rivesi* Dalmasso and *X. americanum* Cobb populations from the Eastern United States' populations. *Journal of Nematology, 14*, 511–516.

Chapter 14
Plant Parasitic Nematodes in Georgia and Alabama

Abolfazl Hajihassani, Kathy S. Lawrence, and Ganpati B. Jagdale

14.1 Introduction

Growers in the State of Georgia incurred losses, including control costs of an estimated \$800 million from diseases in 2015 (Little 2017). The value of the crops used in this estimate was approximately \$5,385 million resulting in a 13.8% relative disease loss across all crops including field (row) crops, fruits and nuts, vegetables, turf, ornamentals and trees (Little 2017). One of the economically important pests that growers are concerned about are plant parasitic nematodes. Nematodes have long been known as soilborne parasites of cultivated crops but in recent years following the methyl bromide phase-out in the United States, there has been an increase in yield losses due to nematode pests. A wide range of plant parasitic nematodes are commonly found in association with crops in Georgia and Alabama (Table 14.1). The most economically damaging species include root knot (*Meloidogyne* spp.), reniform (*Rotylenchulus reniformis*), lance (*Hoplolaimus columbus*), stubby root (*Nanidorus minor*) and ring (*Mesocriconema ornatum* and *M. xenoplax*) nematodes. Other nematode species including root lesion (*Pratylenchus* spp.), sting (*Belonolaimus longicaudatus*) and soybean cyst (*Heterodera glycines*) nematodes are rarely found in fields and considered less economically important in both Georgia and Alabama. In addition to parasitic nematodes, crops are attacked by

A. Hajihassani (✉)
Department of Plant Pathology, University of Georgia, Tifton, GA, USA
e-mail: abolfazl.hajihassani@uga.edu

K. S. Lawrence
Department of Entomology and Plant Pathology, Auburn University, Auburn, AL, USA
e-mail: lawrekk@auburn.edu

G. B. Jagdale
Department of Plant Pathology, University of Georgia, Athens, GA, USA
e-mail: gbjagdal@uga.edu

S. A. Subbotin, J. J. Chitambar (eds.), *Plant Parasitic Nematodes in Sustainable Agriculture of North America*, Sustainability in Plant and Crop Protection,
https://doi.org/10.1007/978-3-319-99588-5_14

Table 14.1 Plant parasitic nematodes associated with different crops in Alabama and Georgia

Nematode	Host/rhizosphere soil	State[a]	References
Belonolaimus longicaudatus	Cotton, peanut, turfgrass	AL, GA	Motsinger et al. (1976a), Johnson (1970) and Norton et al. (1984)
Dolichodorus spp.	Blueberry	GA	Jagdale et al. (2013)
Helicotylenchus clarkei	Peach	AL	Sher (1966)
Helicotylenchus digonicus	Wide host range	AL	Sher (1966)
H. dihystera	Wide host range	AL, GA	Sher (1966), Johnson et al. (1975) and Motsinger et al. (1976a)
H. multicinctus	Tropical ornamentals	AL	Minton et al. (1963)
H. pseudorobostus	Wide host range	AL	Minton et al. (1963)
Helicotylenchus spp.	Switchgrass, blueberry	GA	Mekete et al. (2011) and Jagdale et al. (2013)
Heterodera glycines	Soybean	AL, GA	Motsinger et al. (1976b)
H. cyperi	Yellow nutsedge	GA	Hajihassani et al. (2018a)
Hoplolaimus columbus	Cotton, peanut	AL, GA	Motsinger et al. (1976a) and Norton et al. (1984)
H. galeatus	Wide host range	AL	Norton et al. (1984)
Longidorus sp.	Switchgrass	GA	Mekete et al. (2011)
Meloidogyne incognita	Wide host range including cotton, peanut, corn, soybean, vegetables, pine, blueberries	AL, GA	Motsinger et al. (1974), (1976a), Norton et al. 1984, Nyczepir et al. (1985), Davis and Timper (2000a), (2000b) and Jagdale et al. (2013)
M. arenaria			
M. javanica			
M. hapla			
M. partityla	Pecan	GA	Nyczepir et al. (2002)
M. graminicola	Yellow and purple nutsedge, wheat	GA	Minton and Tucker (1987)
Meloidodera floridensis	Woody plants	AL	Hooper (1958)
Mesocriconema curvatum	Wide host range	AL	Powers et al. (2014)
M. ornatum	Cotton, peanut, mixed forest, blueberries	AL GA	Minton and Bell (1969), Powers et al. (2014), and Jagdale et al. (2013)
M. onoense	Turfgrass	AL	Powers et al. (2014)
M. rusticium	Cotton, mixed forest	AL	Powers et al. (2014)
M. xenoplax	Ornamentals	AL	Powers et al. (2014)
	Peach	GA	Nyczepir et al. (1985)
Mesocriconema sp.	Corn	GA	Davis and Timper (2000b)
Mesoanguina plantaginis	Plantain	AL	Vargas and Sasser (1976)

(continued)

Table 14.1 (continued)

Nematode	Host/rhizosphere soil	State[a]	References
Nanidorus minor	Wide host range including cotton, peanut, corn, sweet corn, onion, broccoli, eggplant	AL, GA	Johnson et al. (1975), Norton et al. (1984), Davis and Timper (2000b), and Hajihassani et al. (2018b)
Paratrichodorus sp.	Switchgrass, blueberries	GA	Mekete et al. (2011)
Pratylenchus brachyurus	Corn, cotton, peanut,	AL, GA	Motsinger et al. (1976a) and Norton et al. (1984)
P. hexincisus	Wide host range	AL, GA	Norton et al. (1984) and Mekete et al. (2011)
P. penetrans	Wide host range	AL	Norton et al. (1984)
P. scribneri	Corn, cotton, peanut, soybean, switchgrass	AL, GA	Norton et al. (1984) and Mekete et al. (2011)
P. vulnus	Peach	AL, GA	Southern Cooperative Series Bulletin (1960) and Fliegel (1969)
P. zeae	Cotton, corn, peanut	AL, GA	Norton et al. (1984)
Pratylenchus sp.	Corn	GA	Davis and Timper (2000b)
Rotylenchulus reniformis	Cotton, soybean	AL, GA	Motsinger et al. (1976a) and Norton et al. (1984)
Trichodorus borneonsis	Soybean, cabbage palm	AL	Rebois and Cairns (1968)
T. primitivus	Corn, cotton, lespendeza, woody ornamentals	AL	Rebois and Cairns (1968)
Tylenchorhynchus claytoni	Field and ornamental crops	AL	Norton et al. (1984)
T. cylindricus	Field and ornamental crops	AL	Norton et al. (1984)
T. martini	Field crops, turfgrass	AL	Johnson (1970) and Norton et al. (1984)
T. nudus	Field, ornamental and native crops	AL	Norton et al. (1984)
Tylenchorhynchus sp.	Corn, blueberry	GA	Davis and Timper (2000a, 2000b) and Jagdale et al. (2013)
Xiphinema americanum	Cotton, peanut, switchgrass	AL	Norton et al. (1984) and Mekete et al. (2011)
X. krugi	Bahia grass, sorghum	AL	Frederick and Tarjan (1975)
Xiphinema sp.	Cotton, peanut, blueberry	GA	Motsinger et al. (1976a) and Jagdale et al. (2013)
X. pacificum	Peach	GA	Nyczepir and Lamberti (2001)

[a]Names of the states are represented by two letter abbreviations: *AL* Alabama, *GA* Georgia

soilborne fungal pathogens including *Fusarium*, *Phytophthora*, *Pythium* and *Rhizoctonia* producing disease complexes. Management of plant parasitic nematodes principally includes the use of chemical and biological nematicides and crop rotation with resistance or tolerant cultivars. Although several species of plant parasitic nematodes cause damage to several different crops, in this chapter major emphasis will be given on the detection, distribution and management of root knot, reniform, lance, stubby root and ring nematodes because of their economic importance in both states of Alabama and Georgia.

14.2 Economically Important Crops in Georgia and Alabama

In the State of Georgia, cotton (*Gossypium hirsutum*), peanut (*Arachis hypogaea*), vegetables (various species), pecan (*Carya illinoinensis*), corn (*Zea mays*), peach (*Prunus persica*) and blueberry (*Vaccinium* spp.) are the most widely and commercially cultivated crops that are considered major contributors to the state's economy (Table 14.2). In Alabama, cotton, hay, corn, peanut, soybean (*Glycine max*), wheat (*Triticum aestivum*) and cucumbers are the most economically important crops

Table 14.2 Major crops produced in Georgia (2017)

Crop	Planted hectares ($\times10^6$)	Harvested hectares ($\times10^6$)	Production in kilogram ($\times10^9$)	Sales in $ ($\times10^6$)
Cotton	0.52	0.51	1.01	794.8
Peanut	0.34	0.33	1.64	780.5
Hay	–[a]	0.25	1.80	187.0
Corn	–	0.099	1.09	178.9
Sweet corn	0.010	0.010	0.204	98.4
Watermelon	0.008	0.008	0.355	74.2
Soybean	0.062	0.061	0.378	61.4
Tobacco	–	0.005	0.012	52.5
Cucumbers	0.004	0.004	0.084	43.8
Bell pepper	0.001	0.001	0.042	37.1
Cabbage	0.002	0.002	0.098	24.9
Squash	0.001	0.001	0.027	23.5
Snap bean	0.005	0.004	0.033	20.1
Wheat	0.064	0.028	0.089	13.9
Cantaloupe	0.001	0.001	0.018	8.9
Rye	0.08	0.0006	0.015	1.7
Pecan	–	–	0.03	–
Peach	–	–	0.01	–

NASS USDA (2018a)
[a]Data not available

Table 14.3 Major crops produced in Alabama (2017)

Crop	Planted hectares ($\times 10^6$)	Harvested hectares ($\times 10^6$)	Production in kilogram ($\times 10^9$)	Sales in $ ($\times 10^6$)
Cotton	0.176	0.174	0.17	265.9
Hay	–[a]	0.348	1.95	208.5
Corn	0.101	0.980	0.99	210.3
Peanut	0.079	0.078	0.32	155.7
Soybean	0.142	0.140	0.43	155.5
Wheat	0.061	0.040	0.21	35.4
Cucumbers	0.022	0.021	0.09	16.6
Oat	0.016	0.004	0.09	2.4
Pecan	–	–	0.01	–
Peach	–	–	0.02	–

NASS USDA (2018b)
[a]Data not available

(Table 14.3). Alabama's crop production valued at $1.01 billion in 2017, was up by 11% compared to 2016 values. In Georgia, the Southern Coastal Plain region supports the largest portion of both field and vegetable crops planted because it is dominated by warm and humid climates, and light sandy or sandy loam soils. Georgia was the first state in the U.S. to produce cotton commercially in 1734. Georgia often ranks second or third nationally in cotton production and hectares grown. Cotton has played a significant role in both the history and economy of the state. Georgia is the biggest peanut-producing state in the country, accounting for more than 45% of the nation's peanut production. The subtropical climate of Georgia is ideal for producing large yields of high-quality peanuts. The State of Georgia is also the nation's largest supplier of pecans, accounting for about a third of the United States pecan production. Pecan trees are commonly found throughout the state. Georgia is the nation's leading producer of fresh market vegetables including cucurbits, onions, leafy greens, bell peppers, tomatoes and sweet corn. Although vegetable crops were traditionally cultivated in Southern Georgia, their production has extended in recent years to areas where predominantly field crops such as corn, potato and others were grown. Georgia vegetables produced a farm gate value of well over $1.1 billion in 2016. In the State of Georgia peach production is a $42.1 million industry (2016 USDA Georgia Agricultural Facts), with production ranking third behind South Carolina and California, respectively. Blueberry production is mostly centered in Southeastern Georgia with a farm gate value in excess of $100 million that accounts for almost one-third of the total fruit and nut crop value for the state. The subtropical climate of Alabama is located in the coastal region with a temperate region in the north. The soils vary dramatically from fine sandy loams to clay loams and crop production occurs across the entire state utilizing the diversity of soil types, climates and growing seasons. Alabama's primary crops are forestry followed by the field crop production of cotton, soybean, corn, peanuts and forages. Alabama is the third largest producer of peanuts in the U.S. and ranks seventh in cotton production.

Soybeans and corn rank 23rd and 28th nationally. Alabama does grow vegetables, melon, potatoes, sweet potatoes, fruits, tree nuts and berries with rankings from 7th for sweet potatoes to 24th for sweet corn. There is a strong movement for the production of specialty high value crops in Alabama which has diversified production and enhanced grower economics. Overall, Alabama crop production was valued at $1.07 billion in 2016 over 8.8 million acres of farm land (2016 USDA Alabama Agriculture Facts).

14.3 Plant Parasitic Nematodes in Georgia and Alabama

14.3.1 Historical Perspective

The history of conducting applied nematology research in the State of Georgia dates back to 1935. In Alabama, the root knot nematode *M. incognita* was first observed as an economic pathogen in cotton in the late 1800s. Since then, considerable effort has been made to obtain information about the effect of nematodes on cultivated crops and strategies to manage these soilborne pests, mostly based on the use of chemical products and resistant cultivars. The primary purpose of all of these efforts has been to manage economically important nematode species to improve the production of crops and to maximize economic welfare of producers. Recent estimates of nematode damage on some major crops grown in Georgia have shown severe reductions in crop value that ranged from 2% to 10% (Table 14.4). Across the entire State of Alabama, it is estimated that 4% and 6% of the cotton crop was lost to the reniform and root knot nematodes respectively, which is estimated at an economic loss of nearly $21.8 million. Specifically, cotton yields over the last 5 years in a reniform nematode infested field with an average at planting of about 5,000 reniform/100 cm^3 of soil, have averaged 50% less compared to an identical field that had no detectable reniform nematodes (Lawrence et al. 2018). Thus, yield losses can be severe in reniform-infested fields. In both Georgia and Alabama, large amounts of traditional and newly introduced chemical nematicides are still being used to

Table 14.4 Estimated values of losses caused by plant parasitic nematodes on some major crops in Georgia

Crop	% Reduction in crop value	Damage in $ (×10^6)	Cost of control in $ (×10^6)	Total in $ (×10^6)
Cotton	10.0	71.3	60.8[a]	132.1
Peanut	2.75	18.8	5.8	24.6
Corn	6.5	16.4	1.3[b]	17.7
Soybean	3.35	4.3	–	4.3
Vegetables[a]	5.0	11.5	–	–

[a]Little (2017)

[b]Vegetable crops include watermelon, cantaloupe, cucumbers, bell peppers, snap beans, squash and cabbage

control the most damaging nematode species, particularly root knot nematodes. Research trends are currently directing efforts towards the use of crop resistance for effective control of nematodes. Additionally, rotation with non-host crops and bio-control have shown potential for plant parasitic nematode management. However, scientific knowledge on the efficacy of biocontrol agents for plant parasitic nematodes lags behind that for other root diseases.

14.3.2 Root Knot Nematodes, Meloidogyne *spp.*

14.3.2.1 Detection and Distribution

Four major species (*M. incognita, M. javanica, M. arenaria* and *M. hapla*) have been reported on numerous crops in Georgia and Alabama (Table 14.1) (Motsinger et al. 1976a; Baird et al. 1996; Powers and Harris 1993; Norton et al. 1984). According to the Nematode Diagnostic Laboratory at University of Georgia that received 6431 soil samples during 2013 through 2017 from different growers located in 126 different counties that covered 4 geographical regions (Coastal Plain, Piedmont, Blue Ridge and Ridge and Valley) of Georgia, root knot nematodes were present in 85.7% of the counties (Fig. 14.1).

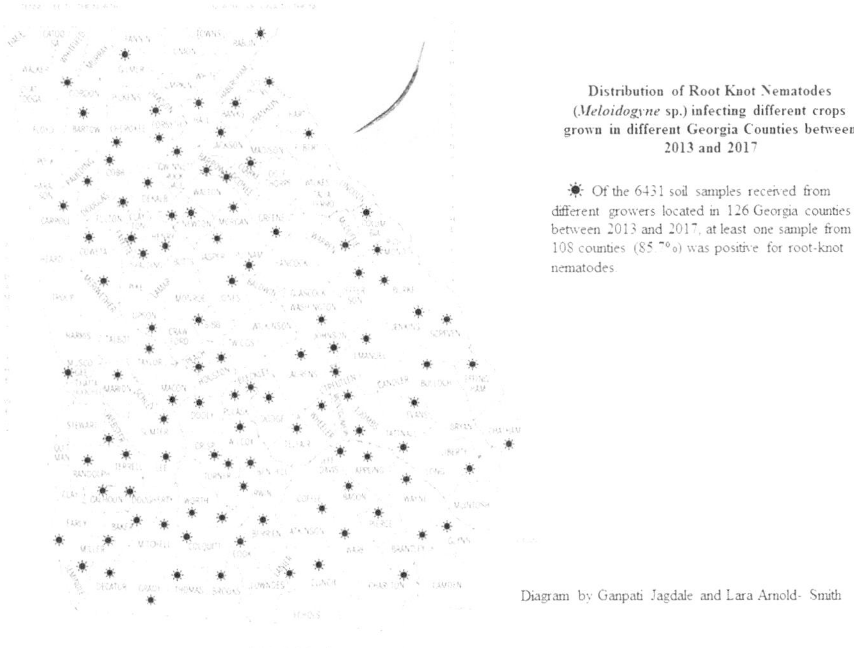

Fig. 14.1 The occurrence and distribution of root knot nematodes on different crops including turfgrasses grown in different Georgia counties

Fig. 14.2 Field symptoms of *Meloidogyne arenaria* damage in peanut, showing a row of completely yellowed and stunted plants (**a**), galls on roots and pegs (**b**) and growth responses of resistant 'GA-14N' (two rows on the left) and susceptible 'GA-06N' (two rows on the right) cultivars treated with a nematicide at the University of Georgia Blank Shank Farm in Tifton, Georgia (**c**). (Photos: Timothy Brenneman and Abolfazl Hajihassani)

Among these species, the southern root knot nematode, *M. incognita*, and the peanut root knot nematode, *M. arenaria*, are widely distributed in field, vegetable and fruit tree growing regions in Georgia. Root knot nematodes feed on roots of plants and produce distinguishing symptomatic galls on the primary and secondary roots that are distinctive and diagnostic in numerous crops including peanut, cotton, soybean, tomato and cucumber (Figs. 14.2, 14.3, and 14.4). In Alabama, *M. incognita* and *M. arenaria* are most often located in the sandy soils of the southern region of this state. A survey of Alabama cotton fields consisting of 969 samples collected in the fall

Fig. 14.3 Field symptoms caused by *Meloidogyne incognita* race 3 in cotton (**a**, **b**), soybean (**c**, **d**) and corn (**e**, **f**) in Alabama showing poor growth of plants and galled roots. (Photos: Kathy Lawrence)

of 1998–2000 found root knot nematodes in 7% of the fields with 57% and 27% of the fields with low or high nematode numbers (Gazaway and McLean 2003). In contrast, a survey of cotton fields in Georgia in 2006 found at least one species of root knot nematode in 69% of the fields planted to cotton. Medium to high populations of *M. incognita* were found in over half of the cotton-producing counties (Kemerait et al. unpubl.). *Meloidogyne* species have been found in 34 out of 102 fields of corn surveyed in 11 counties in Southern Georgia, with *M. incognita* as the most widespread species followed by *M. arenaria* (Davis and Timper 2000a, b). Field studies in

Fig. 14.4 Damage symptoms resulting from *Meloidogyne incognita* parasitism on cucumber grown on fumigated raised beds covered with plastic mulch in Georgia (**a**), heavily galled root systems of cucumber caused by *M. incognita* (**b**). (Photos: Abolfazl Hajihassani)

Georgia have revealed a suppression of about 8.5–35.7% in yield losses due to *M. incognita* parasitism on both susceptible and tolerant cultivars of cotton (Davis and May 2003). *Meloidogyne arenaria* can reduce peanut yields by up to 15% annually in the Southern U.S. Yield losses due to root knot nematodes can be particularly severe when two susceptible host crops are planted in sequence in the same year. This has been observed in many vegetable crops such as cucumber, tomato and squash when grown on raised beds covered with plastic much (Johnson et al. 1996). A preliminary survey of plant parasitic nematodes conducted in fourteen top vegetable-producing counties in Southern Georgia in 2018 showed that root knot nematodes were the dominant parasites. About 67% of surveyed fields grown to multiple vegetable crops including cucumber, tomato, watermelon, cantaloupe, eggplant, peppers, beans, squash, sweet corn and onions were infested by *Meloidogyne* spp. (Marquez and Hajihassani unpubl.).

The root knot nematodes typically become a serious problem in sandy soils, especially during summer and early fall when temperatures are warm and the season has adequate to excessive rainfall. In Alabama, a survey identifying species of root knot nematodes in field crops in 2016 and 2017 found *M. incognita* race 3 as the most prevalent species present. *Meloidogyne arenaria* was present in 3% of the samples (Groover and Lawrence 2018). Although root knot nematode alone is a serious root disease of numerous crops, disease severity and yield loss are often greater in the presence of fungal pathogens. For example, losses in peanut and cucumber due to Cylindrocladium black rot (*Cylindrocladium parasiticum*) and Pythium root rot (*Pythium aphanidermatum*) have been shown to increase substantially in the presence of root-knot nematodes in Georgia, respectively (Dong et al. 2009; Morris et al. 2016). Fusarium wilt is a serious disease complex caused by *F. oxysporum* f. sp. *vasinfectum* and *Meloidogyne* spp. can cause an annual loss of 1.3% or $3.7 million of cotton yield in both Alabama and Georgia (Bell et al. 2017; Lawrence et al. 2018).

14.3.2.2 Management of Root Knot Nematodes

14.3.2.2.1 Cultural Control

Crop Rotation The management of root knot nematodes in Georgia and Alabama has been characterized largely by crop rotation in which the host plants including cotton, soybean, peanut and corn are rotated with poor-hosts or non-host plants. In an ideal rotation, the previous crop suppresses populations of the target nematode and prevents damage to the subsequent crop. Because peanut is a non-host of *M. incognita* and cotton is a non-host of *M. arenaria* (Johnson et al. 1998), rotations with peanut and cotton are highly effective in the management of both nematode species. Davis and Timper (2000a, b) noted that commercial corn hybrids, which are commonly planted in rotation with cotton and peanut in Georgia, supported the reproduction of *M. incognita* (race 3) better than *M. arenaria* suggesting that corn is not a compatible rotation crop for cotton where *M. incognita* is a concern. In Alabama, rotations of cotton, soybean and corn in a *M. incognita* race 3 field found that the nematode populations continued to increase when the host crop was consistent over years. However, even rotations from cotton to a susceptible crop such as corn or soybean, only allowed the *M. incognita* population to increase by 13% and 25%, respectively, compared to continuous cotton (Groover et al. 2017). Since peanut is not a host for the southern root knot, reniform and lance nematodes, it can effectively control these nematode species when planted as rotational crop with susceptible crops in Georgia. Tobacco and many vegetable crops should not be included in rotations with cotton where management of southern and peanut root knot nematodes is the primary concern, even though *M. incognita* races 1 and 2 do not reproduce on cotton. Field and greenhouse research conducted in Georgia have shown that pearl millet hybrids (TifGrain 102) are resistance to various types of plant parasitic nematodes including *M. incognita* race 3, *M. arenaria* race 1, *B. longicaudatus* and *P. brachyurus*. These hybrids can be used in rotations with susceptible crops to reduce the nematode problems in subsequent crops such as peanut and cotton (Timper et al. 2002; Timper and Hanna 2005).

Cover Crops Although cover crops generally increase soil microbial activity, biological diversity and organic matter content, they can also help in suppressing the populations of plant parasitic nematodes and other soilborne pests of cultivated crops. This in turn may reduce the frequency of pesticide applications required to control plant parasitic nematodes. Generally, the use of cover crops for suppression of root knot nematodes should be done with caution because of the broad host range of *Meloidogyne* spp. and susceptibility of certain species/cultivars of cover crops that may increase root knot nematode populations in soil. Greenhouse and field studies in Georgia have revealed that certain summer and winter cover crops including rye (*Secale cereale*; cv. Wrens Abruzzi), pearl millet (*Pennisetum glaucum*; cv. TiftGrain 102), vetch (*Vicia sativa*; cv. Cahaba White), oat (*Avena sativa*), wheat (*Triticum aestivum*) and bahiagrass (*Paspalum notatum*), have a potential to be used as cover crops for the management of root knot nematodes (Johnson et al. 1998; Sumner et al. 1999; Timper et al. 2002, 2006). Timper et al. (2011) reported that incorporating rye residue into soil or scattering on the soil surface had no effect on populations of either *M. incognita* in cotton or *M. arenaria* in peanut. However, the use of high quantities of rye biomass resulted in reduction of *M. incognita* numbers in soil and root gall index in cotton (Timper 2017). As stated above, cover cropping may have suppressive effects on root knot nematodes, but it may support the reproduction of other species of nematodes. For example, oats, wheat and rye may be good hosts for *B. longicaudatus* and *H. columbus* but not for root knot nematodes (Davis et al. 2000). Incorporation of residues of legume cover crops into soil can help to prevent soil erosion, improve water retention in sandy soils and may produce toxic products that can be detrimental to nematodes. For example, preliminary field research in Georgia have shown that incorporation of sunn hemp (*Crotolaria juncea*) residue reduced the root knot nematode population to a depth of 25 cm in soil (Hajihassani et al. unpubl.). Integration of cover crops with other cultural management practices such as tillage or crop rotation with non-hosts may increase the beneficial effects of cover crops in controlling nematodes. Although some *Brassica* species have the potential as winter cover crops and green manure amendments for nematode management, many other species in the Brassicaceae family are known to be susceptible to root knot nematodes. For example, Monfort et al. (2007) reported that there was an increase in the *M. incognita* population density in the rhizosphere of *B. juncea*, *B. oleracea*, *Sinapis alba* or *B. napus* but when the crop residues of these crops incorporated in the soil, the population of nematodes was reduced. This suggests that the efficacy of biofumigation with *Brassica* crops for managing root knot nematodes clearly rely on the plant species used as cover crop and its adaptability to the environment.

14.3.2.2.2 Chemical Control

Precision agriculture has recently become a widely accepted practice in Alabama and Georgia; however, more research is required to fully implement this technique in grower fields. One important aspect of the technology is variable-rate applications of nematicides. In the field, plant parasitic nematodes generally have a patchy and clustered spatial distribution (Lawrence et al. 2008). The distribution varies with nematode species, soil texture and the crop grown. Variable-rate and site-specific application is the application of nematicides only to the areas where the nematode population has reached the economic threshold level and yield enhancement is expected. To implement a successful nematode management program, the nematodes present in the field and their location must be determined (Lawrence et al. 2007, 2008; Moore and Lawrence 2012; Davis et al. 2013). This is accomplished by collecting samples from a uniform systematic grid across the field or through the use of zone sampling (Ortiz et al. 2012). A representative number of soil samples is the key to success for any nematode management program as it becomes essential to decide suitable variable rates of application of nematicides.

The smaller the sample grid size (0.01–0.2 ha) the more detailed the nematode distribution map is generated that results into better placement of the nematicides. However, the more samples the higher the laboratory cost to process them. Zone sampling creates zones or areas of similarity from which samples are collected. Soil texture is one criterion for obtaining points from similar areas. Different nematode genera favor different soil textures so soil texture will influence the damage resulting from infection. Each sample point is geo-referenced using a global positioning system (GPS). This type of sampling is a popular sampling strategy that allows mapping the spatial information for a specific nematode pest. Once the nematode population numbers are located and mapped, nematode contour maps can be developed to graphically represent nematode numbers in a field. The map can be overlaid with yield maps to determine problem areas in the field. Poor crop yields in combination with high nematode numbers are good indications that areas may require nematicide applications. A nematicide prescription map and predetermined application rates are then loaded into the application equipment's computer. The specified amount of nematicides is applied to the selected areas as the equipment moves across the field.

14.3.2.2.3 Resistance

Use of nematode-resistant cultivars not only protects the crop in the field, but also reduces nematode infestations for the subsequent cash crop. *Meloidogyne*-resistant cotton cultivars suppress nematode reproduction compared to the susceptible cultivars but nematode tolerant cotton cultivars will support greater levels of nematode reproduction without affecting yields. Until recently, no commercial cotton cultivar with a high level of resistance to southern root knot nematode was available (Davis and May 2005) but resistant cultivars such as PhytoGen 487 WRF, Deltapine 1747NR B2RF and Stoneville 4946 GLB2 are now commercially available where *M. incognita* is a major problem (Georgia Cotton Commission 2018). Although these cultivars have shown a high level of resistance to the nematode, the infection risk associated with other pathogenic organisms may limit their effective use. For example, DP 1558NR B2RF was affected in some cotton fields in Georgia where the bacterial blight caused by *Xanthomonas citri* pv. *malvacearum* is present (R.C. Kemerait, Univ. Georgia, pers. com.). In peanut, very high levels of root knot nematode resistance have been characterized and introduced from wild species of *Arachis* spp. into newly established peanut cultivars. TifGP-2, Tifguard, Georgia-14N, TifNV-High O/L, NR 0812, and NR 0817 were released as resistant cultivars to *M. arenaria* (Anderson et al. 2006; Holbrook et al. 2008, 2012, 2017; Branch and Brenneman 2015). Field test evaluations in Southern Georgia have shown very high levels of resistance of these improved cultivars to *M. arenaria* in comparison to Georgia-06G, a widely grown and high-yielding susceptible cultivar (T. Brenneman, Univ. Georgia, pers. com.). In addition, Tifguard and its nematode-susceptible sister line, TifGP-2, have high resistance to Tomato Spotted Wilt Virus, making them very suitable for planting in the Southeastern U.S. (Holbrook et al. 2012). In vegetables, some nematode-resistant cultivars are currently available which will reduce production costs and increase marketable yields. Planting resistant cultivars of tomato (multiple cultivars), bell pepper (e.g. Charleston Belle and Carolina Wonder) and sweet potato (e.g. Covington and Evangeline) with resistance to *Meloidogyne* species, might be an effective option in managing root knot nematodes. However, the presence of *Mi* resistance-breaking species such as *M. haplanaria*, *M. hapla* and *M. enterolobii* in some vegetable-growing regions raises concerns about durability of resistance.

14.3.2.2.4 Biological Control

One of the potential biocontrol agents of root knot nematodes is *Pasteuria penetrans*. This endospore-forming, gram-positive bacterium is known as the primary biological agent that causes soil suppressiveness against root knot nematodes. *Pasteuria penetrans* is present in many Georgia peanut fields and can build up its population to levels which are suppressive to nematode populations (Timper 2009; Timper et al. 2016). Field studies in Georgia have shown that increasing *Pasteuria* populations in the soil significantly reduced root knot nematode reproduction (Timper 1999). Studies investigating *P. penetrans* in monoculture rotation systems and their influence in soil suppressiveness have yielded varying outcomes. Rotations including poor hosts for *Meloidogyne* spp. reduced the *P. penetrans* endospore densities compared to a monoculture of peanut (Timper et al. 2001). One of the obstacles of *P. penetrans*-based biological control is the downward movement of spores due to irrigation or rainfall that can result in endospore depletions in the top 15–20 cm of soil (Cetintas and Dickson 2005).

Another biological control option includes various strains of *Bacillus* spp. often targeting *M. incognita* (Kloepper et al. 1992). Gustafson developed BioYield®, a combination of *B. velezensis* strain IN937a and *B. subtilis* strain GB03, in a flowable formulation for management of soilborne pathogens and suppression of *M. incognita.* In Alabama, BioYield® reduced *M. incognita* populations and increased yields in tomato in greenhouse and field trials (Burkett-Cadena et al. 2008). VOTiVO, *Bacillus firmus* GB-126, is marketed by Bayer CropScience as a seed treatment for the control of plant parasitic nematodes on corn, cotton, sorghum and soybean. *Bacillus firmus* GB-126 tests indicated this product reduced egg production in *R. reniformis, Heterodera glycines* and *M. incognita* (Lawerence et al. unpubl.). Induced systemic resistance (ISR) was demonstrated with split-root experiments in the greenhouse and ISR was evident in *H. glycines* split-root assays on soybean but not in *M. incognita* assays on cotton (Schrimsher 2013). The newest biological seed treatment nematicide is Aveo (*B. amyloliquefaciens* strain PTA 4838) by Valent is available for plant parasitic nematode management on corn and soybean."

14.3.3 *Reniform Nematode,* Rotylenchulus reniformis

14.3.3.1 Detection and Distribution

Rotylenchulus reniformis is a major pathogen of cotton in the Southeastern U.S. (Koenning et al. 2004; Robinson 2007). Although first reported on cotton in Georgia in 1940 (Smith 1940) and in Alabama in 1958 (Minton and Hopper 1959), it was not recognized as a serious nematode pest on cotton until 1986, when it caused substantial yield losses in grower fields in South Alabama. The spread of the reniform nematode has been relatively slow across Georgia compared to other parasitic nematodes, in particular the southern root knot nematode. Between 2013 and 2017, the Nematode Diagnostic Laboratory at University of Georgia determined that 25.4% of the counties contained reniform nematodes (Fig. 14.5). A survey of Alabama cotton fields consisting of 969 samples collected in the fall of 1998–2000 found the reniform nematode to be present in 46% of the fields sampled with 44% and 33% of the fields having low and very high populations respectively. Although a damaging pathogen of several crops grown in the region, *R. reniformis* is a primary problem in

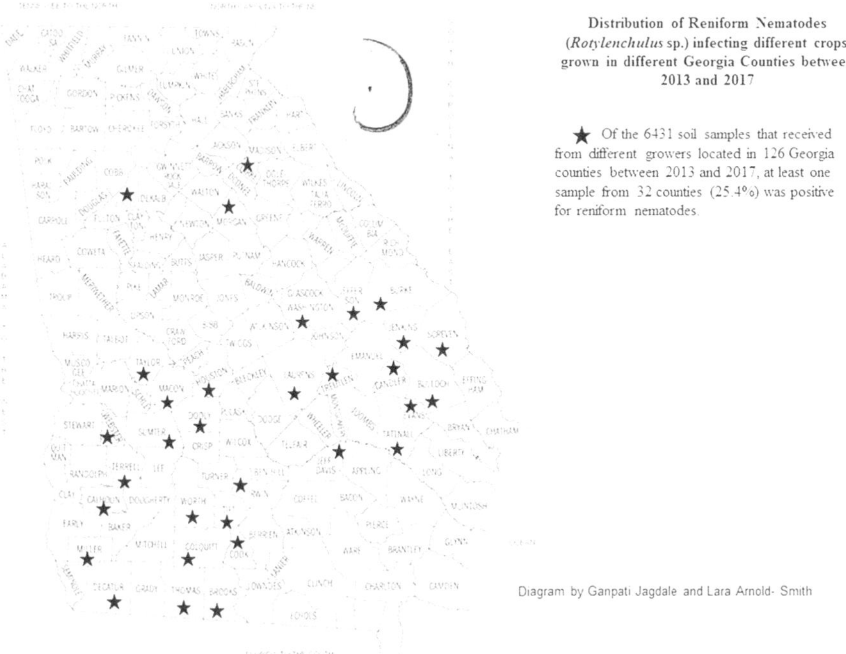

Fig. 14.5 The occurrence and distribution of reniform nematodes on different crops including turfgrasses grown in different Georgia counties

Fig. 14.6 Uneven growth of cotton plants in a reniform nematode infested field in Alabama (**a**), foliar symptoms of interveinal chlorosis associated with reniform nematode infested cotton (**b**), visual cotton yield reductions in areas of high *Rotylenchulus reniformis* numbers near harvest (**c**). (Photos: Kathy Lawrence)

cotton (Fig. 14.6) and is currently present in many of the main cotton-growing counties in Georgia, including nine of the ten counties with the highest cotton production. In a recent (2006) statewide survey of cotton fields, approximately 5% of the fields were infested with *R. reniformis* (Kemerait et al. unpubl.). The nematode can cause serious damages in more restricted areas of the state with heavier soils. A survey conducted in 1989 and 1990 in Alabama found 6.5% of the fields in North and Central Alabama to be infested with *R. reniformis* at populations above the economic threshold. Twelve years later in 2002, *R. reniformis* was found in 46% of the fields in the same regions and half of those fields had populations above the economic threshold (Gazaway and McLean 2003). In Alabama, *R. reniformis* has been shown to prefer finer textured soils and exists above economic thresholds in a wide variety of soil types. The natural migration of the reniform nematode was monitored in Alabama in a silty loam soil, under no-till cotton. The reniform nematode moved 200 cm horizontally and to a 91 cm vertical depth from the initial point of inoculation in one growing season (Moore et al. 2010a). Population density increased steadily in the irrigated trial during both years, exceeding the economic threshold of 1,000 nematodes/150 cm^3.

14.3.3.2 Management of Reniform Nematode

The methods currently used to manage *R. reniformis* in cotton can be economically beneficial if utilized with forethought. Rotation and nematicides are the principle means of *R. reniformis* management. As with many nematode infestations in cotton production systems, nematicide use is the major management tactics for *R. reniformis*. Cultivars resistant and or tolerant to *R. reniformis* have promise to alleviating yield loss, but these are not presently available in current cotton cultivars and the efficacy of tolerant cultivars has been questioned (Robinson et al. 1997; Koenning et al. 2000; Starr et al. 2007). A reniform nematode resistant cultivar named Phytogen will likely be marketed in the near future. Recently, some germplasm lines (LONREN-1, LONREN-2 and BARBREN-713) of upland cotton with high levels of resistance to *R. reniformis* have been developed (Bell et al. 2014, 2015). The BARBREN-713 line also is highly resistant to root knot nematodes. Field trials conducted in Alabama established that *R. reniformis* populations as 50% lower in these resistant lines compared with the susceptible cotton lines at 45 days after planting. However, the use of nematicides did increase yields of both the resistant and susceptible cotton lines (Schrimsher et al. 2014).

14.3.3.2.1 Cultural Control

Crop rotation is recommended as an important tactic for management of reniform nematode. Rotation crops useful for *R. reniformis* suppression include peanut, corn, reniform-resistant soybeans, bermudagrass, bahiagrass and sorghum. In Alabama, crop rotation to non-hosts such as corn or peanuts or highly resistant varieties of soybean, is an effective strategy for the management of *R. reniformis* (Gazaway et al. 2007). Corn, soybean and peanut all reduced initial *R. reniformis* populations compared to continuous cultivation of cotton (Moore et al. 2010b). Cotton yield following 1-year rotations of corn, soybean or peanut yielded 16%, 26% and 17% higher than continuous cotton. Two years of corn, peanuts or soybeans increased cotton yield higher than continuous cotton by 34%, 46% and 40%. All rotations resulted in a net profit over variable costs compared to continuous cotton both with and without a nematicide. The 3-year rotations of corn and soybeans followed by cotton produced the largest increase in net profit over variable costs, both with and without a nematicide. The use of the correct crop rotation for the suppression of the reniform nematode can have a positive impact on cotton yields, even without the use of a nematicide. Many native weed species are host of *R. reniformis* to some degree and can confound the positive effects of crop rotation if not properly controlled (Davis and Webster 2005; Jones et al. 2006; Lawrence et al. 2008; Wang et al. 2003). Davis et al. (2003) have shown that rotations of winter grain crops and soybean cultivars resistant to reniform nematode with cotton are effective for suppression of reniform nematode populations and increasing cotton yield. Although crop rotations with non-host crops are effective in reducing populations and damage incurred by *R. reniformis*, rotations with these crops are often economically prohibitive in many areas where cotton is grown in the United States (Davis et al. 2003; Lawrence and McLean 1999).

Cover crops have not been as beneficial for *R. reniformis* management as they have been for *Meloidogyne* spp. Crimson clover, subterranean clover and hairy vetch were shown to be hosts of *R. reniformis* in greenhouse tests, although field populations did not increase on these cover crops under the natural environmental conditions (Jones et al. 2006). These cover crops may increase initial *R. reniform* populations if the winter is mild and the covers are not terminated before soil temperatures rise. Varieties of radish, black mustard, white mustard, canola, lupin, ryegrass, wheat, oats and rye were poor hosts for *R. reniformis* and did not sustain reniform populations (Jones et al. 2006).

14.3.3.2.2 Chemical Control

An assortment of nematicides have been proven effective for the management of *R. reniformis*, including aldicarb (AgLogic 15G) (Lawrence et al. 2018; Lawrence and McLean 2000), fenamiphos (Nemacur) (Koenning et al. 2007; Lawrence et al. 1990) and terbufos (Counter) (Lawrence et al. 1990). Of the granular pesticides, aldicarb has been the most widely used in cotton production and its continual use has resulted in reports of enhanced degradation by soil microbes thus decreasing its overall efficacy (Lawrence et al. 2005). Fenamiphos is no longer labeled for use in the United States and terbufos was preliminary labeled for use in cotton production in Georgia. Seed applied pesticides such as abamectin, thiodicarb and fluopyram have become widely used in cotton production as a part of Avicta Complete Cotton, Aeris Seed Applied System and COPeO Prime, respectively, and have been reported to provide adequate management of *R. reniformis* (Lawrence and Lawrence 2007; Lawrence et al. 2018). Their protection of the root is limited (Faske and Starr 2007) as is their ability to provide adequate protection against high populations of *R. reniformis* (Moore et al. 2010a, b). The newest seed treatment nematicides on the market in 2018 are Monsanto's tioxazafen (NemaStrike™) and Cortiva's fluazaindolizine (Salibro™). In-furrow spray nematicides are the most recent additions to the nematicide arsenal. Fluopyram combined with imidacloprid (Velum Total) is the most frequently used nematicide in Alabama on cotton. The application of Velum Total resulted in an average 90% decrease in *R. reniformis* eggs/g of root over ten cotton cultivars and increased yield by an average of 23% or 903 kg/ha (Groover et al. 2017).

Oxamyl (Vydate® C-LV) is a foliar applied pesticide that also provides adequate management of *R. reniformis*, often in conjunction with previously mentioned pesticides (Baird et al. 2000; Lawrence and McLean 2000), but has been reported to be less effective in dry conditions (Koenning et al. 2007). Additional options for *R. reniformis* management are biologicals such as *Bacillus firmus* (Poncho®/ VOTiVO®) and *Paecilomyces lilacinus* strain 251 (Nemout) as seed applied formulations (Castillo et al. 2013) that have been reported to have efficacy against the nematode. Furthermore, there are multiple nematophagous fungi with high levels of effectiveness in greenhouse studies (Wang et al. 2004; Castillo et al. 2009) that could prove useful in the future. Overall, the number of pesticides for the management of *R. reniform* is decreasing, resulting in increased challenges for producers.

14.3.4 Lance Nematode, Hoplolaimus *spp.*

14.3.4.1 Detection and Distribution

The lance nematode is a serious parasite of cotton, soybean and corn in parts of Georgia and Alabama (Davis and Noe 2000; Noe 1993). Among multiple species, *H. columbus, H. galeatus* and *H. magnistylus* are considered as the most pathogenic lance species. *Hoplolaimus galeatus* and *H. magnistylus* are the most frequently identified species in Alabama. In Georgia, *H. colombus* has been associated with cotton and soybean, on which tremendous damage and economic yield loss occurs in infested fields. From 2013 to 2017, Nematode Diagnostic Laboratory at University of Georgia found 51.6% of the counties contained lance nematodes (Fig. 14.7). Yield losses due to the nematode have been estimated to be as high as 18% and 48% on cotton and soybean, respectively (Noe 1993); however, losses of more than 50% can occur in sandy soils with high infestations (Fig. 14.8). The economic damage threshold was determined to be 50 nematodes/100 cm^3 of soil. Damaging levels of *H. columbus* has been found in 5% of cotton fields primarily in Georgia's Coastal Plain soils that have relatively high sand contents.

14.3.4.2 Management

Field studies conducted in Georgia (Davis et al. 2000) have shown that removal or destruction of root systems of cotton slightly suppressed populations of *H. columbus* but it had no effect on improvement of the yield of subsequent cotton crops. Control of *H. columbus* on cotton has been achieved primarily through nematicide application. Nematicides are expensive and environmental concerns make their usage problematic. Field research have shown that rotating tobacco with cotton may be effective in suppression of population densities of lance nematode. In soybean,

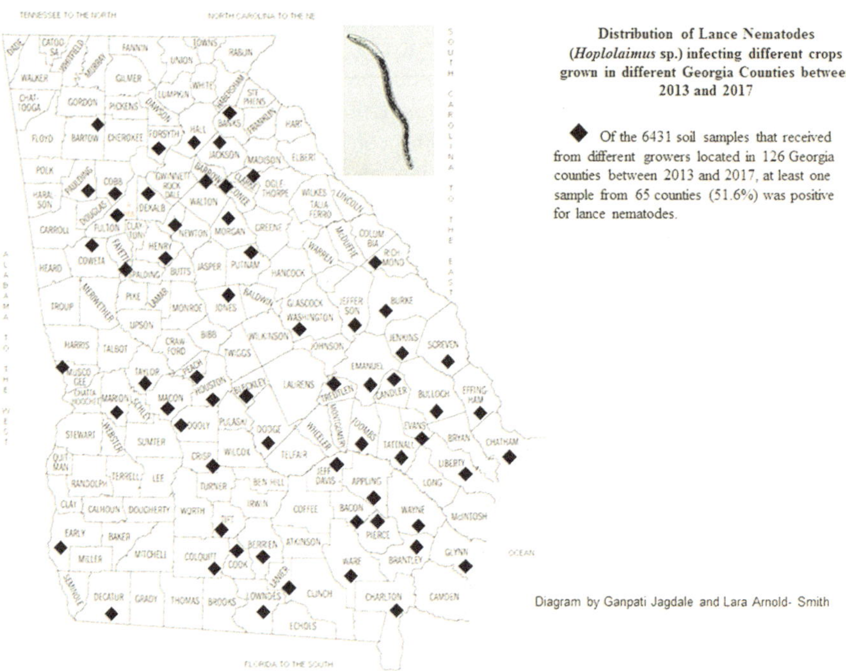

Fig. 14.7 The occurrence and distribution of lance nematodes on different crops including turfgrasses grown in different Georgia counties

Fig. 14.8 Damage symptoms on soybean foliage in a field with low (**a**), moderate (**b**) and high (**c**) population levels of *Hoplolaimus columbus*. (Photos: John Mueller)

management of the lance nematode relies on the use of tolerant cultivars; however, variation in the response of soybean cultivars to *H. columbus* has been reported. Winter wheat and rye planted as cover crops had no impact on *H. columbus* populations (Davis et al. 2000).

14.3.5 Stubby Root Nematode, Paratrichodorus, Trichodorus *and* Nanidorus

14.3.5.1 Detection and Distribution

Stubby root nematodes are among the least studied nematode pests infesting cultivated crops in Georgia and Alabama. From 2013 to 2017, Nematode Diagnostic Laboratory at University of Georgia found 85.7% of the counties contained stubby-root nematodes (Fig. 14.9). Stubby root nematodes cause severe reduction in the growth and yield of multiple field and vegetable crops in the Southeastern U.S. These nematodes feed on the root tips of host crops, thus leading to a stunted, stubby appearance to the root system that can be incorrectly diagnosed as herbicide damage. The shoot of plants may appear stunted with chlorotic foliage (Fig. 14.10). Recent rise in corn acreage in the Southern U.S. has increased the presence of this nematode in the region. The nematode primarily occurred in the Coastal Plain soils of Georgia and Alabama although isolated fields infested with this nematode has been found in Northern Georgia. Severe root pruning to corn roots by the stubby root nematode is most often observed in cool wet springs in the Coastal Plain soils.

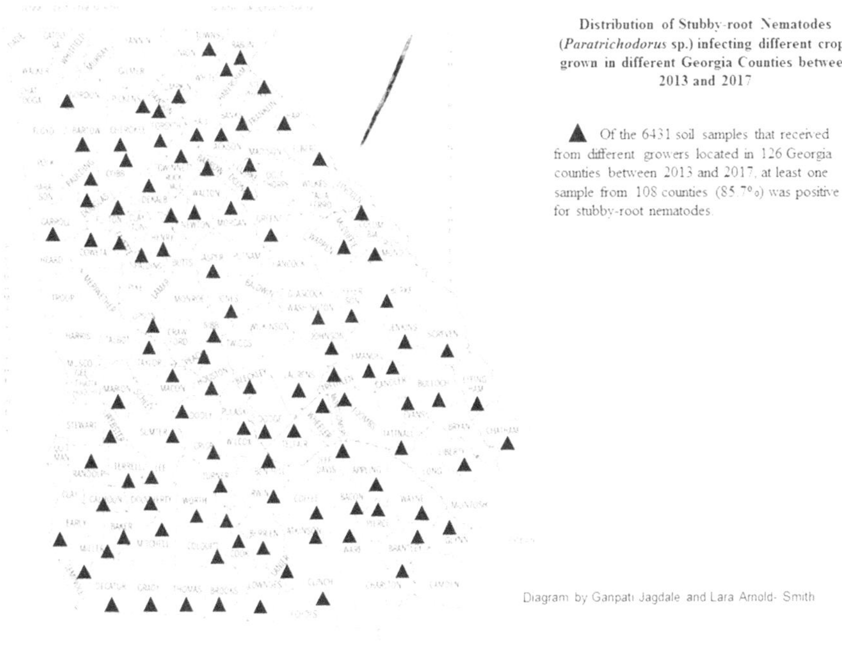

Fig. 14.9 The occurrence and distribution of stubby root nematodes on different crops including turfgrasses grown in different Georgia counties

Fig. 14.10 Field symptoms of *Nanidorus minor* in sweet corn (**a**, **b**), broccoli (**c**, **d**) and onion (**e**, **f**) showing large area of unevenly stunted plants and abbreviated root systems in Georgia. (Photos: Abolfazl Hajihassani)

In the Southern Georgia, *N. minor* is considered a major pest on multiple vegetable crops grown in sandy soils and is responsible for reduction in yield of sweet corn and sweet onion (Hajihassani et al. 2018b).

14.3.5.2 Management

The ability of stubby root nematodes to live deep in the coarse-textured soil profile and to reproduce fast in the presence of host plant roots make control of this nematode particularly challenging. It is known that continuous growing of highly susceptible crops such as corn and certain vegetable crops (e.g. onion, eggplant and sweet corn) can build up *N. minor* population to the economic damaging levels that may necessitate nematicide application on subsequent crops (Hajihassani et al. unpubl.). Tillage tends to reduce numbers of stubby root nematodes as well as rotation to peanut or soybean (Johnson et al. 1974). Cover crops such as pearl millet hybrids (cv. TifGrian 102), cowpea (cv. Mississippi Silver) or seasame (cv. Sesaco 16) tend to keep stubby root nematode populations below the damage threshold and may lessen grower's reliance on chemical control (Timper and Hanna 2005; McSorley and Dickson 1995). Resistant cultivars to the stubby root nematodes are not commercially available in current field and vegetable crops.

14.3.6 Ring Nematodes, Mesocriconema *spp.*

14.3.6.1 Detection and Distribution

Multiple species of ring nematodes (*Mesocriconema* spp.) occur in high population densities in the rhizosphere of crops including blueberry, peanut, soybean, corn, ornamentals, peach, turf grass and vegetables that are grown throughout Georgia. Between 2013 and 2017, the Nematode Diagnostic Laboratory at University of Georgia found ring nematodes in 85.7% of the counties (Fig. 14.11). However, *M. ornatum* and *M. xenoplax* are considered the most damaging species of ring nematode in Georgia. Of these two species, *M. ornatum* is predominantly associated with crops like blueberry, corn, cotton, peanut, soybean, vegetables and turfgrass whereas *M. xenoplax* is mainly associated with peaches, grapes, ornamentals and turfgrasses. Although both *M. ornatum* and *M xenoplax* cause serious damage to many crop species, a major emphasis in this chapter is placed on their impact on blueberries and peaches, respectively, because of their tremendous economic damage to these valuable crops in Georgia.

14.3.6.2 Ring Nematode, *Mesocriconema xenoplax*

In Georgia, peach, *Prunus persica* production is a $31.3 million industry (2012 USDA Georgia Agricultural Facts), but it is on the verge of decline due to the incidence of many diseases like Armillaria root rot and plant parasitic nematodes like ring nematodes, *M. xenoplax* (Savage and Cowart 1942; Miller 1994). Ring nematode is a primary cause of peach tree short life (PTSL) disease that causes premature deaths of peach trees (Nyczepir et al. 1983). Peach tree short life is a disease

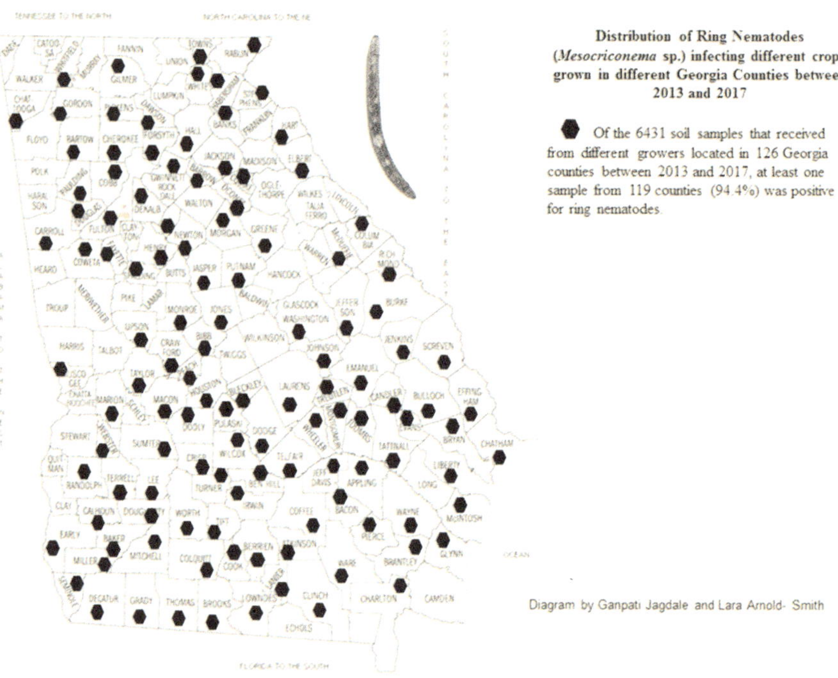

Fig. 14.11 The occurrence and distribution of ring nematodes on different crops including turfgrasses grown in different Georgia counties

complex in which ring nematode infested peach trees become susceptible to combination of factors including cold injuries and bacterial canker disease caused by *Pseudomonas syringae* pv. *syringae* or to each of these individual factors (Brittain and Miller 1978; Nyczepir et al. 1983). According to Nyczepir et al. (1983), *M. xenoplax* infested peach trees died of cold injury, but trees without nematode infestations were resistant to cold injuries. Furthermore, *M. xenoplax* infested trees were more susceptible to bacterial spot disease caused by *Xanthomonas arboricola* pv. *pruni* than uninfected trees (Shepard et al. 1999). The main symptoms of PTSL (Fig. 14.12) include wilting of young leaves, discoloration of cambial tissue and the collapse of new growth above the soil line and eventually death of trees (Nyczepir et al. 1985).

14.3.6.2.1 Management of *M. xenoplax*

The management of ring nematodes is essential for maintaining and optimizing yield of peach orchards. It has been demonstrated that pre-plant fumigation with 67% methyl bromide +33% chloropicrin mixture suppressed the population of *M. xenoplax* in the peach orchards (Nyczepir et al. 2012). Since importation and

Fig. 14.12 Peach tree short life (PTSL) disease caused by ring nematode, *Mesocriconema xenoplax*. (Photos: Andrew Nyczepir)

manufacturing of methyl bromide was banned in the US and Western Europe after January 2005 (Clean Air Act 1990) there was interest in finding alternatives to chemical nematicides to manage ring nematodes infesting peach orchards. Currently the only pre-plant fumigant chemicals available are Telone II (1,3-Dichloropropene) and Vapam® (metam sodium), with Telone II being the one primarily being used and recommended to growers (Horton et al. 2013). Crop rotation with different cover crops has been recognized as one of the best management practices that reduces plant parasitic nematode populations and the associated crop damage (McSorley 2001). Growers in the Southeast generally remove the peach orchard when heavy tree loss from PTSL occurs and often replant these orchards with field crops or small grains instead of peaches. Studies conducted by Georgia scientists on the interaction between small grain crops and *M. xenoplax*, showed that wheat (cv. Stacy) and sorghum (cv. NK2660) plants were poor and nonhost of *M. xenoplax*, respectively (Nyczepir and Bertrand 1990; Nyczepir et al. 1996). They also demonstrated that planting wheat as a groundcover can suppress the populations of *M. xenoplax* and prolonging tree survival on PTSL sites (Nyczepir and Bertrand 2000). Sorghum as green manure was also as effective as methyl bromide in suppressing populations of *M. xenoplax*. According to Nyczepir (2005), rotation of land with wheat/fallow for 3 years prior to re-planting peach orchards can be effective as pre-plant methyl bromide fumigation in suppressing ring nematode populations and increasing tree survival on a PTSL sites. Based on the results of 3-year preplant wheat rotation research, a current recommendation of pre-planting of wheat as rotation crop to prolong tree survival on PTSL sites is available for peach growers in the Southeastern U.S. (Horton et al. 2010). Resistant rootstocks also play an important role in reducing the severity of PTSL. For example, studies conducted in both South Carolina and Georgia showed that peach trees on Guardian rootstock survive better than on Lovell and Nemaguard rootstock when planted in *M. xenoplax* infested fields (Okie et al. 2009). Solarization can influence the population density of *M. xenoplax* in the fields. The effects of solarization, biological control bacteria, *Pseudomonas* spp. and wheat as rotation crop as alternatives to chemical nematicides against *M. xenoplax* in Georgia were evaluated from 2004 to 2011 (Nyczepir et al. 2012). These researchers found that *M. xenoplax* populations were equally suppressed in solar-wheat-treated soil and methyl bromide fumigated plots. Recently, Noe et al. (2015a, b) reported that application of the nematicide fluensulfone (Nimitz) has potential to suppress of *M. xenoplax* population densities on both very susceptible (Nemaguard) and tolerant (Guardian) peach rootstocks to PTSL.

14.3.6.3 Ring nematode, *Mesocriconema ornatum*

Blueberry (*Vaccinium* spp.) is grown in more than 30 states representing over 29,137 ha in the United States (Anon. 2012). The blueberry industry in Georgia continues to grow rapidly, with substantial acreage increasing on a yearly basis. However, although good sites remain for rabbiteye (*V. virgatum*) and southern highbush (*V. corymbosum*) production, the cost of land and site preparation is substantial, especially for southern highbush cultivars that may require added organic matter. Due to the age of the industry in Georgia, many plantings are now reaching the greater than 25-year timeframe and as these plantings decline in productivity, growers often replant these older sites rather than purchase new land. In addition, as newer varieties with desirable traits enter the market, older varieties are often not competitive in yield or quality; therefore, older varieties are often replaced with newer varieties even prior to their natural decline. These replanted sites often exhibit poor plant growth, yellowing, stunting, higher mortality, premature decline (J.P. Noe, Univ. Georgia, pers. com.) and severely reduced yields, symptoms collectively known as Blueberry Replant Disease (BRD; Figs. 14.13a, b), which is considered an emerging threat to the blueberry industries in Georgia (J.P. Noe, Univ. Georgia, pers. com.). In 2008, a preliminary survey of several commercial blueberry fields in Georgia revealed very high ring nematode populations (ca. 1,000 *M. ornatum*/100 cm^3 soil) associated with the rhizosphere of blueberries exhibiting typical BRD symptoms (P.M. Brannen, Univ. Georgia, pers. com.). Major parasitic nematodes frequently associated with commercially grown blueberries in Georgia include ring (*M. ornatum*), dagger (*Xiphinema* spp.), stunt (*Tylenchorhynchs* spp.), spiral (*Helicotylenchus* spp.), lance, root knot and stubby root nematodes (Jagdale et al. 2013). Although the pathogenicity of most of these plant parasitic nematodes to blueberry is unknown, preliminary tests with fumigant nematicides, oxamyl and

Fig. 14.13 Blueberry replant disease caused by ring nematode, *Mesocriconema ornatum* (**a**), blueberry plots infested with *M. ornatum* and treated with methyl bromide (left) and untreated control (right) (**b**). (Photos: Jim Noe and Phillip Brannen)

Telone II showed a strong correlation between increased plant growth, vigor and decreased nematode densities, suggesting detrimental impacts of nematodes (Noe et al. 2012) especially ring nematodes, *M. ornatum* (Jagdale et al. 2013). The widespread occurrence of ring nematodes in blueberry and their demonstrated pathogenicity, indicates that BRD could become a major limitation to continued production on existing farms. The economic impact of BRD could be devastating to growers when establishing new plantings, as the estimated cost of establishing and maintaining blueberry is $93,800/ha for the 4 years normally required before full production (Fonsah et al. 2007). If the farm is infested with ring nematodes, as 52% of the fields sampled in Georgia were (Jagdale et al. 2013), then the grower could lose the entire investment at about the time that the blueberries would normally be coming into production.

14.3.6.3.1 Management of *M. ornatum*

Pre-plant fumigants such as oxamyl and Telone II are available for use against *M. ornatum*, but these products are expensive, pose health risks to the applicator if handled improperly, kill beneficial soil microbiota, highly regulated (U.S. EPA 2009) and they only temporarily suppress nematode populations. Since surviving nematodes will continue feeding and multiplying on new plants, a post-plant nematicide is needed to minimize population densities that increase in blueberry over multiple years; currently, there are no post-plant nematicides registered for use on blueberry. In addition, no nematode-resistant cultivars have yet been identified in blueberry. Although the management of nematodes including *M. ornatum* on blueberry has relied heavily on pre-plant fumigation, there is interest in developing safe alternatives as acceptable post-plant methods of control. In addition, due to increased consumer demand for organic foods including fruits and vegetables, many blueberry growers are also inclined towards production of organic blueberries. Studies on pre-plant fumigation with methyl bromide and solarization of the soil under clear plastic showed that solarization and fumigation reduced population densities of *M. ornatum* by 64% and over 90%, respectively compared with nontreated plots (Noe et al. 2012). Noe et al. (2015b) studied the efficacies of pine bark amendment with and without pre-plant application of soil fumigant against *M. ornatum* under field conditions and showed that the addition of pine bark soil amendment with a robust protocol of pre-plant soil fumigation may provide a more sustainable level of management for blueberry replant disease. Five cultivars each of Rabbiteye (Brightwell, Ochlocknee, Powder Blue, Premiere, Vernon) and southern highbush (Emerald, Farthing, Rebel, Star, Legacy) blueberry types were evaluated for their resistance/tolerance to BRD in fields in Georgia. *Mesocriconema ornatum* population densities increased between May and October for all cultivars, but increases were greatest for highbush cultivars, suggesting that BRD is more severe on southern highbush (Noe et al. 2014).

14.4 Future Research and Challenges

The options available for plant parasitic nematode management include sanitation, resistant and tolerant cultivars, crop rotation, cover crops, conservation tillage, soil amendment, biocontrol and nematicides. In most cases, a stand-alone option for control of plant parasitic nematodes is not sufficient and a combination of management practices will be needed to keep nematode populations below the economic thresholds. With the potential loss or shortage of effective fumigant or nonfumigant nematicides in the future, the need for continued assessment of alternative approaches for environmentally friendly, yet sustainable and effective treatment options has increased. Resistance is the most aggressive, economical treatment to manage plant parasitic nematodes and provides the best opportunity to manage nematodes affecting agricultural crops. Sources of resistance to southern and peanut root knot and reniform nematodes have been identified in some field and vegetable crops. Identifying new sources of resistance are required to develop new cultivars with broad and durable resistance to injurious nematodes. In order to advance breeding for resistance, genetic diversity of nematode populations need to be studied further and new molecular markers for resistance genes needs to be developed in order to expedite the process of introgression of nematode-resistant genes into high-yielding cultivars. The development of cultivars with resistance or tolerance to parasitic nematodes will provide growers with a simple to use, consistently effective and inexpensive tool for nematode management.

References

Anderson, W. F., Holbrook, C. C., & Timper, P. (2006). Registration of root knot nematode resistant peanut germplasm lines NR 0812 and NR 0817. *Crop Science, 46*, 481–482.

Anonymous. (2012). *Noncitrus fruits and nuts: 2011 summary*. Available online at: http://usda01.library.cornell.edu/usda/current/NoncFruiNu/NoncFruiNu-07-06-2012.pdf

Baird, R. E., Davis, R. F., Alt, P. J., Mullinix, B. G., & Padgett, G. B. (1996). Frequency and geographical distribution of plant parasitic nematodes on cotton in Georgia. *Journal of Nematology, 28*, 661–667.

Baird, R. E., Rich, J. R., Herzog, G. A., Utley, S. I., Brown, S., Martin, L. G., & Mullinix, B. G. (2000). Management of *Meloidogyne incognita* in cotton with nematicides. *Nematologica Mediterranean, 28*, 255–259.

Bell, A. A., Robinson, A. F., Quintana, J., Dighe, N. D., Menz, M. A., Stelly, D. M., Zheng, X., Jones, J. E., Overstreet, C., Burris, E., Cantrell, R. G., & Nichols, R. L. (2014). Registration of LONREN-1 and LONREN-2 germplasm lines of upland cotton resistant to reniform nematode. *Journal of Plant Registrations, 8*, 187–190.

Bell, A. A., Robinson, A. F., Quintana, J., Duke, S. E., Starr, J. L., Stelly, D. M., Zheng, Z., Prom, S., Saladino, V., Gutiérrez, O. A., Stetina, S. R., & Nichols, R. L. (2015). Registration of BARBREN-713 germplasm line of Upland cotton resistant to reniform and root knot nematodes. *Journal of Plant Registrations, 9*, 89–93.

Bell, A. A., Kemerait, R. C., Ortiz, C. S., Prom, S., Quintana, J., Nichols, R. L., & Liu, J. (2017). Genetic diversity, virulence, and *Meloidogyne incognita* interactions of *Fusarium oxysporum* isolates causing cotton wilt in Georgia. *Plant Disease, 101*, 948–956.

Branch, W. D., & Brenneman, T. B. (2015). Registration of 'Georgia-14N'peanut. *Journal of Plant Registrations, 9*, 159–161.

Brittain, J. A., & Miller, R. W. (1978). *Managing peach tree short life in the Southeast* (Clemson University Extension Service Bulletin 585). Clemson.

Burkett-Cadena, M., Kokalis-Burelle, N., Lawrence, K. S., van Santen, E., & Kloepper, J. W. (2008). Suppressiveness of root knot nematodes mediated by rhizobacteria. *Biological Control, 47*, 55–59.

Castillo, J. D., Lawrence, K. S., & van Santen, E. (2009). Efficacy of *Arthrobotrys dactyloides*, *Dactylaria brochopaga*, *Fusarium oxysporium*, and *Paecilomyces lilacinus* for biocontrol of reniform nematode (*Rotylenchulus reniformis*). In *Proceedings of the Beltwide Cotton Conferences, 2009* (pp 144–150). National Cotton Council of America. Online. http://www.cotton.org/beltwide/proceedings.htm

Castillo, J. D., Lawrence, K. S., & Kloepper, J. W. (2013). Biocontrol of the reniform nematode by *Bacillus firmus* GB-126 and *Paecilomyces lilacinus* 251 on cotton. *Plant Disease, 97*, 967–976.

Cetintas, R., & Dickson, D. W. (2005). Distribution and downward movement of *Pasteuria penetrans* in field soil. *Journal of Nematology, 37*, 155–160.

Clean Air Act. (1990). Title VI. Stratospheric ozone protection pub L. 101-549, Section U.S. Congress, Washington, DC.

Davis, R., & May, O. L. (2003). Relationships between tolerance and resistance to *Meloidogyne incognita* in cotton. *Journal of Nematology, 35*, 411–416.

Davis, R. F., & May, O. L. (2005). Relationship between yield potential and percentage yield suppression caused by the southern root knot nematode in cotton. *Crop Science, 45*, 2312–2317.

Davis, R. F., & Noe, J. P. (2000). Extracting *Hoplolaimus columbus* from soil and roots: Implications for treatment comparisons. *The Journal of Cotton Science, 4*, 105–111.

Davis, R. F., & Timper, P. (2000a). Resistance in selected corn hybrids to *Meloidogyne arenaria* and *M. incognita*. *Supplement Journal of Nematology, 32*, 633–640.

Davis, R. F., & Timper, P. (2000b). Survey of nematodes associated with corn in Georgia. *Journal of Nematology, 32*, 26 (abstract).

Davis, R. F., & Webster, T. M. (2005). Relative host status of selected weeds and crops for *Meloidogyne incognita* and *Rotylenchulus reniformis*. *Journal of Cotton Science, 9*, 41–46.

Davis, R. F., Baird, R. E., & Mcneill, R. D. (2000). Efficacy of cotton root destruction and winter crops for suppression of *Hoplolaimus columbus*. *Journal of Nematology, 32*, 550–555.

Davis, R. F., Koenning, S. R., Kemerait, R. C., Cummings, T. D., & Hurley, W. D. (2003). *Rotylenchulus reniformis* management in cotton with crop rotation. *Journal of Nematology, 35*, 58–64.

Davis, R. F., Aryal, S. K., Perry, C. D., Sullivan, D. G., Timper, P., Ortiz, B. V., Stevenson, K. L., Vellidis, G., & Hawkins, G. (2013). Utilizing management zones for *Rotylenchulus reniformis* in cotton: Effects on nematode levels, crop damage, and *Pasteuria* sp. *Crop Protection, 50*, 53–60.

Dong, W. B., Brenneman, T. B., Holbrook, C. C., Timper, P., & Culbreath, A. K. (2009). The interaction between *Meloidogyne arenaria* and *Cylindrocladium parasiticum* in runner peanut. *Plant Pathology, 58*, 71–79.

Faske, T. R., & Starr, J. L. (2007). Cotton root protection from plant parasitic nematodes by abamectin-treated seed. *Journal of Nematology, 39*, 27–30.

Fliegel, P. (1969). Population dynamics and pathogenicity of three species of *Pratylenchus* on peach. *Phytopathology, 59*, 120–124.

Fonsah, E. G., Krewer, G., Harrison, K., & Bruorton, M. (2007). Risk-rated economic return analysis for southern Highbush blueberries in soil in Georgia. *HortTechnology, 17*, 571–579.

Frederick, J. J., & Tarjan, A. C. (1975). Morphological variation in *Xiphinema krugi* Lordello, 1955. *Proceedings Soil and Crop Science Society Florida, 34*, 181–185.

Gazaway, G. S., & McLean, K. S. (2003). A survey of plant parasitic nematodes associated with cotton in Alabama. *Journal of Cotton Science, 7*, 1–7.

Gazaway, W. S., Lawrence, K. S., & Akridge, J. R. (2007). Impact of crop rotation and fumigation on cotton production in reniform infested fields. *Proceedings of the National Beltwide Cotton Conference, 1*, 1357–1360. National Cotton Council, Memphis. Available online at: www.cotton.org/beltwide/proceedings

Georgia Cotton Commission. (2018). *Georgia cotton production guide*. Available at: http://www.ugacotton.com/vault/file/2018-UGA-COTTON-PRODUCTION-GUIDE-1.pdf

Groover, W., & Lawrence, K. (2018). *Meloidogyne* spp. identification and distribution in Alabama crops via the differential-host test and molecular analysis. *Proceedings of the National Beltwide Cotton Conference, 1*, 503–505. National Cotton Council, Memphis. Available online at: http://www.cotton.org/beltwide/proceedings/2005-2018/index.htm

Groover, W., Lawrence, K., Xiang, N., Till, S. R., Dodge, D., Dyer, D. R., & Hall, M. (2017). Yield loss of cotton cultivars due to the reniform nematode and the added benefit of velum total. *Proceedings of the National Beltwide Cotton Conference, 1*, 216–219. National Cotton Council, Memphis. Available online at: http://www.cotton.org/beltwide/proceedings/2005-2018/index.htm

Hajihassani, A., Dutta, B., Jagdale, G., & Subbotin, S. (2018a). First report of yellow nutsedge cyst nematode, *Heterodera cyperi*, in Georgia, U.S.A. *Journal of Nematology, 50*, 456–458.

Hajihassani, A., Hamidi, N., Dutta, B., & Tyson, C. (2018b). First report of stubby root nematode, *Paratrichodorus minor*, on onion in Georgia, U.S.A. *Journal of Nematology, 50*, 453–455.

Holbrook, C. C., Jr., Timper, P., Culbreath, A. K., & Kvien, C. K. (2008). Registration of 'Tifguard' peanut. *Journal of Plant Registrations, 2*, 92–94.

Holbrook, C. C., Dong, W. B., Timper, P., Culbreath, A. K., & Kvien, C. K. (2012). Registration of peanut germplasm line TifGP-2, a nematode-susceptible sister line of 'Tifguard'. *Journal of Plant Registration, 6*, 208–211.

Holbrook, C. C., Ozias-Akins, P., Chu, Y., Culbreath, A. K., Kvien, C. K., & Brenneman, T. (2017). Registration of 'TifNV-High O/L' peanut. *Journal of Plant Registrations, 11*, 228–230.

Hooper, B. E. (1958). Plant parasitic nematodes in the soil of southern forest nurseries. *Plant Disease Reporter, 42*, 308–314.

Horton, D., Brannen, P., Bellinger, B., & Ritchie, D. (2010). *Southeastern peach, nectarine, and plum pest management and culture guide* (University of Georgia, Cooperative Extension Service Bulletin #1171), Athens.

Horton, D., Brannen, P., Bellinger, B., & Ritchie, D. (2013). 2013 Southeastern peach, nectarine, and plum pest management and culture guide. University B of Georgia, Cooperative Extension Serviceulletin 1171, Athens, GA.

Jagdale, G. B., Holladay, T., Brannen, P. M., Cline, W. O., Agudelo, P., Nyczepir, A. P., & Noe, J. P. (2013). Incidence and pathogenicity of plant parasitic nematodes associated with blueberry (*Vaccinium* spp.) replant disease in Georgia and North Carolina. *Journal of Nematology, 45*, 92–98.

Johnson, A. W. (1970). Pathogenicity and interaction of three nematode species on six bermudagrasses. *Journal of Nematology, 2*, 36–41.

Johnson, A. W., Dowler, C. C., & Hauser, E. W. (1974). Seasonal population dynamics of selected plant parasitic nematodes on four monocultured crops. *Journal of Nematology, 6*, 187–190.

Johnson, A. W., Dowler, C. C., & Hauser, E. W. (1975). Crop rotation and herbicide effects on population densities of plant parasitic nematodes. *Journal of Nematology, 7*, 158–168.

Johnson, A. W., Dowler, C. C., Glaze, N. C., & Handoo, Z. A. (1996). Role of nematodes, nematicides, and crop rotation on the productivity and quality of potato, sweet potato, peanut, and grain sorghum. *Journal of Nematology, 28*, 389–399.

Johnson, A. W., Minton, N. A., Brenneman, T. B., Todd, J. W., Herzog, G. A., Gascho, G. J., Baker, S. H., & Bondari, K. (1998). Peanut-cotton-rye rotations and chemical soil treatment for managing nematodes and thrips. *Journal of Nematology, 30*, 211–225.

Jones, J. R., Lawrence, K. S., & Lawrence, G. W. (2006). Use of winter cover crops in cotton cropping for management of *Rotylenchulus reniformis*. *Nematropica, 36*, 53–66.

Kloepper, J. W., Rodríguez-Kábana, R., McInroy, J. A., & Young, R. W. (1992). Rhizosphere bacteria antagonistic to soybean cyst (*Heterodera glycines*) and root knot (*Meloidogyne incognita*) nematodes: Identification by fatty acid analysis and frequency of biological control activity. *Plant and Soil, 139*, 75–84.

Koenning, S. R., Barker, K. R., & Bowman, D. T. (2000). Tolerance of selected cotton lines to *Rotylenchulus reniformis*. *Journal of Nematology, 32*, 519–523.

Koenning, S. R., Wrather, J. A., Kirkpatrick, T. L., Walker, N. R., Starr, J. L., & Mueller, J. D. (2004). Plant parasitic nematodes attacking cotton in the United States: Old and emerging production challenges. *Plant Disease, 88*, 100–113.

Koenning, S. R., Morrison, D. E., & Edmisten, K. L. (2007). Relative efficacy of selected nematicides for management of *Rotylenchulus reniformis* in cotton. *Nematropica, 37*, 227–235.

Lawrence, K. S., & Lawrence, G. W. (2007). Performance of the new nematicide treatments on cotton. *Proceedings of the National Beltwide Cotton Conference, 1,* 602–605. National Cotton Council, Memphis. Available online at: www.cotton.org/beltwide/proceedings

Lawrence, G. W., & McLean, K. S. (1999). Plant parasitic nematode pests of soybeans. In L. G. Heatherly & H. F. Hodges (Eds.), Soybean production in the Mid-South (pp. 291–308). Boca Raton: CRC Press LLC, Florida.

Lawrence, G. W., & McLean, K. S. (2000). Effect of foliar applications of oxamyl with aldicarb for the management of *Rotylenchulus reniformis* in cotton. *Annals of Applied Nematology, 32*, 543–549.

Lawrence, G. W., McLean, K. S., Batson, W. E., Miller, D., & Borbon, J. C. (1990). Response of *Rotylenchulus reniformis* to nematicide applications on cotton. *Annals of Applied Nematology, 22*, 707–711.

Lawrence, K. S., Feng, Y., Lawrence, G. W., Burmester, C. H., & Norwood, S. H. (2005). Accelerated degradation of aldicarb and its metabolites in cotton field soils. *Journal of Nematology, 37*, 190–197.

Lawrence, G. W., Lawrence, K. S., & Caceres J. (2007). Options after the furrow is closed. *Proceedings of the National Beltwide Cotton Conference, 1,* 598–601. National Cotton Council, Memphis. Available online at: www.cotton.org/beltwide/proceedings

Lawrence, K. S., Price, A. J., Lawrence, G. W., Jones, J. R., & Akridge, J. R. (2008). Weed hosts for *Rotylenchulus reniformis* in cotton fields rotated with corn in the southeast United States. *Nematropica, 38*, 13–22.

Lawrence, K. S., Hagan, A., Norton, R., Hu, J., Faske, T. R., Hutmacher, R. B., Muller, J., Small, I., Grabau, Z., Kemerait, R. C., Overstreet, C., Price, P., Lawrence, G. W., Allen, T. W., Atwell, S., Idowa, J., Bowman, R., Goodson, J. R., Kelly, H., Woodward, J., Wheeler, T., & Mehl, H. L. (2018). Cotton disease loss estimate committee report, 2017. *Proceedings of the National Beltwide Cotton Conference, 1,* 161–163. National Cotton Council, Memphis. Available online at: www.cotton.org/beltwide/proceedings

Little, E. L. (2017). *Georgia plant disease loss estimates in 2015* (p. 20). University of Georgia Cooperative Extension Service, College of Agricultural and Environmental Sciences, Annual Publication 102–8.

Miller, R. W. (1994). Estimated peach tree losses 1980 to 1992 in South Carolina: Causes and economic impact. In A. P. Nyczepir, P. F. Bertrand, T. G. Beckman (Eds.), Proceedings 6th Stone Fruit Decline Workshop, 26–28 Oct. 1992, (pp. 121–127). Fort Valley, Georgia. USDA-ARS. ARS 122. National Technical Information Service, Springfield, Virginia.

McSorley, R. (2001). Multiple cropping systems for nematode management: A review. *Soil and Crop Science Society of Florida Proceedings, 60*, 132–142.

McSorley, R., & Dickson, D. W. (1995). Effect of tropical rotation crops on *Meloidogyne incognita* and other plant parasitic nematodes. *Journal of Nematology, 27*, 535–544.

Mekete, T., Reynolds, K., Lopez-Nicora, H. D., Gray, M. E., & Niblack, T. L. (2011). Plant parasitic nematodes are potential pathogens of *Miscanthus* × *giganteus* and *Panicum virgatum* used for biofuels. *Plant Disease, 95*, 413–418.

Minton, N. A., & Bell, D. K. (1969). *Criconemoides ornatus* parasitic on peanuts. *Journal of Nematology, 1*, 349–351.

Minton, N. A., & Hopper, B. E. (1959). The reniform and sting nematodes in Alabama. Plant Disease Reporter 43, 47.

Minton, N. A., & Tucker, E. T. (1987). First report of *Meloidogyne graminicola* in Georgia. *Plant Disease, 71*, 376.

Minton, N. A., Cairns, E. J., Minton, E. B., & Hooper, B. E. (1963). Occurrence of plant parasitic nematodes in Alabama. *Plant Disease Reporter, 57*, 946.

Monfort, W. S., Csinos, A. S., Desaeger, J., Seebold, K., Webster, T. M., & Diaz-Perez, J. C. (2007). Evaluating brassica species as an alternative control measure for root knot nematode (*M. incognita*) in Georgia vegetable plasticulture. *Crop Protection, 26*, 1359–1368.

Moore, S. R., & Lawrence, K. S. (2012). *Rotylenchulus reniformis* in cotton: Current methods of management and the future of site-specific management. *Nematropica, 42*, 227–236.

Moore, S. R., Lawrence, K. S., Arriaga, F. J., Burmester, C. H., & van Santen, E. (2010a). Natural migration of *Rotylenchulus reniformis* in a no-till cotton system. *Journal of Nematology, 42*, 307–312.

Moore, S. R., Gazaway, W. S., Lawrence, K. S., Goodman, B., & Akridge, R. (2010b). Value of rotational crops for profit increase and reniform nematode suppression with and without a nematicide in Alabama. *Proceedings of the National Beltwide Cotton Conference, 1*, 260–268. National Cotton Council, Memphis. Available online at: www.cotton.org/beltwide/proceedings

Morris, K., Langston, D. B., Davis, R. F., & Dutta, B. (2016). Evidence for a disease complex between *Pythium aphanidermatum* and root knot nematodes in cucumber. *Plant Health Progress, 17*, 200–201.

Motsinger, R. E., Crawford, J. L., & Thompson, S. S. (1974). Survey of cotton and soybean fields for lance nematodes in east Georgia. *Plant Disease Reporter, 58*, 369–372.

Motsinger, R. E., Crawford, J. L., & Thompson, S. S. (1976a). Nematode survey of peanuts and cotton in southwest Georgia. *Peanut Science, 3*, 72–74.

Motsinger, R. E., Gay, C. M., David, F., & Dekle, J. (1976b). Soybean cyst nematode found in Georgia. *Plant Disease Reporter, 60*, 1087.

NASS USDA. (2018a). *State agriculture overview*. Georgia. Available online at: https://www.nass.usda.gov/Quick_Stats/Ag_Overview/stateOverview.php?state=GEORGIA

NASS USDA. (2018b). *State agriculture overview*. Alabama. Available online at: https://www.nass.usda.gov/Quick_Stats/Ag_Overview/stateOverview.php?state=ALABAMA

Noe, J. P. (1993). Damage functions and population changes of *Hoplolaimus columbus* on cotton and soybean. *Journal of Nematology, 25*, 440–445.

Noe, J. P., Brannen, P. M., Holladay, W. T., & Jagdale, G. B. (2012). Preplant soil treatments to manage blueberry replant disease caused by *Mesocriconema ornatum*. *Journal of Nematology, 44*, 482.

Noe, J. P., Jagdale, G. B., Holladay, W. T., & Brannen, P. M. (2014). Susceptibility of blueberry cultivars to replant disease associated with *Criconemoides ornatum*. *Journal of Nematology, 46*, 212 (abstract).

Noe, J. P, Jagdale, G., Holladay, T., & Brannen, P. M. (2015a). Fluensulfone for management of *Mesocriconema xenoplax* on peach. In *Proceeding of 54th annual meeting of the Society of Nematologists*. July 19–24, 2015. East Lansing.

Noe, J. P., Jagdale, G. B., Holladay, W. T., & Brannen, P. M. (2015b). Management of blueberry (*Vaccinium* spp.) replant disease with pine bark soil amendment and pre-plant fumigation. In *Proceeding of 2015 APS annual meeting*, August 1–5, 2015. Pasadena.

Norton, D. C., Donald, P. L., Kiminski, J., Myers, R., Noel, G., Noffsinger, E. M., Robbins, R. T., Schmitt, D. P., Sosa-Moss, C., & Vrain, T. C. (1984). *Distribution of plant parasitic nematodes species in North America*. Society of Nematologists. 205 pp.

Nyczepir, A. P. (2005). *Nematodes of peach orchards*. Available Online: https://www.clemson.edu/extension/peach/commercial/diseases/files/h7.3.pdf

Nyczepir, A. P., & Bertrand, P. F. (1990). Host suitability of selected small grain and field crops to *Criconemella xenoplax*. *Plant Disease, 74*, 698–701.

Nyczepir, A. P., & Bertrand, P. F. (2000). Preplanting bahiagrass or wheat compared for controlling *Mesocriconema xenoplax* short life in a young peach orchard. *Plant Disease, 84*, 789–793.

Nyczepir, A. P., & Lamberti, F. (2001). First record of *Xiphinema pacificum* from a peach orchard in Georgia. *Plant Disease, 85*, 1119.

Nyczepir, A. P., Zehr, E. I., Lewis, S. A., & Harshman, D. C. (1983). Short life of peach trees induced by *Criconemella xenoplax*. *Plant Disease, 67*, 507–508.

Nyczepir, A. P., Bertrand, P. F., Miller, R. W., & Motsinger, R. E. (1985). Incidence of *Criconemella* spp. and peach orchard histories in short-life and non- short-life sites in Georgia and South Carolina. *Plant Disease, 69*, 874–877.

Nyczepir, A. P., Bertrand, P. F., & Cunfer, B. M. (1996). Suitability of wheat-sorghum, double-crop rotation to manage *Criconemella xenoplax* in peach production. *Plant Disease, 80*, 629–632.

Nyczepir, A. P., Reilly, C. C., & Wood, B. W. (2002). First record of *Meloidogyne partityla* on pecan in Georgia. *Plant Disease, 86*, 441.

Nyczepir, A. P., Kluepfel, D. A., Waldrop, V., & Wechter, W. P. (2012). Soil solarization and biological control for managing *Mesocriconema xenoplax* and short life in a newly established peach orchard. *Plant Disease, 96*, 1309–1314.

Okie, W. R., Reighard, G. L., & Nyczepir, A. P. (2009). Importance of scion cultivar in peach tree short life. *Journal American Pomological Society, 63*, 58–63.

Ortiz, B. V., Perry, C., Sullivan, D., Lu, P., Kemerait, R., Davis, R. F., Smith, A., Vellidis, G., & Nichols, R. (2012). Variable rate application of nematicides on cotton fields: A promising site-specific management strategy. *Journal of Nematology, 44*, 31–39.

Powers, T. O., & Harris, T. S. (1993). A polymerase chain reaction method for identification of five major *Meloidogyne* species. *Journal of Nematology, 25*, 1–6.

Powers, T. O., Bernard, E. C., Harris, T., Higgins, R., Olson, M., Lodoma, M., Mullin, P., Sutton, L., & Powers, K. S. (2014). COI haplotype groups in *Mesocriconema* (Nematoda: Criconematidae) and their morphospecies associations. *Zootaxa, 3827*, 101–146.

Rebois, R. V., & Cairns, E. J. (1968). Nematodes associated with soybeans in Alabama, Florida, and Georgia. *Plant Disease Reporter, 52*, 40–44.

Robinson, A. F. (2007). Reniform in U.S. cotton: When, where, why, and some remedies. *Annual Review of Phytopathology, 45*, 263–288.

Robinson, A. F., Inserra, R. N., Caswell-Chen, E. P., Vovlas, N., & Troccoli, A. (1997). *Rotylenchulus* species: Identification, distribution, host ranges, and crop plant resistance. *Nematropica, 27*, 127–180.

Savage, E. F., & Cowart, F. F. (1942). *Factors affecting peach tree longevity in Georgia* (Georgia Agricultural Experiment Station Research Bulletin 219). Athens.

Schrimsher, D. W. (2013). The studies of plant host resistance to the reniform nematode in upland cotton and the effects of *Bacillus firmus* GB-126 on plant parasitic nematodes. M.S. Thesis. Auburn: Auburn University.

Schrimsher, D. W., Lawrence, K. S., Sikkens, R. B., & Weaver, D. B. (2014). Nematicides enhance growth and yield of *Rotylenchulus reniformis* resistant cotton genotypes. *Journal of Nematology, 46*, 367–375.

Shepard, D. P., Zehr, E. I., & Bridges, W. C. (1999). Increased susceptibility to bacterial spot of peach trees growing in soil infested with *Criconemella xenoplax*. *Plant Disease, 83*, 961–963.

Sher, S. A. (1966). Revision of the Hoplolaimiae (Nematoda). VI. *Helicotylenchus* Steiner. 1945. *Nematologica, 12*, 1–56.

Smith, A. L. (1940). Distribution and relation of meadow nematode, *Pratylenchus pratensis* to Fusarium wilt of cotton in Georgia. *Phythopathology, 30*, 710 (abstract).

Southern Cooperative Series Bulletin. (1960). *Distribution of plant parasitic nematodes in the South*. Bulletin 74/72 pp.

Starr, J. L., Koenning, S. R., Kirkpatrick, T. L., Robinson, A. F., Roberts, P. A., & Nichols, R. L. (2007). The future of nematode management in cotton. *Journal of Nematology, 39*, 283–294.

Sumner, D. R., Minton, N. A., Brenneman, T. B., Burton, G. W., & Johnson, A. W. (1999). Root diseases and nematodes in bahiagrass-vegetable rotations. *Plant Disease, 83*, 55–59.

Timper, P. (1999). Effect of crop rotation and nematicide use on abundance of *Pasteuria penetrans*. *Journal of Nematology, 31*, 575 (abstract).

Timper, P. (2009). Population dynamics of *Meloidogyne arenaria* and *Pasteuria penetrans* in a long-term crop rotation study. *Journal of Nematology, 41*, 291–299.

Timper, P. (2017). Rye residue levels affect suppression of the southern root knot nematode in cotton. *The Journal of Cotton Science, 21*, 242–246.

Timper, P., & Hanna, W. W. (2005). Reproduction of *Belonolaimus longicaudatus*, *Meloidogyne javanica*, *Paratrichodorus minor*, and *Pratylenchus brachyurus* on pearl millet (*Pennisetum glaucum*). *Journal of Nematology, 37*, 214–219.

Timper, P., Minton, N. A., & Johnson, A. W. (2001). Influence of cropping systems on stem rot (*Sclerotium rolfsii*), *Meloidogyne arenaria*, and the nematode antagonist *Pasteuria penetrans* in peanut. *Plant Disease, 85*, 767–772.

Timper, P., Wilson, J. P., Johnson, A. W., & Hanna, W. W. (2002). Evaluation of pearl millet grain hybrids for resistance to *Meloidogyne* spp. and leaf blight caused by *Pyricularia grisea*. *Plant Disease, 86*, 909–914.

Timper, P., Davis, R. F., & Tillman, P. G. (2006). Reproduction of *Meloidogyne incognita* on winter cover crops used in cotton production. *Journal of Nematology, 38*, 83–89.

Timper, P., Davis, R. F., Webster, T. M., Brenneman, T. B., Meyer, S. L. F., Zasada, I. A., Cai, G., & Rice, C. P. (2011). Response of root knot nematodes and Palmer amaranth to tillage and rye green manure. *Agronomy Journal, 103*, 813–821.

Timper, P., Liu, C., Davis, R. F., & Wu, T. (2016). Influence of crop production practices on *Pasteuria penetrans* and suppression of *Meloidogyne incognita*. *Biological Control, 99*, 64–71.

U.S. EPA. (2009). Available online at: http://www.epa.gov/opp00001/reregistration/soil_fumigants/soil-fum-chemicals.html

Vargas, O. F., & Sasser, J. N. (1976). Biology of *Anguina plantaginis* parasitic on *Plantago aristrata*. *Journal of Nematology, 8*, 64–68.

Wang, Q., Klassen, W., Bryan, H. H., Li, Y., & Abdul-Baki, A. A. (2003). Influence of summer cover crops on growth and yield of a subsequent tomato crop in south Florida. *Proceedings of Florida State Horticultural Society, 116*, 140–143.

Wang, K. H., McSorley, R., & Gallaher, R. N. (2004). Effect of winter cover corps on nematode populations levels in north Florida. *Journal of Nematology, 36*, 517–523.

Chapter 15
Important Plant Parasitic Nematodes of Row Crops in Arkansas, Lousiana and Mississippi

Travis R. Faske, Charles Overstreet, Gary Lawrence, and Terry L. Kirkpatrick

15.1 Introduction

This chapter's focus is on the important plant parasitic nematodes and their management on row crops in Arkansas, Louisiana and Mississippi. This region is referred to as the Mid-South. Agronomic crops, production practices and nematode management practice are similar throughout the region.

T. R. Faske (✉)
Division of Agriculture, Lonoke Extension Center, Department of Plant Pathology, University of Arkansas, Lonoke, AR, USA
e-mail: tfaske@uaex.edu

C. Overstreet
Department of Plant Pathology and Crop Physiology, Louisiana State University AgCenter, Baton Rouge, LA, USA
e-mail: COverstreet@agcenter.lsu.edu

G. Lawrence
Department of Biochemistry, Molecular Biology, Entomology and Plant Pathology, Mississippi State University, Starkville, MS, USA
e-mail: GLawrence@endomology.msstate.edu

T. L. Kirkpatrick
Division of Agriculture, Southwest Research and Extension Center, Department of Plant Pathology, University of Arkansas, Hope, AR, USA
e-mail: tkirkpatrick@uaex.edu

S. A. Subbotin, J. J. Chitambar (eds.), *Plant Parasitic Nematodes in Sustainable Agriculture of North America*, Sustainability in Plant and Crop Protection,
https://doi.org/10.1007/978-3-319-99588-5_15

15.2 Economically Important Crops and Importance of Nematodes

The majority of row crop production in Arkansas, Louisiana and Mississippi is concentrated along the Mississippi River Delta (Fig. 15.1). Major row crops in the Mid-South include soybean (*Glycine max*), corn (*Zea mays*), cotton (*Gossypium hirsutum*), rice (*Oryza sativa*), sorghum (*Sorghum bicolor*), sweet potato (*Ipomoea batatas*) and peanut (*Arachis hypogaea*) (Table 15.1). Other areas of row crop production are concentrated near other rivers systems within each state. The total value of production of these row crops in the Mid-South is estimated at 6.9 billion dollars (Table 15.1).

Plant parasitic nematodes are major yield-limiting factors that affect row crop production in Southern United States and in the Mid-South. During the 2010–2014

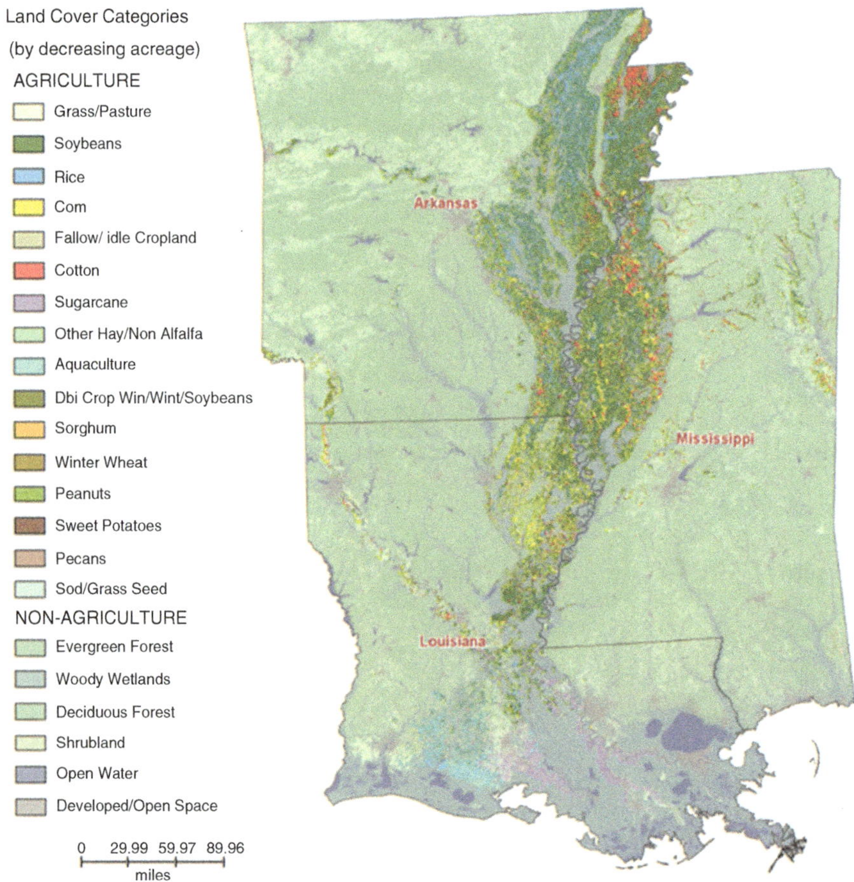

Fig. 15.1 Distribution of agricultural and non-agricultural land coverage categories in 2016 in Arkansas, Louisiana and Mississippi (USDA-NASS 2016d)

Table 15.1 Estimated hectares and crop value of the top seven row crop commodities in 2016 in Arkansas, Louisiana and Mississippi (USDA-NASS 2016a, b, c)

Commodity	Harvested hectares			Total hectares	Value of production (dollars)
	Arkansas	Louisiana	Mississippi		
Soybean	1,254,500	768,900	817,500	2,840,900	2,974,161,000
Rice	615,500	173,200	78,500	867,200	1,427,030,000
Corn	301,500	222,500	291,400	815,400	1,255,337,000
Cotton	151,700	55,400	174,000	381,100	717,802,000
Sorghum	17,800	18,600	4,400	40,800	280,070,000
Sweet potato	1,600	3,800	11,700	17,100	170,621,000
Peanut	9,300	600	15,400	25,300	47,020,000

period, soybean nematodes consistently ranked among the top soybean diseases affecting soybean production in the Southern United States (Allen et al. 2017b). For example, during the 2015 cropping season plant parasitic nematodes contributed to 320,000 ton in grain yield loss, in soybean production in the Mid-South (Allen et al. 2016). Similarly, plant parasitic nematodes were among the top three important yield-limiting factors affecting corn production from 2012 to 2015 in the Southern United States (Mueller et al. 2016). In cotton, plant parasitic nematodes continue to be one of the major yield-limiting factors in the Southern US (Lawrence et al. 2015a, 2016). Some 16,500 ton of cotton were lost in 2015 due to plant parasitic nematodes in the Mid-South (Lawrence et al. 2016). As a general rule, root knot nematodes and reniform nematodes are the most widespread and economically important nematodes on row crops in the Mid-South, although the soybean cyst nematode can also be significant in soybean production, particularly in the more northern (Arkansas, Tennessee, Missouri Bootheel) parts of the region. Research is lacking for many of the plant parasitic nematodes that are also commonly associated with row crops in the Mid-South (Table 15.2).

15.3 Soybean Cyst Nematode, *Heterodera glycines*

The soybean cyst nematode (SCN) is the most important nematode pest of soybean in the United States (Riggs 1977; Allen et al. 2017b). In the Mid-South, yield losses were estimated at two million bushels in 2015 (Allen et al. 2016). *Heterodera glycines* was first reported in 1915 in Japan (Hori 1915) and in the United States in 1955, in North Carolina (Winstead et al. 1955) and now is widely distributed in most soybean-growing areas of the U.S. (Niblack and Riggs 2015). The soybean cyst nematode was first recognized as a problematic pest in 1957 in Arkansas and Mississippi, and in 1967 in Louisiana (Noel 1992). It has been detected in all major soybean-producing counties in the Mid-South (Fig. 15.2a). Dissemination of soybean cyst nematode-infested soil from Japan, as a source of rhizobia inoculum from Asia, is believed to be the source of some of the early infestations in the US.

Table 15.2 Plant parasitic nematodes associated with row crops in Arkansas, Louisiana and Mississippi

Nematode	Crop	State	Reference
Aphelenchoides besseyi	Rice	AR, LA, MS	Martin and Birchfield (1955), Birchfield and Martin (1956), Birchfield et al. (1978), and Norton et al. (1984)
Belonolaimus nortoni	Soybean	AR	Robbins (1982a) and Norton et al. (1984)
Criconemoides annulata	Soybean	LA, MS	Norton et al. (1984)
Helicotylenchus digonicus	Soybean	MS	Rebois and Golden (1978) and Norton et al. (1984)
H. dihystera	Cotton, peanut, soybean	AR, LA, MS	Rebois and Golden (1978), Robbins (1982a), and Norton et al. (1984)
H. multicinctus	Cotton, soybean	LA	Rebois and Golden (1978)
H. pseudorobustus	Soybean	AR, MS	Robbins (1982a) and Norton et al. (1984)
Hemicycliophora triangulum	Soybean	AR	Robbins (1982a)
Heterodera glycines	Soybean	AR, LA, MS	Riggs (1977), Birchfield et al. (1978), Robbins (1982a), Norton et al. (1984), and Robbins et al. (1987)
Hirschmanniella oryzae	Rice	AR, LA, MS	Hollis (1967), Norton et al. (1984), and Wehunt et al. (1989)
Hoplolaimus columbus	Cotton, soybean	LA	Astudill and Birchfield (1980)
H. galeatus	Corn, cotton, grain sorghum, peanut, soybean	AR, LA, MS	Martin and Birchfield (1955), Birchfield and Martin (1956), Birchfield et al. (1978), Rebois and Golden (1978), Robbins (1982a), and Norton et al. (1984)
H. magnistylus	Cotton, soybean	AR, MS	Riggs (1977), Robbins (1982a), Norton et al. (1984), and Robbins et al. (1987, 1989)
Meloidogyne arenaria	Peanut, soybean	MS	Norton et al. (1984)
M. hapla	Soybean	AR	Robbins (1982a)
M. incognita	Corn, cotton, grain sorghum, soybean, sweet potato	AR, LA, MS	Birchfield and Martin (1956), Fielding and Hollis (1956), Birchfield et al. (1978), Robbins (1982a), Thomas and Clark (1983), Norton et al. (1984), Robbins et al. (1989), and Lawrence and McLean (2002)
M. graminicola	Rice	LA	Birchfield et al. (1978)
Mesocriconema onoense	Rice	LA	Birchfield and Martin (1956), Hollis (1967), and Birchfield et al. (1978)
M. ornatum	Soybean	AR, LA, MS	Rebois and Golden (1978) and Robbins et al. (1987)

(continued)

Table 15.2 (continued)

Nematode	Crop	State	Reference
M. xenoplax	Grain sorghum	LA	Wenfrida et al. (1998)
Nanidorus minor	Corn, cotton, soybean	AR, LA, MS	Martin and Birchfield (1955), Fielding and Hollis (1956), Rebois and Golden (1978), and Robbins (1982a)
Paratylenchus projectus	Corn, soybean	AR	Robbins (1982a)
P. tenuicaudatus	Soybean	AR	Robbins (1982a)
Pratylenchus alleni	Soybean	AR	Robbins (1982a)
P. brachyurus	Corn, cotton, soybean, sugarcane	AR, LA, MS	Birchfield and Martin (1956), Fielding and Hollis (1956), Endo (1959), Birchfield et al. (1978), Rebois and Golden (1978), Robbins (1982a), and Robbins et al. (1989)
P. coffeae	Soybean	AR, MS	Rebois and Golden (1978), and Norton et al. (1984)
P. hexincisus	Soybean	AR	Robbins (1982a)
P. neglectus	Corn, soybean	AR	Robbins (1982a)
P. penetrans	Corn, peanut, soybean	AR, LA, MS	Dickerson et al. (1964), Rebois and Golden (1978), and Norton et al. (1984)
P. scribneri	Corn, cotton, soybean	AR, LA, MS	Fielding and Hollis (1956), Rebois and Golden (1978), Robbins (1982a), Norton et al. (1984), and Robbins et al. (1989)
P. vulnus	Soybean	AR	Robbins (1982a) and Norton et al. (1984)
P. zeae	Corn, sugarcane, rice	AR, LA, MS	Martin and Birchfield (1955), Fielding and Hollis (1956), Endo (1959), Birchfield et al. (1978), Rebois and Golden (1978), Robbins (1982a), Norton et al. (1984), Cuarezma-Teran and Trevathan (1985), and Robbins et al. (1989)
Quinisulcius acutus	Soybean	AR, LA, MS	Birchfield et al. (1978), Rebois and Golden (1978), Robbins (1982a), Norton et al. (1984), Cuarezma-Teran and Trevathan (1985), and Robbins et al. (1987)
Rotylenchulus reniformis	Cotton, soybean, sweet potato	AR, LA, MS	Birchfield and Martin (1956), Fielding and Hollis (1956), Birchfield et al. (1978), Robbins (1982a), Thomas and Clark (1983), Norton et al. (1984), Robbins et al. (1989), and Lawrence and McLean (2000)
Scutellonema brachyurus	Soybean	AR, LA	Rebois and Golden (1978) and Norton et al. (1984)
S. bradys	Soybean	AR	Robbins (1982a)
Trichodorus primitivus	Corn	MS	Norton et al. (1984)

(continued)

Table 15.2 (continued)

Nematode	Crop	State	Reference
Tylenchorhynchus annulatus	Corn, grain sorghum, rice, soybean, sugarcane	AR, LA, MS	Fielding and Hollis (1956), Birchfield et al. (1978), Rebois and Golden (1978), Robbins (1982a), Norton et al. (1984), Robbins et al. (1987), Wenfrida et al. (1998), and Bae et al. (2009)
T. canalis	Soybean	AR	Robbins et al. (1987)
T. claytoni	Corn, cotton, soybean	LA, MS	Martin and Birchfield (1955), Rebois and Golden (1978), and Norton et al. (1984)
T. cylindricus	Corn, cotton, soybean	MS	Rebois and Golden (1978) and Norton et al. (1984)
T. ewingi	Soybean	AR	Robbins (1982a) and Robbins et al. (1987)
T. goffarti	Soybean	AR	Robbins (1982a) and Robbins et al. (1987)
T. nudus	Corn, soybean	MS	Rebois and Golden (1978) and Norton et al. (1984)
Xiphinema americanum	Corn, cotton, soybean, sugarcane	AR, LA, MS	Martin and Birchfield (1955), Birchfield et al. (1978), Rebois and Golden (1978), Robbins (1982a), Norton et al. (1984), and Robbins et al. (1987)
X. chambersi	Soybean	AR	Robbins (1982a) and Robbins et al. (1987)
X. rivesi	Soybean	AR	Robbins (1982a) and Robbins et al. (1987)

[*]Names of the states are represented by two letter abbreviations: AR Arkansas, LA Louisiana, MS Mississippi

Heterodora glycines has a host range that includes several genera in the Fabaceae family and a few species outside that family (Riggs 1992). Of the major row crops grown in the Mid-South, soybean is the only crop affected by the soybean cyst nematode. Although soybean cyst nematode was widely distributed in Louisiana in the past, over the past 20 years the nematode has become difficult to find in most fields. This decline is likely related to pathogens or parasites of the nematode in the soil, rather than management practices using host resistance or crop rotation. Since soil temperatures remain fairly warm year-round in Louisiana, microorganisms in the soil could be active all the time.

Populations of soybean cyst nematode differ in their ability to parasitize resistant soybean cultivars. To classify these genetic variants a race classification scheme was developed based on the female ability to develop on four soybean lines; Pickett, Peking, PI 88788 and PI 909763 compared to that of the susceptible standard, cv. Lee (Riggs and Schmitt 1988). Based on the four differential lines, sixteen races are theoretically possible to exist. Race designations are used in the Mid-South with several of the 16 races being reported from field surveys. Races 2–9 and 14 were detected in 1994 in Mississippi, while races 1–5, 9, 10 and 15 were reported in 1988 in Arkansas (Riggs and Schmitt 1988). In a more recent survey in Arkansas, the majority of the soybean cyst nematode populations from the results of a 2015 survey were races 2, 5 and 6, which was similar to the races 2, 4 and 5 reported in 1988 (Riggs and Schmitt 1988; Kirkpatrick 2017). Because a population of soybean

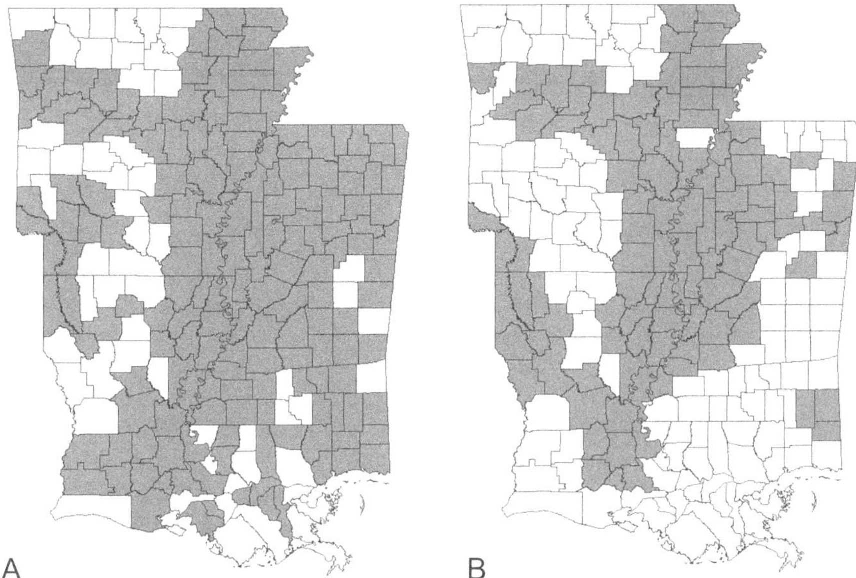

Fig. 15.2 (**a**) Counties (highlighted gray) where *Heterodera glycines* has been detected on soybean in Arkansas, Louisiana, and Mississippi. (**b**) Counties (highlighted gray) where *Meloidogyne incognita* has been detected on row crops in Arkansas, Louisiana, and Mississippi

cyst nematode in a field can vary in genetic diversity, a newer mechanism for describing pathogenic variability, referred to as the HG type scheme, has been adopted (Niblack et al. 2002). As with the race scheme, populations are distinguished based on their ability to reproduce on a set of indicator lines (PI 548402 (Peking), PI 88788, PI 90763, PI 437654, PI 209332, PI 89772 and PI 548316 (Cloud), that represent the major sources of resistance used in the US to develop resistant soybean cultivars.

The soybean cyst nematode occurs across a wide range of temperatures and soil types, but is most problematic in course textured, sandy soils. Crops growing in sandy soils are stressed by low water-holding capacity, plus the stress caused by soybean cyst nematode results in a greater damage potential compared to those growing in finer-textured, clay soils.

15.3.1 Symptoms

Foliar symptoms of infection range from undetectable to stunted, chlorotic plants that may occur in roughly circular or elliptical patterns in a field. Symptomology often depends on the severity of the problem as it relates to the nematode population's ability to reproduce on a soybean cultivar. In general, symptoms are often confused with nutrient deficiencies, although the nematodes are relatively easy to detect visibly without magnification in a field, as white, yellow or brown lemon-shaped females on infected roots. Though damage thresholds vary with soil type, a population density of 500 nematodes/100 cm^3 soil can cause yield loss in the Mid-South.

15.3.2 Management Strategies

Management of the soybean cyst nematode requires several tactics that include cultural practices, the use of resistant cultivars and in some situations, the application of nematicides. Growing crops that are non-hosts to the nematode in rotation with soybean can be an effective means to lower nematode population densities and maintain the nematode densities below an economic threshold (Wrather et al. 1992). Corn, cotton, grain sorghum or peanut are non-host crops and therefore, are a good rotation crop option for the Mid-South. Growing a non-host crop for 2–3 years on a nematode-infested field may be required before a susceptible soybean cultivar can be grown to achieve its full yield potential. The inclusion of a race- or HG-type-specific resistant soybean cultivar is also an economical method for nematode management (Lawrence and McGuire 1987; Wrather et al. 1992) if the appropriate resistance is available in an adapted cultivar. However, the continuous or frequent use of a resistant cultivar may increase the population of individual nematodes that can overcome the host resistance and lead to a "race shift" that eventually renders the resistant cultivar useless. Given that most of the commercially available soybean cultivars adapted to the Mid-South are not resistant to the race or HG types in the region, the use of non-host crops is the best option in soybean cyst nematode management.

Planting early before nematode eggs hatch and juveniles become active in the soil, also known as avoidance, has been suggested as a management tactic. The eggs will not hatch until soil temperatures reach 20 °C (Ross 1988). Early planting before soil temperatures warm-up, allow sufficient time for a soybean plant to become established without nematode damage to its developing root system. In a planting date study in Arkansas, reproduction by SCN was lower on early planted (April) soybeans than late planted (June or July), but grain yield was similar between planting dates (Riggs et al. 2000).

The use of nematicides is another management tactic for soybean cyst nematode management. Fumigant nematicides, including 1, 3 dichloropropene (Telone® II), metam potassium (K-PAM® HL™) and metam sodium (VAPAM® HL), are labeled for use, but are not commonly used in the Mid-South due to cost, the need for special application equipment and rather stringent environmental restrictions. Currently, the vast majority of nematicides used in the Mid-South are applied as seed treatment and are divided between chemical and biological agents. Chemical agents include abamectin (Avicta® 500 FS) and fluopyram (ILeVO® 600 FS), while *Bacillus firmus* (VOTiVO®) and *Pasturia nischizawae* (Clariva® *pn*) are biological agents. Seed-applied nematicides may provide some protection of the developing seedling, but do not provide season-long nematode control. They are more beneficial when paired with moderately resistant cultivars or where there is more than one population of plant parasitic nematode in the field. The decision to use a nematicide should be based on nematode population density, level of cultivar resistance and expected yield benefit.

15.4 Root Knot Nematode, *Meloidogyne incognita*

The southern root knot nematode (*Meloidogyne incognita*) is one of the most important plant parasitic nematode affecting row crop production in the Mid-South and United States. Root knot nematodes were first described on cucumber in 1855 in England (Berkeley 1855) and in 1889 in the Southern U. S. (Neal 1889). Although root knot nematodes are now considered to be indigenous and widely distributed in the U.S. (Chitwood 1949), they were not referenced until 1911 in the Mid-South (Bessey 1911). *Meloidogyne incognita* has a broad host range comprised of thousands of plant species, which include many weed species and row crops. This root knot nematode species attacks most of the major row crops produced in the Mid-South including cotton, corn, grain sorghum, soybean and sweet potato (Fig. 15.2b). *Meloidogyne incognita* Race 3 is the most common biotype, which is probably due to the history of cotton production in the Mid-South (Baker et al. 1985).

Distribution of root knot nematodes within a field is frequently uneven and scattered, particularly in coarse textured, sandy soil. Within these areas, *M. incognita* population density can increase and cause significant damage and symptom development on a susceptible host crop (Thomas and Kirkpatrick 2001). Crops within these areas often suffer the greatest yield losses due to enhanced water stress from nematode infection and the low water holding capacity of sandy soils.

The overwinter survival stage of the root knot nematode is primarily eggs. As a general rule, the total population density of root knot nematode is greatest near harvest in annual row crops. Initially, eggs in the soil or on roots make up the greatest proportion of the total fall population density, but as J2 hatch, the proportion of the population shifts toward J2. Second-stage juvenile survival is short-lived in the absence of a host and the majority die during the winter, so there is a general decline in the total nematode population density. In many cases, the early spring population density is often less than 10% of the total fall population of root knot nematode. In some parts of the Mid-South and in some years, root knot nematodes may survive on some winter weeds or cover crops (Timper et al. 2006). The soil temperature thresholds for J2 infection and female reproduction are 18 °C and 10 °C, respectively (Ploeg and Maris 1999).

15.4.1 Cotton

Meloidogyne incognita is one of the most important, yield-limiting plant parasitic nematode that affects cotton production in the Mid-South. During the 2015 cropping season, yield loss estimates of cotton lint averaged 2.2% in the Mid-South (Lawrence et al. 2016). Over the past 10 years, the estimated yield losses of cotton lint due to *M. incognita* ranged from 2.0% to 2.6% per year for a total lint yield loss of 214,700 ton (NCCB 2017).

15.4.1.1 Symptoms

The degree of symptom development on cotton is positively related to nematode population density. Severely infected cotton plants are stunted, wilt during the heat of the day and may show signs of drought stress or nutrient deficiencies even in the presence of adequate soil moisture and nutrients. The presence of root galls on secondary roots is the best diagnostic symptom (Fig. 15.3a). These galls are visible as early as 45 days after planting on a susceptible cultivar. Root galls are a good indication of nematode presence, but soil sample are better suited to monitor nematode

Fig. 15.3 (**a**) Galls caused by *Meloidogyne incognita* on cotton root system; (**b**) Stunted and yellow plants as a result of a moderate to high population density of *M. incognita*; (**c**) Severely galled soybean root system caused by *M. incognita*; (**d**) Corn root system with clustered and stunted root caused by stubby root nematode. (Photos by T. R. Faske)

population densities and determine if an economic threshold is present. The damage threshold in cotton for *M. incognita* in the Mid-South is 50–100 J2/100 cm^3 soil from soil samples collected in the fall (Robinson 2008).

15.4.1.2 Management Strategies

For the past 20 years, nematode management in cotton has relied heavily on an integrated approach that includes the use of nematicides, host plant resistance (on a very limited basis) and crop rotation. During much of this time, commercially available cotton cultivars, with a suitable level of both nematode resistance and yield potential, were lacking. Thus, the most common nematicides used in the Mid-South were 1, 3-dichloroporpene (Telone® II), aldicarb (Temik® 15G) and oxamyl (Vydate® C-LV). As a general rule, Telone® II was more effective than the non-fumigant nematicides, but they were more expensive and required special equipment for application. Aldicarb provided systemic protection from both early-season insects and nematodes on developing cotton seedlings. Aldicarb was once the most widely used nematicide in the Mid-South, but the use of aldicarb decreased as supplies became limited, because the manufacturer stopped the production of Temik® 15G that was to be phased out by 2018. Recently, however, there has been a renewed interest in the use of aldicarb by cotton producers in the Mid-South, and in 2016, AgLogic™ 15GG was registered for use with the EPA. Currently, the most common nematicides used are those that are applied on the seed coat. There are two groups of seed-applied nematicides: chemical and biological control agents. Abamectin (Avicta® 500 FS) was registered in 2006 as the first chemical seed-applied nematicide in cotton. Abamectin provides some early-season control of *M. incognita* on developing cotton seedlings (Monfort et al. 2006), but seedling protection is limited to a few centimeters from the treated seed as only a small portion of abamectin is transferred along the developing root system (Faske and Starr 2007). Fluopyram is an succinate dehydrogenase inhibitor fungicide that was recently shown to affect *M. incognita* motility and ability to infect tomato seedlings (Faske and Hurd 2015). Fluopyram (COPeO™ Prime) was registered in 2014 for use as a seed-applied nematicide in cotton. Additionally, a formulation of fluopyram + imidacloprid (Velum® Total) was registered in 2015 as a liquid in-seed-furrow spray for use against insects and nematodes in cotton. In field trials, Velum® Total generally provided better suppression of *M. incognita* than other seed-applied nematicides applied on cotton. Fluopyram as COPeO™ Prime preformed similarly to other seed applied nematicides (Lawrence et al. 2015b; Faske et al. 2017). Tioxazafen (NemaStrike™) is currently being evaluated as a seed-applied nematicide and will be marketed for use on cotton.

Of the biological control agents registered for suppression of nematodes, *Bacillus firmus* has been widely used as a seed treatment biological nematicide and is marketed in combination with clothianidin as Poncho®/VOTiVO® for seedling protection against insects and nematodes. Recently, heat-killed *Burkholderia* spp. (BioST® Nematicide 100) has been marketed as a seed treatment for suppression of nematode damage. Although the use of seed-applied nematicides is increasing, all appear to be most effective when used in fields with low nematode population densities, particularly if they are paired with a moderately resistant cultivar.

Nematicides increase production costs and may not be necessary field-wide. Distribution varies both vertically and horizontally within the soil profile (Baker and Campbell 1981). The root knot nematode is most commonly associated with coarse-textured soils, which are often the areas of a field that also suffer the greatest moisture deficit stress and yield loss (Wrather et al. 2002; Monfort et al. 2007). Emerging precision technology now allows soil texture to be estimated and mapped in a field based on apparent electrical conductivity (EC_a), measured with equipment like the Veris 3100 Soil EC Mapping System. Recent studies have shown that areas within the field with the lowest EC_a values, indicating the highest sand content, are high risk zones where nematicide use can have the greatest impact on protecting cotton yield potential (Ortiz et al. 2012; Overstreet et al. 2014). As the use of precision technology including yield monitors, remote sensing, soil EC mapping, etc. increases, so will the opportunities to incorporate site-specific nematicide application as a nematode management tool.

The use of host plant resistance is the most economical and sustainable option for managing plant parasitic nematodes. Resistance suppresses nematode reproduction, which results in a lower nematode population density for the subsequent crop. Pioneering and recent studies have identified a rich source of resistant breeding lines in the germplasm of various *Gossypium* spp. (Robinson and Percival 1997; Robinson et al. 2001). The breeding line Auburn 623 RNR, that was developed from the cross between two moderately resistant parents, Clevewilt 6 and Wild Mexican Jack Jones (Shepherd 1974), was highly resistant to root knot nematodes. This breeding line was later crossed with the recurrent parent cultivar Auburn 56 to develop Auburn 634 RNR (Shepherd 1982), which was back crossed into various recurrent parents with acceptable agronomic characteristics to develop the M-series of breeding lines (e.g. M-120, M-240, M-315) (Shepherd et al. 1996). The mechanisms of resistance in Auburn 623 RNR sources of resistance are not well understood, but resistance is based on both reduced root galling in the host and lower egg production by the nematode (Creech et al. 1995; Jenkins et al. 1995). Studies investigating the inheritance of resistance in Clevewilt 6 indicate that a single recessive gene is involved (Bezawada et al. 2003). A two-gene model for resistance in M-315 was proposed that included a dominant gene from Wild Mexican Jack Jones and an additive gene from Clevewilt 6 (McPherson et al. 2004). The molecular aspects of these genes in M-120 and M-240 were characterized in several studies (Shen et al. 2006; Ynturi et al. 2006; Gutierrez et al. 2010; He et al. 2014). Based on their work, a gene on chromosome 11 (*Mi*-C11) that was present in Clevewilt 6 was primarily responsible for reduction in nematode galling, while a gene on chromosome 14 (*Mi*-C14), present in Wild Mexican Jack Jones was primarily responsible for suppression of nematode reproduction. Five additional sources of resistance were identified from the Yucatan region of Mexico (Robinson and Percival 1997). Though these accessions are not as resistant to *M. incognita* as Auburn 623 RNR, based on nematode biology (reproduction and development), two accession (TX-1174 and

TX-2079) may have genes for resistance that differ from Clevewilt 6 and Wild Mexican Jack Jones (Faske and Starr 2009).

One of the main challenges in breeding for resistance had been integrating resistance from these resistant sources into elite cotton cultivars – a long and slow process. During the mid-1990s resistance to root knot nematode in cotton was moderate at best, with most commercial cultivars containing one resistant gene. By about 2010, some cotton cultivars exhibited much better resistance with a two-gene system. Current commercially available root knot nematode resistant cultivars such as Deltapine DPL1558NR B2RF and Phytogen PHY 427 WRF, are marketed as having two genes and a high level of resistance, while PHY 487 WRF and Stoneville ST 4946 GLB2 have one gene and a moderate level of resistance. It is likely that future resistant cultivars will play a vital role in the management of root knot nematode in cotton.

Crop rotation with a non-host or poor host can be effective at reducing the nematode population density below a damage threshold. Rotation has been used effectively in the Mid-South to manage root knot nematode in cotton. Peanut is a relatively new to the Mid-South production system and is a non-host to the southern root knot nematode, making it a great option as a rotational crop (Kirkpatrick and Sasser 1984). Peanut production has increased in Arkansas and Mississippi, but is still of limited potential in the region due to soil type variability and the relatively low acreage of peanut in relation to that of cotton. Rice is good option as a rotational crop (Bridge 1996) because flooding is an effective tool in nematode control. Unfortunately, most cotton fields are not suitable for rice production because it is so difficult and expensive to maintain adequate flooding levels due to the soil type. Grain sorghum has been recognized as useful rotation crop to manage the southern root knot nematode. Recent studies in the Mid-South have indicated that there is a wide range in host suitability among grain sorghum hybrids (Hurd and Faske 2017), so some grain sorghum hybrids may sustain or possibly increase populations of root knot nematode for the subsequent row crop. Corn is commonly grown in the Mid-South and it too has a wide range in host suitability to root knot nematode. Most corn hybrids are susceptible to root knot nematodes (Davis and Timper 2000). Soybean is a common rotational crop with cotton in the Mid-South. While some soybean cultivars are root knot nematode resistant, most cultivars in all of the maturity groups grown in the Mid-South have little or no resistance (Kirkpatrick et al. 2016).

15.4.2 Soybean

The southern root knot nematode is the most important plant parasitic nematode affecting soybean production in the Mid-South. During the 2015 cropping season, the southern root knot nematode accounted for an estimated average yield loss of 2.0% in Louisiana and Mississippi and 3.6% in Arkansas for a total grain yield loss of 9.4 million bushels (Allen et al. 2016).

15.4.2.1 Symptoms

Above ground symptoms are dependent on nematode population density and crop maturity. Stunted seedlings can be observed at high population densities, while stunted and chlorotic plants are common at mid to late reproductive growth stages where moderate to high population densities occur (Fig. 15.3b). These plants may senesce earlier than non-infected soybean plants. Galls on infected roots are the most diagnostic feature of root knot nematode on soybean. Galling severity depends on population density. Small galls can be observed on soybean at early vegetative stages of growth, but large galls that are easier to identify occur at early and mid-reproductive growth stages (Fig. 15.3c). Severely infected roots may have several galls that coalesce causing the entire root system to appear galled. With this level of severity, many times entire root systems become discolored and necrotic, leaving only a portion of a taproot intact with very few to no secondary roots remaining on the root system. Severely infected plants produce fewer pods and smaller seed per pod, which contributes to lower gain yield. The damage threshold for southern root knot nematode is 60 J2/100 cm^3 soil for soil samples collected in the fall.

15.4.2.2 Management Strategies

The use of resistant cultivars is the most efficient tactic to manage root knot nematodes, because resistant cultivars not only perform better, but may actually lower the overall population density of the nematode (Cook and Evans 1987). The sources of resistances to *M. incognita* in germplasm and breeding lines that are most commonly used to develop resistant cultivars includes Avery, which is a maturity group (MG IV) cultivar, Forrest (MG V), D83-3349 (MG VI), G93-9009 (MG VI), PI 417444 (MG VI), PI 96354 (MG VI) and Gordon (MG VII) (Hartwig and Epps 1973; Boerma et al. 1985; Luzzi et al. 1987; Anand and Shannon 1988; Hartwig et al. 1996; Luzzi et al. 1996). These lines range in resistance to *M. incognita* from partially to highly resistant, with most of the highly resistant germplasm in later soybean maturity groups (MG VI and VII). The inheritance of resistance to *M. incognita* in the cultivar Forrest was reported to be conditioned by a single additive gene (*RMi*1) that confers partial resistance to root galling (Luzzi et al. 1994b); however, horizontal resistance is more common in soybean. A high level of resistance to *M. incognita* in PI 417444 and PI 96354 are conditioned by a few genes that differ from those of Forrest (Luzzi et al. 1994a). The mechanism of resistance in PI 96354 is associated with the inability of J2 to establish a feeding site, or slower development of those individuals that do establish a feeding site. Fewer eggs were produced by survivors on this line than on the susceptible cultivar Bossier (Herman et al. 1991; Moura et al. 1993). Although these PI lines possess unique resistant genes, integrating resistance from these sources into high-yielding cultivars has been a slow process, especially in the early maturity groups (III – V) that are popular in the Mid-South (Kirkpatrick et al. 2016). The majority of the maturity groups grown in the Mid-South are MG IV, followed by MG V and MG III. The availability of elite cultivars with nematode resistance is further complicated by the use of different herbicide resistance traits across the Mid-South.

Due to the lack of available cultivars with good yield potential and a high level of resistance to *M. incognita*, nematode management requires an integrated approach in the Mid-South. Rice is a commonly used in crop rotation in the Mid-South, which is a good option for root knot nematode management as flooded conditions for rice production are unfavorable for nematode survival in the soil. Peanut is a non-host for *M. incognita* and offers an excellent option in fields that are suitable for peanut production. Corn and grain sorghum can increase or sustain a population of root knot nematode depending on host suitability of the cultivar (Davis and Timper 2000; Hurd and Faske 2017). Other cultural practices include subsoil tillage in areas where soil compaction may limit root development (Minton and Parker 1987). Though fumigant nematicides are effective they are generally too expensive to be economically practical in soybean production. Abamectin, fluopyram and *B. firmus*-treated seed provide some suppression of root knot nematode infection on seedling root systems. In general, this suppression of nematode infection is limited with variable responses to yield protection (Hurd et al. 2015, 2017a, b; Jackson et al. 2017). These seed-applied nematicide are best used in fields with a low population density of root knot nematode and paired with at least a moderately resistant cultivar (Jackson et al. 2017). Tioxazafen (NemaStrike™), heat-killed *Burkholderia rinojensis* (BioST® Nematicide 100) and *Bacillus amyloliquefaciens* (AVEO™ EZ Nematicide) are being evaluated as a seed-applied nematicide and field efficacy trials are ongoing to determine the impact of these chemical and biological agents in soybean.

15.4.3 Sweet Potato

The southern root knot nematode (*Meloidogyne incognita*) is one of the most important and widespread plant parasitic nematodes affecting sweet potato production in the Mid-South and worldwide (Overstreet 2013a). Damage from root knot nematode affects both sweet potato quality and yield.

15.4.3.1 Symptoms

The most diagnostic symptom is galls, which appear as spindle-shaped swellings on the fibrous root of sweet potato. Gall size and severity is reflective of nematode population density, but can vary among cultivars. On storage roots, small bumps or blisters can be observed on the root surface. Mature females with egg masses can be detected beneath these raised areas. Cracking on storage roots can be caused both by root knot nematode and fluctuations in soil moisture (Thomas and Clark 1983; Lawrence et al. 1986; Overstreet 2013a).

Sweet potato cultivars with resistance to *M. incognita* are available and provide the most economical approach to management (Overstreet 2013a). Reproduction by *M. incognita* can increase with increasing soil temperatures on resistant cultivars, but not to the same magnitude as that of a susceptible cultivar (Jatalla and Russell

1972). Fumigant nematicides (1,3 dichloropropene and metam sodium) are effective tools and the non-fumigant granular nematicide ethroprop is also registered for use in sweet potato production. Because sweet potato is vegetatively propagated it is important to propagate slips in a root knot nematode-free bed. Infected slips (those with adventitious roots) could transport and distribute root knot nematodes into a new field. Propagation with cuttings would eliminate the risk of root knot nematode infected slips. Other cultural practices including rotation with peanut or some grain sorghum hybrids, may reduce nematode population densities, (Johnson et al. 1996) as well as some fungal diseases (Jenkins et al. 1995).

15.5 Other Root Knot Nematode Species, *Meloidogyne* spp.

Though several other species of *Meloidogyne* affect row crop production in the United States, *M. incognita* is the most common species found on row crops in the Mid-South. Other species have been reported on non-cultivated land or in a field or two in a specific state. In Arkansas, *M. arenaria* and *M. javanica* have been reported on non-cultivated land, while *M. hapla* has been reported on non-cultivated land and on soybean (Robbins 1982b). In Louisiana, *M. javanica* has been detected in a few soybean fields but is not considered a common pest. *Meloidogyne javanica* was found in association with *M. incognita* in some Louisiana soybean fields and the complex contributed to a failure of soybean varieties with resistance to *M. incognita* (E. C. McGawley, pers. comm.). In Mississippi, *M. arenaria* Race 2 has been detected in soybean and *M. arenaria* Race 1 in peanut, but neither species is a common pest. *Meloidogyne javanica* has been detected on vegetables in the southern part of the state, but its impact has not been investigated.

15.6 Stubby Root Nematodes, *Paratrichodorus* spp. and *Trichodorus* spp.

Stubby root nematodes are widespread throughout the U.S. and they are common on row crops in the Mid-South. They have a broad host range that includes hundreds of plant species, but are most damaging to species in the grass family Poaceae.

Population densities of stubby root nematodes within a field are scattered and irregular, both vertically and horizontally. Stubby root nematodes are commonly found in sandy and sandy-loam soils and often occur deeper (*ca.* 30 cm) in the soil profile than other plant parasitic nematodes because they are very sensitive to low soil moisture and mechanical disturbance from tillage. Stubby-root nematode population densities can change quickly during the season with adequate moisture in the root zone. Additionally, population densities can also decrease quickly, making diagnosis difficult without the use of root symptomology.

15.6.1 Symptoms

Stubby root nematodes are ecotoparasitic nematodes. They feed on the meristematic cells of root tips causing root growth to slow and eventually stop, hence the "stubby root" symptom. Unlike other plant parasitic nematodes, stubby root nematodes have an onchiostyle which is a curved, solid stylet. The nematode uses the stylet to puncture the meristem cells, where it secretes salivary materials that are used to construct a feeding tube. The nematode uses the feeding tube to extract nutrients and cellular components before migrating to another cell, leaving the feeding tube behind. As new roots emerge near the root tip they are parasitized by the nematode causing a proliferation of secondary roots near the root tips (Fig. 15.3d). Affected root systems have small, stunted root systems with fewer and shorter secondary roots. Seedlings are especially sensitive to stubby root nematode feeding. On dicots, root symptoms are less obvious and appear as a reduced root system, in contrast to the stubby roots that occur in a grass crop. Foliar symptoms of affected plants are severely stunted and yellow, but continue to develop through reproductive stages of growth.

15.6.2 Management Strategies

Stubby root nematodes have been associated with stunted corn in the Mid-South (Koenning et al. 1999; Faske and Kirkpatrick 2015). Corn is highly susceptible with some variation in host suitability among hybrids (Timper et al. 2007). Given the broad host range of stubby root nematodes, which includes soybean and cotton, nematode population densities are often maintained with the use of these common rotation crops in the Mid-South. Soil tillage can be an effective strategy to reduce nematode population density (Todd 2016). Though nematicides can be effective, they are not always economically beneficial to the farmer. Nematicides registered for use in corn production include fumigants (1,3 dichloropropene), non-fumigant granular nematicides (terbufos) and chemicals applied as a seed treatment; abamectin and *Bacillus firmus*. In general, nematicides applied as a seed treatment are most effective at low population densities of nematodes in the soil.

15.7 Lesion Nematodes, *Pratylenchus* spp.

Several species of *Pratylenchus* have been identified in the Mid-South, but in general, *P. brachyurus* and *P. zeae* are among the most common, especially in corn. Lesion nematodes are migratory endoparasitic nematodes, so a portion of the total viable population in a field may be present in the roots rather than in the soil. Therefore, both a root and a soil sample are needed to determine the total population density of lesion nematode in a field. Lesion nematodes can be found in a range of soil types depending on the species. *Pratylenchus brachyurus* and *P. zeae* were reported to reproduce at a greater rate in silt loam soils than loam or clay soils (Endo 1959).

15.7.1 Symptoms

Symptom development is dependent on nematode species, population density and environmental conditions (soil temperature and moisture). All lesion nematodes cause dark brown lesions on roots, but root lesions vary in size depending on nematode species. Lesion nematode distribution is often aggregated in the field, thus foliar symptoms may occur in irregular patches in the field. Foliar symptoms are non-specific with stunted and yellow plants being the most common descriptions.

15.7.2 Management Strategies

The usefulness of crop rotation in the management of lesion nematodes is species specific. Corn, foxtail millet (*Setaria italica*), grain sorghum, cereal rye (*Secale cereale*), soybean and sudangrass (*Sorghum drummondi*) were good hosts for *P. zeae*, while barley (*Hordium vulgare*), oat (*Avena sativa*) and watermelon (*Citrullus lanatus*) were reported as poor hosts (Endo 1959). Corn, cotton, potato (*Solanum tuberosum*), watermelon and sudangrass, have been reported as good hosts for *P. brachyurus*, while soybean, oat, sweet potato, cereal rye and pearl millet (*Pennisetum glaucum*) are relatively poor hosts (Endo 1959; Timper and Hanna 2005). So, species identification is an important factor when recommending a rotational crop for lesion nematode in the Mid-South. Cover crops resistant to some species of lesion nematode have been effective at reducing nematode population densities, but cover crops resistant to one species of *Pratylenchus* may be susceptible to another, so monitoring nematode population densities is a good practice when using cover crops. Currently, there is no information on the susceptibility of commercially available corn hybrids to lesion nematodes.

15.8 Reniform Nematode, *Rotylenchulus reniformis*

The reniform nematode is recognized as one of the most important plant parasitic nematodes affecting cotton production in the Southern United States. Cotton lint losses in the 2016 cropping season, from three Mid-South states, were estimated at 17,500 ton (Lawrence et al. 2017).

The reniform nematode has become a pathogen of major importance during the past 40 years in the Mid-South. The nematode was first reported in Hawaii in 1940

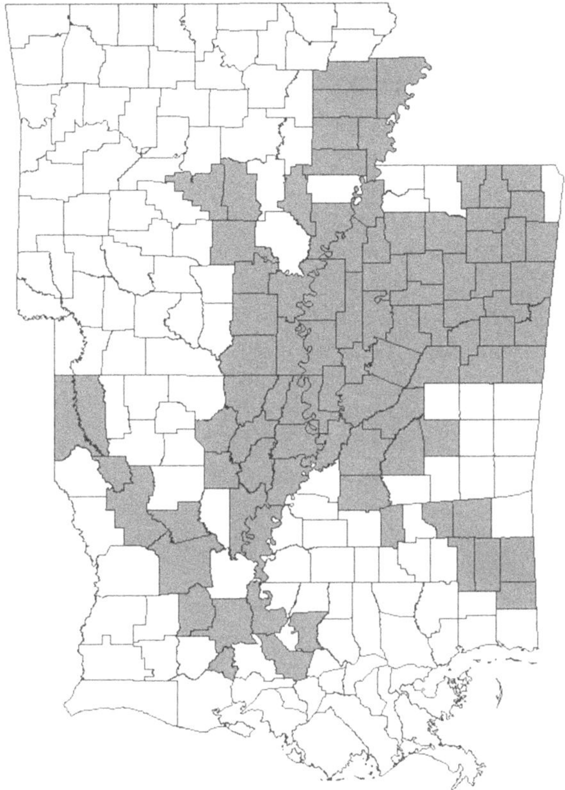

Fig. 15.4 Counties (highlighted gray) where *Rotylenchulus reniformis* has been detected on row crops in Arkansas, Louisiana, and Mississippi

(Linford and Oliveira 1940) and in the Mid-South in Baton Rouge, Louisiana in 1941 (Smith and Taylor 1941). The first occurrence of this nematode in Mississippi was in 1968 and in Arkansas in 1979 (R. T. Robbins, University of Arkansas, pers. comm.). In Louisiana, this nematode was considered as only a minor problem in the 1960s as it was only associated with 800–1000 ha of cotton in two counties (Birchfield and Jones 1961). Reniform nematodes have spread rapidly in the Mid-South since they were first detected. A survey of cotton fields conducted in Louisiana during 1994 and 1995 (Overstreet et al. 2008) indicated that over 200,000 ha were infested with the reniform nematode. McLean and Lawrence (2000) found that reniform nematodes were present in 67% of the fields in Northeast Louisiana. Reniform nematodes are also widely distributed through Mississippi and parts of Arkansas. The distribution of the reniform nematode as of 2017 includes many of the major row crop producing counties or parishes in the Mid-South (Fig. 15.4).

15.8.1 Field Introduction and Distribution

The reniform nematode is very easy to introduce into a production field because of its unique ability to survive in an anhydrobiotic stage (Lawrence and McLean 2001). This is one of the few nematodes that can withstand desiccation for several years, then rehydrate and become active again. Reniform nematodes that are in this dried state can easily be transmitted to fields on any type of farm equipment. Reniform nematodes can spread quickly within fields (Monfort et al. 2008; Xavier et al. 2012a). Once the nematode has entered a field, it can quickly spread through water movement (rainfall events or flood irrigation) or equipment (Overstreet et al. 2008). The spread has been associated with the direction of rows within a field, caused by the general tillage practices in that field (Monfort et al. 2008). In both of these reports, reniform nematode went from a small isolated area to quickly being present over a much greater area of the field.

Although reniform nematodes sometimes appear to be uniformly distributed within a field, populations are in reality described as being a non-clustered horizontal distribution (Lee et al. 2015). Densities are much higher in some locations in a field than others, likely due to a number of factors (Overstreet et al. 2011b). Population densities are strongly influenced by soil texture (Monfort et al. 2008; Xavier et al. 2014) and the nematode is often found in soils with significant silt or clay content. In Louisiana, the highest densities occurred when clay content was between 10% and 20% with lower populations below and above this clay percentage (Xavier et al. 2012c). Monfort et al. (2008) reported that the greatest populations of the reniform nematode occurred when silt content of the soil was between 54% and 60%.

Of particular concern is the distribution of reniform nematode vertically in the soil profile. Reniform nematodes can occur to greater depths in the soil than most other nematode species (Robinson et al. 2005; Robinson et al. 2006). In some cases very low populations can be detected in the upper surface (15–20 cm depth) but very high populations may be present below that depth (Xavier et al. 2012a, b). Rotation with corn or fall or spring tillage contribute to the decline of the nematode at the surface, and since soil samples for nematode analysis are usually taken at this depth, results can sometimes be misleading. Reniform populations deep in the soil profile may quickly rebound when a susceptible crop is grown.

15.8.2 Biology and Hosts

The reniform nematode has a short life cycle requiring only 17–23 days, depending on soil temperature (McGawley and Overstreet 2015). Egg masses typically contain 50–75 eggs and a number of generations of the nematode can develop in a single crop (Overstreet et al. 2009). Reniform nematodes also have a fairly wide host range that includes many broad-leaf weeds (Carter et al. 1995; Robinson et al. 1997). This can make it difficult to reduce or eliminate the nematode once established in a field, since there may be alternative hosts present. Weeds in combination with cotton or soybean have also been found to influence the reniform nematode and may actually suppress population development of this pathogen (Pontif and McGawley 2007, 2008). Common weed species that have been identified as excellent hosts of the reniform nematode include sicklepod (*Senna obtusifolia*), spurred anoda (*Anoda cristata*), entireleaf morning glory (*Ipomoea nil*) and velvetleaf (*Abutilon theophrasti*) which potentially could lead to increased populations (Molin and Stetina 2016).

In the past several years in the Mid-South, a number of researchers have reported that populations of reniform nematodes may not always respond similarly to the same crops or cultivars (Agudelo et al. 2005; Arias et al. 2009; McGawley et al. 2010, 2011). A number of populations have shown differences in how well they reproduce on the same cultivars, and some populations are more pathogenic than others (Xavier et al. 2014; Bhandari et al. 2015). These differences in how the nematode impacts plants from one location to another have serious implications on the use of resistant varieties as a management tool (Agudelo et al. 2005).

15.8.3 Crop Losses

Of the row crops produced in the Mid-South, the reniform nematode is most often associated with cotton (Kirkpatrick and Thomas 2007). Lint loss estimates from reniform nematode to cotton have been reported as high as 50–60% in individual fields (Birchfield and Jones 1961), although based on nematicide trials over the past 40 years, losses are typically more in the 15–30% range (Overstreet 1996). Because reniform nematode may be spread throughout a field, subtle or even serious damage from reniform nematode may go undetected when hot spots are not visible (Lawrence and McLean 2001).

15.8.4 Symptoms and Damage

Typical symptoms of reniform nematode damage on cotton include stunting, delayed flowering and fruit set, uneven plant heights (Fig. 15.5a) and low yield (Lawrence and McLean 2001; Overstreet et al. 2008). Reniform nematode populations vary within a field and shortly after the initial introduction into a field, hots spots or severely stunted areas may be visible. One of the most distinct symptoms of reniform nematode infestation, during the first few years following their initial introduction in a field, is unevenness in plant height (Overstreet et al. 2008). These wavy patterns in plant height are associated with changing populations in the field. Once the reniform nematode has been present in a field for some time, damage may be more uniform, making it more difficult to recognize (Lawrence and McLean 2001). The damage threshold for this nematode varies somewhat between states but levels from 250 to 5000 per 500 cm^3 of soil are considered high enough to cause injury (Greer et al. 2009; Mueller et al. 2012). Factors that impact the level of damage that plants are likely to experience include soil texture, cultivar selection, soil moisture and possibly fertility. Although reniform nematodes can develop in a wide range of soil textures, some soils are much more prone to show damage symptoms (Robinson et al. 1997). Coarse-textured soils will likely sustain the greatest amount of damage and damage may occur with lower population levels of the nematode (Monfort et al. 2008; Overstreet et al. 2011b, 2014). Fine-textured soils with higher

Fig. 15.5 (**a**) Stunted and uneven stand of cotton as result of *Rotylenchulus reniformis* (Overstreet); (**b**) Stunted and uneven stand of cotton as result of a mixed field population of *R. reniformis* and *Meloidogyne incognita*; (**c**) Non-fumigant (left) *vs.* fumigated (right) treated rows in a field with a damaging population of *R. reniformis*; (**d**) Unthrifty soybean growth due to high population density of *R. reniformis*; (**e**) Cracking symptoms of sweet potato storage roots due to early infection by *R. reniformis*; (**f**) Egg masses of *R. reniformis* on sweet potato root system. (Photos by C. Overstreet)

silt or clay content may not be as prone to damage and require higher populations of the nematode to cause injury. Recent studies with site-specific applications of nematicides indicate that soil texture in a field may be much more important than actual nematode populations in determining the level of damage that occurs (Overstreet et al. 2014).

15.8.5 Management Strategies

A phenomenon that has been reported in Louisiana and Texas is the occurrence of soils that suppress reniform nematode (Robinson 2008). Suppressive soils simply mean that nematode populations don't build up as expected on a susceptible host. These soils are believed to have some type of transferable agent, likely some types of biological control organisms. A nematophagous fungus, originally designated as ARF (Arkansas fungus) 18 and recently identified as *Brachyphoris riggsii* (B. Bluhm, pers. comm.), has been found in the Mid-South and was reported to suppress reniform nematode populations in greenhouse experiments (Wang et al. 2004). Alternately, some fungi such as *Rhizoctonia solani,* that causes a seedling disease of cotton called sore shin, have been found to increase infection and subsequent population densities of the nematode (Sankaralingam and McGawley 1994a, b).

The reniform nematode is often found in association with other nematode species (Fig. 15.5b). The interactions between reniform nematodes with the southern root knot nematode (*M. incognita*) have been studied in the Mid-South (Stetina et al. 1997a, b). Based on these studies, root knot nematodes suppressed reniform nematode populations. However, field observations in the Mid-South imply that reniform nematodes appear to become the dominant nematode over time and it actually becomes difficult to find any root knot nematode. This is particularly true when cotton is grown as a monoculture crop. Since some cultivars of soybean or cotton may have some resistance to root knot nematodes and reniform nematode has a shorter life cycle, populations of reniform nematode may simply reach higher levels (Stetina et al. 1997a). However, crop rotations with corn (favors root knot but not reniform nematode) have begun to reverse this trend and more fields now have detectable populations of both nematodes (Overstreet et al. 2011a).

Currently, there are no cultivars of cotton that are resistant to the reniform nematode (Robinson et al. 2004). Breeding efforts have been underway for over three decades to find and incorporate resistance in cotton. Early reports indicated that all of the cotton cultivars and breeding lines in the upland cotton species (*Gossypium hirsutum*) planted in the Mid-South, were susceptible (Birchfield and Brister 1963). Some of the early screening of other *Gossypium* species indicated that resistance to the reniform nematode was present in some species of cotton (Yik and Birchfield 1984); however, incorporating this resistance into *G. hirsutum* proved to be a difficult process. One of the first breeding lines with strong resistance was derived from the cotton species, *G. longicalyx* and released as LONREN (Bell et al. 2014). Unfortunately, LONREN breeding lines showed severe stunting when planted in areas with high population densities of the reniform nematode (Bell et al. 2009). It is likely that a hypersensitive reaction to infection by the nematode in the cotton roots was involved, particularly since LONREN was very effective in killing the nematode as it began to develop with the roots. Germplasm lines with moderate level of resistance were reported from crosses with a germplasm line from Brazil (McCarty et al. 2012). Recently breeding lines derived from *G. barbadense* and referred to as BARREN have been reported (McCarty et al. 2013). The absence of high levels of nematode resistance and poor agronomic performance of BARBREN lines across geographic areas has limited their use for US cotton production.

Although all U.S. cotton is considered susceptible, some cultivars appear to have some level of tolerance. Tolerance implies that the plant may be attacked by the nematode but can still yield well even if it doesn't inhibit nematode reproduction. In a recent study, three cultivars were identified to have some degree of tolerance to the reniform nematode (Stetina et al. 2009). Further research from Mississippi indicated that six of thirteen cultivars tested were considered tolerant of the reniform nematode (Blessitt et al. 2012). These cultivars did not reduce nematode populations at the end of the year, but they did limit economic loss from the nematode. Although tolerant cultivars don't reduce reniform nematode populations they could play a vital role in the overall management of cotton, particularly when combined with other management options.

Aldicarb (Temik® 15G) is a non-fumigant nematicide that was the primary nematicide used in cotton from the early 1980s until recently (Lawrence and McLean 2000; Greer et al. 2009). Aldicarb was effectively used in the Mid-South to reduce nematode losses until 2011 when the product was no longer manufactured. Temik® 15G was applied at low rates (3.4–5.6 kg/ha of formulated material) at the time of planting. The product worked across most soil types and provided early season insect management as well as nematode suppression. The typical response reported in most fields infested with reniform nematode was about 112 kg/ha of lint (Overstreet and McGawley 1994; Overstreet et al. 2002). Aldicarb is available from another company as AgLogic™ 15GG but is not widely available yet. Unfortunately, the long term use of aldicarb in some areas in the Mid-South resulted in reduced benefits of using this product, likely due to accelerated microbial degradation (McLean and Lawrence 2003). Oxamyl (Vydate® C-LV), another carbamate with both insecticidal and nematicidal activity, was also used in cotton throughout the Mid-South until very recently. Oxamyl has been shown to be translocated from leaves to the roots and was available in a liquid formulation that was applied as a foliar spray to cotton to suppress nematode infection. Usually oxamyl was applied at pin-head square and was used in combination with an at-planting nematicide (McLean and Lawrence 2000). Combinations of nematicides such as a fumigant and aldicarb or oxamyl have been reported to provide the greatest yield for cotton (Lawrence et al. 1990).

Seed-applied nematicides came on the market in 2006 with the release of Avicta® Complete Pak for use in cotton which contained a nematicidal component, abamectin, the insecticide thiamethoxam and fungicides azoxystrobin, fludiomonil and mefenoxam. Abamectin provided some protection of the developing seedlings from nematodes when applied on cottonseed (Monfort et al. 2006). The reniform nematode was found to be particularly sensitive to abamectin (Faske and Starr 2006). Although this nematicide was only recommended for use with low to moderate populations of reniform nematode, it quickly became one of the most widely used nematicides. Since this material was already on the seed, it was much more convenient than application of granular or fumigant nematicides. Thiodicarb + imidacloprid (Aeris® seed-applied insecticide/nematicide) was launched in 2008 and serves as another seed treatment nematicide + insecticide in cotton. A formulation of fluopyram + imidacloprid (Velum® Total) applied as an in-furrow spray was registered in 2015 for use in cotton to manage cotton insects and nematodes including the reniform nematode. Although this chemical is an SDHI fungicide, it has been found to be effective against the reniform nematode (Faske and Hurd 2015; Faske et al. 2017). There has been some development of biologicals to manage nematodes in cotton. Poncho®/VOTiVO® is a combination of an insecticide (clothianidin) and a bacterial agent, *Bacillus firmus* that can provide some seedlings protection from early season attack by the reniform nematode. Tioxazafen (NemaStrike™) is being evaluated as a seed-applied nematicide and field efficacy trials are ongoing to determine the impact of this nematicide in cotton production.

Fumigants have been available and used by some cotton producers for many years Fumigant nematicides available today include 1,3 dichloropropene (Telone® II), metam potassium (K-PAM® HL™) and metam sodium (VAPAM® HL) (Greer et al. 2009). Though fumigants are effective, they are also expensive. Fumigants also require special application equipment special permits in some cases and need to be applied prior to planting. Telone® II needs to be applied at least 7 days prior to planting, while VAPAM® HL and K-PAM® HL™ should be applied at least 21 days before planting. Fumigants are applied beneath the row, where they volatilize and move through the soil profile. Although fumigants are very effective against reniform nematode (Fig. 15.5c), whole fields may not require treatment because damaging threshold of the reniform nematode varies among soil texture zones and soil texture zones vary within individual fields. Site-specific application of nematicides has recently been investigated for use in cotton (Overstreet et al. 2014). Fields are divided into management zones (Overstreet et al. 2010) based on apparent electrical conductivity (EC_a), which correlates well with soil texture. This is done with a Veris EC Soil Mapping System which can be used to map the EC_a of a field. The use of verification strips (treated with a nematicide and untreated rows) through the different soil zones can be used to define which zones need to be treated (Overstreet et al. 2010) so, future treatments target only those textural zones where yield loss may occur.

Crop rotation remains one of the most important practices to manage reniform nematode in cotton (Greer et al. 2009). Corn and grain sorghum are excellent rotation crops because they are non-hosts for reniform nematodes and suppress the nematode population density below the damage threshold for the subsequent crop (Stetina et al. 2007; Greer et al. 2009). Two years of corn production is often needed to reduce nematode populations below the damage threshold, particularly if the rotation has followed several years of monoculture cotton production (Stetina et al.

2007). Once the nematode population has dropped, a 1-year corn rotation can reduce damage from reniform in cotton. A few soybean cultivars are resistant to the reniform nematode and can be useful as a rotational crop to reduce the nematode populations in a field. Unfortunately, high populations of reniform nematode can remain deep in the soil profile, allowing population densities to rebound after 1 year of cotton production.

15.8.6 Soybean

The reniform nematode has primarily been an important pathogen of cotton, but in many parts of the Mid-South, soybean is being produced in areas that were previously planted in continuous cotton (Stetina et al. 2014). The reniform nematode has been reported to cause yield losses of 30–60% on soybean (McGawley and Overstreet 2015) and a total yield loss estimate of 3.2 million bushels was reported in 2016 in the Mid-South (Allen et al. 2017a).

15.8.6.1 Symptoms

Reniform nematode damage may not readily identifiable in soybean as it is generally more uniformly distributed in the field (Kirkpatrick et al. 2014). Symptoms include yellowing, stunting, unthrifty growth of plants (Fig. 15.5d) and empty pods (McGawley and Overstreet 2015), although symptoms can vary among cultivars, soil type, nematode population density and environmental conditions. In some cases, the root systems may be stunted with many of the smaller roots appearing discolored or blackened (Overstreet et al. 1992). One of the signs of reniform nematode is abundant egg masses on the root. Because soil is often attached to the egg masses, the root system has a rough or gritty appearance. Symptoms that may show up late in the growing season may be excessive leaf shedding during dry periods. Reniform nematode causes the greatest amount of damage during periods of moisture stress, usually during drought. However, severe damage may also occur under very wet or saturated moisture conditions. Many of the current cultivars of soybean grown in the Mid-South show little or no visible symptoms under adequate moisture conditions and may not show significant yield loss. The lack of visible symptoms and sometimes lack of yield response has made it difficult to determine just how damaging this nematode is to soybeans. Current damage thresholds range from 20 to 4,000/100 cm^3 soil for reniform nematode on soybean across the Mid-South.

15.8.6.2 Management Strategies

Resistance to the reniform nematode has been identified in some soybean cultivars, but very few have a high level of resistance (Robbins et al. 2016, 2017). Historically, the soybean cyst nematode was considered the most important nematode in the Mid-South and Southern U. S. and most of the breeding programs concentrated on this nematode. A widely used source of soybean cyst nematode resistance was the PI 88788 line. Because resistance was originally thought to be linked between soybean cyst and reniform nematodes, it was assumed that all the cultivars that were resistant to soybean cyst were also resistant to reniform nematodes (Rebois et al. 1968, 1970). It is now known that not all the soybean cyst nematode resistant cultivars are effective against reniform. Cultivars that were developed from PI 88788 have only slight resistance to reniform, whereas, those developed from Peking or PI 90763 are highly resistant (Robbins and Rakes 1996). Some of the highly resistant early cultivars included Forrest, Centennial, Sharkey and Stoneville (Robbins et al. 1994). Though today few commercial cultivars have a high level of resistance to reniform nematodes, soybeans in general are not as susceptible to reniform damage as cotton. Resistant cultivars can be used in rotation to suppress reniform population densities in a crop rotation system.

Other rotational crops include peanut and rice, which are considered to be non-hosts for the reniform nematode (Kirkpatrick et al. 2014). These crops can cause a significant drop in nematode population densities in a single season, but often 2 years are needed in fields with a high population density of reniform nematodes. Alternately, crops like cotton and sweet potato are excellent hosts and can sustain or increase reniform population densities for the subsequent crop.

Few nematicides are available to use on soybeans to manage reniform nematode. Fumigants such as 1,3 dichloropropene (Telone® II), metam potassium (K-PAM® HL™) and metam sodium (VAPAM® HL) are registered for use on soybean, but are not commonly used in the Mid-South due to cost of nematicide, product availability and economic benefit to yield. Aldicarb (Temik® 15G) was registered in some states, but not in the Mid-South to manage reniform nematode on soybean. Research is being conducted on the use of site-specific application of nematicides in soybean. Similar to studies in cotton, reniform nematode causes the greatest amount of damage in coarse-textured soils; these are the soil types that are most likely to respond to the application of a nematicide. Seed-applied nematicides like abamectin (Avicta® 500 FS) and fluopyram (ILeVO® 600 FS) are registered for use in soybean. Similarly, the seed-applied bionematicide *Bacillus firmus* (VOTiVO®) is also marketed for reniform suppression. Producers in the Mid-South most commonly use crop rotation to manage reniform nematode on soybean rather than nematicides.

15.8.7 Sweet Potato

The reniform nematode is a common nematode on sweet potato that affects both storage root quality and yield (Overstreet 2013b). Sweet potato is very susceptible to damage by the reniform nematode, but yield loss estimates in the Mid-South are lacking. High yield losses have reported by reniform nematode on sweet potato (Birchfield and Martin 1965), and yield loss estimates of 5–10% have been reported from Louisiana, although these fields contained both reniform and root knot nematodes (Koenning et al. 1999).

15.8.7.1 Symptoms

Symptoms of reniform nematode on sweet potato are often difficult to recognize in the field since there are generally not any distinct foliar symptoms. Some of the earlier cultivars were reported to express some yellowing when grown in reform nematode infested fields (Overstreet 2013b). Infected plants may have discolored fibrous root and mature later than those grown in non-infested fields even when nematode population densities are low. Cracking can occur on the storage roots (Fig. 15.5e) from early infection by reniform nematode. Root knot nematode can also cause cracking of storage roots. Female reniform nematodes are not found in root cracks, whereas *M. incognita* J2 and females can both be found within cracked roots (Overstreet 2009). Cracking is less common with many of the commercially available cultivars as they are less prone to cracking compared to older cultivars. Yield losses in sweet potato are associated with a reduction in the size of storage roots, which reduces the number of marketable sweet potatoes (Abel et al. 2007).

Damage thresholds are lower for older cultivars because of cracking sensitivity, but in general. population levels of reniform nematode that are considered damaging range from 10 to 1000 vermiform/100 cm^3 soil (Smith et al. 2008; Anonymous 2017). Population densities are not uniform within a field, but the nematodes may be distributed throughout the entire field (Burris et al. 2009). Reniform population densities can build up quickly on susceptible cultivars such as Beauegard (Fig. 15.5f).

15.8.7.2 Management Strategies

Nematicides have been one of the most effective methods of management of the reniform nematode (Smith et al. 2008). Fumigant nematicides such as 1,3 dichloropropene (Telone® II), metam potassium (K-PAM® HL™) and metam sodium (VAPAM® HL) are the most common nematicides used to manage nematodes in sweet potato (Overstreet 2013b). Though effective and commonly used to prevent reniform damage, they do increase production cost for sweet potato farmers. All of the fumigants must be applied preplant and require a period of time before sweet potato slips can be transplanted. Aldicarb, a non-fumigant nematicide, was reported to increase the number of USDA number 1 and jumbo storage roots in reniform nematode infested fields in Mississippi (Henn et al. 2006). Fluensulfone (Nimitz®) was recently registered for use on sweet potato, providing growers another option in managing reniform nematode.

Currently, none of the sweet potato cultivars grown in the Mid-South are resistant to the reniform nematode (Smith et al. 2008; Overstreet 2013b); however tolerance has been reported in a few cultivars such as Centennial (Clark and Wright 1983). Unfortunately, growing tolerant cultivars can contribute to the buildup of a nematode population density that can affect the subsequent crop if it is susceptible to the reniform nematode.

Crop rotation is a good strategy for producers with reniform nematodes in production fields. Corn, grain sorghum and peanut can greatly decrease populations in a field (Smith et al. 2008). Other cultural practices include washing storage roots free of soil before planting in the plant bed to avoid the possibility of introducing the nematode into plant beds (Overstreet 2013b). Producers that use vine cuttings (slips) for transplants eliminate the potential dispersal of reniform nematode into non-infested fields.

As discussed in this chapter, plant parasitic nematodes are among the most important yield limiting factors that affect row crop production in the Mid-South. Sustainable agriculture is an integrated system of crop production practices that have specific applications. A long-term goal is addressing the needs of consumers, maintaining production at a profitable level and enhancing the quality of life for farmers. Consequently, an integrated system of tools is used to manage nematodes in the Mid-South. One of the most important factors in nematode management is to understand the nematode species that are affecting crop production, keeping in mind that the species complex in a field can change over time. For example, the soybean cyst nematode was prevalent historically in several counties in Louisiana, but over the past 10 years it is been difficult to find. Conversely, root knot and reniform nematodes are more common. Similarly, the soybean cyst nematode was the most frequently found species of nematode in the 1980s on soybean in Arkansas; however, today (2018), although soybean cyst nematodes are still common inhabitants of soybean fields, the root knot nematode is the most important nematode on soybean in Arkansas.

Host plant resistance and the use of a non-host crop in rotation sequences are among the oldest management tools that are still economical and effective ways to manage plant parasitic nematodes. As discussed in this chapter, resistance and rotational crops are limited in some cropping systems in the Mid-South. In these systems, farmers are more likely to depend on nematicides. Research has provided evidence that site-specific nematicide application where only those areas of the field where soil type and nematode population densities affect yield loss can be an effective approach to nematode management. Reducing the amount of nematicide needed results in a cost savings to the farmer, while also limiting the potential environmental impact of a nematicide across an entire field. Since 2006 there has been an interest in seed-applied nematicides and currently, there are several seed-applied chemical or biological agents being evaluated to suppress plant parasitic nematodes in row crops. Seed-applied nematicides provide a uniform delivery of the nematicide at lower quantities compared to granular applications. Similarly, nematicides which are less toxic to the handler and potentially less toxic to off-target pest, due to lower acute toxicity, are being marketed for use in agriculture.

Because production systems vary across the Mid-South, each system may have a different set of tools to manage nematodes. Similarly, each field likely poses different challenges according to weed, disease, irrigation and nematode presence, so decisions on nematode management should be made on a field by field basis rather than farm-wide. Farms in the Mid-South range from several hundred to several thousand hectares. As cropping systems change so will the population of plant parasitic nematodes, emphasizing the importance of continued research and extension in nematology. There is currently a real concern across the discipline of nematology for the future development of applied, field-oriented nematological expertise. This could have a major impact in future agriculture research and services and the long term goals of sustainable agriculture.

References

Abel, C. A., Adams, L. C., & Stetina, S. R. (2007). Sweet potato yield reduction caused by reniform nematode in the Mississippi Delta. *Plant Health Progress*. https://doi.org/10.1094/PHP-2007-1115-01-RS.

Agudelo, P., Robbins, R. T., Stewart, J. M., & Szalanski, A. L. (2005). Intraspecific variability of *Rotylenchulus reniformis* from cotton-growing regions in the United States. *Journal of Nematology, 37*, 105–114.

Allen, T. W., Bradley, C. A., Damicone, J. P., Dufault, N. S., Faske, T. R., Hollier, C. A., Isakeit, T., Kemerait, R. C., Kleczewski, N. M., Koenning, S. R., Mehl, H. L., Mueller, J. D., Overstreet, C., Price, P., Sikora, E. J., Spurlock, T. N., & Young, H. (2016). Southern United States soybean disease loss estimates for 2015. In *Proceedings of the Southern Soybean disease workers* (pp. 11–15). Pensacola Beach.

Allen, T. W., Bradley, C. A., Damicone, J. P., Dufault, N. S., Faske, T. R., Hollier, C. A., Isakeit, T., Kemerait, R. C., Kleczewski, N. M., Kratochvil, R. J., Mehl, H. L., Mueller, J. D., Overstreet, C., Price, P., Sikora, E. J., Spurlock, T. N., Thiessen, L., Wiebold, W. J., & Young, H. (2017a). Southern United States soybean disease loss estimates for 2016. In *Proceedings of the Southern Soybean disease workers* (pp. 3–8). Pensacola Beach.

Allen, T. W., Bradley, C. A., Sisson, A. J., Byamukama, E., Chilvers, M. I., Coker, C. M., Collins, A. A., Damicone, J. P., Dorrance, A. E., Dufault, N. S., Esker, P. D., Faske, T. R., Giesler, L. J., Grybauskas, A. P., Hershman, D. E., HOllier, C. A., Isakeit, T., Jardine, D. J., Kelly, H. M., Kemerait, R. C., Kleczewski, N. M., Koenning, S. R., Kurle, J. E., Malvick, D. K., Markell, S. G., Mehl, H. L., Mueller, D. S., Mueller, J. D., Mulrooney, R. P., Nelson, B. D., Newman, M. A., Osborne, L., Overstreet, C., Padgett, G. B., Phipps, P. M., Price, P. P., Sikora, E. J., Smith, D. L., Spurlock, T. N., Tande, C. A., Tenuta, A. U., Wise, K. A., & Wrather, J. A. (2017b). Soybean yield loss estimates due to diseases in the United States and Ontario, Canada, from 2010 to 2014. *Plant Health Progress, 18*, 19–27.

Anand, S. C., & Shannon, J. G. (1988). Registration of 'Avery' soybean. *Crop Science, 28*, 1024.

Anonymous. (2017). Population densities (threshold level) of plant parasitic nematodes per pint of soil: Sweet potato. Accessed 10 Oct 2017. Online: http://extension.msstate.edu/sites/default/files/topic-files/sweetpotato/sweetpotato.pdf

Arias, R. S., Stetina, S. R., Tonos, J. L., Scheffler, J. A., & Scheffler, B. E. (2009). Microsatellites reveal genetic diversity in *Rotylenchulus reniformis* populations. *Journal of Nematology, 41*, 146–156.

Astudill, G. E., & Birchfield, W. (1980). Pathology of *Hoplolaimus columbus* on sugarcane. *Phytopathology, 70*, 565.

Bae, C. H., Szalanski, A. L., & Robbins, R. T. (2009). Genetic variation of *Hoplolaimus columbus* populations in the United State using PCR-RFLP analysis of nuclear rDNA ITS regions. *Journal of Nematology, 41*, 187–193.

Baker, K. R., & Campbell, C. L. (1981). Sampling nematode populations. In B. M. Zuckerman & R. A. Rhode (Eds.), *Plant parasitic nematodes* (Vol. 3, pp. 451–474). New York: Academic.

Baker, K. R., Carter, C. C., & Sasser, J. N. (1985). *An advanced treatise on Meloidogyne. Volume II: Methodology*. Raleigh: North Carolina State University.

Bell, A. A., Starr, J. L., Jones, J. E., Lemon, R. G., Nichols, R. L., & Overstreet, C. (2009). Nematode resistance and agronomic performance of LONREN and NEMSTACK line. In *Proceedings of the Beltwide Cotton Conferences National Cotton Council* (p. 178). San Antonio: Cordova.

Bell, A. A., Robinson, A. F., Quintana, J., Dighe, N. D., Menz, M. A., Stelly, D. M., Zheng, X. T., Jones, J. E., Overstreet, C., Burris, E., Cantrell, R. G., & Nichols, R. L. (2014). Registration of LONREN-1 and LONREN-2 germplasm lines of upland cotton resistant to reniform nematode. *Journal of Plant Registrations, 8*, 187–190.

Berkeley, M. J. (1855). Vibrio forming cysts on roots of cucumbers. *Gardener's Chronicle and Agricultural Gazette, 14*, 220.

Bessey, E. A. (1911). *Root knot and its control*. Washington, DC: U. S. Department of Agriculture, Bureau of Plant Industry (89 pp.).

Bezawada, C., Saha, S., Jenkins, J. N., Creech, R. G., & McCarty, J. C. (2003). SSR markers(s) associated with root knot nematode resistant gene(s) in cotton. *Journal of Cotton Science, 7*, 179–184.

Bhandari, B., Myers, G. O., Indest, M. O., & Overstreet, C. (2015). Response of five resistant cotton genotypes to isolates of *Rotylenchulus reniformis* collected from reniform infested fields in Louisiana. *Nematropica, 45*, 252–262.

Birchfield, W., & Brister, L. R. (1963). Susceptibility of cotton and relatives to reniform nematode in Louisiana. *Plant Disease Reporter, 47*, 990–992.

Birchfield, W., & Jones, J. E. (1961). Distribution of the reniform nematode in relation to crop failure of cotton in Louisiana. *Plant Disease Reporter, 45*, 671–673.

Birchfield, W., & Martin, W. J. (1956). Pathogenicity on sugarcane and host plant studies of a species of *Tylenchorhynchus*. *Phytopathology, 46*, 277–280.

Birchfield, W., & Martin, W. J. (1965). Effects of reniform nematode populations on sweet potato yields. *Phytopathology, 55*, 497.

Birchfield, W., Hollis, J. P., & Martin, W. J. (1978). *A list of nematodes associated with some Louisiana plants* (pp. 1–22). Baton Rouge: Louisiana State University Agricultural Experiment Station.

Blessitt, J. A., Stetina, S. R., Wallace, T. P., Smith, P. T., & Sciumbato, G. L. (2012). Cotton (*Gossypium hirsutum*) cultivars exhibiting tolerance to the reniform nematode (*Rotylenchulus reniformis*). *International Journal of Agronomy, 2012*, 893178, 8 p. https://doi.org/10.1155/2012/893178.

Boerma, H. R., Hussey, R. S., Wood, E. D., Barrett, G. B., & Finnerty, S. L. (1985). Registration of 'Gordon' soybean. *Crop Science, 25*, 711–712.

Bridge, J. (1996). Nematode management in sustainable and subsistence agriculture. *Annual Review of Phytopathology, 34*, 201–225.

Burris, E., Burns, D., Smith, T. P., Overstreet, C., & Wolcott, M. C. (2009). GIS/GPS techniques help evaluate soil insect and nematode pest control strategies in sweet potatoes. *Louisiana Agriculture, 52*, 28–29.

Carter, C. H., McGawley, E. C., & Russin, J. S. (1995). Reproduction of *Rotylenchulus reniformis* on weed species common to Louisiana soybean fields. *Journal of Nematology, 27*, 494–495.
Chitwood, G. G. (1949). Root knot nematodes – part I. A revision of the genus *Meloidogyne* Geoldi, 1887. *Proceedings of the Helminthological Society of Washington, 16*, 90–104.
Clark, C. A., & Wright, V. L. (1983). Effect and reproduction of *Rotylenchulus reniformis* on sweet potato selections. *Journal of Nematology, 15*, 197–203.
Cook, R., & Evans, K. (1987). Resistance and tolerance. In R. H. Brown & B. R. Kerry (Eds.), *Principles and practices of nematode control in crops* (pp. 179–231). Marrickville: Academic.
Creech, R. G., Jenkins, J. N., Tang, B., Lawrence, G. W., & McCarty, J. C. (1995). Cotton resistance to root knot nematode: I. Penetration and reproduction. *Crop Science, 35*, 365–368.
Cuarezma-Teran, J. A., & Trevathan, L. E. (1985). Effects of *Pratylenchus zeae* and *Quinisulcius acutus* alone and in combination on sorghum. *Journal of Nematology, 17*, 169–174.
Davis, R. F., & Timper, P. (2000). Resistance in selected corn hybrids to *Meloidogyne arenaria* and *M. incognita*. *Journal of Nematology, 32*, 633–640.
Dickerson, O. J., Darling, H. M., & Griffin, G. D. (1964). Pathogenicity and population trends of *Pratylenchulus penetrans* on potato and corn. *Phytopathology, 54*, 317–322.
Endo, B. Y. (1959). Responses of root lesion nematodes, *Pratylenchus brachyurus* and *P. zeae*, to various plants and soil types. *Phytopathology, 49*, 417–421.
Faske, T. R., & Hurd, K. (2015). Sensitivity of *Meloidogyne incognita* and *Rotylenchulus reniformis* to fluopyram. *Journal of Nematology, 47*, 316–321.
Faske, T. R., & Kirkpatrick, T. L. (2015). *Corn diseases and nematodes. Report no. MP 437* (pp. 1–18). Little Rock: University of Arkansas Cooperative Extension Service.
Faske, T. R., & Starr, J. L. (2006). Sensitivity of *Meloidogyne incognita* and *Rotylenchulus reniformis* to abamectin. *Journal of Nematology, 38*, 240–244.
Faske, T. R., & Starr, J. L. (2007). Cotton root protection from plant parasitic nematodes by abamectin-treated seed. *Journal of Nematology, 39*, 27–30.
Faske, T. R., & Starr, J. L. (2009). Mechanism of resistance to *Meloidogyne incognita* in resistant cotton genotypes. *Nematropica, 39*, 281–288.
Faske, T. R., Allen, T. W., Lawrence, G. W., Lawrence, K. S., Mehl, H. L., Norton, R., Overstreet, C., & Wheeler, T. A. (2017). Beltwide nematode research and education committee report on cotton cultivars and nematicides responses in nematode soils, 2016. In *Proceedings of the Beltwide Cotton Conferences National Cotton Council* (pp. 270–273). Cordova.
Fielding, M. J., & Hollis, J. P. (1956). Occurrence of plant parasitic nematode in Louisiana soils. *Plant Disease Reporter, 40*, 403–405.
Greer, A., Wilson, G., & Kirkpatrick, T. L. (2009). *Management of economically important nematodes of Arkansas cotton* (pp. 1–6). Little Rock: University of Arkansas Cooperative Extension Service.
Gutierrez, O. A., Jenkins, J. N., McCarty, J. C., Wubben, M. J., Hayes, R. W., & Callahan, F. E. (2010). SSR markers closely associated with genes for resistance to root knot nematodes on chromosomes 11 and 14 of upland cotton. *Theoretical and Applied Genetics, 121*, 1323–1337.
He, Y., Kumar, P., Shen, X., Davis, R. F., Becelaere, G. V., May, O. L., Nichols, R. L., & Chee, P. W. (2014). Re-evaluation of the inheritance for root knot nematode resistance in the upland cotton germplasm line M-120 RNR revealed two epistatic QTLs conferring resistance. *Theoretical and Applied Genetics, 127*, 1343–1351.
Hartwig, E. E., & Epps, J. M. (1973). Registration of 'Forrest' soybeans. *Crop Science, 13*, 287.
Hartwig, E. E., Young, D. L., & Gibson, P. (1996). Registration of soybean germplasm line D83-3349 resistant to sudden death syndrome, soybean cyst nematode and two root knot nematodes. *Crop Science, 36*, 212.
Henn, R. A., Burdine, B., Main, J. L., & Baldbalian, C. J. (2006). Evaluation of nematicides for management of reniform nematode and sweet potato yield, 2004. *Fungicide and Nematicide Test, 61*, N010.
Herman, M., Hussey, R. S., & Boerma, H. R. (1991). Penetration and development of *Meloidogyne incognita* on roots of resistant soybean genotypes. *Journal of Nematology, 23*, 155–161.

Hollis, J. P. (1967). Nature of the nematode problem in Louisiana rice fields. *Plant Disease Reporter, 51*, 167–169.
Hori, S. (1915). Phytopathological notes. Sick soil of soybean caused by nematode. *Journal of Plant Protection, 2*, 927–930.
Hurd, K., & Faske, T. R. (2017). Reproduction of *Meloidogyne incognita* and *M. graminis* on several grain sorghum hybrids. *Journal of Nematology, 49*, 156–161.
Hurd, K., Faske, T. R., & Emerson, M. (2015). Evaluation of Poncho/VOTiVO and ILeVO for control of root knot nematode on soybean in Arkansas, 2014. *Plant Disease Management Reports, 9*, N017.
Hurd, K., Faske, T. R., & Emerson, M. (2017a). Efficacy of ILeVO to suppress root knot nematodes on soybean in Arkansas, 2016. *Plant Disease Management Reports, 11*, N034.
Hurd, K., Faske, T. R., & Emerson, M. (2017b). Evaluation of ILeVO to suppress root knot nematode on soybean in Arkansas, 2016. *Plant Disease Management Reports, 11*, N035.
Jackson, C. S., Faske, T. R., & Kirkpatrick, T. L. (2017). Assessment of ILeVO for management of *Meloidogyne incognita* on soybean, 2015. In J. Ross (Ed.), *Arkansas soybean research studies 2016* (pp. 74–77). Fayetteville: University of Arkansas Agriculture Experiment Station.
Jatalla, P., & Russell, C. C. (1972). Nature of sweet potato resistance ot *Meloidogyne incognita* and the effects of temperature on parasitism. *Journal of Nematology, 4*, 1–7.
Jenkins, J. N., Creech, R. G., Tang, B., Lawrence, G. W., & McCarty, J. C. (1995). Cotton resistance to root knot nematode: II. Post-penetration development. *Crop Science, 35*, 369–373.
Johnson, A. W., Dowler, C. C., Glaze, N. C., & Handoo, Z. A. (1996). Role of nematodes, nematicides, and crop rotation on the productivity and quality of potato, sweet potato, peanut, and grain sorghum. *Journal of Nematology, 28*, 389–399.
Kirkpatrick, T. L. (2017). Incidence, population density, and distribution of soybean nematodes in Arkansas. In J. Ross (Ed.), *Arkansas soybean research studies 2015* (pp. 72–74). Fayetteville: University of Arkansas Agriculture Experiment Station.
Kirkpatrick, T. L., & Sasser, J. N. (1984). Crop rotation and races of *Meloidogyne incognita* in cotton root knot management. *Journal of Nematology, 16*, 323–328.
Kirkpatrick, T., & Thomas, A. C. (2007). *Crop rotation for management of nematodes in cotton and soybean*. Little Rock: University of Arkansas Cooperative Extension Service.
Kirkpatrick, T. L., Faske, T. R., & Robbins, B. (2014). *Nematode management, Report no. MP 197* (pp. 1–6). Little Rock: University of Arkansas Cooperative Extension Service.
Kirkpatrick, T. L., Rowe, K., Faske, T. R., & Emerson, M. (2016). Comprehensive disease screening of soybean varieties in Arkansas. In J. Ross (Ed.), *Arkansas soybean research studies 2014, Research Series 631* (pp. 50–51). Fayetteville: University of Arkansas Agriculture Experiment Station.
Koenning, S. R., Overstreet, C., Noling, J. W., Donald, P. A., Boecker, J. O., & Fortnum, B. A. (1999). Survey of crop losses in response to phytoparasitic nematodes in the United States for 1994. *Supplement to the Journal of Nematology, 31*, 587–618.
Lawrence, G. W., & McGuire, J. M. (1987). Influence of soybean cultivar rotation sequences on race development of *Heterodera glycines*, race 3. *Phytopathology, 77*, 1714.
Lawrence, G. W., & McLean, K. S. (2000). Effect of foliar application of oxamyl with aldicarb for the management of *Rotylenchulus reniformis* on cotton. *Journal of Nematology, 32*, 542–549.
Lawrence, G. W., & McLean, K. S. (2001). Reniform nematode. In T. L. Kirkpatrick & C. S. Rothrock (Eds.), *Compendium of cotton diseases* (pp. 42–44). St. Paul: APS Press.
Lawrence, G. W., & McLean, K. S. (2002). Foliar application of oxamyl with aldicarb for the management of *Meloidogyne incognit*a on cotton. *Nematropica, 32*, 103–112.
Lawrence, G. W., Clark, C. A., & Wright, V. L. (1986). Influence of *Meloidogyne incognita* on resistant and susceptible sweet potato cultivars. *Journal of Nematology, 18*, 59–65.
Lawrence, G. W., McLean, K. S., Batson, W. E., Miller, D., & Bordon, J. C. (1990). Response of *Rotylenchulus reniformis* to nematicide application on cotton. *Journal of Nematology, 22*, 707–711.

Lawrence, K., Olsen, M., Faske, T., Hutmacher, R., Muller, J., Marios, J., Kemerait, R., Overstreet, C., Price, P., Sciumbato, G., Lawrence, G., Atwell, S., Thomas, S., Koenning, S., Boman, R., Young, H., Woodward, J., & Mehl, H. (2015a). Cotton disease loss estimates committee report, 2014. In *Proceedings of the Beltwide Cotton Conferences National Cotton Council* (pp. 188–190). San Antonio: Cordova.

Lawrence, K. S., Haung, P., Lawrence, G. W., Faske, T. R., Overstreet, C., Wheeler, T. A., Young, H., Kemerait, B., & Mehl, H. L. (2015b). Beltwide nematode reserach and education committee 2014 nematode research cotton varietal and nematicide responses in nematode soils. In *Proceedings of the Beltwide Cotton Conferences National Cotton Council* (pp. 739–742). Cordova.

Lawrence, K., Hagan, A., Olsen, M., Faske, T., Hutmacher, R., Muller, J., Wright, D., Kemerait, R., Overstreet, C., Price, P., Lawrence, G., Allen, T., Atwell, S., Thomas, S., Edmisten, K., Boman, R., Young, H., Woodward, J., & Mehl, H. (2016). Cotton disease loss estimates committee report, 2015. In *Proceedings of the Beltwide Cotton Conferences National Cotton Council* (pp. 113–115). Cordova.

Lawrence, K., Hagan, A., Norton, R., Faske, T., Hutmacher, R., Muller, J., Wright, D., Small, I., Kemerait, R., Overstreet, C., Price, P., Lawrence, G., Allen, T., Atwell, S., Jones, A., Thomas, S., Goldberg, N., Boman, R., Goodson, J., Kelly, H., Woodward, J., & Mehl, H. (2017). Cotton disease loss estimates committee report, 2016. In *Proceedings of the Beltwide Cotton Conferences National Cotton Council* (pp. 150–152). Cordova.

Lee, H. K., Lawrence, G. W., DuBien, J. L., & Lawrence, K. S. (2015). Seasonal variation and cotton-corn rotation in the spatial distribution of *Rotylenchulus reniformis* in Mississippi cotton soils. *Nematropica, 45*, 72–81.

Linford, M. B., & Oliveira, J. M. (1940). *Rotylenchulus reniformis*, nov. gen. n. sp., a nematode parasite of roots. *Proceeding of the Helminthological Society of Washington, 7*, 35–42.

Luzzi, B. M., Boerma, H. R., & Hussey, R. S. (1987). Resistance to three species of root-knot nematode in soybean. *Crop Science*, 258–262.

Luzzi, B. M., Boerma, H. R., & Hussey, R. S. (1994a). Inheritance of resistance to the southern root-knot nematode in soybean. *Crop Science, 34*, 1240–1243.

Luzzi, B. M., Boerma, H. R., & Hussey, R. S. (1994b). A gene for resistance to the southern root-knot nematode in soybean. *Journal of Heredity, 86*, 484–486.

Luzzi, B. M., Boerma, H. R., Hussey, R. S., Phillips, D. V., Tamulonis, J. P., Finnerty, S. L., & Wood, E. D. (1996). Registration of southern root-knot nematode resistant soybean germplasm line G93-9009. *Crop Registrations, 36*, 823.

Martin, W. J., & Birchfield, W. (1955). Notes on plant parasitic nematodes in Louisiana. *Plant Disease Reporter, 39*, 3–4.

McCarty, J. C., Jenkins, J. N., Wubben, M. J., Hayes, R. W., & LaFoe, J. M. (2012). Registration of three germplasm lines of cotton derived from *Gossypium hirsutum* L. accession T2468 with moderate resistance to the reniform nematode. *Journal of Plant Registrations, 6*, 85–87.

McCarty, J. C., Jenkins, J. N., Wubben, M. J., Gutierrez, O. A., Hayes, R. W., Callahan, F. E., & Deng, D. (2013). Registration of three germplasm lines of cotton derived from *Gossypium barbadense* L. accession GB713 with resistance to the reniform nematode. *Journal of Plant Registrations, 7*, 220–223.

McGawley, E. C., & Overstreet, C. (2015). Reniform nematode. In G. L. Hartman, J. C. Rupe, E. J. Sikora, L. L. Domier, E. L. Davis, & K. L. Steffey (Eds.), *Compendium of soybean diseases* (pp. 96–98). St. Paul: APS Press.

McGawley, E. C., Pontif, M. J., & Overstreet, C. (2010). Variation in reproduction and pathogenicity of geographic isolates of *Rotylenchulus reniformis* on cotton. *Nematropica, 40*, 275–288.

McGawley, E. C., Overstreet, C., & Pontif, M. J. (2011). Variation in reproduction and pathogenicity of geographic isolates of *Rotylenchulus reniformis* on soybean. *Nematropica, 41*, 12–22.

McLean, K. S., & Lawrence, G. W. (2000). A survey of plant parasitic nematodes associated with cotton in Northeastern Louisiana. *Journal of Nematology, 32*, 508–512.

McLean, K. S., & Lawrence, G. W. (2003). Efficacy of aldicarb to *Rotylenchulus reniformis* and biodegradation in cotton field soils. *Journal of Nematology, 35*, 65–72.

McPherson, M. G., Jenkins, J. N., Watson, C. E., & McCarty, J. C. (2004). Inheritance of root knot nematode resistance in M-315 RNR and M78-RNR cotton. *Journal of Cotton Science, 8*, 154–161.

Minton, N. A., & Parker, M. B. (1987). Root knot nematode management and yield of soybean as affected by winter cover crops, tillage systems, and nematicides. *Journal of Nematology, 19*, 38–43.

Molin, W. T., & Stetina, S. R. (2016). Weed hosts and relative weed and cover crop susceptibility to *Rotylenchulus reniformis* in the Mississippi delta. *Nematropica, 46*, 121–131.

Monfort, W. S., Kirkpatrick, T. L., Long, D. L., & Rideout, S. (2006). Efficacy of a novel nematicidal seed treatment against *Meloidogyne incognita* on cotton. *Journal of Nematology, 38*, 245–249.

Monfort, W. S., Kirkpatrick, T. L., Rothrock, C. S., & Mauromoustakos, A. (2007). Potential for site-specific management of *Meloidogyne incognita* in cotton using soil textural zones. *Journal of Nematology, 39*, 1–8.

Monfort, W. S., Kirkpatrick, T. L., & Mauromoustakos, A. (2008). Spread of *Rotylenchulus reniformis* in an Arkansas cotton field over a four-year period. *Journal of Nematology, 40*, 161–166.

Moura, R. M., Davis, E. L., Luzzi, B. M., Boerma, H. R., & Hussey, R. S. (1993). Post-infectinal development of *Meloidogyne incognita* on susceptible and resistant soybean genotypes. *Nematropica, 23*, 7–13.

Mueller, J., Kirkpatrick, T., Overstreet, C., Koenning, S., Kemerait, B., & Nichols, B. (2012). *Managing nematodes in cotton-based cropping systems*. Cary: Cotton Incorporated.

Mueller, D., Wise, K. A., Sisson, A. J., Allen, T. A., Bergstrom, G. C., Bosley, D. B., Bradley, C. A., Broders, K. D., Byamukama, E., Chilvers, M. I., Collins, A., Faske, T. R., Friskop, A. J., Heiniger, R. W., Hollier, C. A., Hooker, D. C., Isakeit, T., Jackson-Ziems, T. A., Jardine, D. J., Kelly, H. M., Kinzer, K., Koenning, S. R., Malvick, D. K., McMullen, M., Meyer, R. F., Paul, P. A., Robertson, A. E., Roth, G. W., Smith, D. L., Tande, C. A., Tenuta, A. U., Vincelli, P., & Warner, F. (2016). Corn yield loss estimates due to diseases in the United States and Ontario, Canada from 2012 to 2015. *Plant Health Progress, 17*, 211–222.

NCCB. (2017). *Cotton disease loss estimate committee. Cotton pest loss data: Disease database*. National Cotton Council of Americal. Cordova. Accessed 13 Oct 2017. Online: http://www.cotton.org/tech/pest/index.cfm

Neal, J. C. (1889). *The root knot disease of the peach, orange and other plants in Florida, due to the work of Anguillula* (pp. 1–51). Washington, DC: Division of Entomology, US Department of Agriculture.

Niblack, T. L., & Riggs, R. D. (2015). Soybean cyst nematode. In G. L. Heartman, J. C. Rupe, E. J. Sikora, L. L. Domier, J. A. Davis, & K. L. Steffey (Eds.), *Compendium of soybean diseases and pest* (pp. 100–104). St. Paul: APS Press.

Niblack, T. L., Arelli, P. R., Noel, G. R., Opperman, C. H., Orf, J. H., Schmitt, D. P., Shannon, J. G., & Tylka, G. L. (2002). A revised classification scheme for genetically diverse populations of *Heterodera glycines*. *Journal of Nematology, 34*, 279–288.

Noel, G. R. (1992). History, distribution, and economics. In R. D. Riggs & J. A. Wrather (Eds.), *Biology and management of the soybean cyst nematode* (pp. 1–13). St. Paul: APS Press.

Norton, D. C., Donald, P., Kimpinski, J., Myers, R. F., Noel, G. R., Noffsinger, E. M., Robbins, R. T., Schmitt, D. P., Sosa-Moss, C., & Vrain, T. C. (1984). *Distribution of plant parasitic nematode species in North America*. Hyattsville: Society of Nematologists.

Ortiz, B. V., Perry, C., Sullivan, D., Lu, P., Kemerait, R., Davis, R. F., Smith, A., Vellidis, G., & Nichols, R. (2012). Variable rate application of nematicides on cotton fields: A promising site-specific management strategy. *Journal of Nematology, 44*, 31–39.

Overstreet, C. (1996). Impact of reniform nematode on cotton production in the U.S.A. *Nematropica, 26*, 216.

Overstreet, C. (2009). Nematodes. In G. Loebenstein & G. Thottappilly (Eds.), *The sweet potato* (pp. 131–155). Dordrecht: Springer.

Overstreet, C. (2013a). Root knot nematode. In C. A. Clark, D. M. Ferrin, T. P. Smith, & G. J. Holmes (Eds.), *Compendium of sweetpotato diseases, pest, and disorders* (pp. 63–67). St. Paul: APS Press.

Overstreet, C. (2013b). Reniform nematode. In C. Clark, D. M. Ferrin, T. P. Smith, & G. J. Holmes (Eds.), *Compendium of sweetpotato diseases, pest, and disorders* (pp. 67–69). St. Paul: APS Press.

Overstreet, C., & McGawley, E. C. (1994). Cotton production and *Rotylenchulus reniformis* in Louisiana. *Journal of Nematology, 26*, 562–563.

Overstreet, C., Whitman, K., & McGawley, E. C. (1992). *Soybean nematodes, Report No. 2147*. Baton Rouge: Louisiana State University, Cooperative Extension Service.

Overstreet, C., McGawley, E. C., & Padgett, B. (2002). Current management strategies employed against the reniform nematode (*Rotylenchulus reniformis*) in cotton production in Louisiana, U.S.A. *Nematology, 4*, 306–307.

Overstreet, C., McGawley, E. C., & Ferrin, D. (2008). *Reniform nematode in Louisiana, Report No. 3095*. Baton Rouge: Louisiana State University, Cooperative Extension Service.

Overstreet, C., Wolcott, M., Burris, E., & Burns, D. (2009). Management zones for cotton nematodes. In *Management zones for cotton nematode National Cotton Council* (pp. 167–176). San Antonio: Cordova.

Overstreet, C., McGawley, E. C., Burns, D., & Wolcott, M. (2010). Using apparent electrical conductivity and verification strips to define nematode management zones in cotton. *Journal of Nematology, 42*, 261–262.

Overstreet, C., McGawley, E. C., & Burns, D. (2011a). Management zone development in cotton against concomitant infestation with *Meloidogyne incognita* and *Rotylenchulus reniformis* for the site-specific application of nematicides. *Nematropica, 41*, 318.

Overstreet, C., McGawley, E. C., Burns, D., & Frazier, R. L. (2011b). Edaphic factors involved in the delineation of management zones for *Rotylenchulus reniformis*. *Journal of Nematology, 43*, 268.

Overstreet, C., McGawley, E. C., Khalilian, A., Kirkpatrick, T. L., Monfort, W. S., Henderson, W., & Mueller, J. D. (2014). Site specific nematode management – development and success in cotton production in the United States. *Journal of Nematology, 46*, 309–320.

Ploeg, A. T., & Maris, P. C. (1999). Effects of temperature on the duration of the life cycle of a *Meloidogyne incognita* population. *Nematology, 1*, 389–393.

Pontif, M. J., & McGawley, E. C. (2007). The influence of morningglory (*Ipomoea lacunosa*), hemp sesbania (*Sesbania exaltata*), and johnsongrass (*Sorghum halepense*) on reproduction of *Rotylenchulus reniformis* on cotton (*Gossypium hirsutum*) and soybean (*Glycine max*). *Nematropica, 37*, 295–305.

Pontif, M. J., & McGawley, E. C. (2008). The influence of leachates from roots of morningglory (*Ipomoea lacunosa*), hemp sesbania (*Sesbania exaltata*), and johnsongrass (*Sorghum halepense*) on eclosion and hatching of eggs of *Rotylenchulus reniformis*. *Nematropica, 38*, 23–35.

Rebois, R. V., & Golden, A. M. (1978). Nematode occurrences in soybean fields in Mississippi and Louisiana. *Plant Disease Reporter, 62*, 433–437.

Rebois, R. V., Johnson, W. C., & Cairns, E. J. (1968). Resistance in soybeans, *Glycine max* L., Merr., to the reniform nematode. *Crop Science, 8*, 394–395.

Rebois, R. V., Epps, J. M., & Hartwig, E. E. (1970). Correlation of resistance in soybeans to *Heterodera glycines* and *Rotylenchulus reniformis*. *Phytopathology, 60*, 695–700.

Riggs, R. D. (1977). Worldwide distribution of soybean-cyst nematode and its economic importance. *Journal of Nematology, 9*, 34–39.

Riggs, R. D. (1992). Host range. In R. D. Riggs & J. A. Wrather (Eds.), *Biology and management of the soybean cyst nematode* (pp. 107–114). St Paul: APS Press.

Riggs, R. D., & Schmitt, D. P. (1988). Complete characterization of the race scheme for *Heterodera glycines*. *Journal of Nematology, 20*, 392–395.

Riggs, R. D., Wrather, J. A., Mauromoustakos, A., & Rakes, L. (2000). Planting date and soybean cultivar maturity group affect population dynamics of *Heterodera glycines*, and all affect yeild of soybean. *Journal of Nematology, 32*, 334–342.

Robbins, R. T. (1982a). Phytoparasitic nematodes associated with soybean in Arkansas. *Journal of Nematology, 14*, 466.

Robbins, R. T. (1982b). Phytoparasitic nematodes of noncultivated habitats in Arkansas. *Journal of Nematology, 14*, 466–467.

Robbins, R. T., & Rakes, L. (1996). Resistance to the reniform nematode in selected soybean cultivars and germplasm lines. *Journal of Nematology, 28*, 612–615.

Robbins, R. T., Riggs, R. D., & Von Steen, D. (1987). Results of annual phytoparasitic nematode surveys of Arkansas soybean fields, 1978–1986. *Journal of Nematology, 19*, 50–55.

Robbins, R. T., Riggs, R. D., & von Steen, D. (1989). Phytoparasitic nematode surveys of Arkansas cotton fields, 1986–88. *Journal of Nematology, 21*, 619–623.

Robbins, R. T., Rakes, L., & Elkins, C. R. (1994). Reniform nematode reproduction and soybean yield of four soybean cultivars in Arkansas. *Journal of Nematology, 26*, 654–658.

Robbins, R. T., Chen, P., Shannon, G., Kantartzi, S., Li, Z., Faske, T., Velie, J., Jackson, L., Gbur, E., & Dombek, D. (2016). Reniform nematode reproduction on soybean cultivars and breeding lines in 2015. In *Proceedings of the Beltwide Cotton Conferences National Cotton Council* (pp. 131–143). Cordova.

Robbins, R. T., Arelli, P., Chen, P., Shannon, G., Kantartzi, S., Fallen, B., Li, Z., Faske, T., Velie, J., Gbur, E., Dombek, D., & Crippen, D. (2017). Reniform nematode reproduction on soybean cultivars and breeding lines in 2016. In *Proceedings of the Beltwide Cotton Conferences National Cotton Council* (pp. 184–197). Cordova.

Robinson, A. F. (2008). Nematode management in cotton. In A. Ciancio & K. G. Mukerji (Eds.), *Integrated management of biocontrol of vegetable and grain crops nematodes* (pp. 149–182). New York: Springerlink.

Robinson, A. F., & Percival, A. E. (1997). Resistance to *Meloidogyne incognita* race 3 and *Rotylenchulus reniformis* in wild accession of *Gossypium hirsutum* and *G. barbadense* from Mexico. *Journal of Nematology, 29*, 746–755.

Robinson, A. F., Inserra, R. N., Caswell-Chen, E. P., Vivlas, N., & Troccoli, A. (1997). *Rotylenchulus* species: Identification, distribution, host ranges, and crop plant resistance. *Nematropica, 27*, 127–180.

Robinson, A. F., Bowman, D. T., Cook, C. G., Jenkins, J. N., Jones, J. E., May, L. O., Oakley, S. R., Oliver, M. J., Roberts, P. A., Robinson, M., Smith, C. W., Starr, J. L., & Stewart, J. M. (2001). Nematode resistance. In T. L. Kirkpatrick & C. S. Rothrock (Eds.), *Compendium of cotton diseases* (pp. 68–72). St. Paul: APS Press.

Robinson, A. F., Bridges, A. C., & Percival, A. E. (2004). New sources of resistance to the reniform (*Rotylenchulus reniformis*) and root knot (*Meloidogyne incognita*) nematode in upland (*Gossypium hirsutum*) and sea island (*G. barbadense*) cotton. *Journal of Cotton Science, 8*, 191–197.

Robinson, A. F., Akridge, R., Bradford, J. M., Cook, C. G., Gazaway, W. S., Kirkpatrick, T. L., Lawrence, G. W., Lee, G., McGawley, E. C., Overstreet, C., Padgett, B., Rodríguez-Kábana, R., Westphal, A., & Young, L. D. (2005). Vertical distribution of *Rotylenchulus reniformis* in cotton fields. *Journal of Nematology, 37*, 265–271.

Robinson, A. F., Akridge, R., Bradford, J. M., Cook, C. G., Gazaway, W. S., McGawley, E. C., Starr, J. L., & Young, D. L. (2006). Suppression of *Rotylenchulus reniformis* 122-cm deep endorses resistance introgression in *Gossypium*. *Journal of Nematology, 38*, 195–209.

Ross, J. P. (1988). Seasonal variation of larval emergence from cysts of the soybean cyst nematode, *Heterodera glycines*. *Phytopathology, 53*, 608–609.

Sankaralingam, A., & McGawley, E. C. (1994a). Interrelationships of *Rotylenchulus reniformis* with *Rhizoctonia solani* on cotton. *Journal of Nematology, 26*, 475–485.

Sankaralingam, A., & McGawley, E. C. (1994b). Influence of *Rhizoctonia solani* on egg hatching and infectivity of *Rotylenchulus reniformis*. *Journal of Nematology, 26*, 486–491.
Shen, X., Bacelaere, G. V., Kumar, P., Davis, R. F., May, O. L., & Chee, P. W. (2006). QTL mapping for resistance to root knot nematode in teh M-120 RNR upland cotton line (*Gossympium hirsutum* L.) of the Auburn 623 RNR source. *Theoretical and Applied Genetics, 113*, 1539–1549.
Shepherd, R. L. (1974). Registration of Auburn 623 RNR cotton germplasm. *Crop Science, 14*, 911.
Shepherd, R. L. (1982). Registration of three germplasm lines of cotton. *Crop Science, 22*, 692.
Shepherd, R. L., McCarty, J. C., Jenkins, J. N., & Parrott, W. L. (1996). Registration of nine cotton germplasm lines resistant to root knot nematode. *Crop Science, 36*, 820.
Smith, A. L., & Taylor, A. L. (1941). Nematode distribution in the 1940 regional cotton-wilt plots. *Phytopathology, 31*, 771.
Smith, T. P., Overstreet, C., Clark, C., Ferrin, D., & Burris, E. (2008). *Nematode management: Louisiana sweet potato production, Report No. 3075*. Baton Rouge: Louisiana State University AgCenter.
Stetina, S. R., Russin, J. S., & McGawley, E. C. (1997a). Replacement series: A tool for characterizing competition between phytoparasitic nematodes. *Journal of Nematology, 29*, 35–42.
Stetina, S. R., McGawley, E. C., & Russin, J. S. (1997b). Relationship between *Meloidogyne incognita* and *Rotylenchulus reniformis* as influenced by soybean genotype. *Journal of Nematology, 29*, 395–403.
Stetina, S. R., Young, D. L., Pettigrew, W. T., & Bruns, H. A. (2007). Effect of corn-cotton rotations on reniform nematode populations and crop yield. *Nematropica, 37*, 237–248.
Stetina, S. R., Sciumbato, G., Young, G. L., & Blessitt, J. A. (2009). Cotton cultivars evaluated for tolerance to reniform nematode. *Plant Health Progress*. https://doi.org/10.1094/PHP-2009-0312-01-RS.
Stetina, S. R., Smith, J. R., & Ray, J. D. (2014). Identification of *Rotylenchulus reniformis* resistant Glycine lines. *Journal of Nematology, 46*, 1–7.
Thomas, R. J., & Clark, C. A. (1983). Population dynamics of *Meloidogyne incognita* and *Rotylenchulus reniformis* alone and in combination, and their effects on sweet potato. *Journal of Nematology, 15*, 204–211.
Thomas, S. H., & Kirkpatrick, T. L. (2001). Root knot nematodes. In T. L. Kirkpatrick & C. S. Rothrock (Eds.), *Compendium of cotton diseases* (pp. 40–42). St. Paul: APS Press.
Timper, P., & Hanna, W. W. (2005). Reproduction of *Belonolaimus longicaudatus*, *Meloidogyne javanica*, *Paratrichodorus minor*, and *Pratylenchus brachyurus* on pearl millet (*Pennisetum glaucum*). *Journal of Nematology, 37*, 214–219.
Timper, P., Davis, R. F., & Tillman, P. G. (2006). Reproduction of *Meloidogyne incognita* on winter cover crops used in cotton production. *Journal of Nematology, 38*, 83–89.
Timper, P., Krakowsky, M. D., & Snook, M. E. (2007). Resistance in maize to *Paratrichodorus minor*. *Nematropica, 37*, 9–20.
Todd, T. C. (2016). Stubby-root nematodes. In G. P. Munkvold & D. G. White (Eds.), *Compendium of corn diseases* (pp. 124–125). St. Paul: APS Press.
USDA-NASS. (2016a). *Mississippi agriculture overview*. United States Department of Agriculture – National Agricultural Statistics Service. Washington, DC. Accessed 11 Oct 2017. Online: https://www.nass.usda.gov/Quick_Stats/Ag_Overview/stateOverview.php?state=MISSISSIPPI
USDA-NASS. (2016b). *Arkansas agriculture overview*. United States Department of Agriculture – National Agricultural Statistics Service. Washington, DC. Accessed 17 July 2017. Online: https://www.nass.usda.gov/Quick_Stats/Ag_Overview/stateOverview.php?state=ARKANSAS
USDA-NASS. (2016c). *Louisiana agriculture overview*. United States Department of Agriculture – National Agricultural Statistics Service. Washington, DC. Accessed 11 Oct 2017. Online: https://www.nass.usda.gov/Quick_Stats/Ag_Overview/stateOverview.php?state=LOUISIANA

USDA-NASS. (2016d). *CropScape – Cropland data layer*. United States Department of Agriculture, National Agricultural Statistics Service. Washington, DC. Accessed 10 Aug 2017. Online: https://nassgeodata.gmu.edu/CropScape/

Wang, K., Riggs, R. D., & Crippen, D. (2004). Suppression of *Rotylenchulus reniformis* on cotton by the nematophagous fungus ARF. *Journal of Nematology, 36*, 186–191.

Wehunt, E. J., Golden, A. M., & Robbins, R. T. (1989). Plant nematodes occurring in Arkansas. *Journal of Nematology, 21*, 677–681.

Wenfrida, I., Russin, J. S., & McGawley, E. C. (1998). Competition between *Tylenchorhynchus annulatus* and *Mesocriconema xenoplax* on grain sorghum as influenced by *Macrophomina phaselolina*. *Journal of Nematology, 30*, 423–430.

Winstead, N. N., Skotland, C. B., & Sasser, J. N. (1955). Soybean-cyst nematode in North Carolina. *Plant Disease Reporter, 39*, 9–11.

Wrather, J. A., Anand, S. C., & Koenning, S. R. (1992). Management by cultural practices. In R. D. Riggs & J. A. Wrather (Eds.), *Biology and management of the soybean cyst nematode* (pp. 125–131). St. Paul: APS Press.

Wrather, J. A., Stevens, W. E., Kirkpatrick, T. L., & Kitchen, N. R. (2002). Effects of site-specific application of aldicarb on cotton in a *Meloidogyne incognita*-infested field. *Journal of Nematology, 34*, 115–119.

Xavier, D. M., Overstreet, C., Kularathna, M., & Martin, C. M. (2012a). Population development of *Rotylenchulus reniformis* in a field over a nine year period. *Nematropica, 42*, 390–391.

Xavier, D. M., Overstreet, C., Kularathna, M. T., Burns, D., & Frazier, R. L. (2012b). Reniform nematode (*Rotylenchulus reniformis*) development across the variable soil texture in a Commerce silt loam field. In *Proceeding of the Beltwide Cotton Conference National Cotton Council* (pp. 242–246). Cordova.

Xavier, D. M., Overstreet, C., McGawley, E. C., Kularathna, M. T., Burns, D., Frazier, R. L., & Martin, C. M. (2012c). Population development of *Rotylenchulus reniformis* in different soil textures within a commerce silt loam field. *Journal of Nematology, 44*, 498.

Xavier, D. M., Overstreet, C., McGawley, E. C., Kularathna, M. T., & Martin, C. M. (2014). The influence of soil texture on reproduction and pathogenicity of *Rotylenchulus reniformis* on cotton. *Nematropica, 44*, 7–14.

Yik, C. P., & Birchfield, W. (1984). Resistant germplasm in *Gossypium* species and related plants to *Rotylenchulus reniformis*. *Journal of Nematology, 16*, 146–153.

Ynturi, P., Jenkins, J. N., McCarty, J. C., Gutierrez, O. A., & Saha, S. (2006). Association of root knot nematode resistance genes with simple sequence repeat markers on two chromosomes in cotton. *Crop Science, 46*, 2670–2674.

Chapter 16
Plant Parasitic Nematodes of Economic Importance in Texas and Oklahoma

Terry A. Wheeler, Jason E. Woodward, and Nathan R. Walker

16.1 Agricultural Crops of Economic Importance in Texas and Oklahoma

The region of Texas and Oklahoma produce a wide range of crops that are grown under a diversity of environments. There were over 66 million hectares involved with farm operations in these two states during 2016. Some of the most economically valuable field crops include cotton ($2.1 billion [value of production or sales] on 1,910,116 ha), hay ($1.37 billion on 3,136,313 ha), corn ($1.26 billion on 910,543 ha), wheat ($973 million on 2,974,439 ha), sorghum ($645 million on 1,157,401 ha), soybean ($127 million on 198,296 ha), peanuts ($120 million on 71,629 ha) and rice ($111 million on 52,609 ha) (Table 16.1). The highest valued vegetables grown in this region include potato ($107 million), cabbage ($31 million), onions ($20 million) and chili peppers ($14 million). Fruits grown in this region include melons ($69 million), grapefruit ($39 million), grapes ($18 million), oranges ($17 million) and peaches ($8 million). There is also a large industry for bedding plants (annuals, $136 million; perennials, $32 million), indoor flowering ($27 million) and foliage plants ($10 million).

T. A. Wheeler (✉)
Texas A&M AgriLife Research, Lubbock, TX, USA
e-mail: ta-wheeler@tamu.edu

J. E. Woodward
Texas A&M AgriLife Extension Service, Lubbock, TX, USA
e-mail: jewoodward@ag.tamu.edu

N. R. Walker
Entomology and Plant Pathology, Oklahoma State University, Stillwater, OK, USA
e-mail: nathan.walker@okstate.edu

S. A. Subbotin, J. J. Chitambar (eds.), *Plant Parasitic Nematodes in Sustainable Agriculture of North America*, Sustainability in Plant and Crop Protection,
https://doi.org/10.1007/978-3-319-99588-5_16

Table 16.1 Rank of select commodities in Oklahoma and Texas by value of production in 2015[a]

Commodity	Value ($) of Production × 1000		Percent of US[b]
	OK	TX	
Cotton	126,546	1,992,429	53.1
Hay	515,320	858,921	8.3
Corn, grain	141,952	1,116,990	2.6
Wheat	471,276	501,615	9.7
Sorghum, grain	73,307	571,616	31.2
Bedding plants, annual		136,112	21.8*
Soybean	102,300	25,116	0.4
Peanuts	6,518	112,992	10.3
Rice		111,154	4.6
Potatoes		106,922	2.7*
Pecans	20,770	73,860	18.3*
Melons	7,290	61,577	8.1*
Grapefruit		38,557	15.6*
Bedding plants, perennial		32,118	5.7*
Cabbage		30,855	6.8*
Flowering plants (indoor)		27,165	3.4*
Sunflower	1,295	21,052	3.9
Canola	20,845		4.6
Onions		19,680	2.1*
Sugarcane		18,768	1.9
Grapes		18,260	0.3*
Rye	17,646		23.4
Oranges		16,509	0.7*
Peppers, chili		14,335	0.2*
Beans		12,779	1.5
Oats	669	10,428	5.2
Squash		10,260	5.4*
Foliage plants (indoor)		10,213	1.9*
Cucumbers		9,122	5.4*
Peaches		8,460	1.3*
Sweet corn		7,138	2.5*
Spinach		5,523	2.1*
Tomatoes		4,860	0.4*

[a]USDA/NASS 2015 State Agriculture Overview for Texas and Oklahoma
[b]Commodities with a * had the % of US value calculated based on total production in the US in 2014

Water stress is a major limiting factor for crop production in the western part of Texas and Oklahoma. In Texas, annual rainfall in the southeastern part of the state, near Houston, averages approximately 129.5 cm, whereas, totals in the northcentral region, near Dallas, average approximately 91.4 cm (Fig. 16.1a). Rainfall amounts for western areas are far less, averaging 48.3 and 22.9 cm for Lubbock (High Plains area) and El Paso (Trans Pecos area) respectively. The eastern part of Oklahoma

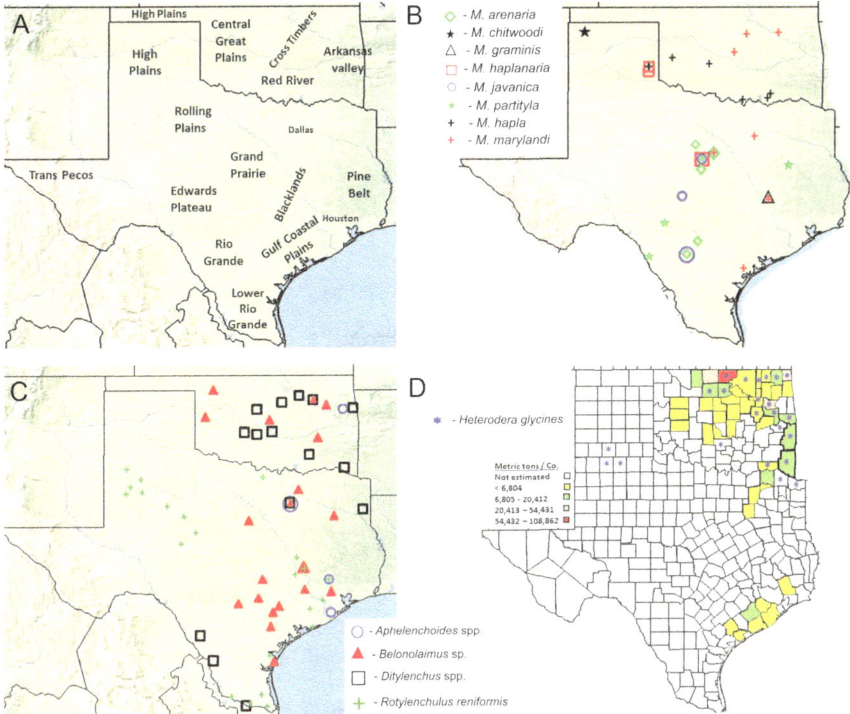

Fig. 16.1 (**a**) Map of geographic regions in Texas and Oklahoma; (**b**) Distribution of root knot nematodes; (**c**) Locations of counties found positive for the foliar nematode *Aphelenchoides* spp. sting nematode *Belonolaimus* sp., stem and bulb nematode, *Ditylenchus* spp. and reniform nematode *Rotylenchulus reniformis*; (**d**) Map of soybean production in 2016 as determined by the U.S. Department of Agriculture National Agricultural Statistics Service. Location of counties that have tested positive for *Heterodera glycines*

averages around 99–119 cm of rain (Arkansas Valley), decreasing to 76–97 cm in the Central Great Plains and down to 38–51 cm in the High Plains.

The most important plant parasitic nematodes associated with crop production in the States of Texas and Oklahoma are the root knot nematodes (*Meloidogyne* spp.). The species of importance can vary depending on the crop. *Meloidogyne incognita* is the root knot nematode species that attacks cotton (*Gossypium hirsutum*), whereas, *M. arenaria*, is the most important species affecting peanut (*Arachis hypogaea*) in Texas and *M. hapla* is the most important species on peanut in Oklahoma. While several root knot nematode species can be found on potato (*Solanum tuberosum*), *M. hapla* is most frequently found in Texas. There are other root knot nematode species that are also present in the region including *M. partityla* on pecan (*Carya illinoinensis*), *M. graminis* on grass species, *M. marylandi* on turf, *M. javanica* and *M. haplanaria* on peanut, *M. chitwoodi* on potato and various species of root knot nematode on soybean. Other nematode species that can be highly damaging, depending on crop and location, include citrus nematode (*Tylenchulus semipenetrans*) on citrus in the Lower Rio Grande Valley, reniform nematode

(*Rotylenchulus reniformis*) on cotton in Texas, soybean cyst nematode (*Heterodera glycines*) on soybean in Oklahoma, sting nematode (*Belonolaimus longicaudatus*) on several crops in both states, stem and bulb nematode (*Ditylenchus dipsaci*) primarily on alfalfa in Oklahoma and the foliar nematode (*Aphlenchoides besseyi*) on rice and ornamentals, bedding plants and indoor flowering plants. The distribution and importance of these nematodes in Texas and Oklahoma are discussed below.

16.2 Plant Parasitic Nematodes That Are Economically Important in Texas and Oklahoma

16.2.1 Root Knot Nematodes, Meloidogyne *spp.*

16.2.1.1 Cotton

The southern root knot nematode, *Meloidogyne incognita*, is found throughout Texas and Oklahoma. Cotton is the most valuable crop in this region, and the highest concentration of cotton is produced in the southern part of the High Plains of Texas (Fig. 16.1a). Approximately 40–50% of the cotton in this area is planted in coarse-textured soils and is infested with *M. incognita* (Starr et al. 1993; Wheeler et al. 2000). Root knot nematode is rarely found in soils with a clay component greater than 40% (Starr et al. 1993). In Oklahoma, *M. incognita* can be found throughout the south and southwestern regions where cotton is grown.

Damage to cotton caused by *M. incognita* in the southern part of the High Plains (in the absence of any management) was estimated at 26% average yield loss, based on 80 field trials conducted over 16 years (Orr and Robinson 1984). Root knot nematodes typically form galls on roots and, consequently, cause plants to have shorter root systems. Interactions with fungi such as *Thielaviopsis basicola,* which is also common in this region, (Walker et al. 1998; Wheeler et al. 2000) can also limit root growth (Ma et al. 2014). This reduction in root length has been associated with a corresponding reduction in water transported through roots (Dorhout et al. 1991). Cotton yields are often limited by insufficient amounts of water in soil. Roots that are inefficient or smaller due to root knot nematode infections will place further stress on plants. However, drier conditions may also inhibit hatching of *M. incognita* eggs and limit movement of second-stage juveniles in soil.

Cotton producers take definite steps to manage root knot nematode, however, management options must fit into their overall production systems. For example, wind erosion is a significant problem in the Southern High Plains and a cover crop of wheat or rye can be planted in the fall or winter to keep the soil from blowing away. Other cover crop species are not often utilized in the western part of Texas and Oklahoma because they can require too much water for establishment and growth. The winter months are relatively dry in the Southern High Plains. From 2007 to 2016, the average accumulation of rainfall from December through March was 7.6 cm in Terry County, TX, which is located centrally within the high root knot nematode-infested region. Recently, across the U. S., there has been a surge in the use of cover crops to improve soil health, and in some cases to assist in nematode management. However, the more arid parts of Texas and Oklahoma are limited to cover crops that require less moisture for plant establishment and growth, or the non-use of cover crops to conserve soil moisture for the cotton crop. Cover crops remove moisture from the soil profile, and consequently in drier years, greatly reduce cotton yield. Deficit irrigation is practiced in 30–40% of the Southern High Plains of Texas, which means that cotton can be irrigated, but there is insufficient irrigation water available to replace water lost through evapotranspiration. The remaining area is termed dryland (rainfed) where *M. incognita* may be present, though often in lower densities than in cotton cultivated on irrigated land. Many dryland hectares are abandoned each year due to lack of rain, therefore, nematode management on non-irrigated land must budget for frequent crop failure.

Crop rotation can be an excellent method of managing root knot nematodes and improving soils for subsequent cotton production. Peanut, as a non-host for *M. incognita*, will provide excellent root knot nematode control in the cotton crop that follows it. Currently in Oklahoma and Texas, less than 10% of the areas infested with *M. incognita* are in rotation with peanuts as water limitations prohibit the production of this crop.

Sorghum is another popular crop that is used in rotation with cotton in Texas and Oklahoma. This crop provides good residue when harvested for grain. The remaining plant residue after harvest does not degrade as quickly as a low-residue crop like cotton. The higher residue left on the soil reduces erosion and improves rainfall retention in soil. However, sorghum cultivars are generally adequate hosts for local populations of *M. incognita* (Orr and Morey 1978). There have been several studies that indicate sorghum is a poor host for *M. incognita* (Aminu-Taiwo et al. 2015; Fortnum and Currin 1988; Ibrahim et al. 1993). Most *M. incognita* host-range studies with sorghum cultivars have been conducted on races that are not found in this region. A survey of *M. incognita* was conducted across the Southern High Plains of Texas and all 50 populations tested were race 4 (Wheeler unpublished). A 3-year study was conducted in Dawson County, TX, on a 2-year cotton and 1-year sorghum rotation compared to continuous cotton. Cotton lint yields in the continuous cotton system averaged 669, 865 and 974 kg/ha at a low, medium and high irrigation rate, respectively (Keeling et al. 2010a, 2011a, 2012a). Cotton lint yields following sorghum in the same field and during the same years, averaged 672, 900 and 1108 kg of lint/ha at a low, medium and high irrigation rate, respectively. The yield improvement in the rotated cotton over continuous cotton was 1%, 4% and 12% for the low, medium and high irrigation rates, respectively (Keeling et al. 2010b, 2011b, 2012b). Root knot nematode density sampled in the fall over those 3 years, averaged 808/500 cm^3 soil in sorghum, 4,473/500 cm^3 soil in cotton following the sorghum crop and 3,649/500 cm^3 soil in continuous cotton (Wheeler unpublished). Generally, irrigated sorghum yields in this region are not considered as profitable as yields from irrigated cotton. Therefore, to maintain profitability, reducing cotton hectares to rotate with sorghum requires better yields from rotated cotton than from continuous cotton.

A better rotation for root knot nematode management is cotton followed by winter wheat, followed by summer fallow. This rotation maximizes retention of rainfall in the soil and greatly increases cotton yields while reducing root knot nematode densities. As irrigation pumping capacities have dropped in Texas and Oklahoma, it has become more popular to leave a portion of a field out of cotton so that irrigation water can be concentrated on only part of that field. The winter wheat/summer fallow fits well into this cropping system. The limited irrigation capacity can be utilized on part of the field during the cotton growing season to produce good cotton yields (typically on less than 18 cm irrigation), and there is limited irrigation available during the winter to grow the wheat crop. The high-residue wheat stubble left after the spring harvest of wheat, allows rain to be retained in the field and keeps the soil from blowing. A 3-year study was conducted on a cotton/wheat/fallow rotation compared to continuous cotton in Dawson County, TX. Using a root knot nematode susceptible cultivar, the yield of cotton following a wheat/fallow rotation averaged 50–57% higher (depending on irrigation rate, Table 16.2) than yield in continuous cotton. Early season galling on cotton was significantly higher in continuous cotton, compared to rotated cotton (Wheeler unpublished), although, by late season, there was no difference in root knot nematode densities in susceptible cotton cultivars in the two cropping systems (Table 16.2). There was, however, a larger reduction in root knot nematode density when a partially resistant cultivar was combined with the rotated (wheat/fallow) cotton crop (Table 16.2).

Commercial, partially resistant root knot nematode cultivars have been available since the 1990s. However, there was almost no cotton planted with root knot nematode resistance in Texas and Oklahoma until 2003 (Table 16.3). Between 2003 and 2016, 0.3% (2007) to 8.9% (2009) of the cotton land in Oklahoma was planted with cultivars that had some resistance to root knot nematode. In Texas from 2003 to

Table 16.2 Effect of crop rotation with winter wheat/summer fallow (W-F) and cotton compared with continuous cotton (CC) on cotton yield and root knot nematode density over a 3-year period

Average irrigation[a] amounts on cotton (cm)		Average lint yield (kg/ha)[b]				Average root knot nematodes/500 cm[b] soil in late summer			
		Susceptible[c]		Partially resistant[c]		Susceptible		Partially resistant	
CC[c]	W-F/C[c]	CC	W-F/C	CC	W-F/C	CC	W-F/C	CC	W-F/C
16.8	15.2	657	1,025	704	1,070	1,667	236	1,967	276
20.6	19.3	772	1,160	892	1,182	2,760	1,780	1,913	704
24.6	23.4	829	1,303	986	1,366	2,098	1,631	3,540	840

[a]After plant establishment, there were three irrigation rates applied in a randomized complete block design (see column 1 and 2) with three replications. The first two columns present preplant and in-season irrigation totals, averaged from 2014 to 2016 for the three irrigation rates

[b]The cotton tests were managed by Dr. Wayne Keeling (Texas A and M AgriLife Research). Yields were reported annually in the AGCARES report (2014–2016) at http://Lubbock.tamu.edu

[c]Cultivars that were planted in a split-plot design with irrigation as the main factor and cultivar as the split plot. There was a susceptible (NG 1511B2RF) and partially resistant (ST 4946GLB2) cultivar planted each year in the test area

Table 16.3 Cotton varieties planted in Oklahoma or Texas that are marketed as resistant or tolerant to root knot nematode[a]

Year	Cotton planted (%) OK	TX	Cultivars marketed as having resistance to root knot nematode
2003	0.73	0.49	[b]ST 5599BR
2004	6.09	1.15	ST 5599BR
2005	0.76	0.63	ST 5599BR
2006	1.69	0.74	ST 5599BR
2007	0.34	0.59	ST 5599BR
2008	4.51	0.49	[b]DP 174RF, ST 5458B2F, ST 5599BR
2009	8.88	3.67	DP 174RF, ST 5458B2F
2010	4.80	1.40	DP 174RF, ST 4288B2F, ST 5458B2F
2011	2.78	5.30	DP 174RF, [b]PHY 367WRF, ST 4288B2F, ST 5458B2F,
2012	3.19	4.77	DP 174RF, PHY 367WRF, ST 4288B2F, ST 5458B2F,
2013	0.63	4.65	DP 174RF, PHY 367WRF, ST 4288B2F, ST 4946GLB2, ST 5458B2F,
2014	0.95	5.67	PHY 367WRF, PHY 417WRF, ST 4288B2F, ST 4946GLB2, ST 5458B2F
2015	0.08	9.41	PHY 367WRF, PHY 417WRF, ST 4946GLB2,
2016	2.03	4.07	DP 1558NRB2RF, PHY 417WRF, ST 4946GLB2
2017	0.00	2.18	DP 1558NRB2RF, DP 1747NRB2XF, PHY 427WRF, ST 4946GLB2

[a]Planted percentages of nematode resistant cultivars to total planted cotton in Oklahoma and Texas, were calculated from the annual "Cotton Varieties Planted", published by the USDA Agricultural Marketing Service, Memphis, TX
[b]*DP* Deltapine, *PHY* Phytogen, *ST* Stoneville

2017, the percentages ranged from 0.5% (in 2003 and 2008) to 9.4% (2015). The reduction in nematode reproduction varies with the cultivar. Cultivars with two resistant genes, (DP 1558NRB2RF (Deltapine, currently a subsidiary of Monsanto), DP 1747NRB2XF, PHY 417WRF (Phytogen, a subsidiary of Corteva Agriscience) and PHY 427WRF, can reduce root knot nematode densities substantially more than a single resistant gene cultivar (DP 174RF, PHY 367WRF, ST 5599BR (Stoneville, currently a subsidiary of Bayer CropScience), ST 5458B2F, ST 4288B2F and ST 4946GLB2) (Wheeler et al. 2016).

Crop rotations with non/poor hosts or fallowing (i.e. no crop grown during the summer months) to reduce root knot nematode densities will only be effective if good weed control is maintained. The southern root knot nematode has a wide host range including many weeds (Rich et al. 2008). Therefore, it is not unusual to see galls on weeds in this region in cotton fields or in weedy, fallow fields (Manuchehri et al. 2015). Weeds that blow into the field can also initiate root knot nematode problems, especially if root knot nematodes can survive in the weed's root system. Significant stunting was found on cotton near the edge of a field, that previously had no history of root knot nematode (Wheeler pers. comm.). The cotton roots were heavily galled with *M. incognita*. Russian-thistle (*Salsola tragus* L.) plants had blown that winter/spring into the field. Root knot nematode counts as high as 36,600/500 cm^3 soil were found in research plots that were placed in that area. That high infestation of root knot nematode built up in a single growing season because infected weeds blew into the field.

Prior to 2011, root knot nematode was controlled primarily with aldicarb® applied in furrows at planting, and occasionally, with oxamyl® after plant emergence (typically around 40–45 days after planting). Aldicarb production by Bayer CropScience (Raleigh, NC) was discontinued in 2011 and although it is currently being manufactured and sold by AgLogic Chemical LLC (AgLogic™ 15G, Chapel Hill, NC), aldicarb has not been distributed for sale in Texas or Oklahoma as of 2017. Seed treatment nematicides have also been utilized to manage root knot nematode in cotton. However, having sufficient moisture to wash the seed applied nematicide off the seed coat and into the soil profile has been difficult in arid environments such as the Southwestern United States. This region has less success with chemical control than in environments where rainfall is more common (Wheeler et al. 2013, 2014). The chemicals, abamectin®, fluopyram® and thiodicarb®, are currently labeled as seed treatment nematicides on cotton and are used to varying extents in Texas and Oklahoma. Fluopyram has also been labeled for application in furrows at-planting of cotton. Texas and Oklahoma are not likely to return to a heavy dependence on chemical control of root knot nematodes, that occurred with aldicarb prior to 2011. The per hectare cost of liquid or granular nematicides has doubled since 2011 and there are now more cultivars available with, at least, partial root knot nematode resistance. The loss of aldicarb in 2011 spurred an increased emphasis on breeding for nematode resistant cultivars. Though these cultivars are not particularly well adapted to the growing conditions in the High Plains, they often provide a yield advantage over susceptible cultivars (Wheeler et al. 2009, 2014). Root knot nematode resistant, cultivar yield response is more consistent than that found for chemical protection (Wheeler et al. 2014), when the amount of water (irrigation and rainfall) is insufficient to properly distribute nematicides around a root profile (Faske and Starr 2007).

16.2.1.2 Peanut

Meloidogyne arenaria, also known as the peanut root knot nematode, is commonly found in South Texas (Atascosa and Frio Counties) and Central Texas (Eastland, Erath and Comanche Counties), as well as Northwestern Rolling Plains (Collingsworth County, TX) (Fig. 16.1b) (Wheeler and Starr 1987; Woodward personal observations). In Oklahoma, *M. hapla* is the primary nematode problem on peanut and has been found in Beckham, Bryan, Caddo and Love Counties (Fig. 16.1b). *Meloidogyne hapla* has also been associated with enhanced pod rot problems, caused by *Pythium* spp. and *Rhizoctonia solani* (Filonow and Russell 1991). *Meloidogyne arenaria* is also found in Oklahoma, but less frequently than *M. hapla*. Other root knot nematode species that have been associated with peanuts include *M. javanica* (Comanche, Frio and Mason Counties in Texas (Fig. 16.1b) (Tomaszewski et al. 1994) and *M. haplanaria* in Collingsworth and Comanche Counties in Texas (Eisenback et al. 2003) (Fig. 16.1b). *Meloidogyne haplanaria* was a newly described species found originally in peanut in Collingsworth County, Texas in 1993. Management of root knot nematodes in peanut can be accomplished by rotation with a non-host. Cotton is a host only for *M. incognita*, making it an excellent rotation crop for all the species that affect peanut. Nematode resistant peanut cultivars have been developed to *M. arenaria* and *M. javanica*. The first root knot nematode resistant cultivar developed was COAN, which was released in 1999 by Texas AandM Experiment Station, followed by NemaTAM in 2002 (Starr and Morgan 2002). The runner cultivar Webb was released in 2013 by Texas A&M AgriLife Research and it combines the high-oleic fatty acid trait with resistance to root knot nematode (*M. arenaria*) (Simpson et al. 2013). As with cotton, the nematicide aldicarb was, at one time, utilized to manage root knot nematode in peanut in Texas and Oklahoma. Other contact nematicides such as ethoprop, were registered for use in peanut, but have also been removed from the market. Several fumigant nematicides such as chloropicrin, dichloropropene and metam-sodium, are available, but are often cost-prohibitive. Currently, there is limited use of fluopyram to manage root knot nematode in Texas and Oklahoma.

16.2.1.3 Grasses and Cereals

There are many root knot nematode related problems with turf and other grass species in Texas and Oklahoma. In most cases, species of *Meloidogyne* have not been identified. However, *M. graminis* was found in bermudagrass (*Cynodon dactylon*) in Collin County, TX (Fig. 16.1b) and *M. marylandi* in turf, zoysiagrass (*Zoysia* sp.) and bermudagrass in multiple counties in Texas (Erath, Dallas, Brazos and Refugio Counties, Fig. 16.1b) (Starr et al. 2007). In Oklahoma, *M. marylandi* has been found in multiple locations in Tulsa and Oklahoma Counties and in a putting green in Stillwater, OK (Walker 2014). *Meloidogyne graminis* was originally identified on bermudagrass in Texas (Orr and Golden 1966), however, it is possible this population was, in fact, *M. marylandi* (Starr et al. 2007). Recently, root knot nematode infestations have become more wide spread in Oklahoma as more golf courses renovate creeping bentgrass greens to ultra-dwarf bermudagrasses. There are other species such as *M. incognita,* that also are common on grass species including corn (*Zea mays*), and thought to occasionally limit yields. Species of grasses with reported root knot nematode problems include bentgrass (*Agrostis* sp.), bermudagrass, annual ryegrass (*Lolium multiflorum*) and wheat (*Triticum* sp.). A more comprehensive list of nematode species affecting turf and their management will be discussed in the section on plant parasitic nematodes on turf.

16.2.1.4 Pecans

The pecan root knot nematode, *M. partityla*, is the most common nematode problem associated with pecans in Texas (Fig. 16.1b, Starr et al. 1996). However, it is also possible for other species of root knot nematode such as *M. incognita,* to be associated with pecans. In Oklahoma, root knot nematode also causes problems on pecans in the southern part of the state, although the species involved has not been identified. Root knot nematode-infested orchards often decline in production, even when managed optimally. There are no real options to reduce nematode damage in pecans, once the nematode becomes established in plants.

16.2.1.5 Vegetables and Fruit Trees

Potatoes and other vegetable crops can be severely impacted by various root knot nematode species. In a survey, during 2002–2003, *Meloidogyne hapla* was the most frequently detected root knot nematode species on Texas potatoes (Powers et al. 2005). Root knot nematode can be easily transmitted in potato planting seed. It is probable that infested planting seed caused the first known occurrence of *M. chitwoodi* in Texas (Szalanski et al. 2001), since the affected field had been created from range land only 2 years prior to the detection of *M. chitwoodi*. *Meloidogyne hapla* was also found to severely damage chili peppers near the state line between Texas and New Mexico in the Southern High Plains (Woodward personal observations). Other vegetable and fruit (non-woody) crops which are commonly impacted by various root knot nematode species include beans (*Phaseolus* spp.), beets (*Beta vulgaris*), blackberry (*Rubus* sp.), cole crops (*Brassica oleracea*) including cabbage, cauliflower, brussel sprouts and broccoli, cucurbits including cantaloupe (*Cucumis melo*), cucumber (*C. sativus*), squash and pumpkins (*Cucurbita* spp.), carrot (*Daucus carota sativus*), peas (*Pisum sativum*), peppers (*Capsicum* spp.), okra (*Abelmoschus esculentus*), sweet potato (*Ipomoea batatas*), tomato (*Lycopersicon esculentum*) and watermelon (*Citrullus lanatus* var. *lanatus*). In Oklahoma, *M. incognita* was a significant problem on tomato grown under hydroponic greenhouse conditions (Walker pers.comm.). Fumigation for nematode control is rarely practiced in this region. Non-fumigant chemical options are often insufficient for controlling root knot nematode in vegetable production systems. In rare occasions, root knot nematode resistant cultivars are available, but usually the best management involves crop rotation with a non-host or fallowing the land prior to planting a susceptible vegetable crop.

16.2.1.6 Woody Perennials

Production of grape (*Vitis* sp.) can be impacted by root knot nematode. The largest area of grape production in Texas is in the Southern High Plains, particularly in Terry and Yoakum Counties. Sandy soils dominate in these counties and are heavily infested with *M. incognita*, because they were planted on land with a long history of cotton production. In Oklahoma, it is likely that multiple species of root knot nematode are capable of infesting vineyards. Root knot nematode infested vineyards are mostly found around Tulsa and Oklahoma City. Management in Texas involves planting rootstock with tolerance to nematodes, primarily 1103 Paulsen and SO4.

In Oklahoma, these rootstocks have performed poorly or inconsistently and rootstocks with *V.* x *champinii* heritage are recommended (Carroll). Peach and plum (*Prunus* sp.) and apple (*Malus domestica*) are other fruit tree species affected by root knot nematodes. Various species of trees also affected by root knot nematodes include ash (*Fraxinus* spp.), catalpa (*Catlapa bignonioides*), elm (*Ulmus* spp.), Ficus (*Ficus* spp.), live oak (*Quercus virginiana*), magnolia (*Magnolia* spp.), marberry (*Ardisia* spp.), mimosa (*Albizzia julibrissin*), mulberry (*Morus* sp.), olive (*Olea europea*), Texas kidneywood (*Eysenhardtia texana*) and wax myrtle (*Myrica cerifera*).

16.2.1.7 Flowers, Ornamental Shrub and Other Plants

Root knot nematode can also cause significant problems on flowers grown in Texas and Oklahoma. Texas A&M AgriLife Extension Service developed a guide for rating the relative sensitivity of various ornamentals to root knot nematodes (Texas Plant Disease Handbook). There are numerous species listed, but annual spring flowers that are considered highly susceptible include morning glory (*Ipomoea purpurea*), zinnia (*Zinnia elegans*) and petunia (*Petunia hybrida*). Highly susceptible annual fall flowers include snapdragon (*Antirrhinum majus*), petunia and calendula (*C. officionalis*). Flowering perennials that are highly susceptible to root knot nematodes include Canna (*Canna* x *generali*), hollyhock (*Althea rosea*) and Shasta daisy (*Chrysanthemum maximum*). Medium to large shrubs that are highly susceptible include Abelia (*Abelia grandiflora*) and Cape jasmine (*Gardenia jasminoides*). Small trees species that are either highly susceptible or susceptible to root knot nematode include fruitless mulberry (*Morus alba*), loquat (*Eriobotryta japonica*) and Japanese magnolia (*Magnolia* spp.).

16.2.2 Foliar Plant Parasitic Nematodes, Aphelenchoides *spp.*

Plant parasitic foliar nematode species belonging to the genus *Aphlenchoides* spp. have been found on different plant species in Texas, and most frequently on rice (*Oryza sativa*) in the coastal prairie region of Brazoria County (Fig. 16.1c). In Oklahoma, it has only been found on phlox. The species *A. besseyi*, has been reported from the rice growing regions of Beaumont, TX and the Lower Rio Grande Valley (Norton 1959). *Aphelenchoides ritzemabosi* is considered a problem on zinnia in Texas (Texas Plant Disease Handbook). *Aphlenchoides* spp. have been associated with other plants including African lily (*Agapanthus africanus*), Australian tree fern (*Sphaeropteris cooperi*), autumn sage (*Salvia greggii*), bean, bermudagrass, blue beard (*Caryopteris* sp.), Boston fern (*Nephrolepis exaltata bostoniensis*), button fern (*Pellaea rotundifolia*), hay scented fern (*Dennstaedtia punctilobula*), Japanese painted fern (*Athyrium niponicum*), Philippine violet (*Barleria cristata*), spleenwort (*Asplenium*), staghorn fern (*Platycerium* sp.) and yarrow (*Achillea ageratifolium millefolium*) (National Plant Diagnostic Network).

16.2.3 Stem and Bulb Nematode

Ditylenchulus dipsaci, the stem and bulb nematode, is an economically important nematode species that is problematic on alfalfa (*Medicago sativa*) in Oklahoma. This nematode has been identified in a few locations in Texas, but is found throughout the eastern and central potions of Oklahoma (Fig. 16.1c). While it is not widespread on alfalfa, it can be devastating in individual fields (Damicone 2013). *Ditylenchus dipsaci* will increase to damaging levels when the winter and early spring weather is cool and wet. The first cutting of alfalfa usually experiences the most severe damage, since hot weather in the summer will limit the nematode buildup. Infected plants are stunted with twisted and crinkled leaves. Severely damaged plants die, resulting in thin stands. The nematode cannot be controlled through chemical means. An integrated approach is recommended, which includes limiting the spread of the nematode into new fields and crop rotation for 2–4 years with non-hosts. Since the nematode can be spread by hay, it is important when cutting an infected field to harvest when the top 5 to 8-cm of soil is dry, and to thoroughly clean harvest equipment free of hay and soil before moving to a new field.

16.2.4 Other Important Plant Parasitic Nematodes of Oklahoma and Texas

16.2.4.1 Plant Parasitic Nematodes on Turf

The most common nematodes found in bentgrass golf course putting greens in Oklahoma are ring (*Mesocriconema* spp.), stubby root (*Paratrichodorus* spp.), stunt (*Tylenchorhynchus* spp.) and spiral (*Helicotylenchus* spp.) nematodes (Walker et al. 2002). Since the removal of fenamiphos® from the market, the frequency and diversity of nematode infestations has increased. In Texas, based on experiences of an extension specialist in turf (W. Crow, University of Florida), the sting nematode (*Belonolaimus* spp.) was considered the most important nematode, although the most frequently found plant parasitic nematodes are the lance nematode (*Hoplolaimus* spp.), stubby root, ring and lesion (*Pratylenchus* spp.) (Crow 2000). The sting nematode was only found in 11% and 1% of bentgrass samples submitted to the Turfgrass diagnostic clinic in Oklahoma and the Plant Diagnostic Clinic in Texas, respectively (Table 16.4). The sting nematode was found frequently in the southeastern part of Texas and in counties in Oklahoma having sandy river bottom soils (Fig. 16.1c). It has been most frequently identified in bermudagrass and creeping bentgrass, as well as on zoysia grass, peanut, soybean and corn. This nematode,

Table 16.4 Percentages of plant parasitic nematodes[a] associated with turf grasses[b] in Oklahoma and Texas from 1995 to 2017[c]

	Bentgrass		Bermudagrass		Bluegrass		Zoysiagrass		Mixed Turf
	OK[d]	TX	OK	TX	OK	TX	OK	TX	TX
Nematodes	% of total number of samples								
Ring	89	22	18	11	27	0	42	11	41
Stunt	80	0	6	0	27	0	8	0	15
Spiral	47	26	16	4	13	100	8	11	7
Root knot	1	0	20	37	0	0	0	11	7
Sting	11	1	4	9	0	0	0	0	0
Lance	11	18	4	3	27	0	0	0	0
Stubby root	12	3	6	2	0	0	25	0	7
Sheath	2	21	0	3	0	0	0	0	2
Lesion	18	0	4	0	7	0	0	0	16
Total number of samples submitted	271	87	50	150	15	1	12	9	61

[a]Ring nematode included *Criconemella* spp., Family Criconematidae, *Criconemoides* spp. and *Mesocriconema* spp.; Stunt nematode included *Tylenchorhynchus* spp. and Family Tylenchorhynchidae; Spiral nematode is *Helicotylenchus* spp.; Root knot nematode is *Meloidogyne* spp.; Sting nematode is *Belonolaimus* spp.; Lance nematode is *Hoplolaimus* spp.; Stubby root nematode included *Paratrichodorus* spp. and *Trichodorus* spp.; Sheath nematode is *Hemicycliophora*; and lesion nematode is *Pratylenchus* spp.

[b]Bentgrass included *Agrostis* spp. and *A. stolonifera*; Bermudagrass included *Cynodon* spp. and *C. dactylon*; Bluegrass included *Poa* and *P. pratensis*, and in Oklahoma also included mixtures of *Poa* and bentgrass; Zoysiagrass included *Zoysia* spp. and *Z. japonica*

[c]The data presented in this table was obtained from the records maintained by the National Plant Diagnosis Network and covered the years from 1995 to 2017

[d]Data was obtained from N. Walker turfgrass diagnostic laboratory in 2011

which requires a high sand content (>80%), is probably the most damaging nematode on turf grasses. Furthermore, soil content of golf course greens can be ideal for the sting nematode as the United States Golf Association requires a sand content of 90% for construction of greens (United States Golf Association 2004). Root knot nematodes are the most economically important nematodes typically found in bermudagrass samples (20% in OK and 37% in TX, Table 16.4). *Meloidogyne marylandi*, which can cause substantial damage on turf, is widespread in Texas (Starr et al. 2007), even though it has only recently been identified in Oklahoma (Walker 2014). The lance and sheath nematodes are found more frequently in bentgrass samples in Texas compared to Oklahoma. The stunt nematode is found more frequently on turf in Oklahoma compared to Texas. In addition, *Peltamigratus christiei* has been reported on warm-season turfgrass species in Oklahoma (Crow and Walker 2003).

Management of nematodes on turf in Oklahoma and Texas is challenging due to heat and drought stress. Recently, several new chemical control options have been introduced to the market, but the optimal choice should be tailored to the specific nematode species present. No single chemical control option is effective against all the nematodes found on turf. It is important to reduce the overall stress placed on turf when damaging levels of nematodes are present. Mowing heights should be raised, turf should be thoroughly irrigated to encourage deep root systems and installation of fans can help reduce turfgrass decline when nematode populations are elevated. Soil fertility should be managed carefully.

16.2.4.2 Citrus Nematode, *Tylenchulus semipenetrans*

The citrus nematode is widely distributed in at least, 93% of citrus orchards in the Lower Rio Grande Valley (Robinson et al. 1987). The citrus nematode can be responsible for a slow decline in plant health. Root growth is significantly reduced as nematode populations increase in infected plants. Under sufficiently high nematode populations, root growth is substantially retarded to cause abnormally small and reduced fruit production. For an efficient management scheme, it is important to start with clean soil, free of the nematode and nematode free rootstock. There are some nematode resistant rootstocks, but typically, susceptible rootstocks are grown in the region (Reynolds et al. 1974). Chemical control, after plant establishment, is limited to oxamyl, which is not always effective (Timmer 1977; Timmer and French 1979). Fluopyram also has a label for citrus, but is recommended for newly established trees or those with root systems distributed around drip irrigation systems. No published research is available yet on performance.

16.2.4.3 Soybean Cyst Nematode, *Heterodera glycines*

The soybean cyst nematode (SCN) has been present in the eastern part of Oklahoma since the 1980s (Tylka and Marett 2014). This nematode has been found in 17 counties in Oklahoma and in 5 counties in Texas (Fig. 16.1d). In Texas, there is almost no soybean production within the counties that were once infested with *H. glycines* (Fig. 16.1d), but have not been positive for SCN since the 1990s (Wheeler personal observations). Both root knot nematode and soybean cyst nematode can cause problems in soybean in Oklahoma. The best management for these two nematode problems comprises crop rotation and use of nematode-resistant soybean cultivars. There does not appear to be SCN type information on soybean cyst nematode for this region. Rotation crops recommended for soybean cyst nematode include alfalfa, canola, corn, cotton, forages, rye, wheat, oats, peanut and sorghum. The appropriate rotation for root knot nematode would depend on the species of root knot nematode present.

16.2.4.4 Reniform Nematode, *Rotylenchulus reniformis*

The reniform nematode was originally found in four counties of the Lower Rio Grande Valley (Norton 1959). However, in 1982 it was found in cotton near New Home, Texas in the High Plains (Robinson 2007). It appears that the cotton producer also farmed in the Lower Rio Grande Valley and, therefore, most likely spread the nematode on infested equipment. The last state-wide survey (Starr et al. 1993) indicated that the reniform nematode was in 12 counties, with a few additional counties that have been identified since then (Wheeler personal observations) (Fig. 16.1c). The reniform nematode is still less frequently found than root knot nematode on cotton in Texas. However, where it does occur, losses are often much higher than those caused by the root knot nematode. It was estimated that reniform nematode losses in cotton fields average 40% (Robinson 2007), which can be contrasted with 26% losses associated with *M. incognita* in cotton in the High Plains of Texas (Orr and Robinson 1984). Reniform nematodes have also been found occasionally on vegetables and citrus in Texas. The relatively slow spread of the reniform nematode is surprising, compared to most other cotton-producing states in the U.S. The damage caused by the reniform nematode increased substantially between 2000 and 2005 in Alabama, Arkansas, Louisiana, Mississippi and Tennessee (Robinson 2007). The reniform nematode is not found west of Texas or in Oklahoma.

The stunting of cotton, caused by reniform nematode, is dramatic when it is distributed in patches in a field. Within 5–10 years, newly infested fields typically become more uniformly infested. Management is primarily with crop rotation using sorghum or corn. After an initial infestation, generally 1 year of rotation with a non-host is sufficient, but over time, it becomes necessary to rotate to a non-host crop for 2–3 years to eliminate the severe stunting seen in cotton. Resistant germplasm has been identified in *Gossypium* species other than *G. hirsutum* (Robinson et al. 2004, 2007) and successfully introgressed into *G. hirsutum*. However, no reniform nematode resistant cultivars have been commercialized. It has been difficult to combine the nematode resistance with adequate yield potential. The variety PHY 417WRF, which has two resistance genes to root knot nematode, allows less reproduction by the reniform nematode than other commercial varieties (Woodward and Wheeler unpublished). This cultivar, while not particularly high yielding in non-reniform nematode fields, consistently yields at least 25% higher than other cultivars in reniform nematode fields (Woodward unpublished).

Chemical control, by fumigation, has been successful at reducing damage caused by reniform nematode. This nematode is often found in soils that have a lower sand content than soils favored by root knot nematodes (Robinson et al. 1987; Starr et al. 1993). These soil types have smaller pores for movement of gas, often resulting in a more limited distribution of fumigant, than through a coarse textured soil. The reniform nematode is also distributed deeper in the soil than the root knot nematode, and therefore, requires deeper fumigation and higher rates. Control of reniform nematode with 1,3 D (dichloropropene) is recommended at a rate of 47 l per hectare at a depth of 51-cm (Wheeler personal observations). In contrast, to control root knot nematode, 28 L/ha of 1,3-D to a depth of 30-cm is usually adequate. The chemical 1,3-D has been used in the High Plains by producers to reduce reniform nematode populations in a few cases. However, it has been difficult to obtain this product in Texas, results of fumigation can be poor, particularly when applied shallow or at rates less than 47 l/ha, and the Texas Department of Agriculture certification to apply soil fumigants has become more difficult to obtain. Furthermore, the cost of the product has discouraged producers from using 1,3-D.

The reniform nematode is so damaging at relatively low densities, that even a 50% reduction in the reniform nematode density can result in no yield improvement (Wheeler et al. 2008). Non-fumigant nematicides such as fluopyram at planting and oxamyl applied around 35–45 days after planting have been used, when available, to control reniform nematode, but it is not clear if either product alone or used in conjunction will be effective. The nematicide aldicarb was heavily utilized in reniform nematode fields at-planting previous to 2011, and was also combined with oxamyl. This combination was somewhat effective. Yield losses are still substantial when non-fumigant pesticides are utilized in reniform nematode fields.

16.3 Conclusions

Nematode problems exist on many crops in Texas and Oklahoma including cotton, peanut, turf grass, citrus, alfalfa, pecans and soybean. The most effective management options for soybean cyst, root knot and reniform nematodes typically involve cultural methods, crop rotation with non/poor hosts and use of cultivars that reduce nematode reproduction. Unfortunately, these options can not be used with nematode problems on turf grass, citrus, alfalfa and pecans. Use of pesticides is practiced most commonly with nematicide seed treatments (cotton and soybean), in-furrow, at-plant nematicide applications (cotton, peanut) and post-plant establishment (turf, citrus and cotton). Crop losses due to nematodes can be severe and often insufficient or uneconomical options exist to substantially reduce these losses.

Acknowledgements We appreciate the assistance of the National Plant Disease Repository for providing certain statistics used in this chapter. We also appreciate the support of Oklahoma State University, Texas A&M AgriLife Research, and Texas A&M AgriLife Extension Service.

References

Aminu-Taiwo, B. R., Fawole, B., & Claudius-Cole, A. O. (2015). Host status of some selected crops to *Meloidogyne incognita*. *International Journal of Agriculture Innovations and Research, 3*, 1431–1435.

Carroll, B. *Rootstocks for grape production*. Oklahoma Cooperative Extension Service HLA-6253. http://pods.dasnr.okstate.edu/docushare/dsweb/Get/Document-3107/F-6253web.pdf. Accessed 14 Dec 2017.

Crow, W. 2000. *Nematodes in Texas golf courses. Texas AandM AgriLife Extension*. Texas A and M System L-5351, pp. 3–10. http://oaktrust.library.tamu.edu/bitstream/handle/1969.1/86880/pdf_1184.pdf?sequence=1. Accessed 14 Dec 2017.

Crow, W. T., & Walker, N. R. (2003). Diagnosis of *Peltamigratus christiei*, a plant parasitic nematode associated with warm-season turfgrasses in the southern United States. Online. *Plant Health Progress*. https://doi.org/10.1094/PHP-2003-0513-01-DG.

Damicone, J. P. (2013). *Alfalfa stem nematode*. Oklahoma Cooperative Extension Service EPP-7648. http://pods.dasnr.okstate.edu/docushare/dsweb/Get/Document-2581/EPP-7648-2013.pdf. Accessed 14 Dec 2017.

Dorhout, R., Gommers, F. J., & Kollöffel, C. (1991). Water transport through tomato roots infected with *Meloidogyne incognita*. *Phytopathology, 81*, 379–385.

Eisenback, J. D., Benard, E. C., Starr, J. L., Lee, T. A., Jr., & Tomaszewski, E. K. (2003). *Meloidogyne haplanaria* n. sp. (Nematoda: Meloidogynidae), a root-knot nematode parasitizing peanut in Texas. *Journal of Nematology, 35*, 395–403.

Faske, T. R., & Starr, J. L. (2007). Cotton root protection from plant parasitic nematodes by abamectin-treated seed. *Journal of Nematology, 39*, 27–30.

Filonow, A. B., & Russell, C. C. (1991). Nematodes and fungi associated with pod rot of peanuts in Oklahoma. *Nematologia Mediterranea, 19*, 207–210.

Fortnum, B. A., & Currin, R. E. (1988). Host suitability of grain sorghum cultivars to *Meloidogyne* spp. *Journal of Nematology (Supplement), 20*, 61–64.

Ibrahim, I. K. A., Lewis, S. A., & Harshman, D. C. (1993). Host suitability of graminaceous crop cultivars from isolates of *Meloidogyne arenaria* and *M. incognita*. *Journal of Nematology (Supplement), 25*, 858–862.

Keeling, W., Bordovsky, J., Reed, J., & Petty, M. (2010a). Cotton variety performance as affected by low-energy precision application (LEPA) irrigation levels at AG-CARES, Lamesa, TX, 2009. In *Agricultural complex for advanced research and extension systems (AG-CARES)* (pp. 1–2). Texas AandM University System Technical Report 10-1. http://lubbock.tamu.edu/files/2011/11/2009AGCARES.pdf. Accessed 14 Dec 2017.

Keeling, W., Bordovsky, J., Reed, J., Petty, M. (2010b). Cotton variety performance in a sorghum/cotton rotation as affected by low-energy precision application (LEPA) irrigation levels at AG-CARES, Lamesa, TX, 2009. In *Agricultural complex for advanced research and extension systems (AG-CARES)* (pp. 3–4). Texas AandM University System Technical Report 11-1. http://lubbock.tamu.edu/files/2011/11/2009AGCARES.pdf. Accessed 14 Dec 2017.

Keeling, W., Bordovsky, J., Reed, J., & Petty, M. (2011a). Cotton variety performance (continuous cotton) as affected by low-energy precision application (LEPA) irrigation levels at AG-CARES, Lamesa, TX, 2010. In *Agricultural complex for advanced research and extension systems (AG-CARES)* (pp. 1–2). Texas A&M University System Technical Report 11-1. http://lubbock.tamu.edu/files/2011/11/2010AGCARES.pdf. Accessed on 14 Dec 2017.

Keeling, W., Bordovsky, J., Reed, J., & Petty, M. (2011b). Cotton variety performance (sorghum-cotton rotation) as affected by low-energy precision application (LEPA) irrigation levels at AG-CARES, Lamesa, TX, 2010. In *Agricultural complex for advanced research and extension systems (AG-CARES)* (pp. 3–4). Texas A&M University System Technical Report 11-1. http://lubbock.tamu.edu/files/2011/11/2010AGCARES.pdf. Accessed 14 Dec 2017.

Keeling, W., Bordovsky, J., Reed, J., & Petty, M. (2012a). Cotton variety performance (continuous cotton) as affected by low-energy precision application (LEPA) irrigation levels at AG-CARES, Lamesa, TX, 2011. In *Agricultural complex for advanced research and extension systems (AG-CARES)* (pp. 1–2). Texas A&M University System Technical Report 12-1. http://lubbock.tamu.edu/files/2012/03/AGCARES20111.pdf. Accessed 14 Dec 2017.

Keeling, W., Bordovsky, J., Reed, J., & Petty, M. (2012b). Cotton variety performance (sorghum-cotton rotation) as affected by low-energy precision application (LEPA) irrigation levels at AG-CARES, Lamesa, TX, 2011. In *Agricultural complex for advanced research and extension systems (AG-CARES)* (pp. 3–4). Texas A&M University System Technical Report 12-1. http://lubbock.tamu.edu/files/2012/03/AGCARES20111.pdf. Accessed 14 Dec 2017.

Ma, J., Jaraba, J., Kirkpatrick, T. L., & Rothrock, C. S. (2014). Effect of *Meloidogyne incognita* and *Thielaviopsis basicola* on cotton growth and root morphology. *Phytopathology, 104*, 507–512.

Manuchehri, M. R., Woodward, J. E., Wheeler, T. A., Dotray, P. A., & Keeling, J. W. (2015). First report of Russian-thistle (*Salsola tragus* L.) as a host for the southern root-knot nematode

(*Meloidogyne incognita*) in the United States. *Plant Health Progress*. https://doi.org/10.1094/PHP-BR-15-0011.

National Plant Diagnostic Network. http://www.npdn.org. Accessed 15 Dec 2017.

Norton, D. C. (1959). *Plant parasitic nematodes in Texas*. College Station: Texas Agricultural Experiment Station.

Orr, C. C., & Golden, A. M. (1966). The pseudo-root knot nematode of turf in Texas. *Plant Disease Report, 50*, 645.

Orr, C. C., & Morey, E. D. (1978). Anatomical response of grain sorghum roots to *Meloidogyne incognita acrita*. *Journal of Nematology, 10*, 48–53.

Orr, C. C., & Robinson, A. F. (1984). Assessment of cotton losses in western Texas caused by *Meloidogyne incognita*. *Plant Disease, 68*, 284–285.

Powers, T. O., Mullin, P. G., Harris, T. S., Sutton, L. A., & Higgins, R. S. (2005). Incorporating molecular identification of *Meloidogyne* spp. into a large-scale regional nematode survey. *Journal of Nematology, 37*, 226–235.

Reynolds, H. W., O'Bannon, J. H., & Nigh, E. L. (1974). *The citrus nematode and its control in the southwest*. USDA-ARS Technical Bulletin 1478.

Rich, J. R., Brito, J. A., Kaur, R., & Ferrell, J. A. (2008). Weed species as hosts of *Meloidogyne*: A review. *Nematropica, 39*, 157–185.

Robinson, A. F. (2007). Reniform in United States Cotton: When, where, why, and some remedies. *Annual Review of Phytopathology, 45*, 263–288.

Robinson, A. F., Heald, C. M., Flanagren, S. L., Thames, W. H., & Amador, J. (1987). Geographical distributions of *Rotylenchulus reniformis*, *Meloidogyne incognita*, and *Tylenchulus semipenetrans*, in the lower Rio Grande Valley as related to soil texture and land use. *Annals of Applied Nematology (Journal of Nematology 19 Supplement), 1*, 20–25.

Robinson, A. F., Bridges, A. C., & Percival, A. E. (2004). New sources of resistance to the reniform (*Rotylenchulus reniformis* Linford and Oliveira) and root-knot (*Meloidogyne incognita* (Kofoid and White) Chitwood) nematode in upland (*Gossypium hirsutum* L.) and sea island (*G. barbadense* L.) cotton. *Journal of Cotton Science, 8*, 191–197.

Robinson, A. F., Bell, A. A., Dighe, N. D., Menz, M. A., Nichols, R. L., & Stelly, D. M. (2007). Introgression of resistance to nematode *Rotylenchulus reniformis* into upland cotton (*Gossypium hirsutum*) from *Gossypium longicalyx*. *Crop Science, 47*, 1865–1877.

Simpson, C. E., Starr, J. L., Baring, M. R., & Wilson, J. N. (2013). Registration of 'Webb' peanut. *Journal of Plant Registrations, 7*, 265.

Starr, J. L., & Morgan, E. R. (2002). Management of the peanut root-knot nematode, *Meloidogyne arenaria* with host resistance. *Plant Health Progress*. https://doi.org/10.1094/PHP-2002-1121-01-HM.

Starr, J. L., Heald, C. M., Robinson, A. F., Smith, R. G., & Krausz, J. P. (1993). *Meloidogyne incognita* and *Rotylenchulus reniformis* and associated soil textures from some cotton production areas of Texas. *Supplement to Journal of Nematology, 25*(4S), 895–899.

Starr, J. L., Tomaszewski, E. K., Mundo-Ocampo, M., & Baldwin, J. G. (1996). *Meloidogyne partityla* on pecan: Isozyme phenotypes and other hosts. *Journal of Nematology, 28*, 565–568.

Starr, J. L., Ong, K. L., Huddleston, M., & Handoo, Z. A. (2007). Control of *Meloidogyne marylandi* on bermudagrass. *Nematropica, 37*, 43–49.

Szalanski, A. L., Mullin, P. G., Harris, T. S., & Powers, T. O. (2001). First report of Columbia root-knot nematode (*Meloidogyne chitwoodi*) in potato in Texas. *Plant Disease, 85*, 442.

Texas Plant Disease Handbook. http://plantdiseasehandbook.tamu.edu/landscaping/shrubs/rating-of-ornamental-plants-to-root knot-nematodes/. Accessed 15 Dec 2017.

Texas Plant Disease Handbook. http://plantdiseasehandbook.tamu.edu/landscaping/flowers/zinnia/. Accessed 15 Dec 2017.

Timmer, L. W. (1977). Control of citrus nematode *Tylenchulus semipenetrans* on fine-textured soil with DBCP and oxamyl. *Journal of Nematology, 9*, 45–50.

Timmer, L. W., & French, J. V. (1979). Control of *Tylenchulus semipenetrans* on citrus with aldicarb, oxamyl, and DBCP. *Journal of Nematology, 11*, 387–394.

Tomaszewski, E. K., Khalil, M. A. M., El-Deep, A. A., Powers, T. O., & Starr, J. L. (1994). *Meloidogyne javanica* parasitic on peanut. *Journal of Nematology, 26*, 436–441.

Tylka, G. L., & Marett, C. C. (2014). Distribution of the soybean cyst nematode, *Heterodera glycines*, in the United States and Canada: 1954–2014. *Plant Health Progress*. https://doi.org/10.1094/PHP-BR-14-0006.

United States Golf Association. (2004). https://www.usga.org/content/dam/usga/images/course-care/2004%20USGA%20Recommendations%20For%20a%20Method%20of%20Putting%20Green%20Cons.pdf. Accessed 14 Dec 2017.

Walker, N. (2014). First report of *Meloidogyne marylandi* infecting bermudagrass in Oklahoma. *Plant Disease, 98*, 1286.

Walker, N. R., Kirkpatrick, T. L., & Rothrock, C. S. (1998). Interactions between *Meloidogyne incognita* and *Thielaviopsis basicola* on cotton (*Gossypium hirsutum*). *Journal of Nematology, 30*, 415–422.

Walker, N., Goad, C. L., Zhang, H., & Martin, D. L. (2002). Factors associated with populations of plant parasitic nematodes in bentgrass putting greens in Oklahoma. *Plant Disease, 86*, 764–768.

Wheeler, T. A., & Starr, J. L. (1987). Incidence and economic importance of plant parasitic nematodes on peanut in Texas. *Peanut Science, 14*, 94–96.

Wheeler, T. A., Hake, K. D., & Dever, J. K. (2000). Survey of *Meloidogyne incognita* and *Thielaviopsis basicola*: Their impact on cotton fruiting and producers' management choices in infested fields. *Supplement to Journal of Nematology, 32*(4S), 576–583.

Wheeler, T. A., Porter, D. O., Archer, D., & Mullinix, B. G., Jr. (2008). Effect of fumigation on *Rotylenchulus reniformis* population density through subsurface drip irrigation located every other furrow. *Journal of Nematology, 40*, 210–216.

Wheeler, T. A., Keeling, J. W., Bordovsky, J. P., Everitt, J., Bronson, K. F., Boman, R. K., & Mullinix, B. G., Jr. (2009). Effect of irrigation rates on three cotton (*Gossypium hirsutum* L.) cultivars in a root-knot nematode (*Meloidogyne incognita*) infested field. *Journal of Cotton Science, 13*, 56–66.

Wheeler, T. A., Lawrence, K. S., Porter, D. O., Keeling, W., & Mullinix, B. G., Jr. (2013). The relationship between environmental variables and response of cotton to nematicides. *Journal of Nematology, 45*, 8–16.

Wheeler, T. A., Siders, K. T., Anderson, M. G., Russell, S. A., Woodward, J. E., & Mulllinix, B. G., Jr. (2014). Management of *Meloidogyne incognita* with chemicals and cultivars in cotton in a semi-arid environment. *Journal of Nematology, 46*, 101–107.

Wheeler, T. A., Siders, K. T., Woodward, J. E. (2016). High Plains root-knot nematode variety trial results. 2016. http://lubbock.tamu.edu/files/2017/01/Rootknot-trials-2016Final.pdf.

Index

A

Anguina spp., 44
Anguina tritici, 48, 250, 277, 330, 352
Aorolaimus leipogrammus, 250
Aorolaimus spp., 250
Aphelenchoides besseyi, 44, 48, 58, 146, 211, 212, 330, 346, 396, 436, 443
Aphelenchoides fragariae, 44, 48, 97, 98, 100, 146, 212, 250, 267, 330, 346
Aphelenchoides myceliophagus, 250
Aphelenchoides parietinus, 44, 48, 250
Aphelenchoides ritzemabosi, 44, 48, 60, 72, 81, 100, 146, 250, 330, 346, 443
Aphelenchoides subtenuis, 250
Atylenchus decalineatus, 44, 60, 71

B

Bakernema inaequale, 175
Belonolaimus euthychilus, 250
Belonolaimus gracilis, 250
Belonolaimus longicaudatus, 3, 44, 210, 212–219, 262, 278, 293, 313, 319, 330, 351, 357, 358, 367, 368, 436
Belonolaimus maritimus, 250
Belonolaimus nortoni, 396
Bentgrass seed gall nematode, 294
Burrowing nematodes, 230–232, 234
Bursaphelenchus xylophilus, 174, 175, 212, 249

C

Cactodera milleri, 60, 63, 175
Cactodera rosae, 175
Cactodera weissi, 60, 63, 175, 250, 277, 334
Carrot cyst nematode, 64
Cereal cyst nematode, 63
Citrus nematode, 223–225, 227, 229, 231, 435, 446
Clover cyst nematode, 34, 66, 318
Corn cyst nematode (CSN), 294, 317, 336, 337
Criconema demani, 250
Criconema fimbriatum, 60
Criconema grassator, 280
Criconema lamellatum, 250
Criconema mutabile, 15
Criconema octangulare, 175
Criconema permistum, 60, 250
Criconema petasum, 60
Criconema princeps, 60
Criconema sphagni, 60, 250
Criconemoides annulatus, 250
Crossonema fimbriatus, 250
Crossonema menzeli, 60, 175
Cyst nematode, 4, 29, 63, 89, 109, 127, 158, 183, 278, 333, 436

D

Dagger nematodes, 41–43, 71, 144, 145, 200–202, 294, 297, 329, 344, 345
Discocriconemella inarata, 100
Ditylenchus destructor, 44, 45, 48, 60, 72, 100, 157, 170, 172–175
Ditylenchus dipsaci, 44–46, 60, 72, 80, 99, 111, 117, 126, 144, 250, 268, 278, 293, 297, 330, 344, 436, 444
Ditylenchus triformis, 250

S. A. Subbotin, J. J. Chitambar (eds.), *Plant Parasitic Nematodes in Sustainable Agriculture of North America*, Sustainability in Plant and Crop Protection, https://doi.org/10.1007/978-3-319-99588-5

Dolichodorus heterocephalus, 44, 58, 60, 212, 250, 294
Dolichodorus marylandicus, 44, 250, 330
Dolichodorus silvestris, 279

G
Geocenamus longus, 60
Geocenamus tenidens, 185
Globodera pallida, 28
Globodera rostochiensis, 9, 27–31
Globodera tabacum, 4, 9–12, 250, 266, 278, 292
Gracilacus acicula, 60

H
Helicotylenchus californicus, 60
Helicotylenchus caroliniensis, 251
Helicotylenchus cornurus, 14, 100
Helicotylenchus crenacauda, 60
Helicotylenchus digonicus, 15, 60, 100, 119, 174, 175, 185, 330, 347, 358, 396
Helicotylenchus dihystera, 14, 100, 147, 185, 251, 294, 330, 358, 396
Helicotylenchus erythrinae, 14, 185, 251, 330
Helicotylenchus exallus, 185, 251
Helicotylenchus hydrophilus, 251
Helicotylenchus microlobus, 15, 175, 185, 196, 251, 330
Helicotylenchus multicinctus, 212, 330, 347, 358, 396
Helicotylenchus paxilli, 212
Helicotylenchus platyurus, 15, 60, 100, 119, 175, 330
Helicotylenchus pseudorobustus, 15, 60, 100, 119, 126, 147, 175, 185, 212, 319, 330, 347, 396
Hemicaloosia graminis, 251
Hemicriconemoides chitwoodi, 251
Hemicriconemoides strictathecatus, 212
Hemicriconemoides wessoni, 212, 251
Hemicycliophora conida, 251
Hemicycliophora gigas, 251
Hemicycliophora gracilis, 44, 251
Hemicycliophora mettleri, 251
Hemicycliophora obtusa, 175
Hemicycliophora parvana, 212, 251
Hemicycliophora robbinsi, 251
Hemicycliophora sheri, 251
Hemicycliophora similis, 44, 60
Hemicycliophora thienemanni, 251
Hemicycliophora triangulum, 396
Hemicycliophora typica, 175
Hemicycliophora uniformis, 60
Hemicycliophora vaccinium, 60, 251
Hemicycliophora vidua, 60, 251
Heteroanguina graminophila, 100
Heterodera avenae, 60, 63
Heterodera carotae, 60, 63, 64, 75
Heterodera cyperi, 251, 358
Heterodera glycines, 34, 60, 63–65, 76, 88–94, 100, 109, 111, 117, 118, 125–129, 131, 132, 134–138, 158, 165–168, 175, 184, 185, 187–192, 197, 212, 251, 260, 261, 291, 292, 306, 308–312, 314, 317, 318, 329, 330, 333–335, 357, 358, 370, 395, 396, 398, 399, 435, 436, 446
Heterodera humuli, 60, 63, 65, 76, 77
Heterodera lespedezae, 251
Heterodera leuceilyma, 212
Heterodera orientalis, 60, 63, 66
Heterodera schachtii, 31, 58, 63, 65, 77, 111, 115, 183, 191, 193, 277
Heterodera trifolii, 34, 63, 66, 100, 175, 316, 333
Heterodera ustinovi, 60, 66, 78, 100
Heterodera zeae, 279, 294, 331, 336, 337
Hirschmanniella gracilis, 60, 175
Hop cyst nematode, 65, 76
Hoplolaimus concaudajuvencus, 212
Hoplolaimus columbus, 252, 263, 264, 331, 357, 358, 368, 375, 376, 396
Hoplolaimus galeatus, 14, 18, 44, 60, 71, 72, 100, 119, 126, 175, 185, 211, 212, 252, 264, 293, 331, 349, 358, 375, 396
Hoplolaimus magnistylus, 319, 375, 396
Hoplolaimus stephanus, 185, 196, 252, 264

L
Lance nematode, 15, 119, 196, 263, 264, 293, 349, 367, 375, 376, 444
Lobocriconema thornei, 60, 72, 175
Longidorus brevannulatus, 44
Longidorus crassus, 252
Longidorus elongatus, 17, 44, 60, 71, 252
Longidorus longicaudatus, 252
Longidorus paralongicaudatus, 252

M
Meloidodera floridensis, 252, 358
Meloidogyne arenaria, 34, 60, 211, 212, 219–221, 223, 252, 256, 285–289, 295, 298, 318, 358, 363, 364, 366–369, 396, 408, 435, 440, 441

Meloidogyne carolinensis, 252, 256, 259
Meloidogyne chitwoodi, 279, 435, 442
Meloidogyne enterolobii, 212, 219–221, 256, 258, 269, 286–288, 298, 370
Meloidogyne floridensis, 211, 219, 220, 222
Meloidogyne graminicola, 34, 358, 396
Meloidogyne graminis, 16, 34, 212, 220, 223, 252, 256, 259, 278, 318, 331, 338, 441
Meloidogyne hapla, 4, 6–9, 34, 61, 64, 67, 78, 100, 115, 126, 147, 169–172, 288, 295, 313, 314, 331, 339, 440, 442
Meloidogyne haplanaria, 212, 220, 222, 370, 435, 440
Meloidogyne incognita, 61, 67, 79, 219, 220, 258, 286, 287, 289, 306, 308, 331, 338, 358, 365, 399, 401, 402, 407, 435, 436
Meloidogyne javanica, 34, 212, 219–223, 253, 256, 285–287, 289, 290, 295, 298, 331, 338, 358, 363, 408, 435, 441
Meloidogyne mali, 279
Meloidogyne marylandi, 259, 445
Meloidogyne megatyla, 253, 256, 259
Meloidogyne microtyla, 61
Meloidogyne naasi, 15, 16, 61, 67, 72, 79, 100, 253, 256, 259
Meloidogyne nataliei, 61
Meloidogyne ottersoni, 158, 175
Meloidogyne ovalis, 174
Meloidogyne partityla, 212, 220, 253, 358, 435, 441
Meloidogyne platani, 285
Meloidogyne querciana, 285
Meloidogyne spatinae, 253
Meloidogyne trifoliophila, 318
Merlinius brevidens, 61, 119, 331, 348
Merlinius joctus, 61
Merlinius lineatus, 185
Merlinius macrodorus, 61
Merlinius tessellatus, 61
Mesoanguina plantaginis, 253, 358
Mesocriconema axeste, 61
Mesocriconema curvatum, 44, 61, 100, 253, 267, 358
Mesocriconema onoense, 358, 396
Mesocriconema ornatum, 44, 61, 212, 253, 267, 331, 351, 357, 358, 379, 382–384, 396
Mesocriconema raskiensis, 185
Mesocriconema reedi, 61
Mesocriconema rusticium, 358
Mesocriconema serratum, 61
Mesocriconema simile, 61, 319, 331, 351
Mesocriconema sphaerocephalum, 254, 267
Mesocriconema xenoplax, 44, 61, 100, 185, 213, 254, 266, 267, 331, 357, 358, 379–381, 397
Michigan grape root knot nematode, 68, 78

N

Nacobbus aberrans, 111
Nacobbus batatiformis, 61, 115
Nagelus aberrans, 185
Nanidorus minor, 43, 61, 71, 100, 175, 213, 254, 264, 265, 331, 350, 357, 359, 378, 379, 397
Needle nematode, 15, 71, 99, 116, 170, 174, 351
Neodolichodorus pachys, 185
Northern root knot nematode (NRKN), 7–9, 35, 78, 169–172, 288, 295, 308, 338
Nothocriconema sphagni, 174, 175

O

Ogma cobbi, 61
Ogma decalineatum, 254
Ogma floridense, 254
Ogma octangularis, 61, 175

P

Paratrichodorus allius, 100, 126, 146, 185, 194, 200, 254
Paratrichodorus atlanticus, 61
Paratrichodorus nanus, 43
Paratrichodorus pachydermus, 43, 61, 70, 80, 331, 350
Paratrichodorus porosus, 43, 61, 254
Paratylenchus dianthus, 331, 352
Paratylenchus goldeni, 254
Paratylenchus hamatus, 61, 72, 185, 331, 352
Paratylenchus nanus, 196, 331, 352
Paratylenchus neoamblycephalus, 101
Paratylenchus projectus, 44, 61, 72, 101, 119, 126, 319, 332, 342, 352, 397
Peltamigratus christiei, 213, 445
Pin nematodes (PN), 195, 196, 352
Potato rot nematode, 48, 72, 170, 172–174
Pratylenchoides laticauda, 61, 70
Pratylenchus agilis, 101, 185, 332, 342
Pratylenchus alleni, 96, 308, 315, 316, 397
Pratylenchus bolivianus, 233
Pratylenchus brachyurus, 213, 233, 234, 254, 263, 315, 332, 341, 359, 367, 397, 409, 410

Pratylenchus coffeae, 213, 230, 233, 234, 254, 263, 315, 332, 342, 397
Pratylenchus crenatus, 61, 96, 101, 175, 280, 297, 315, 332, 342
Pratylenchus hexincisus, 96, 101, 112, 126, 141, 199, 201, 202, 315, 332, 342, 359, 397
Pratylenchus hippeastri, 233
Pratylenchus loosi, 234
Pratylenchus macrostylus, 254, 263
Pratylenchus minyus, 185
Pratylenchus neglectus, 38, 61, 96, 101, 111–113, 159, 175, 185, 194, 201, 297, 315, 332, 342, 397
Pratylenchus penetrans, 5, 7, 38, 68, 96, 111, 141, 158, 222, 254, 280, 315, 341, 359, 397
Pratylenchus pseudocoffeae, 234
Pratylenchus scribneri, 61, 96, 101, 111–113, 126, 141, 142, 159, 164, 175, 186, 194, 199–202, 233, 254, 263, 297, 306, 315, 316, 332, 359, 397
Pratylenchus tenuis, 126
Pratylenchus thornei, 101, 111–113, 175, 297, 332, 342
Pratylenchus vexans, 186
Pratylenchus vulnus, 61, 101, 175, 233, 254, 263, 297, 313, 315, 316, 332, 342, 359, 397
Pratylenchus zeae, 163, 233, 254, 263, 315, 332, 341, 342, 359, 397, 409, 410
Punctodera punctata, 61–63, 66, 67, 78, 186

Q
Quinisulcius acti, 61
Quinisulcius acutoides, 186
Quinisulcius acutus, 61, 101, 102, 119, 186, 332, 348, 397
Quinisulcius capitatus, 61, 254

R
Radopholus similis, 61, 213, 230–233, 235
Reniform nematode, 261, 288, 294, 308, 317, 362, 371–374, 385, 395, 410–421, 435, 446–448
Ring nematode, 266, 267, 293, 297, 350, 351, 360, 379–384, 445
Root knot nematode (RKN), 4, 34, 67, 68, 94–96, 115, 157, 219, 220, 248, 277, 306, 337–339, 362, 395, 435
Root lesion nematode (RLN), 5, 7, 38, 68, 70, 79, 96, 97, 111, 112, 138, 140–144, 159–161, 163, 164, 172, 193, 194, 315, 316, 329, 339, 341–343
Rotylenchulus reniformis, 101, 213, 255, 257, 261, 279, 294, 308, 317, 359, 370–375, 397, 410, 411, 414, 435, 436, 446–448
Rotylenchus buxophilus, 61, 175, 255, 294, 313, 316, 332, 347
Rotylenchus pumilus, 175, 255
Rotylenchus robustus, 61, 332, 347
Rotylenchus uniformis, 44

S
Scutellonema brachyurus, 255, 332, 397
Scutellonema bradys, 397
Sheath nematode, 294, 351, 445
Southern root knot nematode (SRKN), 67, 79, 287, 296, 308, 338, 364, 369, 371, 401, 405–407, 415, 436, 439
Soybean cyst nematode (SCN), 34, 64, 65, 88, 89, 109, 117, 125, 127–129, 131, 132, 134–138, 158, 184, 187–191, 260, 261, 278, 306, 308, 334, 395, 436, 446
Spiral nematode, 99, 141, 145, 147, 196, 294, 319, 346, 347, 445
Stem and bulb nematode, 44–46, 99, 144, 249, 268, 293, 297, 344, 435, 436, 444
Sting nematode, 3, 116, 117, 210, 213, 216, 221, 224, 262, 278, 293, 319, 349, 351, 435, 436, 444, 445
Stubby nematode, 43, 70, 146, 174, 194, 195, 264, 265, 294, 350, 360, 377, 379, 382, 402, 408, 409, 445
Stunt nematode, 15, 19, 71, 148, 195, 294, 319, 329, 348, 445
Sugar beet cyst nematode (SBCN), 31, 65, 77, 96, 115, 191–193

T
Tetylenchus joctus, 44, 71
Tobacco cyst nematode (TCN), 4, 9–12, 266, 292
Trichodorus acutus, 43
Trichodorus aequalis, 43
Trichodorus borneonsis, 359
Trichodorus californicus, 175
Trichodorus elefjohnsoni, 255
Trichodorus obscurus, 43
Trichodorus obtusus, 211, 213, 255, 264, 265
Trichodorus primitivus, 61, 332, 350, 359, 397

Trichodorus proximus, 62
Trichodorus similis, 62
Trophonema arenarium, 62
Trophurus minnesotensis, 126, 148, 186
Trophurus sculptus, 255
Tylenchorhynchus agri, 16, 62, 101
Tylenchorhynchus annulatus, 101, 398
Tylenchorhynchus canalis, 186, 398
Tylenchorhynchus clarus, 14, 62, 333
Tylenchorhynchus claytoni, 3, 13, 14, 18, 62, 71, 148, 186, 255, 348, 359, 398
Tylenchorhynchus cylindricus, 186, 359, 398
Tylenchorhynchus dubius, 13, 18, 19, 44, 62, 333, 348
Tylenchorhynchus ewingi, 398
Tylenchorhynchus goffarti, 398
Tylenchorhynchus latus, 186
Tylenchorhynchus macrurus, 186
Tylenchorhynchus martini, 62, 101, 102, 319, 359
Tylenchorhynchus maximus, 14, 44, 62, 101, 175, 186, 255
Tylenchorhynchus nudus, 14, 18, 62, 101, 102, 186, 359, 398
Tylenchorhynchus parvus, 62
Tylenchorhynchus robustus, 186, 195
Tylenchulus palustris, 279, 293
Tylenchulus semipenetrans, 213, 223–230, 435, 446
Tylenchulus sp., 255

V

Vittatidera zeaphila, 317

W

Wheat seed gall, 48

X

Xenocriconemella macrodora, 62, 255, 333
Xiphinema americanum, 41–43, 62, 71, 80, 101, 119, 127, 144, 175, 186, 200, 279, 294, 297, 333, 345, 359, 398
Xiphinema bakeri, 255
Xiphinema chambersi, 101, 144, 175, 255, 333, 345, 398
Xiphinema diversicaudatum, 62, 71
Xiphinema krugi, 255, 359
Xiphinema pacificum, 359
Xiphinema rivesi, 41, 43, 62, 71, 80, 294, 297, 333, 345, 398
Xiphinema vulgare, 213